Navigation: Science and Technology

Volume 10

This series *Navigation: Science and Technology (NST)* presents new developments and advances in various aspects of navigation - from land navigation, marine navigation, aeronautic navigation to space navigation; and from basic theories, mechanisms, to modern techniques. It publishes monographs, edited volumes, lecture notes and professional books on topics relevant to navigation - quickly, up to date and with a high quality. A special focus of the series is the technologies of the Global Navigation Satellite Systems (GNSSs), as well as the latest progress made in the existing systems (GPS, BDS, Galileo, GLONASS, etc.). To help readers keep abreast of the latest advances in the field, the key topics in NST include but are not limited to:

- Satellite Navigation Signal Systems
- GNSS Navigation Applications
- Position Determination
- Navigational instrument
- Atomic Clock Technique and Time-Frequency System
- X-ray pulsar-based navigation and timing
- Test and Evaluation
- User Terminal Technology
- Navigation in Space
- New theories and technologies of navigation
- Policies and Standards

This book series is indexed in **SCOPUS** and **EI Compendex** database.

More information about this series at http://www.springer.com/series/15704

Yu Lu

BDS/GPS Dual-Mode Software Receiver

Principles and Implementation Technology

中国工信出版集团 電子工業出版社 PUBLISHING HOUSE OF ELECTRONICS INDUSTRY http://www.phei.com.cn

Springer

Yu Lu
Nurlink Technology Co. Ltd
Beijing, China

ISSN 2522-0454 ISSN 2522-0462 (electronic)
Navigation: Science and Technology
ISBN 978-981-16-1074-5 ISBN 978-981-16-1075-2 (eBook)
https://doi.org/10.1007/978-981-16-1075-2

Jointly published with Publishing House of Electronics Industry
The print edition is not for sale in China (Mainland). Customers from China (Mainland) please order the print book from: Publishing House of Electronics Industry.
ISBN of the Chinese edition: 978-7-121-28525-7

This Springer imprint is published by the registered company Springer Nature Singapore Pte Ltd.
The registered company address is: 152 Beach Road, #21-01/04 Gateway East, Singapore 189721, Singapore

Preface

The Global Navigation Satellite System (GNSS) is a system that uses space satellites for positioning and navigation purposes. In the early stages of GPS development, this technology was the domain of only a small number of professionals, and was most widely adopted in fields such as the military, national defense, precision guidance, land mapping, professional measurement, and atmospheric research. However, with the arrival of the 21st century, this technology is becoming increasingly indispensable in daily life, and is used for such purposes as vehicle navigation, mobile navigation, anti-loss watches for children, and location-based services.

Currently, major GNSS systems include GPS in the United States, GLONASS in Russia, the BeiDou system in China, and the Galileo system in the European Union, among which GPS is the most mature and has the highest market acceptance. The development of GLONASS stagnated after the disintegration of the Soviet, but has been revitalized and is proceeding rapidly in the 21st century. Although China's BeiDou system had a late start, it has been developing at high speed. A regional service system covering China and parts of the Asia-Pacific region was completed in December 2012, and an international service system with global coverage is planned for completion by 2020.

In this context, scientists and technicians in China require more technical literature about the research and design of the BeiDou receiver. However, since the BeiDou system is still in the process of development and requires improvement in certain aspects, mature technical literature remains scarce. The publication of this book responds to this issue to a certain extent, and provides a reference for those who are engaged in the research and design of BeiDou receivers and their algorithms.

The preliminary content of this book is based on my *GPS Global Positioning Receiver—Principles and Software Implementation*, which was released by the Electronics Industry Publishing House in 2009. After the completion of this text, I spent some time working on system design and algorithm research for the BeiDou receiver, but focused mainly on project implementation. Later, I conducted research on the implementation of the dual-system software receiver for GPS and BeiDou, and gained a deeper understanding of the theoretical issues that I had not completely grasped before. In 2011, Song Mei, Editor at the Electronics Industry Publishing House, invited me to write a professional text about the BeiDou receiver. However, I felt that my technical knowledge was inadequate. Moreover, the BeiDou system was still undergoing improvement and evolution, and much of the technology was immature and unstable. Therefore, I did not accept Ms. Song's invitation, as I was concerned about technical accuracy, and did not wish to mislead readers. Later, with repeated encouragement from Ms. Song, I finally decided to write the book. My reason was that in a rapidly developing field like the BeiDou navigation system, it is unrealistic to wait for technology to mature and stabilize. Staged summaries and discussions, which the book contains, are also meaningful.

There are nine chapters in this book. Chapter 1 explains the space and time system in the GNSS system, and introduces the basic principles of positioning, offering readers (especially beginners) a basic understanding of the system's framework.

Chapter 2 details the history of the GPS and BeiDou systems, and describes the current state of the two GNSS systems and their future trajectories.

Chapter 3 explains the format of GPS and BeiDou signals and navigation messages. This chapter is the basis for an understanding of subsequent baseband signal processing, and is also the theoretical basis for implementing data demodulation, bit synchronization, carrier synchronization, and sub-frame synchronization in the design of the receiver.

Chapter 4 elaborates on the theoretical principles of signal acquisition and tracking. Several technical solutions for signal acquisition that are used in engineering practice are analyzed in detail, as well as the principles of carrier tracking and Pseudo-Random code tracking. In the last section of the chapter, GPS and BeiDou sub-frame synchronization technology is also analyzed and compared.

Chapter 5 explains the most commonly used measurements of GNSS receivers, including pseudo-range measurement, carrier phase measurement, and Doppler measurement. The mathematical principles of observations and measurements, and the noise features are also analyzed in this chapter, paving the way for a subsequent understanding of the PVT solution.

Chapter 6 analyzes the calculation of a satellite's position and velocity, including GPS and BeiDou GEO/MEO/IGSO satellites, and provides an overview of the extended ephemeris technology adopted in mainstream commercial receivers in the market today.

Chapter 7 offers a detailed explanation of the theory and specific methods of positioning, velocity, and time solution, and also analyzes several commonly used Kalman filtering models in receivers. This chapter also includes part of the state parameter estimation theory, as well as the principles of the least-squares solution and the Kalman filtering solution.

Chapter 8 explains the RF front-end in the receiver, with emphasis on its principle and the selection of performance parameters. Examples are given to illustrate the key arrangement of the RF front-end, offering a well-rounded comprehension of the receiver.

Chapter 9 details the source code of the GPS and BeiDou dual-system software receivers designed according to the theoretical principles discussed in this book. At the same time, the GPS and BeiDou dual-mode IF signals obtained in real situations are processed and integrated into intuitive data curves and graphs to indicate signal acquisition, signal tracking, position velocity, and time resolution. It is recommended that readers read the chapter with the software receiver source code for a thorough comprehension of the theories introduced in the former chapters. Since some of the formulae in the theoretical part of the book are slightly cumbersome, they are listed as appendices for reference.

In the field of engineering technology, it is easier to know than to do. True knowledge can only be achieved through concrete action, especially in terms of GNSS receiver technology. In view of this, in the process of compiling material for this book, I heavily emphasized the source codes of the GPS and BeiDou dual-mode software receivers. These source codes, written in C and Matlab programming languages, enable the implementation of functions such as signal acquisition, carrier and pseudo-random code tracking, sub-frame synchronization, telegram demodulation, satellite ephemeris position calculation, and PVT solution (least-squares and Kalman filtering). Readers can also make situational modifications to the codes to suit their own needs, and thus implement their own software receivers. All source code can be downloaded for free from the website http://www.gnssbook.cn/. I have posted my e-mail address and WeChat account on this website for the purpose of academic communication.

In the process of writing this book, I received support from my colleagues at the Institute of Microelectronics of the Chinese Academy of Sciences, from whose sharp minds I have drawn great knowledge and inspiration. In addition, Song Mei, Editor at the Electronics Industry Publishing House, gave me the utmost encouragement, without which this book could not have been written. Huang Jian, a Masters student at the Institute of Microelectronics of the Chinese Academy of Sciences, participated in the proofreading of the first draft of the book and the compilation of the GUI interface in the source code of the software receiver. I would hereby like to express my gratitude to the above-mentioned people. I would also like to thank my wife, He Yan, and my children, who are my driving force. Due to my limited knowledge and experience, and the rapid development of the GPS and BeiDou systems, mistakes and imperfections are inevitable in this book.

I welcome corrections and critique from readers.

Beijing, China Yu Lu
January 2016

Introduction

This book offers a detailed explanation of the principles of GPS and BDS dual-system receivers from the perspective of electronic engineering and communication technology. While describing the historical evolution of GPS and BDS systems, the book also elaborates on almost all of the signal processing theories applied to GPS and BDS receivers, from signal tracking and acquisition to satellite position velocity calculation, model analysis of observed measurements, and positioning navigation solutions. The author's research and development experience in this field over many years is also integrated in the discussion of the theories. While elaborating on theoretical knowledge, the book explains the GPS and BeiDou dual-system software receiver by closely combining the theories, and gives the source code for the software receiver, which realizes all of the theoretical points explained herein. Readers can run the corresponding program while reading the theoretical part of the book, and then analyze the results and modify the source code according to their own needs. This will allow for a better understanding of the design theory of GPS and BDS dual-system receivers and lay a solid foundation for further research.

The content of this book is substantial. Theoretically and practically useful, it is suitable for engineers, technicians, and research institutes working on satellite navigation and positioning in fields such as electronics, aerospace, surveying and mapping, automatic control, geography, transportation, agriculture, forestry, remote sensing, and planning, as well as senior undergraduate and graduate students in electronics majors. It can be used both for academic reference and as a supplementary textbook.

Contents

Chapter 1
Positioning, Coordinate System, and Time Standard

The Global Navigation Satellite System (GNSS) was invented to provide positioning and navigating service for end users through radio-navigation signals sent by navigational satellites from space. In general, GNSS refers to all navigational satellite systems, including global, regional, and augmentative. Currently there are four main global GNSS systems, namely Global Positioning System (GPS) in the United States, GLONASS in Russia, BeiDou Navigation Satellite System (BDS) in China, and Galileo in the European Union. Regional systems include China's BeiDou-1, Japan's Quasi-Zenith Satellite System (QZSS), and the Indian Regional Navigation Satellite System (IRNSS). Others, like WASS and EGNOS, are augmentation systems. In terms of maturity and popularity, the USA's GPS unquestionably ranks top. The other global GNSS systems differ from GPS in radio frequency, signal modulation, and navigation message, but their basic fundamentals for realizing positioning and navigation are basically the same. In this sense, getting acquainted with GPS is a good start point for beginners in the field of satellite navigation.

Starting from the basics, the first section of this chapter explains the fundamentals of the GPS system in a comprehensible way, proving how simple they are. We avoid using complicated theoretical derivation and equations, opting instead for simple, daily language in order to retain readers' interest and enthusiasm for the GPS receiver.

One purpose of the GPS receiver is positioning, so the receiver itself must be part of a coordinate system in order to provide a meaningful position. Therefore, in the second section of this chapter, we briefly explain a couple of different coordinate systems, as well as conversions among commonly-used systems. GNSS has high requirements on time, as precise time measurement has a direct impact on positional accuracy. Some scholars even claim that precise time synchronization must be achieved for GNSS before it can be used for positioning. Hence, in the third section of this chapter, we discuss some of the time systems that are commonly used around the world, and analyze the relationships between them.

Y. Lu, *BDS/GPS Dual-Mode Software Receiver*, Navigation: Science and Technology 10, https://doi.org/10.1007/978-981-16-1075-2_1

1.1 The Question Raised

1.1.1 Basic Purpose and Basic Positioning System

It is commonly agreed that the fundamental purpose of a GPS receiver is positioning, namely to answer the question: *Where am I?* Therefore, we will start with positioning.

We live in a three-dimensional world, so the question above should be considered in a three-dimensional way. However, it is easier to begin with two-dimensional positioning, as the situation is more complicated in a three-dimensional space, and too simple in a single dimension.

Consider this: Mr. A lives in a two-dimensional world, and his position is indicated by two-dimensional Cartesian coordinates (seen in Fig. 1.1).

If Mr. A wants to know his coordinate (x, y), the simplest way is to measure the distance between him and the two coordinate axes with a ruler. But this method has major limitations. Firstly, Mr. A must know the position of the coordinate axes. However, in real life, the concept of coordinate axes is abstract. For example, under circumstances where virtual coordinate axes extend through mountains and oceans, direct physical measurement is unattainable. Secondly, when Mr. A's position is too far away from the axes—say 1000 km—it would not be possible to find a long enough ruler.

To address these problems, Mr. A doesn't choose the coordinate axes as his measurement reference. Instead, he chooses two points with known coordinates on the two-dimensional plane, P_1 (x_1, y_1) and P_2 (x_2, y_2) (seen in Fig. 1.2. In this case, as long as he knows the distance between himself and P_1 (x_1, y_1) and P_2 (x_2, y_2)—symbolically S_1 and S_2—he can work out his own coordinate (x, y) using the two equations below:

$$S_1 = \sqrt{(x - x_1)^2 + (y - y_1)^2} \tag{1.1}$$

$$S_2 = \sqrt{(x - x_2)^2 + (y - y_2)^2} \tag{1.2}$$

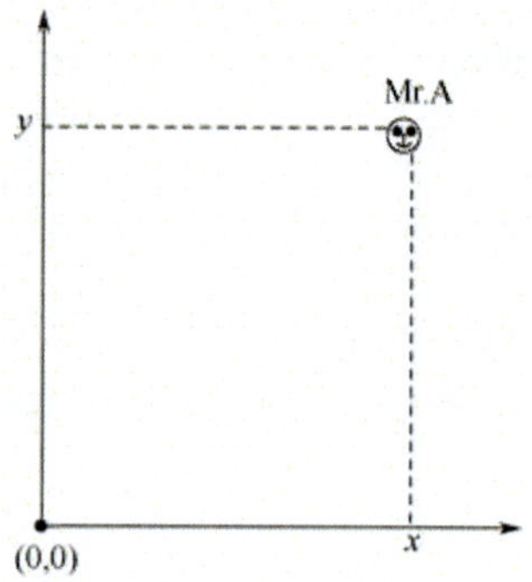

Fig. 1.1 Mr. A lives in A two-dimensional world

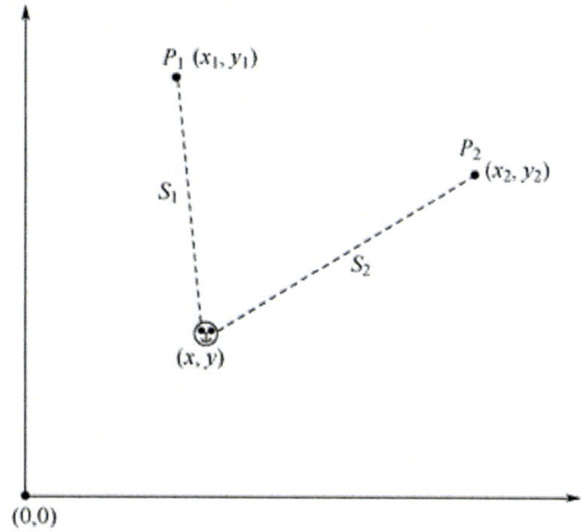

Fig. 1.2 Mr. A uses two reference points for positioning

Equations (1.1) and (1.2) are non-linear, so the iteration method can be applied. In the physical sense, the two equations indicate the track of two circles whose intersection is Mr. A's exact position. Of course, theoretically, there will be two intersections, but one of them is often impossible, for instance, if an intersection is in the sky, as Mr. A can only stay on the surface of the earth.

Instead of measuring coordinates directly, this alternative method measures the distance between Mr. A and a specific reference point with a known position. There are numerous ways of measuring distance, from simple methods like measuring with a ruler to intricate approaches like triangulation, as well as distance measuring equipment like laser range finders and ultrasonic range finders. Here we will use an alternative method that it is currently applied in pseudo-range measurement among GNSS receivers. Its basic principle is to identify the distance by recording the propagation time of electromagnetic waves between the user and the reference point.

The propagation velocity of electromagnetic waves in free space, which equals the velocity of light under normal circumstances, is constant and usually represented by c. Thus, if we know the propagation time of electromagnetic waves from the reference point to the user, we also know the distance between them, namely S_1 and S_2 on the left side of Eqs. (1.1) and (1.2), through multiplying time by the velocity of light.

One way to do this is to let the reference point flash once at a known time—t_s. When Mr. A sees the flash at t_r, he can use equation $S = c\,(t_r - t_s)$ to acquire the propagation length, i.e. his distance from the reference point.

c stands for the velocity of light, which is 2.997×10^8 m/s. As Mr. A can see flashes from two reference points, he will get two sending times—t_{s1} and t_{s2}—and two receiving times—t_{r1} and t_{r2}. To make things simpler, point P_1 and P_2 can have the same sending time, namely $t_{s1} = t_{s2}$, hence we use t_s to represent them both in the analysis below. However, the receiving times differ for P_1 and P_2 unless Mr. A stands in a position where P_1 and P_2 share the same distance from it. In other words,

P_1, P_2, and Mr. A's position form an isosceles triangle, otherwise t_{r1} and t_{r2} will always be different.

Now, Mr. A has successfully turned a measurement of distance into a measurement of time, so the next question is how to record t_{r1} and t_{r2}?

The answer is quite simple. All we need to do is to put three clocks at P_1, P_2 and Mr. A's position respectively, and calibrate them to the same time. Then, we let P_1 and P_2 flash simultaneously, for example, at 7:00:00. If Mr. A sees the flash from P_1 and P_2 at 7:00:01 and 7:00:02 respectively, he will know that his distance from P_1 equals the propagation distance of the flash in 1 s, and that his distance from P_2 equals the propagation distance of the flash in 2 s.

This method seems workable, as Mr. A simply needs a clock to realize the distance measurement, not a ruler. This is the fundamental theory behind GPS.

This way of recording time appears simple. However, problems emerge after deliberation.

The first problem is this: how can Mr. A distinguish the P_1 flash from the P_2 flash if they are identical? The second problem is ensuring that t_{r1} and t_{r2} are recorded precisely. To thoroughly understand this problem, we must consider the absolute value of the velocity of light. Light travels at 300,000 km per second. Therefore, if Mr. A's clock is slightly inaccurate and the time error is 1 ms, then the 1 ms error will also appear in the recorded values of t_{r1} and t_{r2}, causing a 300 km error in the measurement of distance $\left(2.997 \times 10^8 \times 1\,\text{ms} = 299.7\,\text{km}\right)$. One millisecond seems minor in daily life, but it can lead to significant errors in situations like this.

The first problem is easy to solve. We can use a red flash for P_1 and a blue flash for P_2 to distinguish them. In a broader sense, it means that a unique identification (ID number) must be assigned to every reference point.

However, the second problem is more difficult to solve. To do so, it is necessary to understand how a clock works.

1.1.2 *The Workings of a Clock*

Throughout the history of human civilization, despite changes in the world and the development of technology, time has been measured through counting cyclical events. In ancient times people looked to the rise and fall of the sun to delineate a day and to record time. In the time of Galileo and Huygens, pendulum clocks recorded time through counting the harmonic motion period of simple pendulum movement, while the electronic age gave rise to clocks with electronic oscillators, which were more stable. The clock signals that are extensively applied in modern digital circuits work on a certain kind of cyclical event.

Quartz crystal oscillators are widely used to set the frequency in modern clocks. For instance, the frequency of crystal oscillators in quartz clocks is 32,768 Hz. To count seconds, a counter must be placed in the clock to count the oscillation periods of the 32,768 Hz crystal oscillator. When the number of periods reaches 2^{15}, a signal of 1 s is generated. However, only when the crystal oscillator oscillates precisely at

32,768 Hz does the signal stand exactly for 1 s, as 32,768 Hz is the nominal value of this oscillator, not the measured value. The two paramount evaluation indexes of a crystal oscillator are the *Accuracy* and *Stability* of the oscillation frequency. A difference between the nominal and measured value of the oscillation frequency is almost inevitable, and this difference is called frequency accuracy. For example, if the nominal value of the frequency of a crystal oscillator is 1 MHz, and the measured value is 999,999 Hz, then the frequency accuracy will be 1 Hz. For greater precision, we use the term *Relative Frequency Accuracy*. Its value is indicated as $\frac{\Delta f}{f_0}$; Δf is the difference between the nominal and measured value, and f_0 is the nominal value of the oscillator frequency. It is obvious that frequency accuracy is a ratio—a dimensionless quantity—and can be represented as ppm, 1 ppm = 10^{-6}. In the example above, the relative frequency accuracy is 10^{-6} (1 ppm). The difference between the nominal and measured value of the frequency varies with time. Its derivative with respect to time indicates the frequency stability of the crystal oscillator. Frequency stability correlates with many elements, including temperature, voltage, aging, and external dynamic stress.

The below equation is used to calculate the value of relative frequency accuracy:

$$F = \frac{f - f_0}{f_0} \tag{1.3}$$

In Eq. (1.3), f_0 is the nominal value and f is the measured value.

If we count the oscillation period of a crystal oscillator with a nominal frequency as f_0, and the counting time depends on the time set, which is represented by N here, then theoretically the time recorded by the clock should be

$$T_0 = \frac{N}{f_0} \tag{1.4}$$

However, considering frequency accuracy, the actual time recorded should be

$$T_0 = \frac{N}{f} \tag{1.5}$$

Suppose that $\Delta f = f - f_0$, then the time error should be

$$\Delta T = T - T_0 = \frac{N\Delta f}{f f_0} \approx \frac{T_0 \Delta f}{f_0} = T_0 F \tag{1.6}$$

We use an approximation in Eq. (1.6) because under normal circumstances, $\Delta f \ll f_0$.

From Eq. (1.6) we can infer that $\frac{\Delta T}{T_0} = F$, which means that a time error is bound to occur as long as the value of F is not 0 and the error value is proportional to the value of F. If 1 s = F, then $\Delta T = F$. In the physical plane, this conclusion suggests that after we record the nominal frequency value of a crystal oscillator for 1 s, the difference between the length of time we get and the length of 1 s in theory is the relative frequency accuracy of this oscillator. The application of this conclusion in GPS system receivers will be elaborated in subsequent chapters.

Now, the problem facing Mr. A is that even if he tunes his clock to the same time on the clocks at P_1 and P_2, influenced by relative frequency accuracy as discussed above, he cannot expect accurate time information from his own clock after a while, unless he tunes his clock constantly, which is impossible.

As we are looking for solutions to the second problem, a third issue arises: how can we guarantee time consistency of clocks at P_1 and P_2 if absolute consistency of time is unachievable? This problem arises because the forgoing discussion is based on the hypothesis that P_1 and P_2 indeed flashes simultaneously so that they can share the same sending time, otherwise the subsequent discussion will be impossible.

The third problem can be solved this way: there are merely two reference points, so we can allocate a large budget to sophisticated technology to make two extremely precise clocks. In fact, the satellites that work as reference points in the GPS system are equipped with atomic clocks with rubidium or cesium oscillators, whose relative frequency stability can reach 10^{-12}–10^{-14}. At the same time, the ground control station monitors atomic clocks on the satellites at all times, and makes adjustments when necessary.

However, what if Mr. A has an atomic clock as well? Can we solve the second problem this way? Theoretically, the answer is yes, but in practice it is no, because end users cannot afford receivers with atomic clocks due to the high cost of raw materials and sophisticated technology.[1] The reason why reference points can be used to configure atomic clocks is that there are only a handful of reference points, but there are tens of thousands of users.

A solution has not yet been found for the second problem.

1.1.3 A Modified System

Putting the second problem aside for the moment, suppose that Mr. A's clock is accurate enough to record the receiving time of the flash so that he can determine his position using Eqs. (1.1) and (1.2), yet imperfections still exist. Now, the method we use for positioning is as follows: let P_1 and P_2 flash at fixed instant, such as every second sharp. For example, P_1 and P_2 flash simultaneously at 7:00:00, 7:00:01, and 7:00:02, and the color of the flash emitted from the same reference point every second does not change.

There are two problems. First, Mr. A can only get his position once per second. Thus, if he misses the flash of this second, he has to wait for another one at the next

[1]With the progress of technology and the reduction of costs, it is not impossible to consider the use of an atomic clock in a GPS receiver. Some companies are already developing and producing Chip-Scale Atomic Clocks, which are one centimeter in size and consume tens of milliwatts of power, and can already be used in small receivers. These clocks may be considered in future receiver designs.

second. The ideal positioning is that every time Mr. A looks at his clock, he can also know his real time position. Second, if Mr. A is far away from the reference point (for example, more than 300,000 km) it will take the flash more than 1 s to reach him, which means that he cannot distinguish the flash sent 1 s ago from the one that was sent 10 s ago because they are identical. Consequently, the distance cannot be accurately calculated, and positioning is out of the question. We may call this problem *Ambiguity of a Whole Second*, originating from the sameness of every flash.

To address this problem, we need to modify the basis of our positioning system. First, we should increase the frequency of the flashes at the reference points from once per second to 10^6 per second, namely once per μs, while P_1 and P_2 still send a red flash and a blue flash concurrently every μs. Then, we modulate the information about the sending time carried by every flash, so that Mr. A can get the sending time of this flash according to the modulated time information when it arrives. To read this information, Mr. A must equip himself with a flash receiver as well as a clock, which may appear as a function of an existing instrument to decode the modulated information.

Figure 1.3 shows the improved flash-sending model at each reference point. The red and blue flashes are sent concurrently, and the sending time is modulated on every flash. Something that needs to be stressed here is that strict synchronicity exists only at reference points, not at the receiving point, i.e. Mr. A's receiver. Obviously, the reason is that the distance between Mr. A and the two reference points is different,hence the propagation time of the flashes will not be the same. That is why the synchronicity disappears at receiving ends.

What impact do the two modifications have on positioning?

First, every time Mr. A looks at the time on his clock, his receiver is bound to receive a red flash and a blue flash as long as he can wait one more 1 μs. Even if the

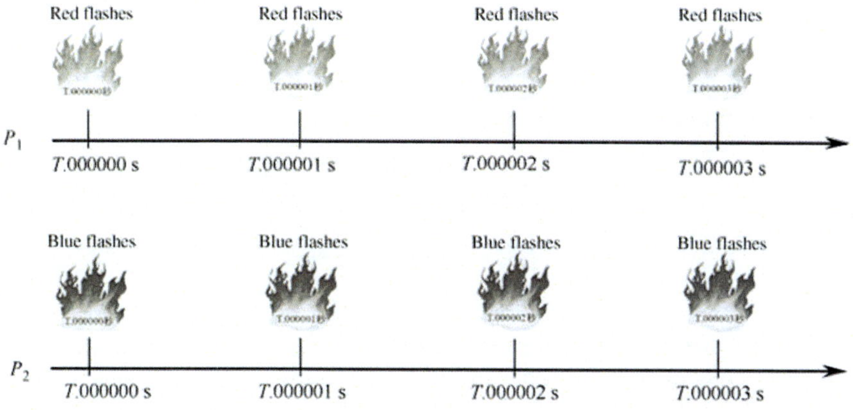

Fig. 1.3 The red and blue flashes are emitted 1 μs apart, and the time of modulator emission is also adjusted at each flash

two flashes do not reach Mr. A in strict synchronicity, the time error will be less than 1 μs.

We already know that the distance error brought by 1 μs is roughly 300 m. If the 300 m error is tolerable, we can say that the two flashes arrive simultaneously and the arriving time is what is shown on Mr. A's clock. As mentioned earlier, two flashes with the same arriving time have different sending times.

If Mr. A's receiver receives a red flash and a blue flash at t_r, then he can work out the respective sending times of both flashes from the modulated information on them. Now, Mr. A has three observed times: t_{s1} and t_{s2}, the sending time of the red and blue flash, together with his local time t_r. t_{s1} and t_{s2} are always accurate, as they are provided by the atomic clocks at the reference points. Let us assume that Mr. A's clock is also an atomic one with extreme precision, which guarantees the accuracy of t_r. Then, the propagation time of the two flashes will be $(t_{s1} - t_r)$ and $(t_{s2} - t_r)$, so we can write

$$S_1 = c(t_{s1} - t_r) = \sqrt{(x - x_1)^2 + (y - y_1)^2} \tag{1.7}$$

$$S_2 = c(t_{s2} - t_r) = \sqrt{(x - x_2)^2 + (y - y_2)^2} \tag{1.8}$$

1 μs is so short that almost any time Mr. A reads his receiver, he can get the value of t_{s1}, t_{s2}, and t_r to calculate his position. And because every flash carries the information of its sending time, the problem of whole second ambiguity does not exist either.

However, we should keep in mind that the feasibility of this modified system depends on the premise that Mr. A is equipped with an atomic clock, which is impractical. So, the second problem put forward in Sect. 1.1.1 is still unsolved.

Now if Mr. A uses a common clock with limited precision, then the local time provided by the receiver every time he looks at it will be inaccurate, represented by t_r', as opposed to t_{s1} and t_{s2}, which are precise because they are decoded from the modulated signal carried by the flashes instead of directly shown on local clocks. If the difference between t_r' and t_r is b, i.e. $t_r' = t_r - b$, then we can get the following equations from Eqs. (1.7) to (1.8):

$$\rho_1 = c\left(t_{s1} - t_r'\right) = \sqrt{(x - x_1)^2 + (y - y_1)^2} + cb \tag{1.9}$$

$$\rho_2 = c\left(t_{s2} - t_r'\right) = \sqrt{(x - x_2)^2 + (y - y_2)^2} + cb \tag{1.10}$$

We use ρ instead of S in these two equations, as ρ is not the real distance. Its difference from the real distance is b, a time constant. Thus, we call ρ the pseudo-range. There are three unknown quantities (x, y, b) in merely two equations, for which they cannot be solved.

After observing Eqs. (1.9) and (1.10), we can see that each equation includes parameters for a specific reference point. There are two reference points, P_1 and P_2, for Mr. A, so we can write two equations accordingly. Consequently, one more equation is attainable if there is one more reference point, and the value of (x, y, b) is easy to work out.

If we add a new reference point P_3 (x_3, y_3) that emits green flashes concurrently with P_1 and P_2, then Mr. A's receiver will theoretically receive three flashes of different colors per μs (seen in Fig. 1.4).

Correspondingly, we can write three equations:

$$\rho_1 = c\left(t_{s1} - t_r'\right) = \sqrt{(x - x_1)^2 + (y - y_1)^2} + cb \quad (1.11)$$

$$\rho_2 = c\left(t_{s2} - t_r'\right) = \sqrt{(x - x_2)^2 + (y - y_2)^2} + cb \quad (1.12)$$

$$\rho_3 = c\left(t_{s3} - t_r'\right) = \sqrt{(x - x_3)^2 + (y - y_3)^2} + cb \quad (1.13)$$

Hence, Mr. A can find the value of (x, y, b). The second question mentioned in Sect. 1.1.1 is solved. Both positioning information and the difference between t_r' and t_r are found. Mr. A can use the difference to correct local time t_r'. This is called precise timing, which is an unexpected by-product.

As well as finding his position, Mr. A can also work out his local time. It can be seen from Eqs. (1.11) to (1.13) that the accuracy of positional measurement and that of time measurement are tightly coupled. The accuracy of pseudo-range ρ_i directly determines the positional accuracy and time accuracy. There is a linear correlation between time accuracy and positional accuracy, with the coefficient being $1/c$. For example, if the positional accuracy is 30 m, the time accuracy will be $30/c = 100$ ns. If the positional accuracy is 300 m, the time accuracy will be $300/c = 1$ μs.

The method of using three reference points for positioning is presented in Fig. 1.5. Now Mr. A can find his position (x, y), as well as the difference b between t_r and the accurate time.

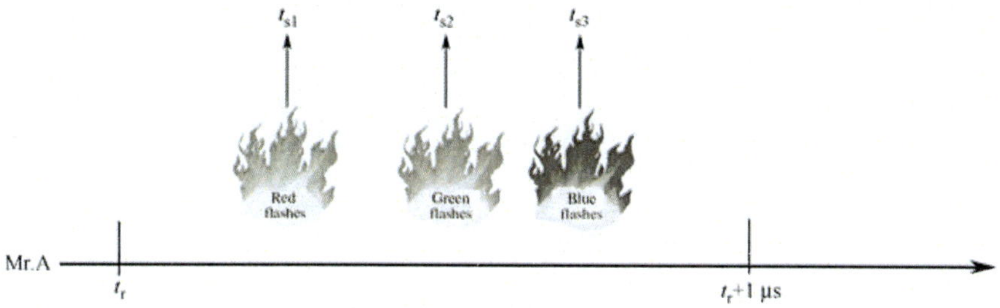

Fig. 1.4 Mr. A receives three flashes at t_r

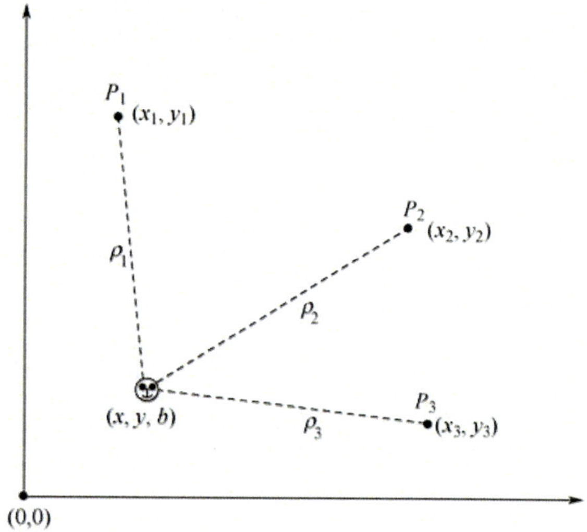

Fig. 1.5 Mr. A now needs three reference points to find his location and realize the timing function

1.1.4 Summary of the Modified System

In Sect. 1.1.3, we improved the original positioning system and obtained a surprisingly good result. Mr. A can now perform positioning calculations at any time, and can also work out the exact time. Of course, the improved system is more complex. As for the reference point, it flashes 10^6 times per second, and the sending time must be modulated on every flash. Mr. A needs a receiver that can receive flashes from at least three reference points, and can demodulate flash information. The advantage of the modified system is that a precise local clock is no longer needed. Instead, an inexpensive clock can be used to find out the correct local time.

This system is based on a two-dimensional world, but it is not difficult to apply it to the three-dimensional world in which we are living. We add one axis into Mr. A's coordinates, giving (x, y, z, b). Correspondingly, there are four minimum required reference points, and the number of required equations is also four.

$$\rho_1 = c\left(t_{s1} - t_r'\right) = \sqrt{(x - x_1)^2 + (y - y_1)^2} + cb \tag{1.14}$$

$$\rho_2 = c\left(t_{s2} - t_r'\right) = \sqrt{(x - x_2)^2 + (y - y_2)^2} + cb \tag{1.15}$$

$$\rho_3 = c\left(t_{s3} - t_r'\right) = \sqrt{(x - x_3)^2 + (y - y_3)^2} + cb \tag{1.16}$$

$$\rho_4 = c\left(t_{s4} - t_r'\right) = \sqrt{(x - x_4)^2 + (y - y_4)^2} + cb \tag{1.17}$$

Equations (1.14)–(1.17) are the pseudo-range equations of GPS. The above illustration involves reading the observed quantity of the pseudo-range in GPS receivers. The methods of observation will be discussed in Chap. 5. In the design of the receiver, more than four satellite observations will be obtained, so Eqs. (1.14)–(1.17) will become overdetermined equations. In Chap. 7, we will explain how to gain a positioning result from more than four equations as accurately as possible by using the methods of least square method and Kalman filtering.

The following are the basic requirements discussed in the sections above that make this positioning system feasible.

i. Several reference points are needed, the coordinates of which are known.
ii. These reference points transmit certain signals that have unique identification information (ID) to distinguish themselves from the signals of other reference points.
iii. The reference points continuously transmit signals that can be received by other receivers.
iv. The signals transmitted by every reference point are synchronized in time.
v. After receiving a signal from a reference point, the user can know the exact transmission time of the received signal from its modulation information.
vi. The user has a clock, which doesn't need to be precise.

These requirements may seem insignificant to beginners, but we hope you will keep them in mind. Only when you finally grasp and appreciate GPS design will you fully understand the overall process of GPS positioning and the principles of GPS receivers, remembering the requirements.

These requirements show that GPS has a high demand for time synchronization, which necessitates a strictly defined time standard as a reference. Besides, to get an effective positioning result, it must operate within a strictly defined coordinate system. Therefore, some of the most common coordinate systems and time standards will be briefly introduced in the following two sections, as they form the basis for subsequent chapters.

The positioning principles of GPS, BeiDou, GLONASS and Galileo are basically the same. As long as you have grasped the basic principles of GPS positioning, the positioning principles of several other GNSS systems are not difficult to understand.

1.2 Common Coordinate Systems

1.2.1 Earth Centered Inertial Frame

The inertial coordinate system must be stationary or moving at a constant speed with no acceleration, so that Newton's law of motion can be applied. The origin of the inertial coordinate system can be at any point, and its three coordinate axes can be in any three mutually orthogonal directions. The coordinate axes in different directions define different inertial coordinate systems. The conversion between different coordinate systems is detailed in Appendix B.

The selection of an inertial coordinate system should be considered if: the origin of the inertial system coincides with the centroid of the Earth; the Z-axis and the Earth's rotational axis coincide; the X-axis and the Y-axis form the Earth's equatorial plane; and the X-axis points to the vernal equinox, which is on the intersection line of the Earth's equatorial plane and revolution track, so the Y axis, the X axis, and the Z axis form a right-handed coordinate system. Since the spring equinox doesn't change with the Earth's self-rotation, the X-axis can be considered as a fixed one. Any polar motion can be considered to be fixed in a short time. Therefore, this inertial system is called the Earth Centered Inertial Frame, abbreviated to the ECI frame. Figure 1.6 is a schematic diagram of the ECI coordinate system. The shaded area shows the plane of the Earth's orbit. There is an angle between the Earth's rotation plane and the ecliptic, known as the obliquity of the ecliptic. The two intersections between the two planes are called the autumnal equinox and the vernal equinox. The Earth passes the autumnal equinox on September 23 and the vernal equinox on March 21 every year. Both are the positions on the Earth's orbit in which days and nights are of equal length.

The Earth orbits the sun with a one-year orbital period, and the sun orbits the Milky Way. The Earth's axis also moves in space, including complex precession and nutation. Figure 1.7 is from the official IERS website (https://www.iers.org/), and shows the polar motion of the Earth's rotation axis between 2003 and 2013. The unit of the X and Y axes is an arcsecond. Simultaneously, due to the gravity of the Earth as an irregular sphere and the gravity of the moon, the position of the Earth's centroid is also disturbed. Therefore, strictly speaking, the Earth Centered Inertial Frame is not a real inertial system. It can only be considered to be an inertial system for a short period of time.

Fig. 1.6 The ECI coordinate system

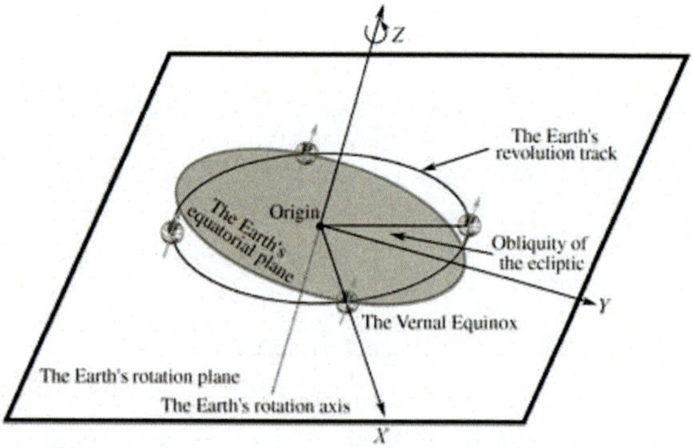

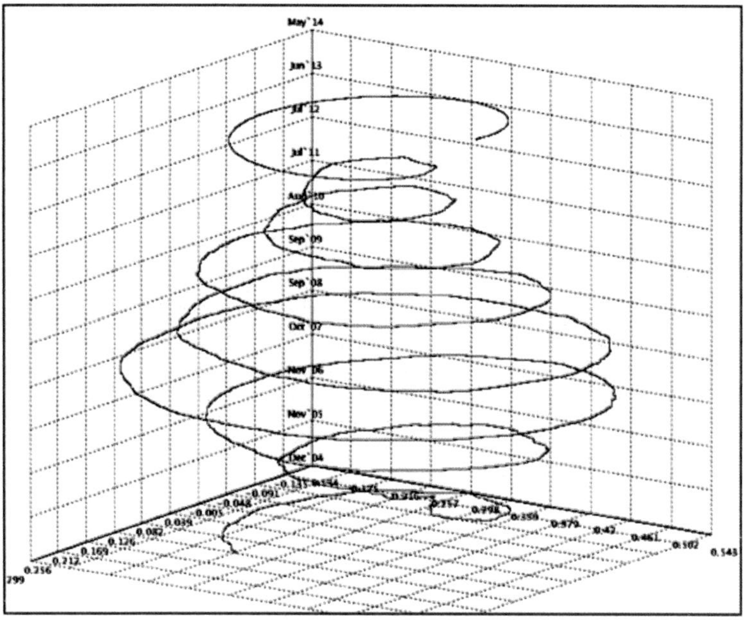

Fig. 1.7 The polar motion of the Earth's rotation axis from 2003 to 2013

1.2.2 The Geodetic Frame

The Geodetic Frame is the latitude, longitude, and height (or altitude) coordinate system that is often used in daily life, also known as the LLH coordinate system. Before defining the Geodetic Frame, we should explain the term geoid.

The shape of the Earth is roughly a spheroid. The length between two poles is slightly shorter than the diameter of the plane in which the Equator is located. According to the relevant parameters of the WGS84 coordinate system,[2] the semi major-axis of the Earth is 6,378,137 m and the semi minor-axis is 6,356,752 m. The semi minor-axis is nearly 20 km shorter than the semi major-axis. The ellipsoid of the Earth rotates around the semi minor-axis, so it is a slightly flattened ellipsoid. The center of the ellipsoid coincides with the Earth's center, so only two pieces of information are needed to define the ellipsoid coordinate system: the length of the semi major-axis and the semi minor-axis, marked as a and b respectively. From these two quantities, two other important quantities can also be derived:

$$\text{Eccentricity}(e)\colon e^2 = \left(a^2 - b^2\right)/a^2$$

[2]The ellipsoid model defined in other coordinate systems is slightly different from WGS84, but the difference is not very big, so taking WGS84 as an example does not cause serious confusion.

$$\text{Flattening}(f)\text{: } f = (a - b)/a$$

The relationship between flattening and eccentricity is $e^2 = 2f - f^2$. According to the widely used WGS84 coordinate system, the eccentricity of the Earth's ellipsoid is 0.08181919, while the flattening is 0.00335281.

If the surface of the Earth were smooth, then the natural surface would be the spherical surface of the revolving ellipsoid. However, in reality, the Earth's surface is uneven and abnormal, and the highest peak (Mount Everest) is nearly 20 km higher than the lowest area (the Mariana Trench). In this circumstance, defining the geoid is a very complicated process. In general, performing least-squares fitting on a huge amount of data is required to obtain the definition. This data often includes geographic mapping and gravity fields, which give different definitions of geoids in different periods. Updates to measuring instruments and methods have greatly improved this process. Since the start of the aviation era, remote sensing telemetry technology based on an Earth-orbiting satellite has provided more effective measurements for the geoid.

This book focuses on the design of the GNSS receiver, so we don't need to spend too much time and effort on the definition and measurement of the geoid. Readers should simply view it as a geopotential surface that refers to the average global sea level. The surface has the same potential gravitational energy everywhere on its face, and the direction of the gravitational vector is perpendicular to the tangent plane there. The concept of the geoid will be used when defining elevation.

Figure 1.8 helps us understand the definition of the Geodetic Frame. Point *P* is a point somewhere on the geoid. The plane formed by point *P* and the Earth's rotation axis is perpendicular to the equatorial plane. It is called the meridian. The meridian and the Equator have an intersection line *OQ*. The well-known prime meridian passes through the meridian of the Greenwich Observatory in England. The geodesic coordinate of point *P* consists of three coordinate quantities, of which the definitions are as follows (Fig. 1.9).

- Latitude: The angle between *OQ* and the normal point/vector on the surface of the ellipsoid over point *P* is usually expressed as Ø. Latitude near the equator is small, and gradually becomes larger as it moves towards the poles. The range of latitude is [−90°, +90°]. It is positive in the northern hemisphere and negative in the southern hemisphere, so the north pole is +90°, and the south pole is −90°.
- Longitude: The angle between *OQ* and the meridian measured in the equatorial plane is generally expressed as λ. The range of longitude is [−180°, +180°], and the east of the prime meridian is positive.
- Height: To make a projection from *P* to the geoid. The distance from *P* to the projection point is the height, which is usually expressed by *h*.

On most parts of the Earth's surface, the normal direction of point *P* in Fig. 1.8 does not pass through the center of the Earth, but with a small difference, except when crossing point *P* at latitude 0° or ±90°. Therefore, the demonstration in the figure is only approximate, and should not be used in practical application.

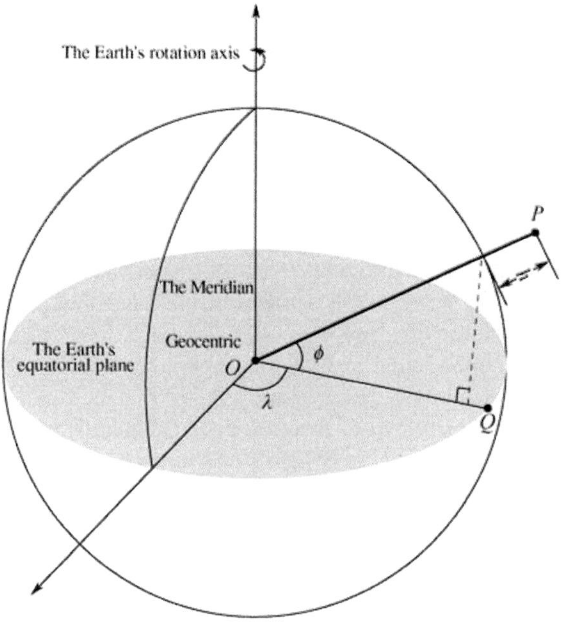

Fig. 1.8 The geodesic coordinate system

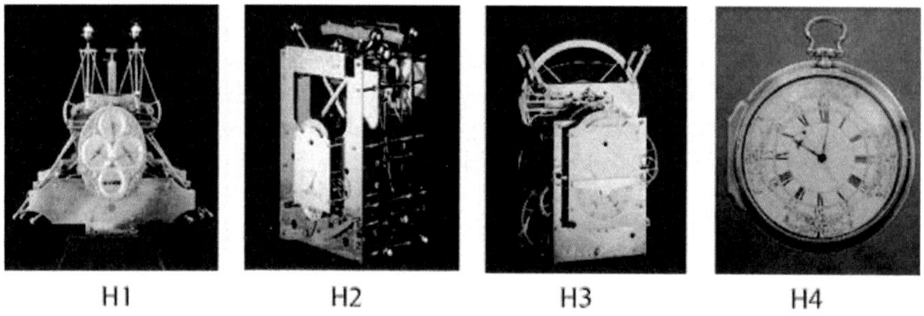

Fig. 1.9 John Harrison's four navigational clocks finally provided a solution to longitude measurement. (Pictures from the internet)

The curve of the same latitude on the surface of the Earth is called the parallel, and all parallels are circles. The higher the latitude, the smaller the radius of the parallel. The curve of the same longitude is called the meridian. All meridians are elliptical circles, and they stretch from the North Pole to the South Pole. Historically, latitude has been measured by gauging the elevation angle of the sun with a nautical instrument such as a sextant, and correcting the nautical calendar. Accurate measurement of longitude has experienced several obstacles, because the division of the longitude line is artificial. Unlike latitude, whose lines have obvious natural differences in geography, topography, and celestial movements, longitude cannot be gauged by directly measuring the angle of the sun. In the Age of Discovery that began in the fifteenth century there were urgent demands for the positioning of fleets in ocean voyages. A preponderance of shipwrecks in the seventeenth and eighteenth centuries prompted an interest in achieving accurate maritime positioning. The most famous shipwreck occurred on 2nd November 1707, when a fleet led by Admiral Cloudesley Shovell of the Royal Navy defeated the French Naval fleet and sank on its return due to fog. Nearly 2000 members of the navy perished in the maritime space near the Isles of Scilly (latitude: 49.9334° N, and longitude 6.325° W). In 1714, British Parliament passed the Longitude Act to legislate the search for a solution to longitude measurement. Many methods proposed, including the star map method, the lunar distance method, and the clock method. Eventually, the problem was solved by a British craftsmen named John Harrison. He constructed a series of highly accurate nautical clocks—known as H1, H2, H3, and H4—to maintain precise sailing time, and the position of the Greenwich Observatory in England was determined as the prime meridian. This was to secure Britain's hegemony over the ocean for the next nearly two hundred years.

The geodesic frame is not a Cartesian coordinate system but a polar coordinate system, so the method in Appendix B cannot be used when performing coordinate transformation. Likewise, the geodetic frame has some inconveniences. For example, there will be different distances for the same longitude span at different latitudes. For instance, longitude 1° near the Equator has a span of about 110 km, while latitude 60° only has a span of around 55 km.

The geodesic frame and navigation techniques have some interesting relationships. For example, a nautical mile is defined as the arc length of one minute of the Earth's ellipsoid, which is about 1.855 km. Therefore, it can be more easily calculated that the length of the Earth's Equator is about $360 \times 60 = 21{,}600$ nautical miles. Today, the speed of a ship is often expressed in knots. One knot is one nautical mile/hour. Because longitude lines are large circles, the geometric distance differs by 60 nautical miles if the latitude is one degree different in the north–south direction. A ship sailing 60 knots in the north–south direction indicates that the latitude has changed by one degree. This unit setting allows the ship to calculate its latitude and longitude coordinates by its speed during navigation. This technique is known as Dead Reckoning, or DR technology. Modern DR is often undertaken by inertial sensors or other speed sensing devices.

In the GNSS receiver, the position coordinates of the satellite are represented in the ECEF. The calculated user position coordinates are also represented in the ECEF, but in many circumstances it is more common to use the geodetic frame. Therefore, mutual conversion between the ECEF and the geodetic coordinate frame is an important issue. The ENU coordinate system (explained later) needs to know the longitude and latitude of the user before it can be determined.

1.2.3 The ECEF Coordinate System

The full name of the ECEF coordinate system is the Earth-Centered-Earth-Fixed coordinate system. It refers to all coordinate systems whose origins are the center of mass, and whose coordinate axes are rigidly attached to the Earth. However, in the field of GNSS, the ECEF system refers to a rectangular coordinate system with the center of mass as its origin, with the X axis pointing to the intersection point of the prime meridian plane (whose outline is 0° longitude) and the Equator, i.e. 0° latitude, and the Z axis, which coincides with the rotation axis of the Earth, pointing to the North Pole. The Y, X, and Z axes form a right-handed coordinate system, as shown in Fig. 1.10.

Because the Earth revolves around both itself and the Sun, the ECEF coordinate system is not inertial. Unlike the ECI inertial coordinate system, ECEF has an angular velocity while revolving, and can be written as:

$$\omega_{ie} \approx \frac{(1+365.25)\times 2\pi}{365.25\times 24\times 3600} = 7.292115\times 10^{-5}\,\text{rad/s} \tag{1.18}$$

Equation (1.18) is interpreted this way: it takes one year, namely 365.25 days, for the Earth to revolve once around the Sun, and it revolves (365.25+ 1) rounds in

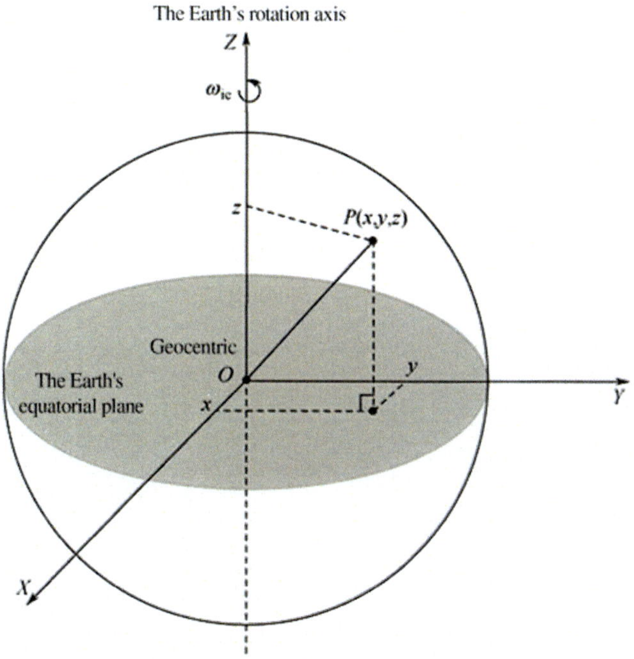

Fig. 1.10 The ECEF coordinate system

a year, of which the extra round comes from the revolution of the Earth. Then, we divide the radian with the time it takes, i.e. $365.25 \times 24 \times 3600$ s, to get the angular velocity. The angular velocity ω_{ie} is a vector, and since it rotates around the Z-axis of the ECEF coordinate system, it is written as $[0, 0, \omega_{ie}]^T$.

Even though the angular velocity of the ECEF system is small in value, it should not be neglected under certain circumstances. For example, in the following chapter, the navigation positioning algorithm will explain why the rotation angle of the Earth must be taken into consideration during the period when satellite signals travel from space to the ground; the positions of satellites in the positioning system need to be adjusted accordingly, otherwise positioning errors will occur.

Without exception, the parameters for the ECEF coordinate frame will be indicated in every interface control document (ICD) of the GNSS system. Table 1.1 shows the main differences in the coordinate frames of four GNSS systems.

Minor differences exist in the various coordinate systems in terms of origins, semi-major axes of the Earth's ellipsoid, flattening of the ellipsoid, and gravitational constants, among which the semi-major axes and the flattening of the ellipsoid define the shape and eccentricity of the Earth's ellipsoid. Given that the ECEF coordinate system is non-inertial, there must be an angular velocity of the Earth's rotation. Besides, the gravitational constant is closely related to the calculation of a satellite's orbit period, the principles of which will be elaborated in Chap. 6.

Inside the receiver, the position of satellites functioning as reference points is indicated inside the ECEF system, along with the position of the user. However, the ECEF coordinate is somewhat unintelligible to common users if the output is delivered directly, as most people are accustomed to positions in the form of longitude, latitude, and height. Thus, the conversion between ECEF and the geodesic coordinate system is significant.

Further analysis suggests that we can use the following equations to realize the conversion:

$$x = (R_N + h)\cos(\phi)\cos(\lambda) \tag{1.19}$$

$$y = (R_N + h)\cos(\phi)\sin(\lambda) \tag{1.20}$$

$$z = [R_N(1 - e^2) + h]\sin(\phi) \tag{1.21}$$

in which

$$R_N = \frac{a}{\sqrt{1 - e^2 \sin^2(\phi)}} \tag{1.22}$$

The significance of R_N is explained in Fig. 1.11. The elliptic plane in the figure is the meridian plane that point P is on. If we draw a vertical line from point P to the elliptic curve, and it intersects with the ellipse and Y axis at points N and T, then

Table 1.1 Definition of the main parameters of the ECEF frame in the four GNSS systems

	Difference term				
GNSS	Coordinate frame	Semi−major axis/m	Oblateness	Angular velocity of the Earth's rotation/(radian/s)	Gravitational constant (GM)
GPS	WGS−84	6,378,137	1/298.257,223,563	$7.292{,}115{,}146{,}7 \times 10^{-5}$	$398{,}600.5 \times 10^{9}$
BDS	CGCS 2000	6,378,137	1/298.257,222,101	$7.292{,}115{,}146{,}7 \times 10^{-5}$	$398{,}600.4418 \times 10^{9}$
GLONASS	PZ−90	6,378,136	1/298.257,839,303	$7.292{,}115 \times 10^{-5}$	$398{,}600.44 \times 10^{9}$
Galileo	GTRF	6,378,137	1/298.257,222,101	$7.292{,}115{,}146{,}7 \times 10^{-5}$	$398{,}600.4418 \times 10^{9}$

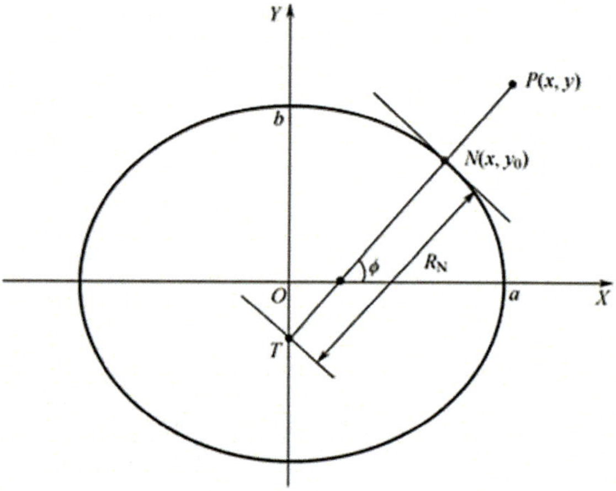

Fig. 1.11 The Significance of R_N

the value of R_N is the linear distance between points N and T. However, the ellipse in the figure is at variance with reality, as the eccentricity of the Earth's meridian plane is so small in value that the plane can be deemed a circle. The eccentricity is exaggerated here to clarify the figure. The intersection angle of line TN and the X axis is latitude Ø. Its derivation is too complicated, so it is defined in Appendix D.

Equations (1.19)–(1.21) show the relationship between velocity $(\dot{x}, \dot{y}, \dot{z})$ and $(\dot{\emptyset}, \dot{\lambda}, \dot{h})$ in the ECEF system. If we take derivatives of time on both sides of Eqs. (1.19)–(1.21), we get:

$$\dot{x} = (R_M + h)[-\sin(\phi)\cos(\lambda)]\dot{\phi} + (R_N + h)[-\cos(\phi)\sin(\lambda)]\dot{\lambda} + [\cos(\phi)\cos(\lambda)]\dot{h} \tag{1.23}$$

$$\dot{y} = (R_M + h)[-\sin(\phi)\sin(\lambda)]\dot{\phi} + (R_N + h)[\cos(\phi)\cos(\lambda)]\dot{\lambda} + [\cos(\phi)\sin(\lambda)]\dot{h} \tag{1.24}$$

$$\dot{z} = (R_M + h)[\cos(\phi)]\dot{\phi} + [\sin(\phi)]\dot{h} \tag{1.25}$$

Then, we rewrite Eqs. (1.23)–(1.25) in matrix form:

$$\begin{bmatrix} \dot{x} \\ \dot{y} \\ \dot{z} \end{bmatrix} = \begin{bmatrix} -\cos(\lambda)\sin(\phi) & -\sin(\lambda) & -\cos(\lambda)\cos(\phi) \\ -\sin(\lambda)\sin(\phi) & \cos(\lambda) & -\sin(\lambda)\cos(\phi) \\ \cos(\phi) & 0 & -\sin(\phi) \end{bmatrix} \begin{bmatrix} (R_M + h) & 0 & 0 \\ 0 & (R_N + h)\cos(\phi) & 0 \\ 0 & 0 & -1 \end{bmatrix} \begin{bmatrix} \dot{\phi} \\ \dot{\lambda} \\ \dot{h} \end{bmatrix} \tag{1.26}$$

R_M in Eq. (1.26) can be defined as

$$R_M = \frac{a(1-e^2)}{\left[1-e^2\sin^2(\phi)\right]^{3/2}} \tag{1.27}$$

Theoretically, if we know how to convert (Ø, λ, h) to (x, y, z), we can also perform the conversion backwards. However, it is difficult to achieve analytical solutions to Eqs. (1.19)–(1.21), so we usually adopt the iterative method listed below to make the conversion possible.

First, we initialize relative parameters:

$$\begin{aligned} \mathrm{h} &= 0 \\ R_N &= a \\ p &= \sqrt{x^2+y^2} \end{aligned}$$

Then, we iterate using the equations below until the solutions converge:

$$\begin{aligned} \sin(\phi) &= \frac{z}{(1-e^2)R_N+h} \\ \phi &= \mathrm{atan}\left(\frac{z+e^2R_N\sin(\phi)}{p}\right) \\ R_N &= \frac{a}{\sqrt{1-e^2\sin^2(\phi)}} \\ h &= \frac{p}{\cos(\phi)} - R_N \end{aligned}$$

In theory, from the above iteration process, we can get the value of Ø directly from arcsin $[z/(1-e^2)R_\mathrm{N}+h]$, but practice proves that Ø will converge much faster through calculating $\mathrm{atan}(\frac{z+e^2R_N\sin(\phi)}{p})$. The iteration method shown above allows the accuracy of Ø to reach centimeter level after five iterations.

1.2.4 The ENU or NED Coordinate System

The mechanism of the ENU coordinate system is indicated in Fig. 1.12. We take point P on the surface of the Earth as the origin of the ENU system. Then, we draw a tangent plane of the Earth's ellipsoid through P and make the Y axis point due north, the X axis due east, and the Z axis in the direction of the norm. To distinguish the ENU system from the ECEF system, which is also shown in the figure, we mark axes X, Y, Z of the former as X_{ENU}, Y_{ENU} and Z_{ENU}.

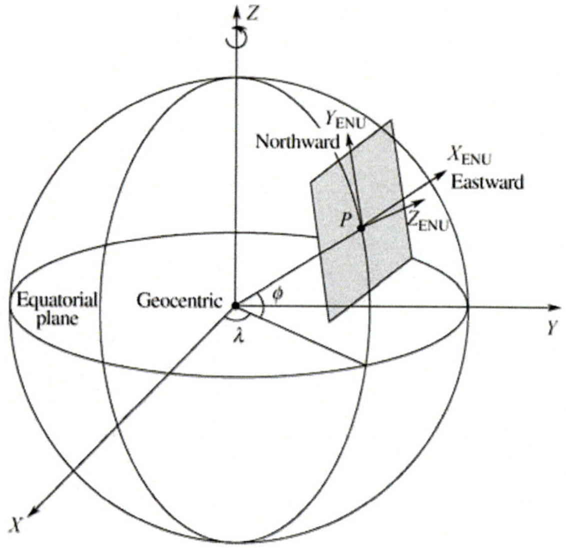

Fig. 1.12 The relationship between the ENU and ECEF coordinates

By definition, the position of the ENU system is decided by that of point P and will move with it. This is why some scholars also call it the Earth-fixed Coordinate System.

In real life, people are used to marking the position of a certain point with regard to themselves using terms *north, south, east,* and *west,* which are also applicable in describing positions in a local ENU system. For example, due north of point P refers to the direction along which the Y axis extends in the ENU system. Similarly, due east is where axis X points. In the Earth-centered inertial coordinate system, the *true north* of a certain point located in Shanghai, China, is completely different from that of a certain point in San Francisco, USA. Thus, confusion will arise if we talk about directions without knowing the position of relative reference points.

From Fig. 1.10, it would not be difficult to figure out a way to convert the ECEF system to the ENU system. First, we rotate the ECEF system around the Z axis by $(\lambda + \pi/2)$ degrees. The rotation matrix can be written as

$$\boldsymbol{R}_1 = \begin{bmatrix} \cos(\pi/2+\lambda) & \sin(\pi/2+\lambda) & 0 \\ -\sin(\pi/2+\lambda) & \cos(\pi/2+\lambda) & 0 \\ 0 & 0 & 1 \end{bmatrix} = \begin{bmatrix} -\sin(\lambda) & \cos(\lambda) & 0 \\ -\cos(\lambda) & -\sin(\lambda) & 0 \\ 0 & 0 & 1 \end{bmatrix} \tag{1.28}$$

Then, we rotate the coordinates around axis X by $(\pi/2 - \phi)$ degrees, the rotation matrix of which is

$$\boldsymbol{R}_2 = \begin{bmatrix} 1 & 0 & 0 \\ 0 & \cos(\pi/2-\phi) & \sin(\pi/2-\phi) \\ 0 & -\sin(\pi/2-\phi) & \cos(\pi/2-\phi) \end{bmatrix} = \begin{bmatrix} 1 & 0 & 0 \\ 0 & \sin(\phi) & \cos(\phi) \\ 0 & -\cos(\phi) & \sin(\phi) \end{bmatrix} \tag{1.29}$$

So, the rotation matrix converted from the ECEF system to ECU coordinates is

$$\boldsymbol{R}_{\mathrm{e2t}} = \boldsymbol{R}_2 \boldsymbol{R}_1 = \begin{bmatrix} -\sin(\lambda) & \cos(\lambda) & 0 \\ -\cos(\lambda)\sin(\phi) & -\sin(\lambda)\sin(\phi) & \cos(\phi) \\ \cos(\lambda)\cos(\phi) & \sin(\lambda)\cos(\phi) & \sin(\phi) \end{bmatrix} \tag{1.30}$$

The subscript e in Eq. (1.30) stands for the ECEF system, and subscript t (the first letter of *tangent plane*) represents the Earth-fixed Coordinate System.

NED coordinates are similar to the ENU system, as they are converted from the latter by interchanging its X and Y axis, and reversing the direction of its Z axis. In this case, directions N, E, and D can still form a right-handed coordinate system. The conversion can be written as

$$\begin{bmatrix} n \\ e \\ d \end{bmatrix} = \begin{bmatrix} 0 & 1 & 0 \\ 1 & 0 & 0 \\ 0 & 0 & -1 \end{bmatrix} \begin{bmatrix} e \\ n \\ u \end{bmatrix} \tag{1.31}$$

Therefore, the conversion from the ECEF system to NED coordinates becomes

$$\begin{bmatrix} 0 & 1 & 0 \\ 1 & 0 & 0 \\ 0 & 0 & -1 \end{bmatrix} \boldsymbol{R}_{\mathrm{e2t}} = \begin{bmatrix} -\cos(\lambda)\sin(\phi) & -\sin(\lambda)\sin(\phi) & \cos(\phi) \\ -\sin(\lambda) & \cos(\lambda) & 0 \\ -\cos(\lambda)\cos(\phi) & -\sin(\lambda)\cos(\phi) & -\sin(\phi) \end{bmatrix} \tag{1.32}$$

No matter whether the ECEF system is converted to the ENU system or to NED coordinates, the conversion matrix $\boldsymbol{R}_{\mathrm{e2t}}$ is a unitary one, namely

$$\boldsymbol{R}_{\mathrm{e2t}} \boldsymbol{R}_{\mathrm{e2t}}^{\mathrm{T}} = \boldsymbol{I} \tag{1.33}$$

In physics, Eq. (1.33) can be analyzed in two steps. $R_{\mathrm{e2t}}^{\mathrm{T}} = R_{\mathrm{e2t}}^{-1}$ means the conversion from the ENU system to the ECEF system, so the first step is to convert vectors from the ENU system to the ECEF system and left-multiplying matrix. R_{e2t} means the conversion from the ECEF system to the ENU system. The whole process brings no change to the original vectors.

The ENU system is broadly applied in GNSS receivers, and its most popular use is in calculating a satellite's elevation angle and azimuth. Since the position coordinates of GNSS satellites and the calculated position coordinates of the user are all in the ECEF system, we can assume that the vector of the relative position between GNSS satellites and the user is $(\Delta x, \Delta y, \Delta z)_{\mathrm{ECEF}}$, then convert it to the ENU system via the R_{e2t} matrix:

$$\begin{bmatrix} \Delta e \\ \Delta n \\ \Delta u \end{bmatrix} = \boldsymbol{R}_{\mathrm{e2t}} \begin{bmatrix} \Delta x \\ \Delta y \\ \Delta z \end{bmatrix} \tag{1.34}$$

The elevation angle α of the satellites is defined as the intersection angle between the relative position vector $(\Delta x, \Delta y, \Delta z)_{\mathrm{ENU}}$ and the tangent plane between the east axis and the north axis. The angle can be calculated using the equations below:

$$\alpha = \arcsin\left[\frac{\Delta u}{\sqrt{\Delta e^2 + \Delta n^2 + \Delta u^2}}\right] \tag{1.35}$$

or

$$\alpha = \arctan\left[\frac{\Delta u}{\sqrt{\Delta e^2 + \Delta n^2}}\right] \tag{1.36}$$

The azimuth β is defined as the intersection angle between the projection vector of the relative position vector on the tangent plane and the north axis, i.e.

$$\alpha = \arctan\left[\frac{\Delta e}{\Delta n}\right] \tag{1.37}$$

Matrix R_{e2t} is also useful in the conversion of the velocity vector between the ECEF and ENU systems, which is written as $[\dot{x}, \dot{y}, \dot{z}]^T$ in the ECEF system and $[v_n, v_e, v_d]^T$ in the ENU system. So,

$$\begin{bmatrix} v_n \\ v_e \\ v_d \end{bmatrix} = \begin{bmatrix} -\cos(\lambda)\sin(\phi) & -\sin(\lambda)\sin(\phi) & \cos(\phi) \\ -\sin(\lambda) & \cos(\lambda) & 0 \\ -\cos(\lambda)\cos(\phi) & -\sin(\lambda)\cos(\phi) & -\sin(\phi) \end{bmatrix} \begin{bmatrix} \dot{x} \\ \dot{y} \\ \dot{z} \end{bmatrix} \tag{1.38}$$

By comparing Eqs. (1.38) and (1.26), we know that

$$\begin{bmatrix} v_n \\ v_e \\ v_d \end{bmatrix} = \begin{bmatrix} (R_M + h) & 0 & 0 \\ 0 & (R_N + h)\cos(\phi) & 0 \\ 0 & 0 & -1 \end{bmatrix} \begin{bmatrix} \dot{\phi} \\ \dot{\lambda} \\ \dot{h} \end{bmatrix} \tag{1.39}$$

Equation (1.39) explains the relation between the velocity vectors of the north and east and the changing rate of the longitude and the latitude. The inverse transformation of Eq. (1.39) is

$$\begin{bmatrix} \dot{\phi} \\ \dot{\lambda} \\ \dot{h} \end{bmatrix} = \begin{bmatrix} \frac{1}{(R_M+h)} & 0 & 0 \\ 0 & \frac{1}{(R_N+h)\cos(\phi)} & 0 \\ 0 & 0 & -1 \end{bmatrix} \begin{bmatrix} v_n \\ v_e \\ v_d \end{bmatrix} \tag{1.40}$$

1.2.5 The Body Frame Coordinate System

Body frame refers to coordinates that are strictly affixed to moving objects. Here, this means all objects that require navigation, like moving aircraft, cars, and vessels. Generally, the origin of a body frame is a fixed point in or on the moving object, for example, its center of mass. The axes of the frame point to the three orthogonal directions of the moving object, and are usually represented by $[u, v, w]$. Theoretically, we can set any directions for the axes as long as they are orthogonal. However, in practice, for the sake of convenience we will make axis u extend along the forward direction of the moving object, axis v towards its right, and axis w in a downward direction. The three axes will then form a right-handed coordinate system, seen in Fig. 1.13.

Since the body frame is strictly affixed to the moving object, it will doubtlessly move with it (see Fig. 1.13). This pattern is similar but not identical to the ENU system. For instance, when a moving object rotates around its mass point at a certain position on the Earth's surface, the ENU system stays the same while the body frame rotates with it.

The body frame is closely related to attitude-control requirements. The translation of a moving object decides the position of its center of mass, while the rotation of the object around its center of mass decides its attitude, including its yaw, pitch, and roll. We can rotate the body frame around axis u to get its roll, rotate it around axis v to get its pitch, and rotate it around axis w to get its yaw. The definitions of the three types of attitude are presented in Fig. 1.14.

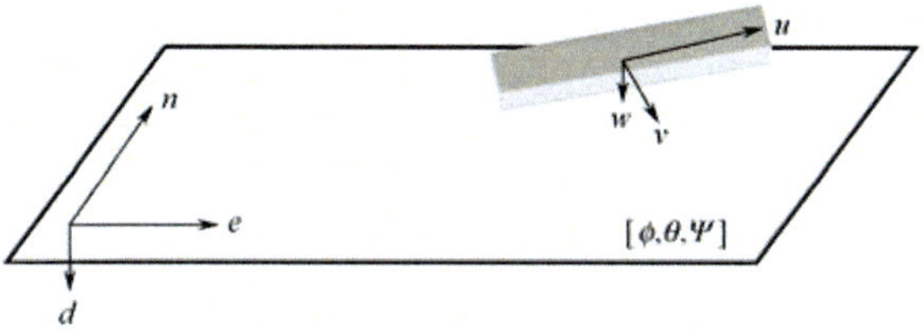

Fig. 1.13 An illustration of a body coordinate system as it changes with the motion of the carrier

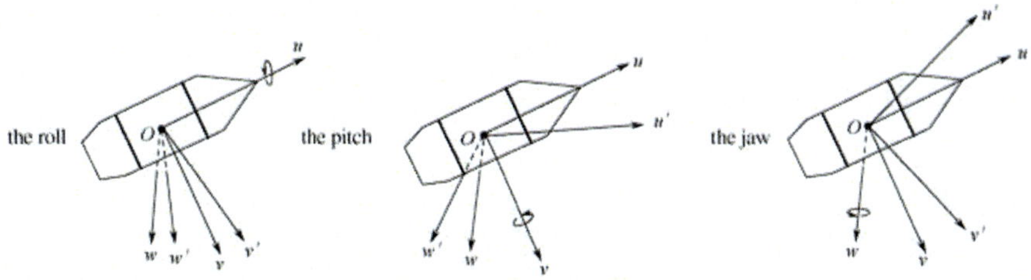

Fig. 1.14 Definition of the pitch, roll, and yaw

The attitude of a moving object cannot be determined by observing a single GNSS, but by two or more GNSS antenna arrays, and this measurement is called GNSS attitude determination. Attitude measurement of moving objects using GNSS is a GNSS attitude determination problem. Many scientific breakthroughs have been made on this issue, and there is a wide array of relevant literature on the subject for readers who are interested.

In a navigation system, the required position and attitude information of a moving object is always linked to a certain navigational coordinate system, which generally means a fixed reference coordinate system. Meanwhile, the body frame, relative to the reference coordinates, changes with the movement of the object. Thus, to realize a conversion from the body frame to the reference coordinates, we must know the position and attitude of the moving object, which is usually achieved by relative sensors fixed on the object, like an Inertial Measurement Unit (IMU). We can tell from Fig. 1.14 that the navigation coordinates are of the NED kind, hence the position and attitude of the moving object are also indicated in that coordinate system. Since this book only deals with the contents of independent GPS receivers, the body frame coordinate system is only briefly explained. Interested readers can refer to publications on inertial navigation and integrated navigation.

1.3 The Time System

The seven basic units in the International System of Units (abbreviated as SI) are *metre (m), kilogram (m), second (s), ampere (A), kelvin (K), candela (cd),* and *mole (mol).* Time is one of the seven units, and is hitherto the most accurate measurement of quantity in physics. As previously mentioned, although the accuracy of time measurement has increased with time, its basic principles have not changed very much. They consist of recording the passage of time through counting a periodic phenomenon, like sunrise and sunset, the turning over of a sand clock, or the swinging of a pendulum, and the oscillation of atomic clocks. Different periodic phenomena lead to different time systems. Since Sect. 1.1.1 explains the importance of accurate time synchronization and measurement to the functioning of GPS system, this section will elaborate on several major time systems in history and at present.

1.3.1 Solar Time and Sidereal Time

In ancient times, people recorded time by referring to the rising and setting of the Sun. One rise and set suggested the passage of one solar day—the period in which the Earth rotates once. However, if we compare two random solar days, it is easy to see that they are not identical in length, which means that solar days are not uniform. This is because the Earth's orbit around the Sun is not a perfect circle, but an ellipse. According to Kepler's second law, the velocity of the Earth's movement is faster on certain spots along the orbit and slower on others. Meanwhile, there is an angle of intersection between the Earth's equatorial and ecliptic planes, which means that the rotation axis of the Earth is not perpendicular to the ecliptic plane—another factor that will make the length of solar days uneven. This gave rise to the hypothesis of a *flat sun*, around which the Earth rotates at a constant speed in a hypothesized round orbit, with its rotation axis perpendicular to the ecliptic plane. Under this circumstance, a complete rotation period of the Earth is called a *mean solar day*.

A sidereal day is defined as the time in which the Earth rotates once, relative to a distant star. The length of a sidereal day is not constant in the strict sense, which is why the *mean sidereal day* was introduced to the system. The Earth revolves around the Sun as it rotates, thus in the period of one year (here referring to the period in which the Earth travels once around the Sun), the Earth will rotate one extra time relative to a distant star. This means that a mean solar day is longer than a mean sidereal day by approximately four minutes. The accurate value is:

$$\text{The difference} = \frac{3600 \times 24\,\text{s}}{365.25} \approx 236\,\text{s} = 3\,\text{m}\,56\,\text{s}$$

So, the approximate length of a mean sidereal day is 23 h 56 min and 4 s. In fact, it takes half of the sidereal day (around 11 h 58 min and 2 s) for a GPS satellite to travel around the Earth.

The time systems based on mean solar days and mean sidereal days are called *Solar Time* and *Sidereal Time* respectively.

The mean solar time using the Greenwich Meridian as the baseline is called *Universal Time* (UT). It is also a time system based on the Earth's rotation. In the early twentieth century, astronomers found out that the rotation velocity of the Earth was not constant. Even though factors affecting its velocity are still unclear, the Earth is rotating more and more slowly. Astronomers have also noticed that the movement of the Earth's poles and the change of seasons also affect the accuracy of UT. Taking these things into consideration, they introduced three time standards—UT0, UT1, and UT2—to the recording of UT. UT0 is the mean solar time recorded by multiple observatories. Based on UT0, UT1 corrects errors brought about by the polar motion of the Earth, and UT2 adds the effect that the change of seasons has on the Earth. In brief, the relationships among the three versions of UT can be written as

$$\text{UT1} = \text{UT0} + \Delta\lambda \tag{1.41}$$

$$\text{UT2} = \text{UT1} + \Delta\text{TS} \tag{1.42}$$

In Eq. (1.41), $\Delta\lambda$ stands for the correction of polar motion, while in Eq. (1.42), ΔTS is the correction of the Earth's rotation velocity error caused by the change of seasons.

Even after correction, UT2 cannot completely offset the slowing down of the Earth's rotation and other irregular variations. Therefore, UT2 is still an uneven time system. The 1 s obtained by this definition cannot meet the growing requirements for time accuracy, and will unquestionably be replaced by more accurate and uniform atomic time.

1.3.2 Dynamical Time

Dynamical Time is an important concept in astronomy. It is mainly used to describe the movement of celestial bodies in certain coordinate systems and gravitational fields, affected by some basic factors like general relativity and the inertial system. Approximately speaking, the nearest inertial system to us is the coordinate system whose origin is at the barycenter of the solar system, known as Barycentric Dynamical Time (TDB).[3] It considers the effect of relativity, and is continuous as well as consistent. If we place a clock on the surface of the Earth, because of the rotation of the Earth in the gravitational field of the solar system, the difference between the time on that clock and TDB can reach a maximum of 1.6 ms.

While describing the movement of near-earth objects (NEOs), we should include another form of dynamical time, namely *Terrestrial Dynamical Time* (TDT). This originated from *Ephemeris Time*, which describes the time scale of NEO ephemeris. Before the emergence of accurate atomic time, 1 s in ephemeris time was defined by American astronomer Simon Newcomb as the $\frac{1}{31556925.9747}$ of the time scale of the year 1900. As people discovered the negative effect of the Earth's rotation on ephemeris time, it was replaced by TDT. On 1st January 1977, a deviation of 32.184 s was added to TDT based on International Atomic Time (TAI):

$$\mathrm{TDT} = \mathrm{TAI} + 32.184\ \mathrm{s} \tag{1.43}$$

1.3.3 Atomic Time and Coordinated Universal Time (UTC)

Universal time based on the Earth's rotation can only reach an accuracy of 10^{-9}. With growing requirements for greater accuracy and stability, timing systems that were created based on the rotation of the Earth were no longer sufficient. In the 1950s, an atomic time system was established based on the transition mechanism among atomic energy levels. The updated definition of a *second* is the time it takes for the transition radiation to oscillate 9,192,631,770 times in the zero-magnetic field between two hyperfine levels of a cesium atom in its ground state at sea level. Atomic time is accurate, consistent, and free from the influence of the Earth's rotation and revolution. TAI is the time system built through integrating data from multiple atomic clocks around the world. The starting point of atomic time should have overlapped

[3]The acronym here and the later TDT is from the word in French.

with UT2 at 00:00 on 1st January 1958, but the former fell 0.0039 s behind the latter. This was recorded as a historical fact.

Due to the high accuracy and stability of the frequency of atomic clocks, the corresponding derived atomic time also has very high accuracy and stability in terms of time. Currently, the frequency stability of a rubidium atomic clock can reach 10^{-12}–10^{-13} in a day. Accordingly, the cumulative clock bias in a year is around 31 μs. Meanwhile, the stability of a hydrogen atomic clock can be as high as 10^{-14}, which means that the annual cumulative clock bias is an astonishing 3.1 μs.

However, the near perfection of atomic time has brought about a problem. With the passage of time, a greater difference will appear between atomic time and universal time. The reason for this is that the slowing of the Earth's rotation will cause UT2 to fall behind atomic time, and the gap will keep expanding. Figure 1.15 shows IERS's record of the change in the Earth's rotation period from 1973 to 2008. The cause of its loss of pace is still unclear, and hypotheses (such as the tide effect between the Earth and the moon, the melting of icebergs, and the accretion of the centrosphere) are yet to be verified.

To shrink the gap between the two time systems, the concept of *Coordinated Universal Time* (UTC) was put forward. The basic timing unit was still seconds in atomic time, but the concept of *leap seconds* was brought in. A leap second can be both positive and negative, and reins the time difference between UTC and UT1 within 0.9 of a second. However, special attention is needed here to distinguish UTC from UT, as UTC is an atomic clock in nature while UT is a system based on the Earth's rotation.

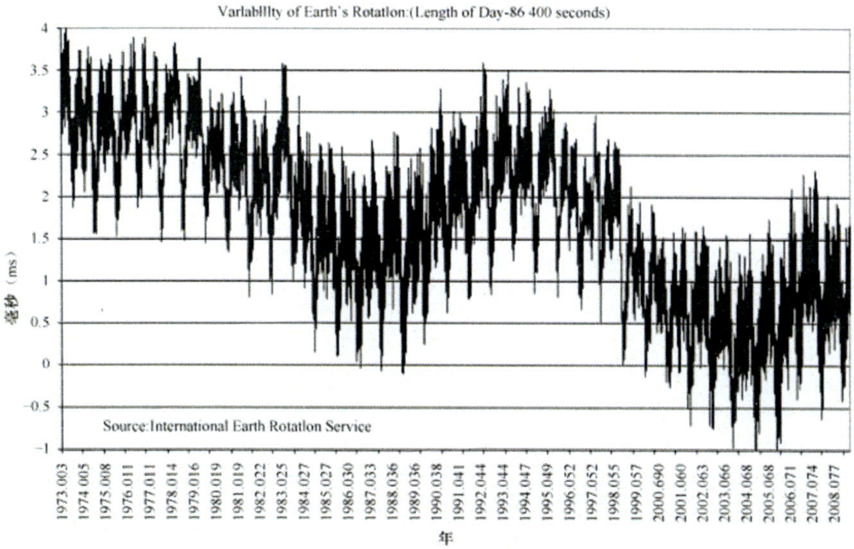

Fig. 1.15 Changes in the Earth's rotation period between 1973 and 2008

The International Earth Rotation Service (IERS) oversees the maintenance and release of leap seconds. Normally, the release date is 30th June or 31st December every year. If a leap second appears, then the length of the minute containing it is 61 s instead of 60 s. Therefore, in the long run, the UTC curve should resemble the curve of a step function instead of a smooth one. Thus far, UTC has proven to be a significant time standard influencing multiple sectors of daily life, as almost every country in the world uses it as the standard time. It is also an important task for GPS receivers to convert GPS time to UTC time.

IERS (https://www.iers.org) is an international organization that provides Earth orientation parameters, having established a celestial agreement and terrestrial reference system. As well as the leap seconds mentioned earlier in this section, it is also in charge of the detection and release of the polar motion issue in the ECI system mentioned earlier, along with related data. IERS was co-founded by the International Astronomical Union (IAU) and the General Assembly of the International Union of Geodesy and Geophysics (IUGG) in 1987, and started functioning the following year. It currently has branches in the USA, Europe, and Australia.

1.3.4 GPS Time (GPST)

GPS time (GPST) is the time reference for the GPS system. It is a form of atomic time by nature, and must be understood thoroughly if one is to grasp the principles of the GPS receiver. The time reference for GPS time comes from measuring the frequencies of a series of atomic clocks, including those installed in ground control stations and carried by space satellites. The whole-second carry time of GPST synchronizes with UTC despite a minor difference of approximately 10 ns. However, unlike UTC, GPST is consistent, and is free from the concern of leap seconds. From this we can infer that there will be jumps in the time difference between GPST and UTC.

The recording of GPS time started at the interlocking of midnight on 5th January 1980 and dawn on 6th January 1980, which was 00:00:00 on 6th January 1980 in UTC. From then on, the time difference between the two systems has been growing (its historical record is shown in Table 1.2. By the time this book is published (June, 2016), the difference will already have reached 16 s, with the latest jump second having occurred on 1st July 2012, after which,

$$\text{GPST} \approx \text{UTC} + 16 \tag{1.44}$$

An approximate equals sign is used in Eq. (1.44) because, apart from the whole-second difference between GPST and UTC caused by the accumulation of leap seconds, there is still a difference of less than 1 μs within each second, whose value can be computed with UTC time parameters broadcasted by GPS satellites.

Table 1.2 Historical records of GPST and UTC deviations

Date	Julian day	GPST−UTC
Jan. 1st 1980	JD 2 444 239.5	0
Jul. 1st 1981	JD 2 444 786.5	1
Jul. 1st 1982	JD 2 445 151.5	2
Jul. 1st 1983	JD 2 445 516.5	3
Jul. 1st 1985	JD 2 446 247.5	4
Jan. 1st 1988	JD 2 447 161.5	5
Jan. 1st 1990	JD 2 447 892.5	6
Jan. 1st 1991	JD 2 448 257.5	7
Jul. 1st 1992	JD 2 448 804.5	8
Jul. 1st 1993	JD 2 449 169.5	9
Jul. 1st 1994	JD 2 449 534.5	10
Jan. 1st 1996	JD 2 450 083.5	11
Jul. 1st 1997	JD 2 450 630.5	12
Jan. 1st 1999	JD 2 451 179.5	13
Jan. 1st 2006	JD 2 453 736.5	14
Jan. 1st 2009	JD 2 454 832.5	15
Jul. 1st 2012	JD 2 456 109.1	16

A new concept, the *Julian Day* (JD), is presented in Fig. 1.2. It is a type of long-term day recording model that excludes the recording of years and months, measuring only the number of days since the Greenwich mean noon (UT 12:00) of 1st January 4713 BCE. This model is mainly adopted by astronomers as a singular calendar in the astronomical field. Unlike today's concept of a day, which is the time from one midnight to another, a Julian day starts on noon of one day and ends at noon on the next day. This is because astronomical observations are mostly carried out at night, and the Julian day model is more convenient for recording astronomical events. One Julian period[4] is 7980 years. The first year of the recording of Julian days (4713 BCE) is also the first year of the latest Julian period.

[4]The derived Julian Cycle is related to three factors: the Solar Cycle, Metonic Cycle, and Indiction Cycle. The Solar Cycle lasts for 28 years, the Metonic Cycle lasts for 19 years, and the Indiction Cycle lasts for 15 years. The small common multiple of 28, 19, and 15 is $28 \times 19 \times 15 = 7980$. The Solar Cycle and Metonic Cycle are determined by the orbit of the Sun and the Moon respectively, while the Indiction Cycle was the period of taxation issued by the ancient Roman Emperor.

However, Julian days are too lengthy to be used in practical situations. Instead, the concept of a Modified Julian Day (MJD) was introduced. The last MJD started on midnight on 17th November 1858, with the corresponding Julian day being 2,400,000.5, so we have

$$\mathrm{MJD} = \mathrm{JD} - 2\,400\,000.5 \tag{1.45}$$

As for GPST, the corresponding Julian day for its starting point, 00:00:00 on 6th January 1980, is JD 2,444,244.5.

Meanwhile, the relationship between GPST and TAI is much simpler, as they are both forms of atomic time and possess continuity and evenness. The time difference between them is fixed at 19 s, namely

$$\mathrm{TAI} = \mathrm{GPST} + 19\ \mathrm{s} \tag{1.46}$$

The basic unit for measuring time in GPST is also seconds, but GPST considers continuous time to be periodic. The period is one week, that is, $7 \times 24 \times 3600 = 604,800$ s. The count of seconds in GPST is cleared to zero at midnight on Saturday or at dawn on Sunday every week. Its week count is incremented by 1, and the second count is re-accumulated until the beginning of the next week. So when referring to GPST, besides giving the second count in that week, we must also give the number of weeks of the current GPST. In the GPS navigation message, 10 bites are used to record the number of weeks, so the recorded number of weeks is the remainder of the absolute number of weeks of GPST divided by 1024; when the number reaches 1024, it will overflow and be cleared. Since the beginning of GPST, the number of weeks has been cleared once, between midnight on 21st August 1999 and early morning on the 22nd.

The radio navigation signals that are transmitted by each GPS satellite have a continuous timestamp, which is consistent with the conclusions we obtained in Sect. 1.1.4. Whenever a user receives the GPS satellite signal, they can get the transmission time of the current signal through a simple calculation. This timestamp is sent as a GPST and includes the number of GPS weeks and the count of seconds in the current week. The navigation signals transmitted by all GPS satellites are strictly synchronized, which is essential for the implementation of GPS positioning. The synchronization relationship between different GPS signals and BeiDou satellite signals is also detailed in Sect. 3.4.

1.3.5 BeiDou Time (BDT)

BeiDou Time (BDT) is the time base of the BeiDou satellite navigation system. Similar to GPST, it is a form of atomic time whose time reference is derived from the time measurement of a series of atomic clocks, including observations from ground-based monitoring stations and atomic clocks carried by BeiDou satellites. Since the BeiDou satellite navigation system is still in a process of continuous promotion and deployment, many aspects have yet to be perfected, so the detailed information about BDT is difficult to find from published literature. Some of the conclusions about BDT in this book are based on very limited literature, or are even inferred from analogies with GPST and common sense. Therefore readers must take note that mistakes and omissions are inevitable.

According to version 1.0 of the BeiDou Interface Control Document (ICD) published on 27th December 2012, in BDT, time is accumulated as seconds of SI units, and no leap seconds are involved. So, like GPST, BDT is continuous. The starting epoch of BDT was UTC 00:00:00 on 1st January 2006, and the corresponding Julian day is 2,453,736.

BDT records time the same way as GPST, i.e. by counting the number of weeks and the number of seconds in a week. The value of the latter is 604,800 s, so starting from 00:00:00 on 1st January 2006, BDT adds one week to every 604,800 s counted, and clears the count of seconds in the week. Since BDT is also a continuous time system, its difference from UTC will inevitably widen. From 1st January 2006 to 2013, there were two adjustments in leap seconds, so as of May 2013, the time difference between UTC and BDT was already 2 s.

Given that the number of weeks and the count of seconds in a week at the starting epoch of BDT is

$$\text{GPST} = [1356\,\text{weeks},\ 14.000\,\text{s}]$$

we can infer that the relation between BDT and GDST is

$$\text{BDT week number} = \text{GPST week number} + 1356 \tag{1.47}$$

$$\text{BDT week seconds count} = \text{GPST week seconds count} + 14 \tag{1.48}$$

The issue of overflow must be considered in the calculation of Eq. (1.48), which means that when the value (the second count in a week in GPST + 14) exceeds 604,800, we should increase the number of BDT weeks by 1.

As a conclusion of this chapter, Fig. 1.16 indicates the interrelation of several time systems.

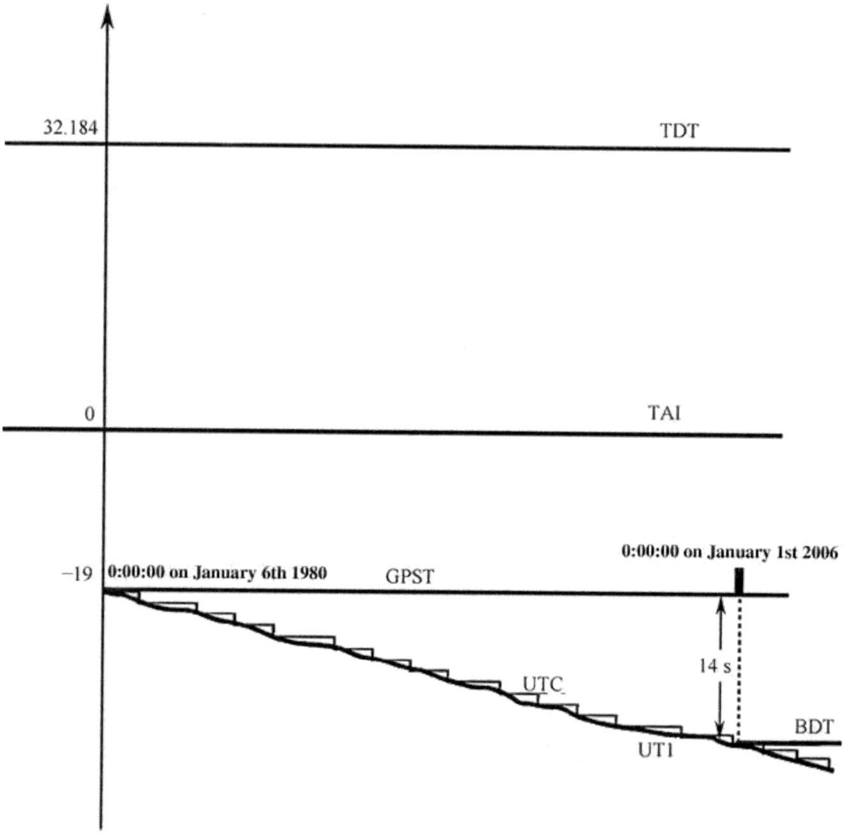

Fig. 1.16 The interrelation of several time systems

References

1. Chong C (2010) Questions and answers on common knowledge of satellite navigation. Electronic Industry Press, Beijing
2. Sobel D (2007) Longitude: a true story of a lonely genius solving the biggest problem of his times (trans: Xiao M). Shanghai People's Publishing House, Shanghai
3. Baolin L, Wei F (1998) The god of battlefield positioning—the military application of GPS. Lantian Publishing House, Beijing
4. Zaocheng Z, Weiqun Z et al (2009) The basic principle of the atomic clock and time frequency measurement technology. Shanghai Science and Technology Literature Press, Shanghai
5. Farrell JA (2008) Aided navigation. GPS with high rate sensors. McGraw Hill
6. Jiyu L (2003) The principle and method of GPS satellite navigation and positioning. Science Press, Beijing
7. Kaplan ED (2006) Understanding GPS principles and applications, 2nd edn. Artech House Publishers
8. Navstar GPS space segment/navigation user interfaces, IS-GPS-200G. Sept 5, 2012
9. China Satellite Navigation System Management Office. "Beidou satellite navigation system space signal interface control document" (open service signal), Version 2.0. Dec 2013

10. Misra P, Enge P (2008) Global positioning system—signal, measurement and performance, 2nd edn. Electronic Industry Press, Beijing
11. GPS website: https://www.gps.gov
12. Beidou website: https://www.beidou.gov.cn
13. Galileo website: https://www.gsa.europa.eu
14. GLONASS website: https://glonass-iac.ru/en/
15. IERS website: https://www.iers.org

Chapter 2
Brief Introduction to the GPS and BeiDou Satellite Navigation Systems

This chapter first gives a brief overview of the GPS system from the macro perspective, then gives a similar overview of China's independently planned, designed, and implemented BeiDou Satellite Navigation System. Emphasis is placed on the comprehensive description of the satellite constellation, signal system, carrier frequency band, performance index, and application scope of the GPS and BeiDou systems from the perspective of top-level design. At the same time, some historical development events relating to the satellite navigation system are described and remarked upon. This will provide readers who are new to the field with a preliminary understanding of the system's composition as well as the historical evolution of the field, so as to grasp the characteristics of the existing system and the direction of future development. As a major event in satellite navigation in the twenty-first century, the modernization of the GPS system will be described in detail in this chapter, and the impact of this event on future GPS receivers will be analyzed. At the same time, the performance evolution trend of future GNSS receivers will be analyzed in combination with the deployment and implementation of China's BeiDou Satellite Navigation System.

2.1 The History of the GPS System

The basic principle of GPS positioning is radio navigation, i.e. to navigate and position by measuring the electrical signal parameters (such as amplitude, frequency, and carrier phase) of electromagnetic waves during spatial propagation.

On 4th October 1957, the Soviet Union launched the first artificial Earth satellite, Sputnik-I, which was the first step in using space satellites to achieve radio navigation. Sputnik-I was a polished metal sphere with a diameter of 58 cm and a mass of approximately 83.6 kg. Four transmitting antennae were installed on the exterior to

Y. Lu, *BDS/GPS Dual-Mode Software Receiver*, Navigation: Science and Technology 10, https://doi.org/10.1007/978-981-16-1075-2_2

broadcast continuous radio wave signals of 20.005 and 40.002 MHz. Its orbit was elliptical, with an eccentricity of 0.05201, and its semi-major axis was 6955.2 km. According to the basic properties of the elliptical curve, it can be inferred that the orbital perigee distance was about 6593 km from the center of the Earth, and the apogee was about 7317 km from the center of the Earth, considering that the radius of the Earth is about 6378 km, so Sputnik-I basically travelled between 215 and 939 km from the ground plane. At the same time, Kepler's law can be used to calculate an orbital period of 92 min and 12 s and a flight speed of around 8 km/s.

Sputnik-I was not launched for navigation but for atmospheric observations. By measuring its aerodynamic drag, the atmospheric concentration of the outer atmosphere could be inferred. Information about the ionosphere could be gained through observations of radio waves, and a series of interesting physical phenomena were later observed (Fig. 2.1).

A few days after the launch of Sputnik-I, William Guier and George Weiffenbach—two physicists at the Applied Physics Laboratory (APL) at Johns Hopkins University—discovered the Doppler shift phenomenon on Sputnik-I, which was caused by radio waves.

Weiffenbach was completing his doctoral thesis on microwave spectroscopy, so they used a 20 MHz receiver to pick up the 20.005 MHz radio wave from Sputnik-I. The different frequency signal was within the range of sonic frequencies that can be heard directly by the human ear. They later spent an entire night recording all of the radio signals as the satellite rose above the ground and then fell below it. When they analyzed the frequency of the complete radio signal, they discovered a change in the radio frequency during the whole satellite operation. From that they quickly realized

Fig. 2.1 Sputnik-I: the first artificial earth satellite

that the curve can reflect changes in relative motion speed between the satellite and the receiver. These changes are directly related to the satellite's flight track and the geographic location of the receiver. Combined with dual-frequency signal data from the later Sputnik-II, Guier and Weiffenbach worked out the six orbital parameters and three system parameters of the satellite operation, including the free electron concentration of the atmospheric ionosphere.

Dr. Frank McClure, then deputy director of the Applied Physics Laboratory, was aware of Guier and Weiffenbach's work. He came up with an inverse proposition that if the orbital parameters of the satellite are known, can the user's position be calculated based on the Doppler shift?

Guier and Weiffenbach quickly completed a theoretical analysis of this proposition. It showed that it was not only feasible, but also that the positioning accuracy would be much higher than expected, because the unknown quantity is lower than the positive proposition. The TRANSIT Navigation System was later developed from their conclusion.

The TRANSIT Navigation System, also known as the Navy Navigation Satellite System (NNSS), was originally designed to solve the positioning problem of the Polaris Missile—a nuclear-armed submarine-launched ballistic missile. As the precise location of the nuclear submarine itself could not be determined, the accuracy of the target position of a missile attack could not be guaranteed. The Polaris nuclear submarine was an extremely important part of the ternary system of the US strategic nuclear force, so this was a major strategic problem.

It is worth mentioning that although the TRANSIT Navigation System was designed to solve a strategic problem for the US military, when the United States later decrypted navigation equipment and computer programs in 1967, some civilian marine equipment and ships also began to use the TRANSIT Navigation System for positioning.

The TRANSIT Navigation System was developed through a series of experimental satellite launches, such as the TRANSIT −1A and −1B, and the subsequent TRANSIT −5A, −5B, −5E, and −5C. Seven and a half years were spent on development, after which the satellites were finalized and put into use. The final TRANSIT satellite runs on a circular polar orbit with an orbital height of about 1075 km, an orbital period of about 107 min, and a space flight speed of about 7.3 km/s. As Fig. 2.2 shows, since the satellite's orbit passes through the Earth's polar axis, the orbital distribution of the constellation resembles a huge birdcage with the Earth inside it.

The transmission carrier frequency of the TRANSIT satellite was made up of the 150 MHz and the 400 MHz radio navigation signals. They were used to measure the Doppler count of the received radio signal. The user could only receive a signal from one TRANSIT satellite at any time, which is completely different from the later GPS system. Whenever a TRANSIT satellite appears above the horizon, the receiver begins to obtain navigation signals from the satellite, including six orbital parameters and orbital disturbances. From these, the user can calculate the satellite's position at any time. The 150 and 400 MHz dual-frequency signals are used to eliminate the effects of ionospheric delay, further improving the accuracy of positioning.

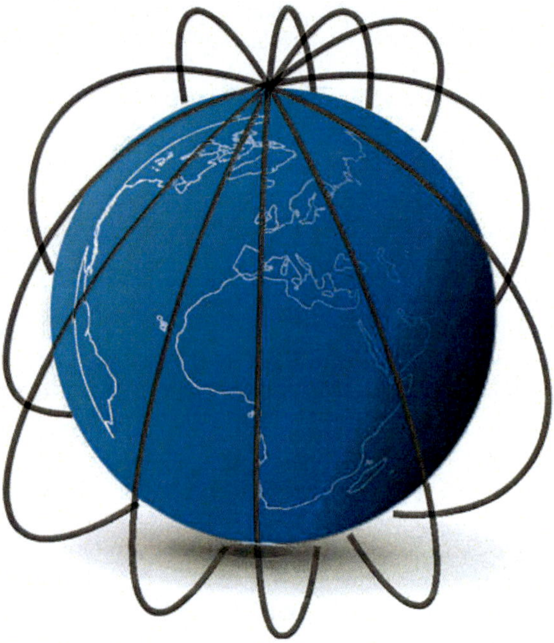

Fig. 2.2 Distribution of the constellation orbits of TRANSIT satellites

Positioning using the TRANSIT Navigation System is carried out as follows: when the TRANSIT satellite passes over the receiver, the receiver first locks the radio navigation signal broadcasted by the satellite, then demodulates the satellite's ephemeris data and the track disturbance in the navigation message. It then calculates the satellite's position at any time, and the Doppler count of the satellite at different orbital positions are calculated at the same time. The Doppler count here includes information about the user's position, which can be calculated by the hyperbolic method in radio navigation. Under the conditions of the era, the process was complicated and involved intensive data processing, requiring a sophisticated computer. Figure 2.3 is a schematic diagram of the positioning of the TRANSIT receiver by measuring the Doppler count of satellite signals at different times.

Figure 2.4 shows the Doppler count $D(t_i, t_{i+1})$ of satellite signals that the receiver obtains from four time records ($i = 0 \ldots 3$). $T = t_{i+1} - t_i$ is the time difference between the two observations. If the technical details are not calculated, $D(t_i, t_{i+1})$ is

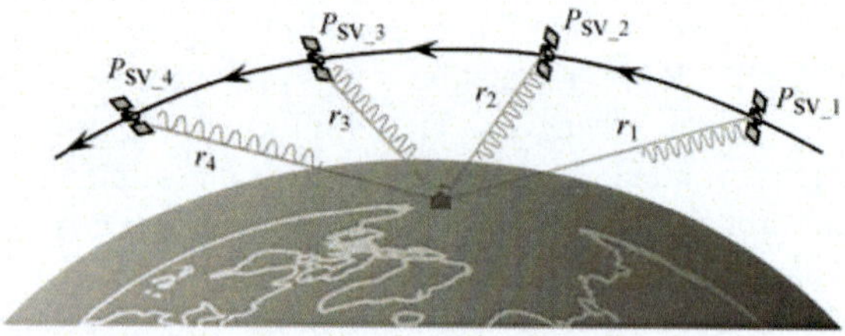

Fig. 2.3 Positioning the TRANSIT receiver by Doppler counting

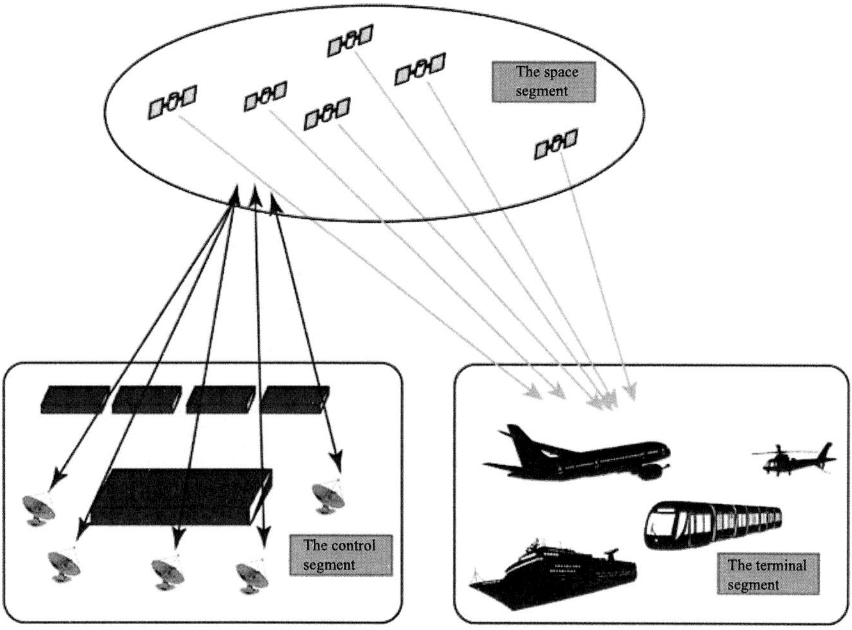

Fig. 2.4 The general structure of the GPS system

$$D(t_i, t_{i+1}) = \frac{1}{\lambda}\int_{t_i}^{t_{i+1}} \dot{r}(t)\mathrm{d}t = \frac{1}{\lambda}\left[r(t_{i+1}) - r(t_i)\right] \quad (2.1)$$

$D(t_i, t_{i+1})$ indicates the difference in distance from the satellites to the receiver in t_i and t_{i+1}. Therefore, as long as the Doppler count value is accumulated three times, the longitude, latitude, and frequency difference of the receiver can be corrected to achieve positioning. In fact, because the receiver can obtain many Doppler counts during a satellite's flight, the receiver processes these redundant observations using the least-squares method to gain an optimal user position solution. If the reader is familiar with the GPS carrier phase observation, the $D(t_i, t_{i+1})$ and GPS carrier phase measurements are similar.

The positioning principle of the TRANSIT system determines the positioning accuracy of objects that are stationary or slow moving. For an object moving at high speed, there will be errors in positioning accuracy This is because the distance vector between the receiver and the satellite is changing at two observations. This change has two causes, one of which is generated by the orbital flight of the satellite, and the other by the user's motion. The former is composed of satellite ephemeris data. It can be calculated more accurately, and the latter needs to calculate the position according to the user's motion speed. In this process, the user's speed error will correspond to the estimated user position error.

In general, for slow-moving ships, the TRANSIT system can achieve a positioning accuracy of about 100 m. For stationary carriers, such as offshore oil drilling platforms, the TRANSIT system can provide a 10 m accuracy through long-term repeated data processing. Given the technical conditions of the time, this accuracy index has been quite good. The main problem with the TRANSIT system is that it cannot be continuously positioned because the satellite at the top of the day must wait for a period of time before it can receive another satellite signal. The interval may be several minutes or several hours, depending on the latitude, but the average waiting time is 90 min. Therefore, the TRANSIT system is more suitable for offshore ships. Its demand for positioning updates is infrequent, but it is not suitable for applications requiring frequent or continuous positioning. However, the significance of the TRANSIT system is to enable the R&D department to gain initial experience in satellite positioning and to verify the feasibility of positioning by satellite systems, paving the way for the development of GPS systems.

The TRANSIT system is the first space radio navigation system in the world that has been able to provide globally accurate positioning of satellites that require only one precise orbit. It is also worth mentioning that the principle of positioning of the TRANSIT system through Doppler counting is still applicable to GPS satellites. In fact, when the GPS constellation has not yet achieved full global coverage, in some areas where GPS constellations are poorly distributed, researchers have tried to use Doppler counting of GPS satellites by using the positioning principle of the TRANSIT system. Positioning is performed as described in Ref. [1]. Interested readers can also achieve positioning today by observing the Doppler count of only one GPS satellite.

In the 1960s and 1970s, a series of major events took place involving political, military, and engineering technologies. These events affected the evolution of GPS. The Cuban missile crisis of 1962 aggravated the US-Soviet nuclear arms race during the Cold War. The United States decided to establish a ternary system of strategic nuclear forces in its air, land and sea forces. The US Air Force's Intercontinental Ballistic Missile (ICBM) and the US Navy's submarine-launched ballistic missiles (SLBM) caused an urgent demand for high-speed, real-time, all-day, global coverage navigation and positioning systems. These technical requirements far exceeded the positioning capabilities provided by the TRANSIT system.

The Naval Research Lab developed the Timation satellite in 1967, based on the TRANSIT system. 'Timation' was not an existing word in English, but a new term created by combining the first three letters of 'time' and the last five letters of 'navigation', to define a satellite and time system. In fact, the original purpose of the Timation satellite was to solve a prominent problem in passive ranging positioning technology, namely high-precision clock synchronization and time keeping issues. The Timation-I (launched in 1967) and Timation-II (launched in 1969) satellites used a special quartz crystal as the clock source, but it was not effective. The Timation-III satellite (launched in 1974) made a breakthrough—using an atomic clock for time-keeping within the system. A series of experiments performed by the Timation satellite verified the feasibility of a high-precision time synchronization system based on space satellites. This conclusion was one of the most important foundations for the normal operation of the GPS system.

Compared with the US Navy, the US Air Force (USAF) had greater demands for high-performance navigation and positioning systems, because the US Navy's TRANSIT system was clearly unable to meet the requirements of highly dynamic aircraft such as warplanes and missiles. In the 1960s, the US Air Force backed a series of navigation-related research projects, including MOSAIC, Program 57, and 621B. Some of these plans are still State secrets, and it is difficult to find detailed descriptions from publications. However, some scattered conclusions can still be drawn. A plan that was closely related to GPS was the 621B program, which developed and validated a complex pseudo-random code-based ranging code scheme that is used in today's GPS systems.

A similar navigation project that was performed almost simultaneously during the same period was the US Army's SECOR positioning satellite system, which consisted of three known ground stations and a space satellite. The ground stations sent positioning signals to the location to be determined by the satellite signal transponder. The positioning signal was used to position the fourth unknown location. From this perspective, the system was similar to the forwarding satellite positioning system. The system launched 13 satellites in the 1960s, and undertook a series of experiments. These results and conclusions, together with the Timation system and the 621B project, constitute the basis of the GPS system.

From these developments, we can see the development motivations and goals of the GPS project, as well as the expected application scenarios. GPS was firstly designed for military purposes, so it is not surprising that the US Department of Defense (DoD) led the project.

In 1973, the US Department of Defense approved the overall structure of the GPS project. Its original name was the Defense Navigation Satellite System (DNSS), which was later changed to NavSTAR-GPS (an abbreviation of Navigation by Satellite Timing and Ranging—Global Positioning System).[1] Later this was is simplified to 'GPS'.

The GPS project is jointly managed by the US Air Force, the US Space and Missile Center (SMC), and the NavSTAR-GPS Joint Project Office (JPO), of which JPO consists of the Air Force, Navy, Army, Marine Corps, Department of Transportation, and member representatives of the Defense Map Agency. There are also representatives from NATO and Australia. The JPO plays an important role in the advancement of the project. It is responsible for the development of the ground-based monitoring station and area, as well as the updating of space satellites. It also oversees the development of military receivers and PPS service terminals.

Below are some basic design features of the GPS system.

- 24 h of uninterrupted service;
- Global coverage;

[1]Or 'Navigation Signal Timing and Ranging'.

- 3D positional positioning and speed information; precise timing function;
- Continuous, real-time positioning;
- Unlimited number of users;
- Passive positioning for easy radio silence;
- Limited civil service accuracy; higher military service accuracy.

Although the original purpose of GPS was military, decision makers soon became aware of the huge market and business opportunities in the civil field, so added the possibility of applying for civilian purposes from the start. However, in order to prevent civilian GPS receivers from being used for military purposes, especially against the national interests of the United States, the positioning accuracy of civilian services is determined by a signal format that is lower than the accuracy of military services. Selective availability, also known as the SA policy, further limits the positioning accuracy of civil services. After considering military and civilian needs, the US Department of Defense divides the services provided by GPS into the following two categories:

- Standard Location Service (SPS): Serving non-authorized users, including applications for civil, commercial, and academic research;
- Precision Location Service (PPS): Serving authorized users.

The PPS service is controlled by a specially encrypted cryptographic technology that is only available for DoD authorized users. The majority of civil users can only use the SPS service, which is affected by the SA policy in the early stage of the system. As a result, the positioning accuracy has been criticized, as it is only around 100 m. Also, there is a lack of dual-frequency observation to eliminate the influence of the ionospheric delay. The US government announced the cancellation of SA interference in May 2000. In 2007, the US Department of Defense announced that future GPS-III satellites will not have the ability to impose SA interference, which augurs the permanent abolition of SA policies.

The passive positioning method is designed to ensure radio silence in military applications. Active positioning requires the end user to communicate with the satellite via the uplink, thus inevitably exposing its existence. From this perspective, the GPS receiver only passively receives and processes signals from the satellites, answering the question "where am I?" This is different from the BeiDou-1 system in China. The BeiDou-1 receiver is an active positioning method, so it not only must know its own position, but also let others know where it is. The following chapter will elaborate on the BeiDou system.

The satellite orbits have been selected well. Global satellite orbits are generally divided into low, medium, and stationary. Low-orbit satellites have orbital altitudes below 2000 km, and have both advantages and disadvantages for navigation positioning. The advantages are that the height of the satellite is low, the distance between it and the user receiver is relatively short, and the path loss of the navigation signal is relatively small, meaning that a low-power signal transmitter can be used. The satellite is fast, and the spatial constellation distribution changes rapidly. It is therefore advantageous for localization algorithms that utilize satellite geometric spatial correlation. However, the shortcomings are also obvious. First, the satellite's transit time is short, and the receiver needs to continuously capture new satellites, resulting in a complex receiver control algorithm. The low orbit leads to an obvious atmospheric drag effect, and the satellite orbit is disturbed. The number of low orbit satellites required to achieve global coverage is much larger. One obvious example is a globally covered Motorola's Iridium system, which needs 66 satellites. The geostationary orbit satellite is a geosynchronous satellite located 36,000 km above the Equator. Although it only needs a small number of satellites to achieve global coverage, the higher orbital altitude determines that the spaceborne signal transmitter must apply more power. Since all satellites are located at the Equator, the positioning accuracy cannot meet the requirements of users in high latitudes. Readers will understand this after learning about the geometric precision factor in the GPS positioning algorithm.

Therefore, the GPS system finally selected a medium-orbit satellite with an orbital altitude of about 20,000 km. The satellite transit time is several hours, depending on the location. Moreover, the number of global-coverage satellites is moderate, at twenty-four. At present, in order to improve the system's performance, the number of satellites in orbit at any one time can be as high as 30, and the path loss of the medium-orbit satellite is also within the acceptable range. This does not impose excessive demands on the power of the spaceborne signal transmitter.

According to Ref. [2], the implementation of GPS projects can be divided into the following four stages.

1. Pre-research

From 1960 to 1972 there were experiments with various types of satellite positioning and navigation systems, accumulating experience, and solving a series of technical engineering problems, including the signal system, carrier selection, time synchronization, precise orbit determination, and terminal positioning equipment. The operations of the TRANSIT system, the Timation system, the 621B plan, and the SECOR system described above all occurred in this stage.

2. Conceptual Verification

This phase, from 1973 to 1979, verified the feasibility of GPS by launching test satellites. The early Block-I satellite of the GPS system began to be launched in this stage. However, before the launch of the first Block-I satellite, the United States also launched two experimental navigation technology satellites (NTS). The navigation signal used on the satellite NTS-2 was the GPS signal that was later adopted. The NTS-2 satellite was launched from Vandenberg Air Force Base in June 1977, but failed after eight months of operation. On the basis of the NTS satellite, 11 Block-I satellites were subsequently launched. The last Block-I satellite failed in 1995. At this stage, a large amount of design verification work was also carried out on the control section and user terminal.

3. Comprehensive Development and Design

From 1979 to 1985, 11 Block-I satellites were launched, and the provider and commercial contracts for the Block-II satellites were also determined. The developers of the user terminal receivers were also initially identified. The terminal prototype was tested in a large number of terrestrial and marine environments. Twenty-eight Block-II satellites began development at this stage.

4. Deployment and Practice

This phase began in 1986 and continues to the present day. On 14th February 1989, the first Block-II satellite was pushed into orbit in space by the Delta-II rocket. Nearly two and a half years later, in June 1991, a hybrid constellation of Block-I satellites and Block-II satellites was able to provide global two-dimensional positioning capabilities. In December 1993, the JPO announced that it had achieved three-dimensional positioning on a global scale. A spatial constellation of all 24 satellites was realized in April 1994.

It is worth mentioning that in 1991, during the first Gulf War, the GPS system had not yet achieved global coverage of satellite constellations, but was nevertheless used in battlefield command, deployment and rescue of combat units, and precision guided weapons. It played an important role in the war, and greatly enhanced military capabilities, showing the huge potential of satellite navigation systems in the field of warfare. As the US government cancelled the SA interference in 2000, GPS technology and end products began to enter daily life. Today, GPS applications are found in vehicle navigation, field rescue, outdoor sports, person tracking, seismic monitoring, atmospheric monitoring, and location based services.

2.2 The Construction of the GPS System

While seemingly complex, the GPS system can be divided into three parts: the space segment, the control segment, and the terminal segment, as shown in Fig. 2.4.

1. The Space Segment

The space segment is composed of 24 GPS satellites that are orbiting the Earth. At the start of the GPS constellation design, to ensure that there were more than four visible satellites at any time on any position on the Earth, the initial plan was to have a minimum of 24 satellites. However, the number of on-orbit GPS satellites in the Space network is always greater than 24, and the current number of is maintained at 29 ~ 31. The US officially refers to the 24 satellites that make up the GPS constellation as the "core constellation" or "basic constellation". The deployment of the core constellation is designed to offer basic positioning within the GPS system on a global scale. GPS satellites provide better, more reliable positioning results.

GPS satellites orbit the earth in a nearly circular elliptical orbit. The orbital eccentricity is between 0.002 and 0.01. There are slight differences in the orbital eccentricity of the various satellites. Twenty-four GPS satellites are distributed on six orbital planes, with a 55° angle between each orbital plane and the Earth's equatorial plane. The elevation of the adjacent orbital planes is 60°, with four to six GPS satellites per orbital plane. The spacing between GPS satellites is not evenly distributed. The long semi-axis of the orbit is around 26,560 km, and it takes the satellite about 11 h and 58 min to travel around the Earth. This is exactly half the length of the stellar day. Consider the Earth's rotation period of 24 h; from the perspective of the ECI coordinate system, the GPS satellite has travelled two rounds when the Earth completes one rotation, so the same GPS satellite orbits cover the same position on the Earth at roughly the same time on two consecutive days. Strictly speaking, observers at the same location on Earth every day will find that the same GPS satellite appears from the horizon four minutes earlier than the previous day, which is due to the time difference between the Earth's rotation and the two orbital periods of the GPS satellite. This also determines that observers at the same location on Earth will observe almost identical GPS satellite constellation distributions at the same time in the next two days.

The distribution of the GPS satellite constellation is shown in Fig. 2.5a. Figure 2.5b shows the distribution of GPS satellites on the six orbital planes in December 2007, where the number on the orbit indicates the satellite's PRN number. A total of 30 satellites are used for networking on the six orbital planes. Readers should note that the status of the satellite constellation here is only for a specific period of time; the information will be adjusted as the satellite is updated and maintained, or when it fails.

The six orbital planes of the GPS constellation are represented by the letters A, B, C, D, E, and F. The satellites in the orbit are represented by the numbers 1 ~ 4.

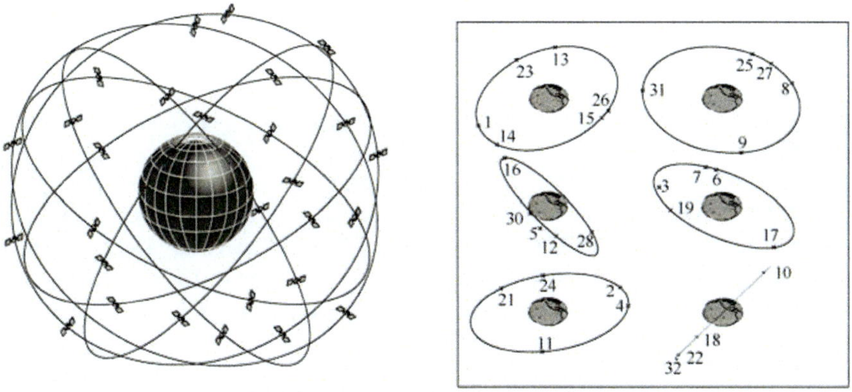

(a) Distribution of GPS satellite constellations

(b) Distribution of satellites on the GPS orbital plane (December 2017)

Fig. 2.5 Satellite constellation map

Each GPS satellite can be represented by two characters. One is the track number, and the other is the satellite number within the orbit. For example, A1, B3, and E4 represent satellite 1 on orbital plane A, satellite 3 on orbital plane B, and satellite 4 on orbital plane E. The satellite number should not be confused with the satellite PRN number. The former only indicates the slot number in the same orbital plane, which is not the satellite's PRN number.

GPS satellites have been through two generations since their inception, namely Block-I and Block-II. Block-II includes the improved Block-IIA, Block-IIR, Block-IIR (M), and Block-IIF series.

Block-I launched its first satellite in 1978, and by 1985 it had launched 11 in total. The initial design life of the Block-I satellite was five years, but the actual operating life ranged from four to 13.5 years. The satellite with the longest life span was the third—the GPS I-3 or GPS SVN-3, where SVN-3 was the spacecraft number of the satellite that was launched in October 1978 and failed in May 1992, after 13.5 years of operation. The last Block-I satellite in operation as of March 1996 was the GPS I-10, the 10th Block-I satellite. It was launched in September 1984, and its actual working time was 11.5 years. The purpose of the Block-I satellite was to verify the feasibility of the entire system and to identify problems in its workings, providing valuable engineering information for the development of Block-II satellites. As of November 1995, all services provided by the Block-I satellites were terminated.

Like the Block-I satellite, the Block-II/IIA satellite was developed by Rockwell International in the United States. The Block-II series was improved on the basis of Block-I. One of the design goals was to provide up to 14 days of positioning service without any communication between the satellite and the control section. Of course, the quality of service gradually deteriorated over time. This series of satellites was equipped with a three-axis position stabilizer, which ensured good positional control. The solar panels on both sides provided 710 W of power. The satellite was also equipped with a hydrazine propulsion system to provide a certain orbit correction capability. The clock reference was delivered by two rubidium and cesium atomic clocks. The Block-II satellite had a mass of 1660 kg, and nine were launched by the Delta-II rocket between February 1989 and October 1990. It had a design life of 7.5 years, but the actual working life can be longer. The last Block-II satellite stopped service in March 2007, and currently there are none in the GPS constellation.

Compared with the Block-II satellite, the service time of the Block-IIA satellite without communication connection increased to 180 days. The total mass also increased from 1660 kg to 1816 kg. Nineteen Block-IIA satellites were launched between November 1990 and November 1997. As of June 2013, there were still eight Block-IIA satellites operating normally, taking up A3, A5, C2, C6, D4, E5, E5, and F5 of the GPS satellite's orbital plane respectively.

The Block-IIR satellite was developed by Lockheed Martin in the United States. The letter 'R' in its name stands for 'replenishment'. Its main purpose was to replace the Block-II/IIA satellites that had reached the end of their operational lives. The Block-IIR satellite did not receive many performance updates. The total mass of the satellite increased to 2217 kg, the power of the solar panels rose to 800 W, and the predicted working life increased to 10 years. From July 1997 to November 2004, 12 Block-IIR satellites were launched. As of June 2013, they were still in normal operation.

The Block-IIR(M) was also developed by Lockheed Martin in the United States. The letter 'M' in the suffix stands for 'modernized'. As its name suggests, the series was developed to fit with the GPS modernization plan. Compared with the previous series of Block-IIR satellites, the biggest improvements were the increase of the L2C signal on the L2 carrier, and the addition of new military code signals. The solar panel power, satellite quality, and predicted working life were the same as those of Block-IIR. Eight Block-IIR(M) satellites were launched from September 2005 to August 2009. As of June 2013, seven were working normally and one (GPS SVN-49) was unusable. The Block-IIR and Block-IIR(M) (12 + 8 satellites) together constitute the body of the current GPS constellation.

The Block-IIF satellite was developed by Boeing Corporation in the United States, with the letter 'F' standing for 'follow-on'. Boeing expects to manufacture a total of 12 Block-IIF satellites. The first was launched in May 2010. By June 2013, four had been put into orbit. Compared to the previous Block-II/IIA/IIR/IIR(M) series, the Block-IIF satellite has a longer predicted working life (12 years). It uses a hybrid helium atomic clock to provide an extremely accurate clock reference, and has a third civil signal (L5) that provides more flexible power control for military signals to effectively counter blockage and interference from hotspots. As of June 2013, three Block-IIF satellites were working normally, with SVN66 (corresponding to PRN27) in the testing stage.

Table 2.1 shows the corresponding list of GPS satellites' PRN numbers, SVN numbers, and satellite types as of June 2013. The characters in the table indicate the working status of the satellite, with 'H' indicating a healthy state. Among the 32 satellites, there are eight Block-IIA, 12 Block-IIR, eight Block-IIR(M), and four

Table 2.1 List of GPS satellite types, PRN numbers, and SVN numbers (June 2013)

PRN	1	2	3	4	5	6	7	8	9	10	11	12	13	14	15	16
SVN	63	61	33	34	50	36	48	38	39	40	46	58	43	41	55	56
IIA			H	H		H		H	H	H						
IIR		H									H		H	H		H
IIR (M)					H		H					H			H	
IIF	H															
PRN	17	18	19	20	21	22	23	24	25	26	27	28	29	30	31	32
SVN	53	54	59	51	45	47	60	65	62	26	66	44	57	49	52	23
IIA										H						H
IIR		H	H	H	H	H	H					H				
IIR (M)	H												H	U	H	
IIF								H	H		T					

Note H—Healthy, T—in Test, U—Unavailable or unstable

Block-IIF. It is worth noting that the information given in this table comes from the US Coast Guard's website (USCG). The information in the table will change over time. Old satellites will gradually be replaced by new models. PRN numbers will change due to satellite adjustment, so readers should make adjustments according to the latest information on the official website when using the table.

The main task of the GPS satellite in the space segment is to broadcast the navigation radio signal with which users on Earth can position themselves. The satellite itself has a certain degree of fault tolerance and self-detection logic, and keeps the same time as the ground control station. Satellite maintenance and status monitoring can be achieved through communication, and the uploading of periodic navigation messages is part of these maintenance and monitoring tasks. Satellite maintenance and monitoring operations generally do not interrupt the broadcast of navigation signals, but Block-IIA satellites generate intermittent interruptions of navigation signals ranging from six to 24 s when uploading navigation messages. However, this is rare.

2. The Control Segment

The control segment is a set of global networks that are responsible for tracking the status of GPS satellites, detecting the quality of satellite signals, and analyzing their positioning performance to send control commands and data to the satellites. In physics, this mainly includes:

- a master control station (MCS);
- an alternate master control station (alternate MCS);
- four ground antenna stations;
- a global monitor station network (Monitor Stations).

The master control station is located at Schriever Air Force Base, Colorado, and the alternate master control station is located at Vandenberg Air Force Base, California. The master control station completes the core mission of the GPS control system, and is equivalent to the nerve center of the entire control segment. The input of the master control station comes from the navigation message from the monitoring station network. The spatial position of the GPS satellite is obtained through complex calculations. The status of the GPS satellite and its running orbit parameters are monitored, and the abnormal satellites are detected and processed over time. The navigation message is handed over to the ground antenna station to upload, and the signal from the on-board atomic clock is monitored to ensure synchronization between the GPS satellites. For the master control station to complete its task, it has to be supported by the global monitoring station network.

The communication between the GPS monitoring station and the GPS satellite is one-way. The navigation messages transmitted by the transit GPS satellites are tracked, the atmospheric data and the navigation code transmitted by the ranging code and the carrier observation satellite are obtained, and the information is then sent to the master control station. The communication link between the monitoring station and the master control station is the US Defense Satellite Communications System

(DSCS). Since GPS satellites are medium-orbit satellites with a global trajectory, the monitoring station must consist of multiple global networks to ensure real-time tracking of all satellites.

At the start of the operation of the GPS system, there were only six monitoring stations, all operated by the US Air Force. They were located in Hawaii and Colorado Springs in the USA, Kwajalein in the South Pacific, and Ascension Island, Diego Garcia, and Cape Canaveral in the Indian Ocean. Since 2006, 11 monitoring stations have been deployed in the United States (two in Alaska and Washington DC) under the responsibility of the National Geospatial-Intelligence Agency (NGA), as well as in Ecuador (Quito), Australia (Adelaide), the United Kingdom (Hermitage), Bahrain (Manama), South Korea (Osan), South Africa (Pretoria), New Zealand (Wellington), Argentina (Buenos Aires), and Tahiti Island (Papeete). When all 17 monitoring stations are put into operation, any GPS satellite can be tracked by at least three monitoring stations in real time. This means that the integrity and accuracy of a satellite's orbital parameters and ephemeris data can be guaranteed, ensuring optimal operation for the whole system.

The ground antenna station is responsible for data transmission with GPS satellites. This data transmission is bidirectional, and is realized by S-band signal communication. The ground antenna station and the master control station also communicate through the defense satellite communication system, sending ephemeris data, clock correction data, and some control commands from the master control station to the satellite. There are four ground antenna stations around the world, located in Kwajalein and the Asuncion Islands in the South Pacific, Diego Garcia in the Indian Ocean, and Florida.

Figure 2.6 shows the global distribution of the GPS control segment, including that of the master control station, alternative master control station, monitor station, and ground antenna.

3. The Terminal Segment

The terminal segment is also known as the user segment, and includes GPS users from all over the world, covering a wide variety of applications. From national organizations and government agencies to a company or individual, GPS receivers can be designed and developed according to individual needs. On this basis, a series of applications and services can be developed, so in this sense the terminal segment is the most dynamic and creative part of the GPS system.

Since its inception, the GPS system has been designed to provide global users with all-weather, continuous real-time, three-dimensional positioning and navigation services. The work and infrastructure of the space segment and control segment are built for this purpose. For the end user, positioning navigation and subsequent applications are realized by the GPS receiver. Therefore, the terminal segment plays the final and critical role in the GPS system, without which it is meaningless for the space segment and the control segment to exist.

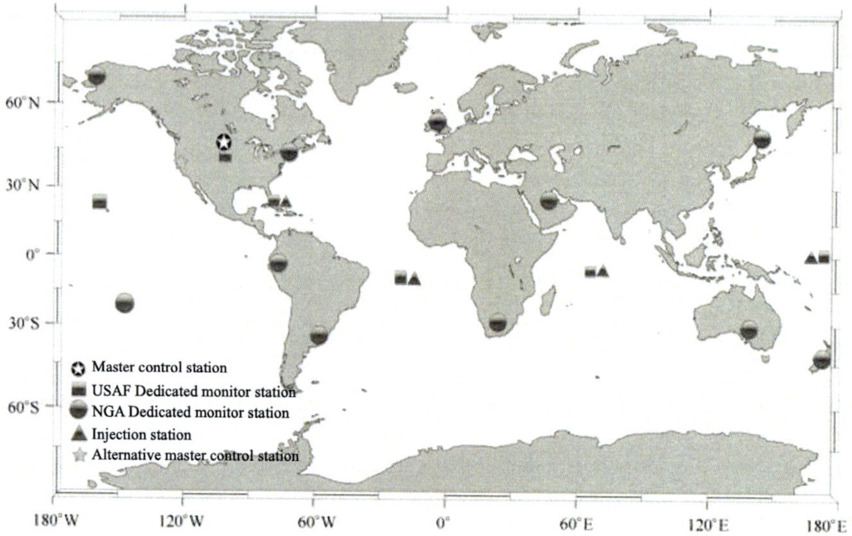

Fig. 2.6 Global distribution of GPS control segments

Although the specific implementation of the terminal segment is diverse, the infrastructure will include GPS receivers, antennae, and associated accessory equipment. From the specific implementation form, it can be divided into the hardware GPS receiver and software GPS receiver; from the application category, it can be divided into the navigation receiver and professional surveying receiver. From the accuracy of the service provided, it can be divided into precise positioning service receivers (PPS) and standard positioning service receivers (SPS). Although the classifications are different, the basic structure and principle of GPS receivers are almost the same. The hardware architecture includes antennae, RF front-ends, signal processing units, microprocessor units, human-machine interfaces, and related ancillary facilities. Modern software receivers may not include all of the above hardware units, but a large number of signal processing and positioning solution algorithms are implemented by software programming, which is similar in terms of the nature of signal processing.

Early GPS receivers were introduced in the 1980s. They were designed primarily to validate GPS concepts and system functions. Being limited to the technology and manufacturing processes of that time, GPS receivers were very cumbersome and limited in functionality. Figure 2.7 shows the world's first human-operable GPS receiver AN/PSN-8, developed and produced by Rockwell Collins in 1988. The receiver weighed 17 lb, was battery powered, and consumed a high amount of energy. At a price of about $40,000, it was limited to device technology, and could only be designed in a single channel to process multiple satellite signals in a time-sharing manner. Although the product was designed to be used by individual soldiers in the

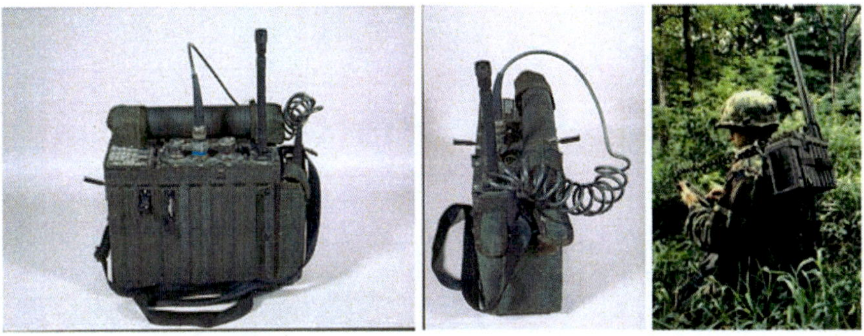

Fig. 2.7 AN/PSN-8—the world's first manpower-operated GPS receiver

field, it was often used on off-road vehicles or helicopters because of its size and weight.

After the AN/PSN-8 was produced, the US military initially estimated the demand to be around 900. However, the total demand in the following four years far exceeded this figure, reaching 75,000. The subsequently improved version was also used in the United States for Conventional Air-Launched Cruise Missiles (CALCM) and Standoff Land Attack Missiles (SLAM). This shows the great potential of the GPS receiver for military applications.

In 1983, when the Soviet Union shot down a South Korean 007 passenger airliner, the then President Ronald Reagan announced that the USA would provide GPS positioning services to civilian and commercial users around the world to improve navigation accuracy and safety for aircraft. Since then, a large number of commercial companies have been involved in the development and production of GPS receivers. Various types of models, different performances, and GPS receivers for different applications have mushroomed. Along with the continuous improvement of semiconductor device technology and advancement of packaging technology, miniaturized, low-power, low-cost GPS chips and complete machines are becoming better known.

Although modern GPS receivers have different shapes, they mainly contain the same parts, as shown in Fig. 2.8.

The antenna is the first point at which the receiver obtains the satellite signal, so its performance directly affects the subsequent performance parameters. Since the polarization of GPS satellite signals is right-handed circular polarization (RHCP), under normal conditions the receiver antennae are all right-handed circularly polarized designs. A left-handed circularly polarized antenna is used only in certain special applications, such as when receiving reflected signals. RF bandwidth is also an important factor in selecting an antenna. The proper RF bandwidth minimizes noise while ensuring that the signal is transmitted without loss. If the bandwidth is too narrow, the signal energy is lost. If the bandwidth is too wide, the signal energy can be completely passed, but at the same time, more noise is mixed into the RF front-end, so too large or too small a bandwidth is not suitable. In general, for receiver antennae that process

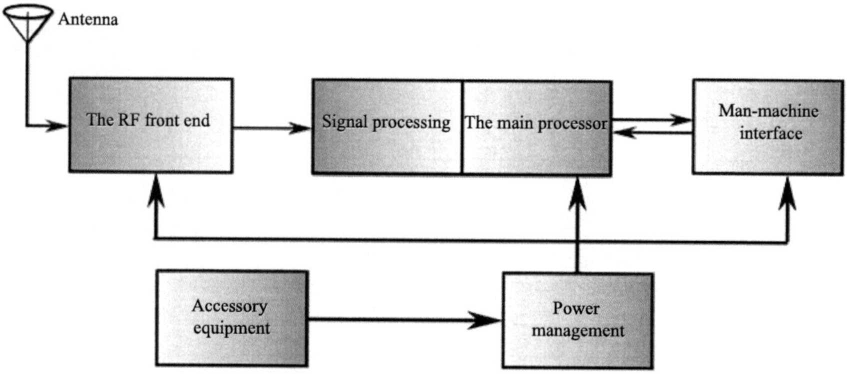

Fig. 2.8 The main functional unit modules of a modern GPS receiver

L1 C/A code signals, the RF bandwidth should be around 2.046 MHz. For dual-frequency receiver antennae that process L1 and L2 P(Y) codes, the RF bandwidth at L1 frequency must be guaranteed to be 2.046 MHz while the RF bandwidth at the L2 frequency is 20.46 MHz. This dual-frequency GPS antenna is designed to be more complex and costly. The antenna of the high-precision receiver must also ensure that accurate RF signals of different frequency bands reach the phase center of the antenna. Some of them also need to suppress multipath signals, such as choke coil antennae. A low-noise amplifier can also be designed inside the antenna (LNA). In addition to ensuring the gain of the antenna, it also helps to reduce the impact of subsequent RF devices on the total noise figure of the system. Such an antenna is called an active antenna, and vice versa, a passive antenna. When using an active antenna, it is important to note that the LNA inside the antenna must be fed in a suitable way, otherwise it will not work properly. Some special applications require suppression of radio frequency interference, which is attenuated or suppressed by special analog or digital signal processing. This type of antenna is called an anti-jamming antenna, and is used in military applications. In physics, there are various types of antennae, such as the helical antenna, the microstrip antenna, the patch antenna, and the back cavity planar spiral antenna. The appropriate physical form and mounting dimensions must be considered when installing the device.

The function of the RF front-end is to amplify and filter the RF signal received by the antenna, filter out the out-of-band noise signal, and then mix it with the locally generated carrier signal to obtain the intermediate frequency (IF) signal. At this time, further amplification or filtering can be performed. After the signal is amplified to the appropriate level, it is sent to the ADC for digital sampling and quantization. The purpose of sampling is for subsequent digital signal processing. Generally speaking, the RF front-end will output the sampled quantized digital signal, but some RF front-ends will also output the analog IF signal, and the user can select the appropriate ADC to complete the sampling and quantization. The main parameters of the RF front-end include RF bandwidth, noise figure, system gain (usually expressed in dB, i.e.

decibels), RF signal frequency, AD sampling frequency, IF signal center frequency, quantization bit width, and power consumption.

The signal processing unit and the main processor unit in Fig. 2.8 are combined. The main reason is that the signal processing unit can be implemented by a specific dedicated hardware chip, such as in the implementation of most commercial receivers, or a code running on a piece of software, such as a software radio receiver solution. However, from the perspective of signal processing, the specific software or hardware implementation does not change the core algorithm principle of signal processing. It requires complex baseband algorithms and navigation and positioning algorithms, which are often inseparable from the main processor's instruction program, execution, and software scheduling. The signal processing unit and the main processing unit are at the core of the entire receiver system and the system overall, so they are closely related to the implementation of performance indicators and functions.

The signal processing part is responsible for capturing and tracking the GPS signal, implementing carrier synchronization, bit synchronization, and sub-frame synchronization, and then performing demodulation of the navigation message, extraction of the observation measurement, and further completing satellite ephemeris data collection and navigation positioning solutions, as well as processing the entire signal. The process includes intensive mathematical operations and complex software logic controls. In most cases, a large number of floating-point operations are required, so the demands for computing and processing power on the main processor are very high. A large number of operations and complex software processing often mean an increase in power consumption. It is a good alternative to implement computationally intensive processing tasks using hardware logic, but this will inevitably require certain cost and size of the hardware. The use of efficient and sophisticated baseband algorithms plays a decisive role in reducing system power consumption, and the size and cost of chips. In applications where product performance and power consumption must be considered, the final solution is often a compromise between the two.

After the main processor completes the positioning and navigation solution, the location information must be transmitted to the user through a certain interface mode, or displayed through a device such as an LCD or an LED. Modern receivers can send positioning results through RS232, RS485, USB, network card interface, SPI, IIC, CAN bus, and other physical interfaces. Some receivers also have real-time data storage function. The positioning results are placed in non-volatile storage media, such as a flash memory card, USB stick, or portable hard drive, for later analysis and processing.

The power supply unit is the power source for the whole system operation, and it also ensures the normal operation of the system. The power supply can be implemented by built-in, external, or battery-powered sources. It can be divided into linear voltage regulation and switching mode. The requirements for the power supply vary depending on the application. For example, a portable device must be powered by a battery, and a battery charge and discharge management circuit must be provided accordingly. When used outdoors, the power consumption of the system directly affects the length of battery life, so low-power design and complex power management circuits must be considered for this application. The power performance parameters of the receiver mainly include power supply voltage, current capacity, ripple parameters, and conversion efficiency.

Auxiliary equipment mainly includes cables, such as RF signal cables, data interface cables, and power cables. Some devices also require display devices, and systems that provide data-assisted (A-GPS) and differential GPS (D-GPS) functions, as well as a wireless link device, such as a digital radio or GPRS/CDMA/3G modem.

2.3 The GPS Modernization Plan

After more than 20 years of operation, problems began to arise in the GPS system. Due to the advancement of new technologies and the development of devices, the system functions and hardware and software facilities of the original system were unable to meet the needs and application environment of the new era. Calls were for updates and improvements to the system, and in 2000, US Congress officially approved the GPS modernization plan.

GPS modernization, initially dominated by the US government and the military, involved the modernization of the space segment and the control segment. However, updates to the terminal segment inevitable followed due to the progress of space satellites and the activation of new navigation signals. However, unlike the space segment and the control segment, the modernization of the terminal segment was spontaneously promoted by civil society, commercial companies, and academia.

The modernization of the space segment included the development and deployment of next-generation GPS satellites, namely the Block-III series, which added new civilian signals L2C, L5, and L1C together with M-code (a new military signal) based on the existing L1 C/A and L2 P(Y) signals.

The L2C and L5 signals were proposed for three main reasons:

1. Accurate correction of the ionospheric group delay required two signals of different frequencies.
2. The demand for the availability of positioning signals continued to rise, especially in safety-related applications, while the original signals experienced deficiency in this domain.
3. The requirements for the accuracy and rapidity of positioning results also placed higher demands on GPS signals.

As a first step in GPS modernization, the US government abolished the SA policy in May 2000, which resulted in a positioning accuracy of 10 m for civil single-frequency receivers—an order of magnitude higher than the accuracy before the SA policy was abolished. The initial purpose of the SA policy was to artificially reduce the positioning accuracy of unauthorized users, or to control that of the unauthorized GPS applications. However, upon realizing that the policy had an adverse effect on the promotion of GPS in the civilian market, and that non-authorized users could

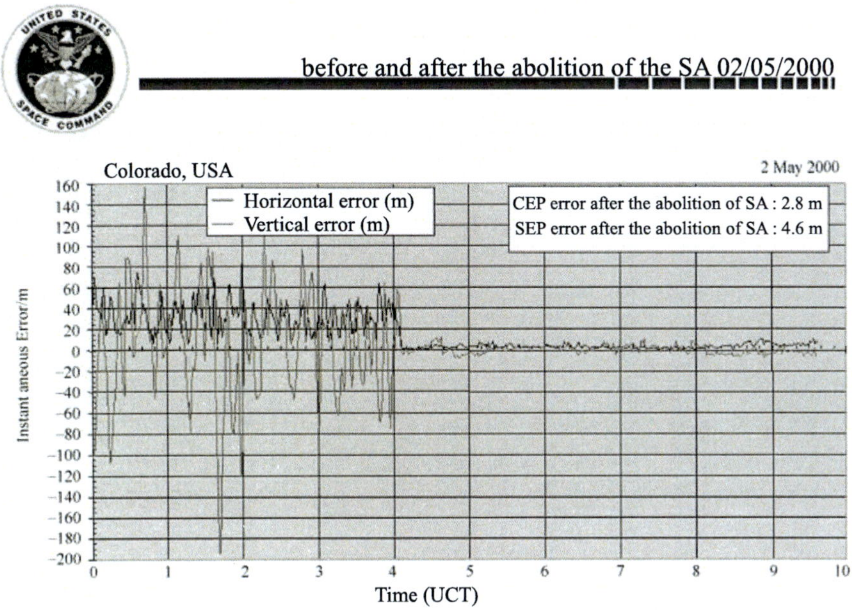

Fig. 2.9 Comparison of GPS positioning errors before and after the abolition of the SA policy

offset the impact of SA on positioning accuracy at a small cost using differential GPS technology, the US government ultimately decided to abolish SA policy.

Figure 2.9 shows the GPS positioning error recorded at the time before and after the abolition of SA. The unit of the ordinate is meters, and that of the abscissa is time. The light gray curve in the figure indicates the positioning error in the vertical direction, and the dark gray curve indicates that in the horizontal direction. The data was recorded by Rob Conley using a Trimble SV6 receiver at a GPS support center in Colorado Springs, USA. The time of recording was around 04:05:00 on 2nd May 2000 in UTC time, when the SA policy lost its validity. The figure clearly shows that the horizontal and the vertical position error had been significantly reduced since the abolition. Further statistical analysis of the data shows that the positioning error before the abolition was within a circle with a radius of 45.0 m (95% probability), yet it shrank to 6.5 m (95% probability) after the abolition.

In 2008, Lockheed Martin won a contract for the modernization of the GPS space segment. It was responsible for the development, testing, and deployment of the Block-III satellite, and planned to launch the first and second Block-III satellites in 2014 and 2015 [3]. The Raytheon Company won the contract for the modernization of the control segment in February 2010. The future control segment is called the *Next Generation GPS Operational Control System*, referred to as OCX.

As the modernization of the GPS space segment continues, in the foreseeable future the space constellation will include Block-IIA, Block-IIR, Block-IIRM, Block-IIF and Block-III satellites. Figure 2.10 is a schematic diagram of the navigation

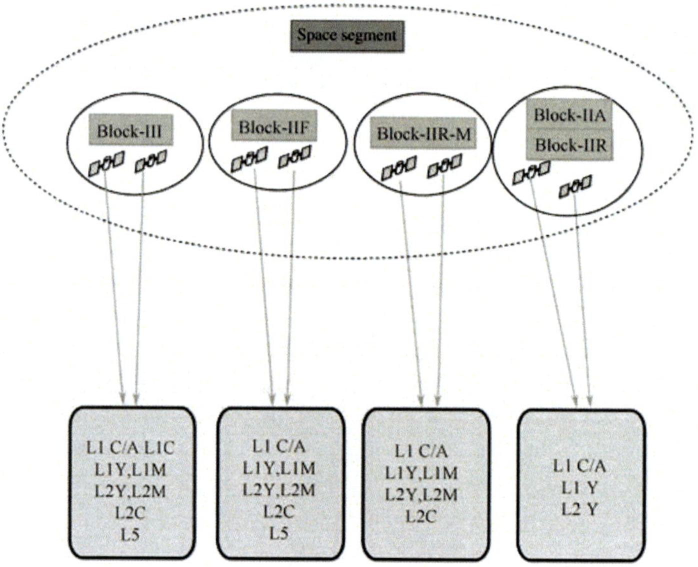

Fig. 2.10 Types of navigation signals emitted after the modernization of the GPS space segment

signals transmitted by various types of satellites in the space segment after GPS modernization.

Lockheed Martin is the contract supplier for the future Block-III satellite. According to its plan, in the future GPS-III satellites will be divided into three types: IIIA, IIIB, and IIIC, manufactured on its A2100S platform. The A2100S platform is a highly reliable mainstream satellite platform developed by LM, and the first A2100 satellite has been running stably for more than 13 years. The satellite platform adopts a box-board structure and modular design to support multiple communication loads, and can adapt to various launching platforms. The Block-IIIA satellite is scheduled to launch in 2014. In addition to raising the number of L1C signals, it also increases the signal strength received by users on the Earth's surface, and leaves room for compatibility with subsequent IIIB and IIIC. A "hot-spot" function aiming for local signal enhancement will be added to Block-IIIC satellites. With the help of spot beams, it will improve the signal intensity in hot-spot regions on the Earth to enhance the anti-interference ability of GPS receivers in the region.

As described in Sect. 2.2, satellites included in the existing GPS space segment are Block-IIA, Block-IIR, Block-IIR-M, and Block-IIF, where Block-IIA and Block-IIR transmit L1 C/A signals and L1 P(Y), L2 P(Y) signals, yet only the L1 C/A signal is for civil use. A new military code signal and L2C signal is included in the navigation signals transmitted by Block-IIR-M, of which the L2C signal is the second civilian signal. The third civilian signal (L5) is transmitted by Block-IIF. In the future, the fourth civilian signal (L1C) is expected to be added to Block-IIIA, but the activation of this signal will depend on the implementation of the OCX [4]. Figure 2.10 shows the types of navigation signals that will be transmitted by various types of GPS satellites in the future.

1. The L2C signal

The carrier frequency of the L2C signal is L2 (1227.6 MHz), which, together with the original L1 frequency C/A signal, constitutes a dual-frequency navigation signal for both business and civilian use. The direct advantage of this type of signal is that ionospheric delays can be eliminated for civilian users through integrating data from dual-frequency observation, thereby improving the positioning accuracy. The ionospheric delay is the delay occurring when the radio navigation signal passes through a dispersion medium such as the ionosphere. The delay is inversely proportional to the square of the carrier frequency, so the impact of the ionosphere can be completely eliminated by integrating observations of the L1 and L2 frequencies. Prior to this, civil receivers could only correct the ionospheric delay using a general model, such as the Klobuchar model in Appendix E, or applying differential corrections obtained from the data link. Therefore, the ionospheric error is a major factor affecting single-frequency GPS receivers with an independent positioning function, which was also a serious problem for the civilian market of GPS receivers before modernization.

There are two pseudo-random codes for the L2C signal—the CM code and the CL code. 10,230 chips form one code period of the CM code, and 767,250 chips of the CL code. The chip rate of the CM code and the CL code are both 511,500 bps, for which the CM code repeats one period every 20 ms while the CL code repeats one period every 1500 ms. The CM code and the CL code modulate the carrier in a time division multiplexed manner, so the bit rate of the final combined pseudo-code is 1.023 Mbps.

The L2C signal is still modulated in BPSK, and the bit rate of its navigation telegram is 25 bps. Modulate the telegram into CM code in the form of 1/2 forward error correction (FEC) code, and its final information bit rate will be 50 bps. To distinguish it from the original navigation telegram, the L2C navigation message is called CNAV, in which the letter 'C' indicates the civil signal. The format of the CNAV message is still a 300-bit sub-frame. However, since the CNAV message is more streamlined and efficient, it takes less time for a L2C receiver to collect all ephemeris data and realize first positioning than the L1 C/A receiver.

Unlike the CM code, the CL code does not modulate the message bits, for obvious reasons: the CM code is used to achieve fast capture of the signal, and the coherent integration is longer necessary for the CL code because there is no data bit jumping, making the acquisition of weak signals easier. Besides, in terms of signal tracking, since there is no data bit hopping, the carrier tracking loop can be a pure phase-locked loop instead of a Costas loop, and the performance of carrier tracking can be improved by 6 dB. This is why the CL code has better performance in signal tracking.

For more on the L2C signal format and CNAV, see Ref. [5].

2. The L5 signal

Before the plan for GPS modernization was put forward, the satellite navigation community agreed that the civilian market was in urgent need of a civil navigation signal with another frequency point other than the L1 frequency. That signal had to meet the following three conditions:

(a) The impact of ionospheric delay can be eliminated through dual-frequency observation.
(b) It has enough robustness to serve in safety sectors such as aerospace and aircraft navigation.
(c) It performs better in signal acquisition and tracking, and has better positioning accuracy.

At first, the Federal Aviation Administration (FAA) opposed the idea of using the L2C signal as the second civilian navigation signal, mainly because its frequency was not within the band allowed by the Aviation Radio Navigation Service (ARNS). The ARNS band is the aeronautical radio navigation band allocated by the ITU for safety services, and it must not be occupied by any other types of services.

Figure 2.11 shows the frequency range of ARNS and RNSS specified by the ITU (International Telecommunication Union). The figure shows that the ARNS band is divided into two segments. The first half is 1559 ~ 1610 MHz, including the frequency of L1 signals transmitted by GPS and GLONASS and the frequency of B1 signals transmitted by China's BeiDou satellites. The second half of the ARNS is 1151 ~ 1214 MHz. It should be noted that the frequency of the L2C signal (1227.6 MHz) is not within the ARNS frequency range. Instead, it falls in the wireless positioning

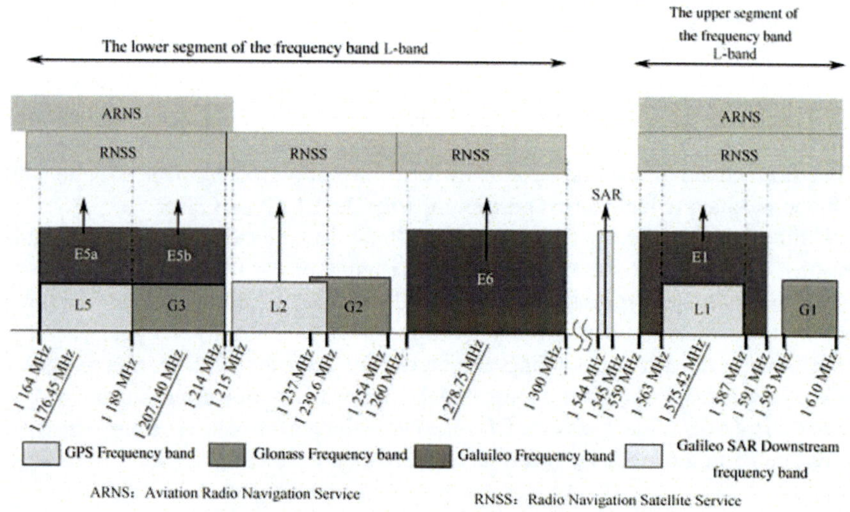

Fig. 2.11 The RNSS and ARNS frequency range stipulated by the ITU

service frequency band, which may be affected by electromagnetic waves emitted by high-power ground radar. This means that signal interference may occur, which may even lead to positioning failure when severe.

Later, after careful investigations with concerned parties, the United States Department of Defense and the Department of Transportation (DOT) decided to add a new civil navigation signal with a frequency of 1176.45 MHz while retaining the L2C signal. The frequency falls within the ARNS band and accords with requirements of the US Federal Aviation Administration (FAA). The signal, of which the frequency is named L5, is the third civil navigation signal.

Block-IIF and subsequent models of GPS satellites transmit L5 signals. The L5 signal is divided into a data channel and a pilot channel, which are respectively modulated on the in-phase and quadrature carriers through BPSK (10). The pseudo-code rate is 10.23 Mbps, the pseudo-code period is 1 ms, and the whole period includes 10,230 chips. The navigation message rate is 50 bps, which produces a 100-bps bit stream, later modulated into the data channel, in 1/2 FEC encoding. The pilot channel has no modulated data bits, which means that it allows longer coherent integration and can provide more accurate phase tracking results and carrier phase measurements. Besides, the integration of measurements of L1, L2C, and L5 signals offers a greater range of combinations of wide lanes and narrow lanes, which is more conducive to the rapid solution of carrier phase ambiguity.

The chip length of the L5 signal is only 1/10 of that of the L1 C/A signal, which, together with the chip length of P(Y) code, is the shortest chip length of any civil navigation signal so far. It is common knowledge that the shorter the chip length, the higher the accuracy of the theoretical positioning accuracy, due to the higher measurement accuracy of the pseudo-range observation. Therefore, the L5 signal can provide higher theoretical positioning accuracy than the current L1 C/A code signal. At the same time, in the multipath environment, shorter pseudo-code chips contribute to better multipath suppression performance. Longer pseudo-code periods can also elevate the spread-spectrum gain as well as better autocorrelation and cross-correlation properties, allowing for more accurate acquisition of weaker signals from occluded satellites without being interfered by cross-correlation peaks of stronger signals from other satellites. All of these advantages make the L5 signal the most likely to be widely used in safety applications in the future.

More on the L5 signal format and CNAV modulated on it can be found in Ref. [6].

3. The L1C signal

The L1C signal has been proposed to facilitate compatibility and interoperability between GPS and other satellite navigation systems in the world. It is expected that the L1C signal will start to be transmitted by Block-III in 2015. By 2026, there will be 24 satellites transmitting L1C signal in Space.

The frequency of the L1C signal is 1 575.42 MHz, which is the same as the frequency of the existing L1C/A signal. As seen in Fig. 2.11, the L1C signal also falls in the ARNS band. In order not to overly disturb the existing L1C/A signal spectrum, the MBOC modulation mode is adopted for the L1C signal. The carrier component is also comprised of two parts: the in-phase component and the quadrature component. The first is used as an exclusive channel for pilot frequency, while the modulation of navigation messages is completed by the latter, which is used as the information channel. The BOC (1, 1) mode is adopted for the pilot channel, and the BOC (6, 1) mode for the information channel. The pseudo-code rate of the L1C signal is 1.023 MHz, and its pseudo-code period is 10,230 chips. Its cycle length is 10 ms.

In the future, there will be more than four civil navigation signals on the L1 frequency, including L1 C/A for GPS, L1C, the E1 signal of the European Galileo system, and the B1-C signal of China's BDS. This means that L1 frequency GNSS receivers in the future will receive and process navigation signals from multiple systems using only one RF front-end device. Therefore, maintaining compatibility and interoperability among multiple systems will be an important task for these receivers.

In addition to the new civil navigation signals introduced above, a new military code signal (the M-Code) will also be transmitted by the Block-III satellites. An innovative spot beam antenna will be adopted for the military code signal. It can change the beam of the transmitting antenna that is sent to specific regional hotspots when necessary, so as to achieve local signal enhancement. Its maximum gain can reach 20 dB, which means that the signal power of the M-Code received by users on the surface of the Earth is no longer fixed. This technology is called 'combat zone signal enhancement'. The M-Code is designed to be autonomously positioned. Here, 'autonomous' means that the receiver only needs to process the M-Code signal to achieve positioning, unlike the P(Y) code, which requires the receiver to acquire and track the C/A code before realizing its positioning. The advantage is that the M-Code can be successfully located even when the C/A code signal is turned off or suffering severe interference. Since this book focuses on the development of civilian signals, the M-Code will not be detailed any further.

Figure 2.12 is a schematic diagram of the civil navigation signal provided by the space segment after GPS modernization. Once the modernization is completed, there will be more navigation signals available for civil users, and the design and implementation of the receiver will be diversified. There will be more products for different markets and applications, and one or more combinations of navigation signals will be designed for different products to achieve specific performances. All of this was unimaginable before GPS modernization.

2.4 An Overview of BDS

Satellite navigation systems have great national strategic significance and can generate enormous domestic and international economic interests. This means that all major industrialized countries around the globe are all implementing or planning

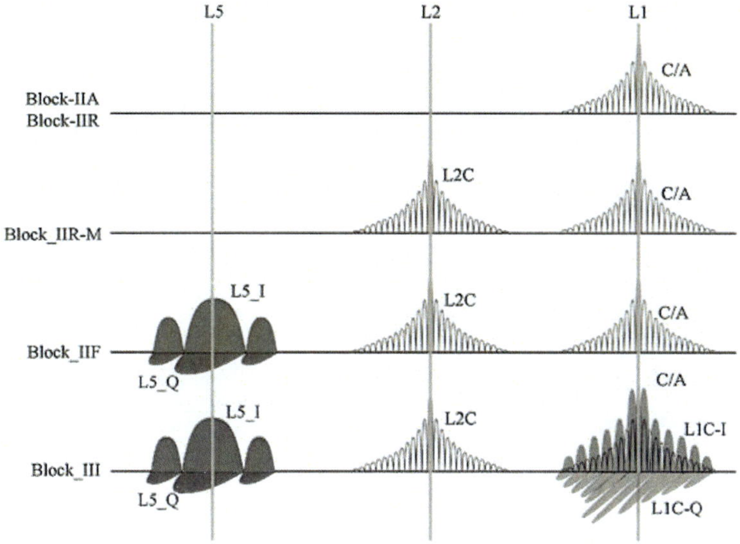

Fig. 2.12 Civil navigation signals transmitted by various types of GPS satellites after GPS modernization

to implement their own versions. As one of the world's most influential political, military, and economic powers, China is no exception. BDS (officially named BeiDou Navigation Satellite System) is a self-developed and independently operated global satellite navigation system currently under implementation in China. It is known internationally as BeiDou or Compass. BDS is one of the world's four major GNSS systems, together with America's GPS, Russia's GLONASS, and the EU's Galileo. The logo of BDS is shown in Fig. 2.13.

According to Ref. [7], the development of BDS has been carried out in the following three steps.

Fig. 2.13 The BDS logo

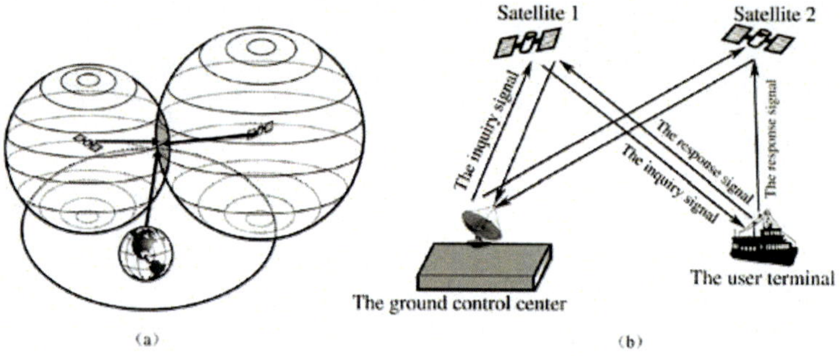

Fig. 2.14 Positioning principle of BDS-1

1. The construction of the experimental BeiDou navigation system;
2. The provision of regional service;
3. The provision of global service.

The experimental BeiDou navigation system constructed in the first step was called 'BDS-1', and the system constructed in the second and third steps was the 'BDS-2'. At the time of publication, the second step has been completed, and the third step is under progressive implementation.

China began the construction of the experimental BeiDou navigation system in 1994. On 31st October and 21st December 2000, it launched two BeiDou satellites. On 25th May 2003, the third BeiDou satellite was sent into Space, forming a constellation consisting of two geostationary satellites (located at 80° and 140° east longitude respectively) and an on-orbit backup satellite (located at 110.50° east longitude). On the ground, there is a central control system, a calibration system, and various user terminals, which together form BDS-1.

BDS-1 adopts the active positioning method with a three-ball intersection as its basic principle. In Fig. 2.14a, the three spheres are centered on satellite 1, satellite 2 and the Earth's center of mass. First, the ground control center sends an inquiry signal to users in the service area through two satellites in the geostationary orbit. The user will respond to one of the satellites if there is a request for the positioning service. At the same time, the user will also send a response signal to the other satellite. Then, the two response signals will be sent to the ground control station via the satellite transponder, and the ground control station will calculate the distance between the two satellites based on the two observations. It will then obtain the elevation according to the digital map database or the user's own altimeter. Thereby, three equations related to the user's position are generated, from which the user's latitude and longitude can be calculated. Figure 2.14b is a schematic diagram of the positioning principle and information flow of BDS-1.

The implementation of BDS-1 solved a major problem in China's satellite navigation system. In the 1990s, when China's domestic capital reserves were limited and its technology was underdeveloped, the inception of BeiDou was a major achievement for the satellite navigation industry. It made China the third country in the world to establish a satellite navigation system after the United States and the Soviet Union.

Since the BDS-1 user terminal needs to establish a two-way communication with the satellites, BDS-1 can not only conduct conventional positioning but can also deliver short messages. That is, the user terminal can perform two-way message communication through the satellite relay and the ground control center. A message with no more than 120 Chinese characters can be transmitted each time, which allows the user and others to know their location without turning to additional communication links—a unique function that is not yet available in the GPS and GLONASS systems. It is also this function that allowed BDS-1 to play an important role in the relief work after the Wenchuan Earthquake in 2008 [8].

1. Deficiencies of BDS-1

BDS-1 has been widely applied in China and its surrounding areas since its completion. It has also played an important role in mapping, telecommunication, water conservancy, transportation, fisheries, exploration, forest fire prevention, earthquake relief, and national security. However, it has also shown deficiencies, which are listed below.

(1) BDS-1 is not a global satellite navigation and positioning system

Since there are only two geostationary orbit satellites in the BDS-1 space constellation, global signal coverage cannot be achieved. What's more, the positioning of BDS-1 relies on the elevation database, which means that it is impossible to realize positioning in areas where elevation information is unattainable. Besides, BDS-1 can only effectively cover the area between 5° ~ 55° north latitude and 70° ~ 140° east longitude.

(2) BDS-1 is not suitable for high-dynamic users

According to the positioning principle of BDS-1, the BeiDou system must go through the process of inquiry-response-calculation-results return before the positioning result is obtained. The whole process takes about 1 s, so the real-time performance of BDS-1 is poor. This means that it is difficult to apply in high-dynamic applications such as aircraft navigation and precision weapon guidance.

(3) BDS-1 offers inadequate concealment of the user's position

This is caused by the positioning mechanism. When the user requires positioning, the receiver must send information to the control center. Therefore, radio silence will be compromised in this active positioning mode, which is disadvantageous in the military field.

(4) The population of BDS-1 users is limited

BDS-1 is an active positioning system with two-way ranging, so the number of users depends on the maximum data communication bandwidth of the satellite link. It is also related to the maximum rate of channel blocking, interrogation, and user response. On the contrary, in the passive positioning mode that is adopted in GPS and GLONASS, the navigation signal is broadcasted, and the user only needs to passively receive and process it. Therefore, the user population of the two systems is unlimited.

(5) The user terminal receiver is expensive, making it less competitive in the market

The upward communication link in BDS-1 requires a transmitter to be equipped in every terminal receiver, so they are not able to compete with GPS receivers in terms of size, weight, power consumption, and price. For this reason, it comes off worse in the market.

2. BDS-2

BDS-2 is designed to build a navigation system that is independent, open, inclusive, technologically advanced, stable, reliable, and globally covered. It can provide all-weather, all-day positioning, navigation, and time services with high precision and high reliability. It can also deliver short messages.

According to Ref. [9], in the future BDS will cover the space segment, the control segment, and the user terminal segment, similar to other GNSS systems. The space segment contains 35 satellites, including five geostationary orbit (GEO) satellites, three inclined geosynchronous orbit (IGSO) satellites, and 27 medium earth orbit (MEO) satellites. The orbital information of five GEO satellites at 58.75° east longitude, 80° east longitude, 110.5° east longitude, 140° east longitude, 160° east longitude, IGSO, and MEO are shown in Table 2.2.

The future distribution of the BeiDou satellite constellations is shown in Fig. 2.15. The large circular orbit parallel to the Earth's equatorial plane is the orbit of five GEO satellites, and the three large circular orbits at oblique angles to the equatorial plane are the orbits of the IGSO satellite, while the 27 MEO satellites move on the remaining three smaller elliptical orbits.

BDS will have basic functions such as passive positioning, speed measurement, and timing. The positioning accuracy is 10 m (confidence 95%), the speed measurement accuracy is 0.2 m/s, and the timing accuracy is 20 ns. Ref. [9] lists the signal bands for the following three BeiDou satellites.

- B1:1 559.052 ~ 1 591.788 MHz;
- B2:1 166.22 ~ 1 217.37 MHz;

Table 2.2 IGSO and MEO satellite orbit information for BDS

Relative information	MEO	IGSO
Number of satellites	27	3
Number of orbits	3	3
Orbital altitude	About 21,500 km	About 36,000 km
Orbital inclination	55°	55°

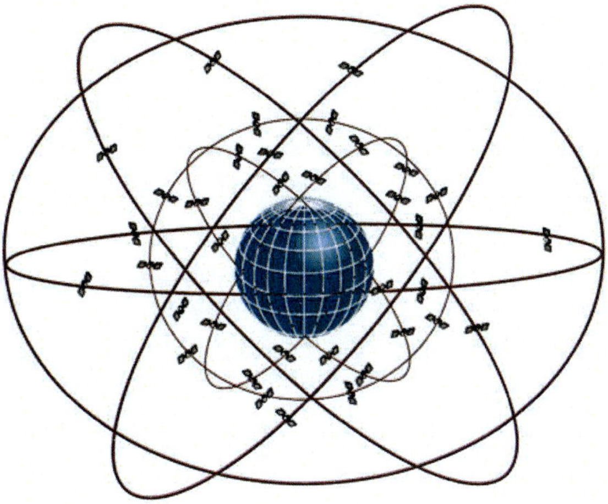

Fig. 2.15 Spatial constellation distribution of future BDS

- B3:1 250.618 ~ 1 286.423 MHz.

The basic parameters of the BeiDou navigation signal are shown in Table 2.3.

The B1 frequency in the BeiDou navigation signal is the same as the L1 frequency of GPS. In the future, it can be interoperated more successfully with navigation signals of GPS L1 C/A L1C, Galileo E1, and other systems such as GLONASS. The B2 frequency falls within the ARNS band, so it can be applied more effectively to the safety field. The B3 frequency is authorized, and is not available to unlicensed users such as the mass commercial market. Table 2.2 shows that in the future, all open service (OS) BeiDou navigation signals will have information channels and pilot

Table 2.3 Basic parameters of the BeiDou navigation signal

Signal component	Carrier frequency (MHz)	Pseudo-code rate (cps)	Data rate (bps/sps)	Modulation method	Service type
B1-C_D	1575.42	1.023	50/100	MBOC (6, 1, 1/11)	OS
B1-C_P			No		
B1$_D$		2.046	50/100	BOC (14, 2)	AS
B1$_P$			No		
B2a$_D$	1191.795	10.23	25/50	AltBOC (15, 10)	OS
B2a$_P$			No		
B2b$_D$			50/100		
B2b$_P$			No		
B3	1268.52	10.23	500	QPSK (10)	AS
B3-A_D		2.5575	50/100	BOC (15, 2.5)	AS

Note AS—Authorized service; OS—Open service

channels, which will help to improve the signal acquisition and tracking performance of future BeiDou receivers.

Similar to GPS, the ground control section is also included in the construction of BDS-2. It is composed of a main control station, a time synchronization/injection station, and a monitoring station, and is responsible for the operation control of system navigation. The main control station is the operation control center of BDS. Its main tasks include:

- Collection of navigation signal monitoring data transmitted by each time synchronization/injection station and monitoring station, data processing, and generation of satellite navigation messages;
- Task planning and scheduling, as well as system operation management and control;
- Comparison of satellite time and ground time, delivery of navigation message parameters to the satellite;
- Monitoring of satellite payload and analysis of abnormal situations.

The time synchronization/injection station is responsible for measuring satellite time synchronization and transmitting navigation message parameters to the satellite. The monitoring station continuously observes the satellite navigation signals and provides real-time observation data for the main control stations. The function and architecture of the ground control segment of the BeiDou system are very similar to those of GPS. Readers can use relevant knowledge of the GPS ground control section for reference.

On 27th December 2012, the Chinese government released the BeiDou Navigation Satellite System Signal In Space Interface Control Document (Version 1.0), marking the completion of the regional service system of the BDS-2. From then on, the BDS-2 provided navigation and positioning services to China and its surrounding regions and countries. The space constellation of the BDS-2 consists of 14 satellites, including five GEO satellites, five IGSO satellites, and four MEO satellites. The orbital height of the GEO satellite is 35,786 km, which is fixed at 58.75°, 80°, 110.5°, 140°, and 160° east longitude. The orbital height of the IGSO satellites is 35,786 km, and the orbital inclination is 55°. They are distributed in three orbital planes. The elevation point of the orbital plane is 120° away from the right ascension; the orbital height of the MEO satellite is 21,528 km, the orbital inclination is 55°, and the satellite period is 12 h, 55 min, 23 s, which is slightly longer than the GPS satellite's operating cycle. Four MEO satellites are located within two orbital planes. Table 2.4 shows the transmission record of the BeiDou navigation satellite.

It is worth mentioning that the 12th, 13th, 14th, and 15th BDS-2 satellites were all successfully launched though the "Two Satellites in One Rocket" approach, using the LM-3B carrier rocket. It was the first time that China used this method to launch medium and high-orbit satellites. According to Ref. [10], the "Two Satellites in One Rocket" and "Multiple Satellites in One Rocket" approaches will continue to be adopted in the launch of MEO satellites, while the GEO and IGSO satellites will be launched individually.

Table 2.4 Transmission record of the BeiDou navigation satellites

Satellite type	Launch date	Carrier rocket	Orbit
BDS-1-1	2000.10.31	CZ-3A	GEO
BDS-1-2	2000.12.21	CZ-3A	GEO
BDS-1-3	2003.05.25	CZ-3A	GEO
BDS-1-4	2007.02.03	CZ-3A	GEO
BDS-2-1	2007.04.14	CZ-3A	MEO
BDS-2-2	2009.04.15	CZ-3C	GEO
BDS-2-3	2010.01.17	CZ-3C	GEO
BDS-2-4	2010.06.02	CZ-3C	GEO
BDS-2-5	2010.08.01	CZ-3A	IGSO
BDS-2-6	2010.11.01	CZ-3C	GEO
BDS-2-7	2010.12.18	CZ-3A	IGSO
BDS-2-8	2011.04.10	CZ-3A	IGSO
BDS-2-9	2011.07.27	CZ-3A	IGSO
BDS-2-10	2011.12.02	CZ-3A	IGSO
BDS-2-11	2012.02.25	CZ-3C	GEO
BDS-2-12, 2-13	2012.04.30	CZ-3B	MEO
BDS-2-14, 2-15	2012.09.19	CZ-3B	MEO
BDS-2-16	2012.10.25	CZ-3C	GEO

Up until now, 16 BDS-2 satellites have been launched in total. Two are experimental satellites, and the other 14 in-orbit satellites provide positioning services. For BDS-2 to achieve global coverage by 2020, 24 more MEO satellites still need to be launched. According to *BeiDou News* [10], major changes will be applied to subsequent MEO satellites. At present, the four MEO satellites in the regional BDS-2 system not only provide conventional navigation and positioning services, but also undertake an important task, which is to test and verify the subsequent MEO satellites in terms of orbit, launch, measurement and control, and large-system interface. Major changes will be made to later MEO satellites in terms of platform and load, which include increasing the predicted lifetime of the satellites, improving autonomous survivability, designing new structural configurations to accommodate the "Multiple Satellites in One Rocket" approach, and providing space time benchmarks with higher accuracy.

The service diagram of the regional BDS-2 system is shown in Fig. 2.16. It covers most areas between 55° south latitude and 55° north latitude, 70° east longitude and 150° east longitude, as shown in the wire frame in the figure.

The smooth implementation of the regional BDS-2 service system marks a major step for China in becoming a nation with advanced satellite navigation technology. It also lays a solid foundation for the smooth progress of the third step of the BeiDou project. However, there are still some important tasks to accomplish.

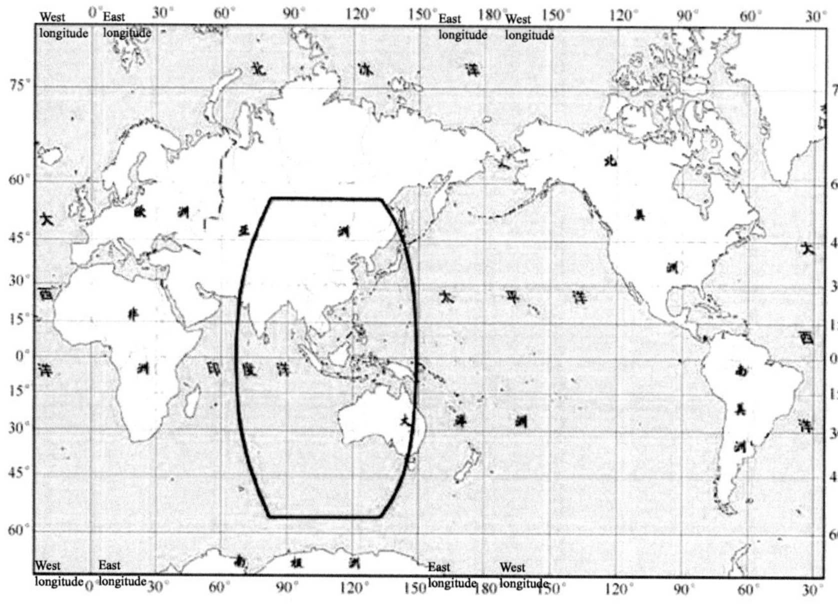

Fig. 2.16 Coverage of the regional BDS-2 service

(a) Ensuring the solid and stable operation of the regional BDS-2 service system

Existing BeiDou satellites must be monitored and evaluated to ensure the safe, stable and reliable operation of the satellite system. In order to further improve the reliability of the BeiDou constellation, several backup satellites are needed to ensure that service quality in the relevant area will not degrade or be affected when an inactive satellite appears.

(b) Preparing for a global BDS-2 system

The signal system for global BDS in the future will be different from the current one. Therefore, it is necessary to evaluate and test the new signal system through practical engineering. At the same time, existing satellites must be modified to increase service life and self-sustaining ability.

(c) Production and launch of satellites in the global BDS-2 system

The intensive launch period for BDS-2 satellites has not yet begun, but reasonable plans still need to be made for the production and launch of satellites.

(d) Key technology and marketing of the BeiDou navigation receiver

While BDS is achieving global coverage, the Chinese government needs to learn from the lesson of the Russian GLONASS system. The rapid deployment of GLONASS in the 1990s aroused great interest and expectation among international users at that time, as the system's positioning accuracy was even more accurate than the GPS civilian positioning with the SA policy. However, with the degrading of its economy, Russia lacked the financial resources to support continuous updates and improvements to the GLONASS space constellation. As a result, the number of available satellites declined, resulting in the continuous decline of service quality, and a loss of trust in GLONASS from users and receiver manufacturers. This lack of confidence made reconstruction difficult. Therefore, strong policy support and capital investment from the Chinese government is required for BDS before a virtuous cycle of "government input—market application—financial return" is formed.

The Chinese government's approach to implementing BDS is thorough and intense, and the construction of its terminal segment needs similar attention. Domestic and international satellite navigation receiver manufacturers have already launched navigation chips or receivers based on BDS. Since the service of the BeiDou navigation system is oriented towards global users, the competition that BDS receiver equipment faces in the market will also be global. This is both an opportunity and a challenge. On the one hand, domestic BDS users can choose products from global GNSS receiver suppliers; on the other hand, BDS receiver manufacturers in China can also sell their products on the global market.

This book focuses on the navigation signals of the regional BDS-2 service system. If not specifically explained, the term "BeiDou" refers to the regional service system of BDS-2.

References

1. Jiang Z (1994) Research on GPS doppler positioning method. China Navig 2
2. Navstar GPS User Equipment—Introduction, Public Release Version, NAVASTAR-GPS JPO USAF, September 1996
3. Lockheed Martin Press Release. US Air Force Awards Lockheed Martin GPS III Flight Operations Contract, 2008-05-31
4. Shaw M (2011) GPS modernization: on the road to the future, GPS IIR/IIR-M and GPS III. UN/UAE/US Workshop on GNSS Applications, Dubai, UAE, January 16, 2011
5. Navstar GPS Space Segment/Navigation User Interfaces, IS-GPS-200G, September 5, 2012
6. Navstar GPS Space Segment/User Segment L5 Interfaces. IS-GPS-705A, June 8, 2010
7. China Satellite Navigation System Management Office. Beidou Satellite Navigation System Development Report V2.2, December 2013
8. The use of Beidou satellite navigation system in disaster relief. Xinhua News, May 18, 2008
9. International Committee on Global Navigation Satellite Systems. Current and planned global and regional navigation satellite systems and satellite-based augmentations systems. New Work, 2010
10. China Beidou Satellite Navigation System website. http://www.beidou.gov.cn
11. Wu Y (1980) Development and applications of the transit navigation satellite system. Foreign Space Technol 4
12. Chen W (1979) The first-generation navigation satellite system of the United States: the transit. Foreign Space Dyn 5
13. Stansel TA Jr (1978) The current situation and future of the transit (trans: Wang G). J Navig 25(1)

14. Beard RL, Murray J, White JD (1986) GPS clock technology and the navy PTTI programs at the U.S. naval research laboratory
15. Guier WH, Weiffenbach GC (1998) Genesis of satellite navigation. Johns Hopkins Apl Tech Digest 19(1)
16. Dick SJ, Launius RD (2007) Societal impact of spaceflight, Chapter 17. National Aeronautics and Space Administration (NASA), Office of External Relations, History Division, Washington, D.C.
17. GPS Wing Reaches GPS III IBR Milestone, Inside GNSS, November 10, 2008
18. http://space.skyrocket.de/doc_sdat/navstar-2a.htm
19. Global Positioning management System IIR. Lockheed Martin Space Systems Company
20. Krebs G (2012) GPS-2R (Navstar-2R). Gunter's space page. Retrieved 11 July 2012
21. http://en.wikipedia.org/wiki/List_of_GPS_satellites
22. http://www.gps.gov/systems/gps/space/
23. http://www.boeing.com/boeing/defense-space/space/gps/
24. http://www.navcen.uscg.gov/?Do=constellationstatus
25. Global Positioning System Standard Positioning Service Performance Standard, Department of Defense, USA. Public Release, 4th edn, 2008
26. Adams TK (2006) The army after next: the first postindustrial army. Greenwood Publishing Group
27. ION Navigation Museum URL, http://www.ion.org/museum/item_view.cfm?cid=7&scid=9&iid=10
28. http://www.gps.gov/systems/gps/modernization/sa/data/
29. McDonald KD (2002) The modernization of GPS: plans, new capabilities and the future relationship to Galileo. J Glob Position Syst 1(1):1–17
30. European Space Agency. http://www.navipedia.net
31. http://www.gps.gov/systems/gps/modernization/civilsignals/
32. Li J (2007) A comparative analysis of Beidou satellite navigation and positioning system and global positioning system. Beijing Surv Mapp 1
33. Dai S, Ma C, Liao J (2010) An analysis and research of Beidou navigation and positioning system. Comput Digit Eng 3
34. Tang J, Yu L, Wang S (2008) An analysis of the application status of Beidou satellite navigation and positioning system. GNSS World China 2
35. China National Administration of GNSS and Applications (2009) Compass view on compatibility and interoperability. ICG Working Group A Meeting on GNSS Interoperability, July 30, 2009
36. China Satellite Navigation System Management Office (2013) Beidou satellite navigation system open service performance specification version 1.0, December 2013
37. Diggelen FV (2009) A-GPS, Assisted GPS, GNSS, and SBAS. Artech House

Chapter 3
GPS and the Signal Format and Navigation Message of BDS-2

The GPS receiver works in the passive positioning mode. The positioning mode of China's BDS-2 is very similar to that of GPS, which means that the GPS/BeiDou receiver only passively receives navigation signals from satellites, and calculate user positioning through a series of software/hardware signal processing modules and corresponding algorithms. So, understanding the structure of the satellite and satellite signals is of great significance for grasping the principle and performance of each module inside the receiver.

Code division multiple access (CDMA) is adopted for the navigation signals of GPS and BDS, and signals transmitted by multiple satellites in each system share the same carrier frequency. A pseudo-random code is an identification through which multiple satellite signals sharing the same carrier frequency can be distinguished from each other. It also broadens the initial signal bandwidth, which is key to detecting and processing weak signals from the distant surface of the Earth. In addition, the tracking loop inside the receiver can provide pseudo-range measurement after tracking the pseudo-random code phase, based on which position calculation is achieved. Therefore, the pseudo-random code is of great significance to satellite navigation signals.

A detailed introduction to the GPS and BeiDou navigation signals is presented in this chapter, including carrier components, pseudo-random codes, and navigation messages. With the modernization of GPS and the globalization of BDS-2, there will be more variations in navigation signals in the future. However, due to the limitations of space, this chapter will only cover the L1 C/A code of GPS and the D1 and D2 code of the B1 frequency of BDS-2. With an understanding of these basic signal structures and properties, readers will more easily grasp the principles of other similar satellite navigation signals, laying the foundation for further research.

Y. Lu, *BDS/GPS Dual-Mode Software Receiver*, Navigation: Science and Technology 10, https://doi.org/10.1007/978-981-16-1075-2_3

3.1 The GPS Signal

3.1.1 The Generation Mechanism of the GPS Signal

Figure 3.1 shows the generation mechanism of the L1 and L2 signals transmitted by GPS satellites. In the figure, 10.23 MHz is the reference clock frequency generated by the atomic clocks carried by GPS satellites. Generally, this frequency is recorded as f_0, and all the clocks in the GPS signal generation are generated from it. The 10.23 MHz clock is multiplied by 120 to obtain a carrier with an L2 frequency of 1fv

227.6 MHz. Then, the carrier signal is transformed into an L2 signal through BPSK modulation. At the same time, the 10.23 MHz clock is multiplied by 154 to generate a carrier with an L1 frequency of 1575.42 MHz, and the in-phase and quadrature components will be obtained by the carrier signal through phase shifters. Then, the modulo-two addition of the P(Y), C/A code and the navigation message are respectively modulated to obtain the L1 signal. The 10.23 MHz is also the generator clock of the pseudo-code of the P(Y) code, which drives the C/A code generator with a 1.023 MHz clock divided by 10. As seen in Fig. 3.1, the BPSK signal modulated at the L2 frequency can be a P(Y) code, a modulo-two addition of the navigation telegram and the P(Y) code or C/A code, which is decided by a selector. Currently, the L2 signal generally selects the modulo-two addition of the P(Y) code and the navigation message.

The 10.23 MHz clock on the GPS satellite drives the P(Y) code generator, so the chip width of the P(Y) code is approximately 0.1 μs, while that of the C/A code is approximately 1 μs, which is 10 times the width of the P(Y) code. The navigation

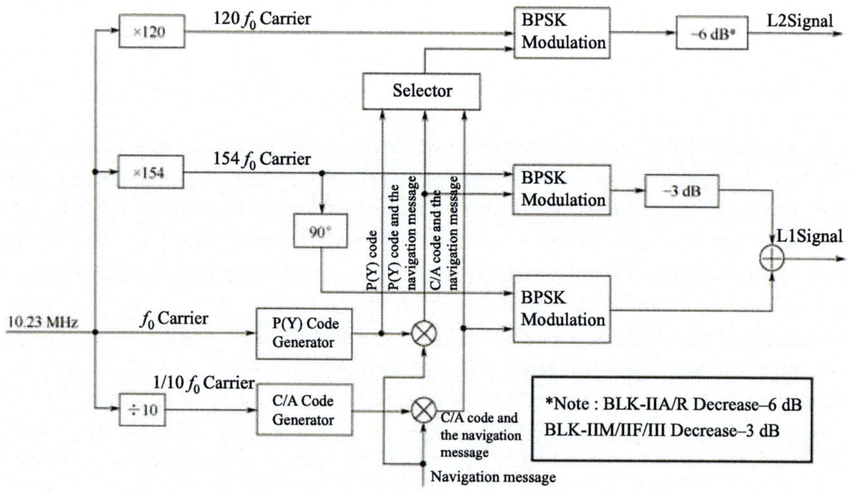

Fig. 3.1 Generation mechanism of the L1 and L2 signals in the GPS system

signal modulated by the P(Y) code provides high-precision location service (PPS), while the navigation signal modulated by the C/A code provides standard location service (SPS).

According to NavSTAR GPS Space Segment/Navigation User Interfaces [1], the P code generator is driven by the 10.23 MHz clock, and generates P codes with long periods (about 10^{14} chips). This process is repeated every seven days. The US government encrypts P-codes through an "anti-spoofing" (AS) policy. The primary purpose of this act is to protect authorized users from false GPS signals while restricting unauthorized users from using PPS services. The encryption process is performed by a special "Y code". The Y code chip rate is the same as the P code, but the user can only have access to the code with a key. The encrypted signal is generally called the P(Y) code signal. Through the AS policy, the US government can control the coverage of the PPS service. The PPS service is exclusive to authorized users. Unauthorized users, such as the civilians, can only use the SPS service provided by the C/A code. As seen in Fig. 3.1, the SPS service only provides the C/A code signal at the L1 frequency. Due to the lack of dual-frequency observation to eliminate the effect of the ionospheric delay on positioning accuracy, the positioning accuracy of SPS is slightly lower than PPS.

The P(Y) code period is very long, making direct capture quite difficult. However, the period of the C/A code is short, so it is usually captured first. After the preliminary information of the C/A code (such as the code phase and the transmission time) is obtained, the capture of the P(Y) code is performed. In this sense, the C/A code is also referred to as the "coarse code". Newer technologies have been developed to capture P(Y) codes directly. The L1M and L2M signals that emerged in GPS modernization can also directly capture and locate the P(Y) code without the transitional C/A code.

In fact, influenced by the relativistic effect, the output frequency of the atomic clock is not exactly 10.23 MHz; a manual offset is included. According to the time-expansion notion within the special theory of relativity, since the GPS satellites moves at a high speed, the time per day is shortened by about 7.2 μs if a clock operating at the orbital speed of the GPS satellite is recorded. Besides, according to the gravitational field theory of general relativity, the gravitational field is relatively weak for satellites more than 20,000 km away from the Earth, which leads to the acceleration of the clock. The time of one day recorded by the clock will increase by 45.8 μs daily. The final result is an increase of approximately 38.6 μs per day. If the on-board GPS clock is not adjusted, the position error caused by the time error will reach 11 km after one day. So the GPS clock was manually slowed down by $38.6 \times 10_{-6}/(24 \times 3600) \approx 4.467 \times 10^{-10}$ Hz, and the actual atomic clock output is 10.229,999,995,43 MHz. It is worth noting that the frequency correction here is to fix the reference frequency of the atomic clock, and is unrelated to the position of the satellite. If the flight speed and altitude of the satellite remain unchanged, a simple frequency correction like this can solve the problem. However, since the satellite's orbit is an ellipse, its flight speed and gravitational potential will vary with position, hence the impact of the relativistic effect on the satellite at different locations in the flight orbit varies. In order to ensure the accuracy of the pseudo-range observation, a relativistic correction for satellites in different positions will be included in the GPS satellite's broadcast ephemeris. When measuring the pseudo-range, we should take the correction into consideration to elevate the precision of pseudo-range observation.

If readers have different opinions on the correlation between clock accuracy and positioning accuracy, it is time for us to look back at the impact of the SA policy implemented earlier in the GPS system on the positioning accuracy. The intention of the SA policy was to purposefully limit the positioning accuracy of unauthorized users through GPS satellite signals, i.e. adding controllable errors to the positioning measurements. To achieve this policy, an artificial "jitter" amount was added to the satellite clock, which would affect the C/A, P(Y) code and carrier phase observation, and ultimately affect user positioning accuracy. Authorized users can eliminate the impact of the jitter with a key. Unauthorized users who do not have a key can also eliminate the adverse effect by means of difference and the like. According to the relevant literature, the main error of the positioning results of the autonomous GPS receiver came from the SA policy during its implementation, where the horizontal and vertical errors were 100 m and 156 m (95% confidence) respectively. The change in positioning accuracy before and after the SA policy was abolished can be clearly seen in Fig. 2.9.

The P(Y) code and C/A code generator in Fig. 3.1 generate 0 and 1 logic signals. The navigation message is also displayed in 0, 1 sequences. The modulated BPSK signal is obtained through modulo-two addition of the codes and the message. Here, the modulo-two addition of the binary numbers is equivalent to the multiplication of decimal numbers. The operation rules are as shown in Table 3.1.

BPSK is the acronym for binary phase shift keying. When the modulation signal is 0, the carrier phase is unchanged; when it is 1, the carrier phase is inverted. The modulated GPS signal is the composite signal generated by the modulo-two addition of the navigation message and the pseudo-code. The frequencies of the modulated carrier are 1575.42 MHz (L1) and 1227.6 MHz (L2). For the P(Y) code signal, there are 154 L1 or 120 L2 carrier cycles within one chip width, while for C/A code signals, there are 1540 L1 carrier cycles within one chip width. The final composition of the L1/L2 navigation signal is shown in Fig. 3.2.

In Fig. 3.2, the C/A code of the GPS L1 frequency is set as an example. The navigation message bit, pseudo-random code, carrier signal, and BPSK signal are all included. The pseudo-code period of the C/A code is 1 ms, so the pseudo-random code is repeated every 1 ms, and the carrier portion is synchronized with it. Also, it can be seen from the figure that the clock of the navigation message at the bit transition time

Table 3.1 Modulo-two addition

Operation number 1	Operation number 2	Result of the modulo-two addition
0	0	0
0	1	1
1	0	1
1	1	0

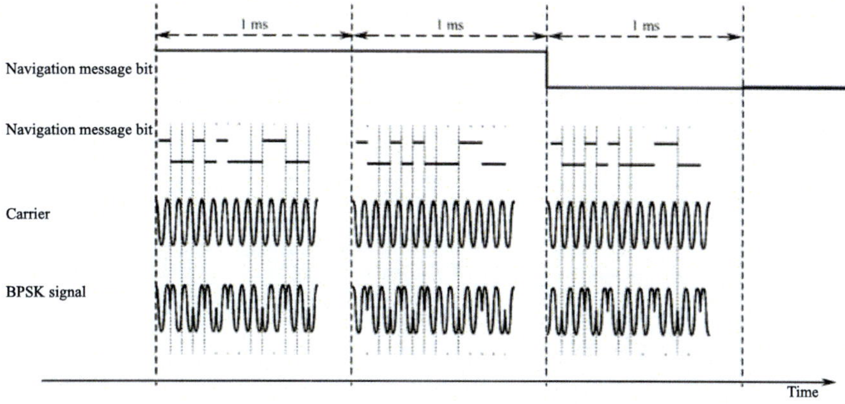

Fig. 3.2 Composition of the GPS navigation signal

is synchronized with that of the pseudo-random code, and all these synchronization relationships are guaranteed by the GPS onboard clock and subsequent clock circuits. In fact, there are 1540 carrier cycles inside a C/A code chip (for L1 carrier signals). Due to limitations of space, Fig. 3.2 does not adequately reflect this proportion. It can be clearly seen that the carrier phase of the BPSK signal in the lower part of the figure changes with the modulo-two addition result of the C/A code and the navigation telegram bits. The GPS signal of the P(Y) code can be similarly understood according to Fig. 3.2.

Before GPS modernization, each GPS satellite transmitted navigation signals of two frequencies, L1 and L2, where L1 carried both C/A and P(Y) code signal, and L2 only carried the P(Y) code signal, as shown in Fig. 3.3.

P(Y) signal: Encrypted and only available for authorized users, modulated on both L1 and L2 carriers with a chip rate of 10.23 MCP.

C/A signal: Open to the public, modulated only on the L1 carrier with a chip rate of 1.023 MCP.

The navigation message rates on L1 and L2 are both 50 bps, and the width of each information bit is 20 ms, so there are 20,460 C/A chips or 204,600 P(Y) chips inside each navigation telegram. The rate of P(Y) chip hopping is 10 times that of C/A chip hopping. Guaranteed by the precisely assembled on-board clock circuit, the hopping of the navigation telegram and the pseudo-code is strictly synchronized. Furthermore, thanks to the monitoring and correction of all of the GPS satellites' on-board clocks by the GPS control segment, the hopping clocks of respective navigation message generated by all GPS satellites are also synchronized with the clocks of the respective pseudo-code generators, which is a basic premise for the GPS system to achieve positioning.

The GPS navigation signal consists of three components: the carrier wave, pseudo-code, and navigation message. Its equation can be written as:

$$s_{\mathrm{L1}}(t) = \sqrt{2P_{\mathrm{C}}}D(t)c(t)\cos[\omega_{\mathrm{L1}}t + \theta_{\mathrm{L1}}] + \sqrt{2P_{\mathrm{Y1}}}D(t)y(t)\sin[\omega_{\mathrm{L1}}t + \theta_{\mathrm{L1}}] \quad (3.1)$$

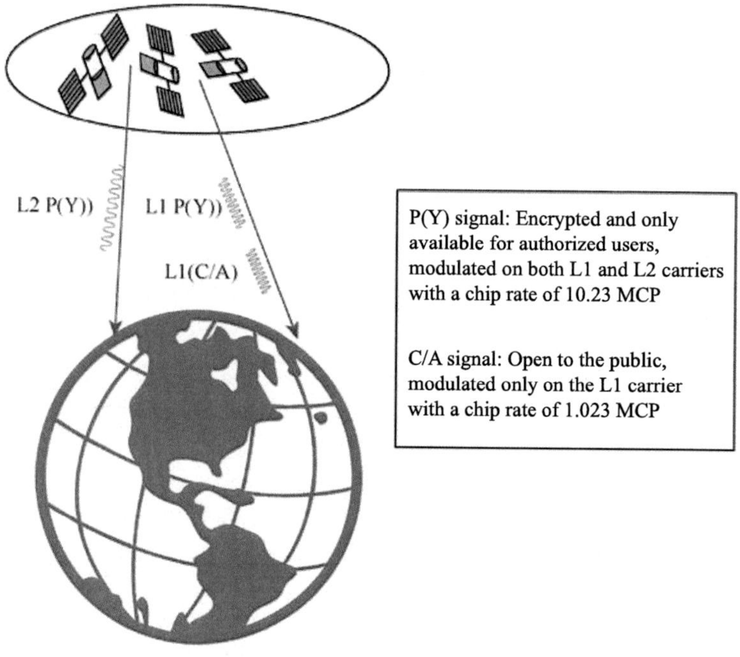

Fig. 3.3 Navigation signals transmitted on the GPS L1 carrier and L2 carriers

$$s_{L2}(t) = \sqrt{2P_{Y2}}D(t)y(t)\sin[\omega_{L2}t + \theta_{L2}] \tag{3.2}$$

where, *c(t)* and *y(t)* are C/A code and P(Y) code respectively, *D(t)* is the navigation message data bits, and ω is the carrier frequency, whose subscripts L1 and L2 represent the L1 carrier and L2 carrier. P_C, P_{Y2}, and P_{Y1} are the powers of different signal components. The C/A and the P(Y) code signal are modulated on the L1 carrier. Although modulated on the same frequency carrier, the two signals are 90° out of phase and remain “orthogonal”, thus enabling the receiver to obtain and distinguish the two navigation signals.

The chip rate of *c(t)* is 1.023 MHz, and the width of each chip is about 1 μs, so the error of one chip represents about 300 m in distance. The pseudo-random code corresponding to *c(t)* is sometimes referred to as a “coarse code”. Its period is 1023 chips, which is 1 ms in length. The chip rate of *y(t)* is 10.23 MHz, which is 10 times the C/A code rate, so the error of the corresponding chip represents 30 m in distance. Thus, it can be deducted that the receiver using the *y(t)* code is theoretically nine times higher than the C/A code in distance positioning accuracy. In reality, the positioning accuracy of the receiver will be affected by many factors, and the pseudo-code chip width is merely one of them.

D(t) represents the modulated navigation telegram bit. As its rate is 50 bps, the length of one bit is 20 ms. The navigation message is intended to provide satellite ephemeris and almanac data, which are used to calculate the position and velocity of the satellite. Parameters like satellite clock correction parameters, ionospheric and tropospheric delay parameters, UTC time parameters, and satellite health are also indicated in navigation messages. Apart from these parameters, the navigation message is also helpful for obtaining the signal's transmission time. As seen in Chap. 1, it is a very important step for the receiver to acquire the sending time of the signal, because the position calculation of the GPS satellite is directly determined by its signal transmission time. Furthermore, in order to obtain the pseudo-range observation, it is also necessary to know the signal transmission time of the satellite at a certain time point.

After the navigation message and the pseudo-random code are multiplied, the original signal bandwidth is stretched from 100 Hz to approximately 2 MHz (for C/A codes) and 20 MHz (for P(Y) codes). The available spectrum representation of the GPS signal is shown in Fig. 3.4.

The main purposes of the GPS signal using the spread spectrum signal are as follows.

(a) The pseudo-random code is the identification of different satellite signals.

Equations (3.1) and (3.2) show that all satellite signals share the same carrier frequency, which means that the spectrum of all satellite signals is mixed in the spectrogram. The signal obtained by the receiver is also the coexistence signal of multiple satellites. However, serious co-channel interference is not aroused thanks to the fact that each satellite has its own unique pseudo-random code, which has a strong autocorrelation.

(b) Broadening the frequency band lowers the signal-to-noise ratio of the received signal.

This can be seen from the Shannon formula in information theory:

$$C = B\mathrm{lb}(1 + S/N) \tag{3.3}$$

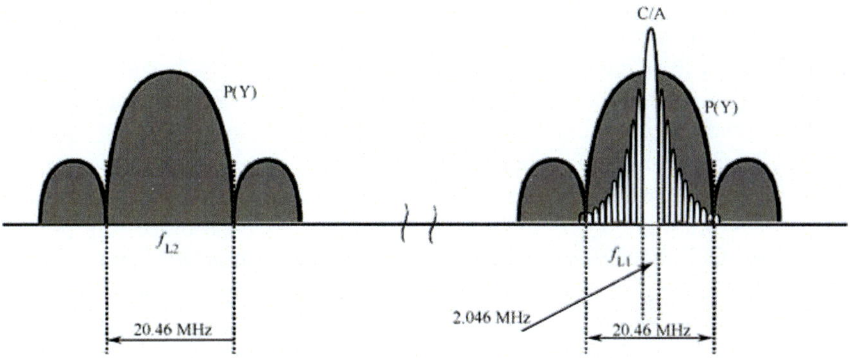

Fig. 3.4 Illustration of the GPS signal frequency spectrum on L1 and L2

where, C is the channel capacity, B is the bandwidth of the channel, and S/N is the signal-to-noise ratio. It is clear that if the value of C is constant, and if B is widened, then S/N can be quite low. That is, the power of the received signal can be low or even lower than the noise power.

(c) The chip phase of the pseudo-random code provides the necessary ranging signal for positioning at the front end of the GPS receiver.

The code phase is directly related to the distance of the satellite signal transmission. The receiver obtains the pseudo-code phase by detecting the position of the correlation peak, thereby obtaining the pseudo-range observation and realizing positioning.

In current major satellite navigation systems, the carrier modulation method of the navigation signal may be different, but almost all of the ranging signals are generated by the combination of the pseudo-random code and navigation message. CDMA is adopted for the various satellite signals of the GPS, Galileo and BeiDou systems, in which the pseudo-random code is indispensable. For Russia's GLONASS systems, FDMA is used for different satellite signals, so the differentiation of satellite signals depends on different carrier frequencies. However, the combination of the pseudo-random code and navigation message is still useful in the system. Different from the former three, the navigation signals of all GLONASS satellites use one set of pseudo-random codes, in which case only the second and third properties the pseudo-random code mentioned above are embodied. It is worth mentioning that since 2008, the Russian government has been considering transmitting CDMA signals on new GLONASS satellites. It is expected that the newly launched K series and later satellites will transmit two public CDMA signals, but there has been no further news yet.

There are expected to be three more civilian signals in the process of GPS modernization, namely L1C, L2C, and L5. The L2C frequency is the same as the L2 frequency point of the existing P(Y) code. L5 is the frequency of ARNS, a newly added frequency band. The L1C frequency point is the same as the existing L1 frequency point, the better to interact with the existing C/A code signal and signals of other GNSS systems. Since the Galileo system, BeiDou system, and GLONASS system have all launched new public service signals, the civilian satellite navigation signals of the future will be more diverse, but their basic structures will remain similar. Therefore, grasping the contents of this section will facilitate readers' understanding of relevant knowledge.

Table 3.2 shows the basic characteristics of the L1C, L2C, and L5 signals, from which we can see that the three signals use forward error correction and pilot channel mechanisms to ensure better data demodulation and baseband processing performance. TMBOC modulation is used for the L1C signal to minimize the interference between the distribution of the signal spectrum and the existing signals. Similar to the L1 C/A signal, BPSK modulation is adopted for the L2C and L5 signals, so the existing processing method can be transplanted easily into the processing of the L2C and L5 signals in the future.

Table 3.2 The basic characteristics of the L1C, L2C, and L5 signals after GPS modernization

Signal type	Carrier frequency (MHz)	Multi–access mode	Modulation mode	Pseudo–code rate (Mcps)	Modulation content	Navigation message rate
L1C	1575.42	CDMA	TMBOC	1.023	Data I+ Pilot Q	50 bps/100 sps FEC
L2C	1227.6	CDMA	BPSK	1.023	CM+CL time share	25 bps/50 sps FEC
L5	1176.45	CDMA	BPSK	10.23	Data I+ Pilot Q	50 bps/100 sps FEC

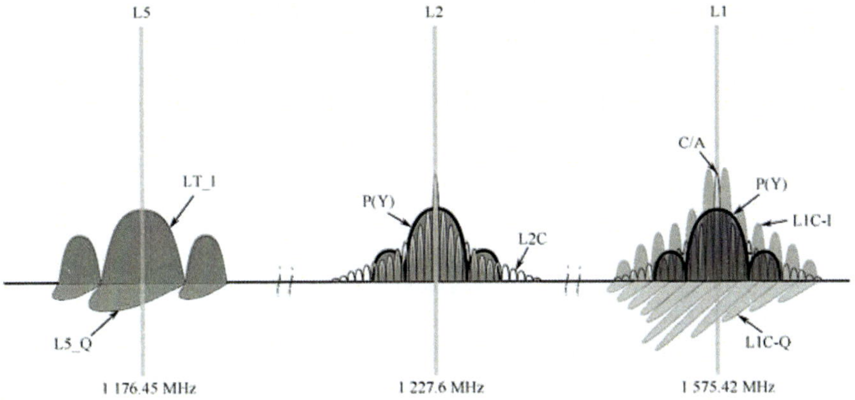

Fig. 3.5 The spectrum distribution of navigation signals after GPS modernization

Fig£ure 3.5 shows the spectrum distribution of several GPS navigation signals that the receiver can use after GPS modernization, including the signal spectrum of L1 C/A, P(Y), L1C, L2C, and L5. The new M-Code signal is not included in the figure. Since pilot and data channels are applied to L5 and L1C signals, what is shown in the figure is the orthogonal spectral components of the two channels. For technical details on L1C, L2C and L5, please refer to its interface control document [1–3].

3.1.2 *The C/A Code Generator*

The pseudo-random code of the GPS signal varies among signals, such as the C/A code on L1, the P(Y) code on L2 and L1, the CM and CL codes in the L2C signal, the I5 and Q5 codes on L5, and the L1CD and L1CP codes on L1C. Although the code length and bit content of these pseudo-random codes are different, their basic mechanisms are very similar. With an in-depth understanding of the principle of one of the codes, readers will quickly grasp the others. Here, the most common C/A code in the GPS receiver is taken as an example for detailed explanation. Readers can refer to the technical documents for other pseudo-random codes for further study and research.

The C/A code at the L1 frequency is essentially a Gold code. This Gold code is obtained by the bitwise addition of the preferred pairs of two m-sequences with different code words controlled by the synchronous clock. Unlike ordinary addition, the addition method adopted here is modulo-two addition, which has been described in detail in Fig. 3.1. The cross-correlation function between each code group of the Gold code maintains cross-correlation between the original two m-sequences, and the maximum cross-correlation value does not exceed that between the original two m-sequences. When the relative phase relationship between two m-sequences is changed, a new Gold code will be generated. So, the advantage of the Gold code is that it can generate multiple independent code groups according to the preferred pairs of two m-sequences, which is highly significant for the CDMA system. For the specific application of GPS systems, it means that the system can accommodate enough satellites, and the signals broadcast by different satellites can be distinguished from each other by their respective independent Gold codes (C/A codes).

The C/A code used by GPS satellites is a balanced Gold code. Here, "balanced" means that the number "1" in the overall sequence is at most one more than "0", which means there is basically no DC component, so the carrier suppression can be better when modulating the carrier. If the balance of the pseudo-code is off, leakage of the coded clock component will occur in the balanced modulator of the modulation circuit. This will cause the spread spectrum signal at the transmitting end to lose signal concealment, and will also be a waste of transmission power. For the receiver, the carrier component is not sufficiently suppressed, so it will enter the subsequent signal processing unit as a narrowband signal, increasing the internal interference of the system. Therefore, the balance of the pseudo-code must be considered when selecting the C/A code for the GPS satellite signal.

Many factors should be taken into consideration in the selection of C/A codes. One is the autocorrelation that is as sharp as possible and a cross-correlation that is as weak as possible for the code, which will be described in more detail in later chapters. The second is the code length, because the longer the code period, the greater the spread spectrum gain and the better the autocorrelation and cross-correlation properties will be. However, if a code period is excessively long, the difficulty in signal acquisition will increase, which will lead to a significant problem, namely that the uncertainty of the pseudo-code phase will increase, and thereby the signal capture time will expand. These two factors were taken care of in the initial GPS signal system through the coexistence of the C/A code and P(Y) code, as the C/A code has a short period of 1,023 chips, which equals the time length of 1 ms, so it can be captured quickly. However, the period of P(Y) code is seven days, which makes it hard to capture, although it can contribute an excellent cross-correlation suppression and autocorrelation function. Therefore, the C/A code is generally captured first, and basic time information and phase are obtained. The information is captured later by the P(Y) code.

The schema for the generation of a C/A code is given in ICD-GPS-200, the official GPS document, as shown in Fig. 3.6.

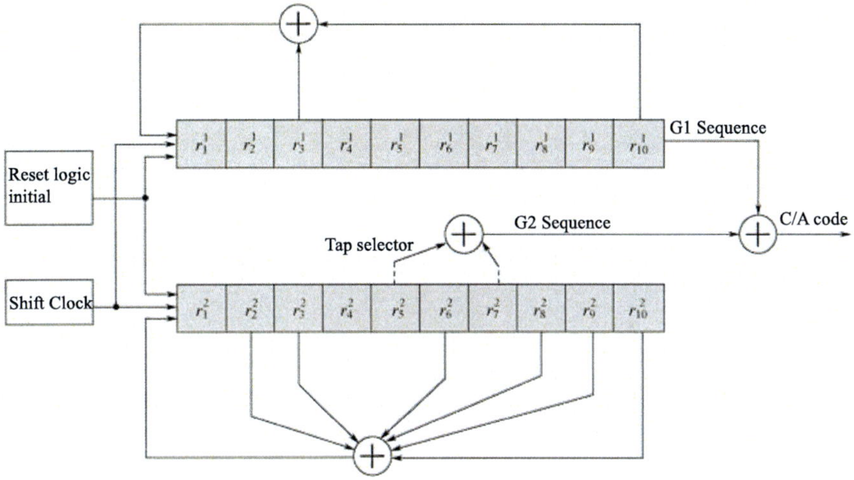

Fig. 3.6 Generation of the C/A code through a tap selector

As can be seen in the figure, the C/A code consists of two m-sequences, called the G1 sequence and G2 sequence respectively. The C/A code used in each satellite ranging signal is generated through the modulo-two addition of the G1 sequence and G2 sequence after a certain tap selection. The bit number of linear shift register bits per m-sequence is ten. According to the nature of the m-sequence, the period of the G1 and G2 sequence are both 210–1 = 1023 chips, which is also the period of the Gold code obtained through the modulo-two addition of the two sequences.

In Fig. 3.6, there are two linear-shift register groups and 10 registers in each group. The upper half of the register set corresponds to the G1 sequence, and the lower half of the register set corresponds to the G2 sequence. The frequency of the operating clock of the shift register is usually 1.023 MHz. It is set by the actual pseudo-code rate, which refers in the satellite to the octave of the 10.23 MHz clock output from the atomic clock and in the user receiver, refers to the pseudo-code generated by the local pseudo-code clock generator. It is usually outputted by a numerically controlled oscillator (NCO) controlled by a pseudo-code loop whose frequency value is fine-tuned around 1.023 MHz. The period of the Gold code is 1,023 chips, so driven by the 1.023 MHz clock, the length of a pseudo-code period is 1 ms. That is, all pseudo-code chips of one cycle are generated every 1 ms, and the length of each chip is approximately 1 μs (≈0.977 517 μs).

The generator polynomial of the m-sequence is an important basis for theoretical analysis of this sequence. The linear feedback tap configuration in Fig. 3.6 shows that the generator polynomial of G1 and G2 is

$$P_{G1} = 1 + x^3 + x^{10} \tag{3.4}$$

$$P_{G2} = 1 + x^2 + x^3 + x^6 + x^8 + x^9 + x^{10} \tag{3.5}$$

In Fig. 3.6, the content of the linear shift register of G1 is represented by r_i^1, where $i = 1, \ldots, 10$. When the rising edge of each clock comes, the content of the second to 10th registers is updated by the content of the previous register. The content of the first registers is updated by the feedback value F, and the whole process can be expressed by the following formula:

$$\begin{cases} F = r_3^1 \oplus r_{10}^1 \\ r_1^1 = F \\ r_i^1 = r_{i-1}^1, \quad i = 2, \cdots, 10 \end{cases} \tag{3.6}$$

Similarly, the content of the linear shift register of G2 is represented by r_i^2, and the update process of the register contents of G2 can be written as

$$\begin{cases} F = r_2^2 \oplus r_3^2 \oplus r_6^2 \oplus r_8^2 \oplus r_9^2 \oplus r_{10}^2 \\ r_1^2 = F \\ r_i^2 = r_{i-1}^2, \quad i = 2, \ldots, 10 \end{cases} \tag{3.7}$$

The initial phases of the registers of G1 and G2 are all set as "1", namely 0x3FF. Here, 10 "1"s are represented by hexadecimal numbers. The initial phase setting of the registers is critical, as an unsuitable setting will not lead to the desired result. For example, if the initial phases are all set as "0", then the feedback value F will always be "0". As a result, the obtained m-sequence is also composed of "0"s.

The output C/A code is the result of the modulo-two addition of the last register content of G1 and the contents of several registers of G2, which is

$$C_{C/A}(k) = r_{10}^1(k) \oplus \left[r_{S_1}^2(k) \oplus r_{S_2}^2(k)\right], \quad k = 1, \ldots, 1023 \tag{3.8}$$

where, the subscripts s_1 and s_2 are the tap positions of the G2 linear shift register. Changes to the generated C/A code can be achieved by changing the value of s_1 and s_2.

Unlike generating a C/A code through changing the tap position, another method generates different C/A codes by changing the delay phase of the G2 sequence. The block diagram for this method is shown in Fig. 3.7.

Compared with Fig. 3.6, this method eliminates the tap selection logic and increases the delay phase unit. The G1 and G2 sequences are still generated according to Eqs. (3.4) and (3.5), namely, the feedback tap settings of the G1 register and the G2 register remain unchanged. After the G2 sequence is generated, it needs to delay the *Di* chips and then perform the modulo two-addition with the G1 sequence. The result is the C/A code stream, where *Di* is the phase delay amount, and *i* corresponds to the 37 C/A codes of GPS from 1 to 37. Each C/A code corresponds to a different *Di* value.

The initial phase of the registers of G1 and G2 in the delay phase method are the same as those in the tap selection methods, which is 0x3FF, or all "1"s. The content

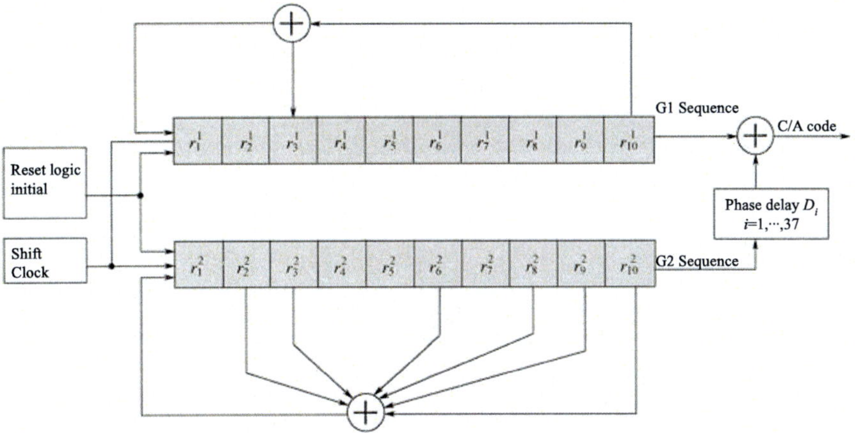

Fig. 3.7 Changing the C/A code through adjusting the phase delay of the G2 sequence

of the generated C/A code can be expressed as

$$C_{\text{C/A}}(k) = r_{10}^1(k) \oplus r_{10}^2(k + D_i), \quad k = 1, \ldots, 1023 \tag{3.9}$$

During the exploration of the specific implementation of the delay phase method, a third way of generating C/A codes emerges. This method is enlightened by a property of the m-sequence, which is that the modulo-two sum of an m-sequence and its own time-shifted sequence is itself; the only change is the phase. Therefore, the output of the G2 shift register corresponding to different phase delays in the phase delay method can also be obtained by changing the initial phase of G2. This method is called the initial phase method, as shown in Fig. 3.8.

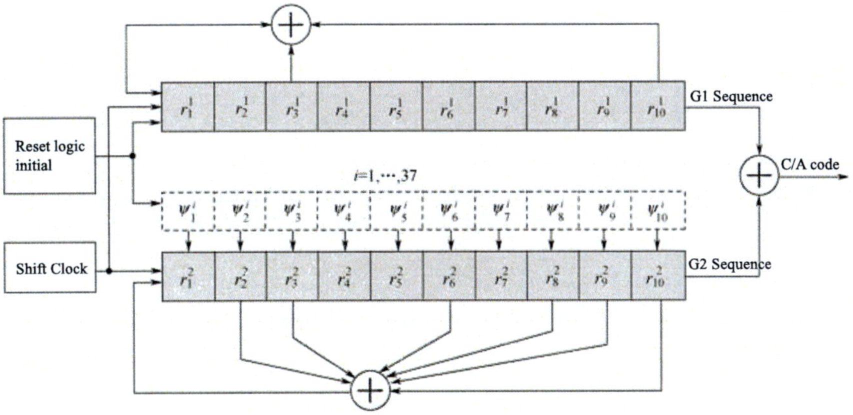

Fig. 3.8 Generation of the C/A code through changing the initial phase of the G2 register

The initial phase of the linear shift register of G1 in Fig. 3.8 is still 0x3FF, but the initial phase of G2 is different according to the C/A code. Therefore, $\Psi_1^i \ldots \Psi_{10}^i$ is used in the figure to represent different initial settings, and their superscript i, whose value ranges from 1 to 37, represents the i-th C/A code.

In the initial phase method, the content of the register groups of G1 is still updated according to Eq. (3.6), while that of the G2 register groups is updated according to Eq. (3.10), where the only step added is the initialization of the phase of the G2 register when compared with Eq. 3.7.

$$\begin{cases} \text{Initialize: } [r_i^2] = [\Psi_i^j], \quad i = 1, \ldots, 10 \\ F = r_2^2 \oplus r_3^2 \oplus r_6^2 \oplus r_8^2 \oplus r_9^2 \oplus r_{10}^2 \\ r_1^2 = F \\ r_i^2 = r_{i-1}^2, \quad i = 2, \ldots, 10 \end{cases} \tag{3.10}$$

The content of the C/A code generated by the initial phase method can be expressed by Eq. (3.11):

$$C_{\mathrm{C/A}}(k) = r_{10}^1(k) \oplus r_{10}^2(k), \quad k = 1, \ldots, 1023 \tag{3.11}$$

The C/A codes generated by the above three methods are the same. So one of them can be selected according to the specific situation. For the convenience of engineers, this book summarizes the specific configuration information of the three methods in Table 3.3, including the tap selection of the tap setting method, the D_i value of the delay phase method, and the setting of $\Psi_1^i \ldots \Psi_{10}^i$ in the initial phase method (represented by hexadecimal and binary numbers). Table 3.3 shows the information of PRN1–PRN37 of GPS signals. Currently, PRN1–PRN32 is used by the GPS space segment, while PRN33–PRN37 is reserved for ground services, such as pseudo-satellite applications.

3.1.3 *Autocorrelation and Cross-Correlation of C/A Codes*

The autocorrelation function measures the similarity of a signal with itself after being offset on the time axis by a certain length of time. For a completely random function, since its value at the current time point and the next one is completely uncorrelated, its autocorrelation function should be 0 if the time offset is not 0. A typical example is a white noise signal. When it comes to the autocorrelation function of the C/A code, it is generally defined as a time-averaged autocorrelation function which can be written as:

$$R_{i,i}(\tau) = \frac{1}{T}\int_0^T c_i(t)c_i(t+\tau)\mathrm{d}t, \quad \tau \in (-T/2, T/2) \tag{3.12}$$

Table 3.3 Summary of three methods for generating C/A codes

C/A code number	Setting of the tap selector (s1, s2)	Phase delay *Di* (Unit: chips)	Initial G2 phase[a]	
			(In hexadecimal notation)	(In binary notation)
1	2⊕6	5	0x320	1100100000
2	3⊕7	6	0x390	1110010000
3	4⊕8	7	0x3C8	1111001000
4	5⊕9	8	0x3E4	1111100100
5	1⊕9	17	0x25B	1001011011
6	2⊕10	18	0x32D	1100101101
7	1⊕8	139	0x259	1001011001
8	2⊕9	140	0x32C	1100101100
9	3⊕10	141	0x396	1110010110
10	2⊕3	251	0x344	1101000100
11	3⊕4	252	0x3A2	1110100010
12	5⊕6	254	0x3E8	1111101000
13	6⊕7	255	0x3F4	1111110100
14	7⊕8	256	0x3FA	1111111010
15	8⊕9	257	0x3FD	1111111101
16	9⊕10	258	0x3FE	1111111110
17	1⊕4	469	0x26E	1001101110
18	2⊕5	470	0x337	1100110111
19	3⊕6	471	0x39B	1110011011
20	4⊕7	472	0x3CD	1111001101
21	5⊕8	473	0x3E6	1111100110
22	6⊕9	474	0x3F3	1111110011
23	1⊕3	509	0x233	1000110011
24	4⊕6	512	0x3C6	1111000110
25	5⊕7	513	0x3E3	1111100011
26	6⊕8	514	0x3F1	1111110001
27	7⊕9	515	0x3F8	1111111000
28	8⊕10	516	0x3FC	1111111100
29	1⊕6	859	0x257	1001010111
30	2⊕7	860	0x32B	1100101011
31	3⊕8	861	0x395	1110010101
32	4⊕9	862	0x3CA	1111001010
33	5⊕10	863	0x3E5	1111100101
34	4⊕10	950	0x3CB	1111001011

(continued)

Table 3.3 (continued)

C/A code number	Setting of the tap selector (s1, s2)	Phase delay *Di* (Unit: chips)	Initial G2 phase[a]	
			(In hexadecimal notation)	(In binary notation)
35	1⊕7	947	0x25C	1001011100
36	2⊕8	948	0x32E	1100101110
37	4⊕10	950	0x3CB	1111001011

[a]LSB Corresponds to Ψ_1^i, MSB Corresponds to Ψ_{10}^i

where, *i* represents the C/A code of the *i*-th PRN, and the subscript (*i, i*) stands for the autocorrelation function of the PRN code. This is convenient for the following presentation of the cross-correlation function, as the cross-correlation function between the C/A codes of the *i*-th and *j*-th PRNs can be represented by subscripts (*i, j*) by analogy. T is the period of the C/A code. Assuming that the chip length is T_c and the chip number in one cycle is N, then $T = NT_c$. For n-order linear feedback shift registers, $N = 2^n - 1$; for C/A codes, $n = 10$, which is the number of bits in the G1 and G2 shift register banks. According to the previous section, T_c of the GPS C/A code ≈0.977 517 μs, $N = 1023$, then $T = 1$ ms, so the C/A code period is 1 ms.

Equation (3.12) is the integral averaging of the product of the C/A code after a delay of τ, since the C/A code is a periodic function, which is

$$c_i(t) = c_i(t + NT), \quad N = 0, 1, 2, \ldots \tag{3.13}$$

When substituting Eq. (3.13) into (3.12), we can see that $R_{i,i}(\tau)$ is also a periodic function with a period of T, which is the same as the C/A code period. This is also the reason why the value of τ in Eq. (3.12) is in $(-T/2, T/2)$.

The calculation of the C/A code autocorrelation function is directly shown in Fig. 3.9. The uppermost waveform signal in the figure is $c_i(t)$, and the waveform signal below it is $c_i(t)$ after τ, a period of delay. When the value of τ is positive, the signal waveform moves to the right, and when it is negative, the signal waveform

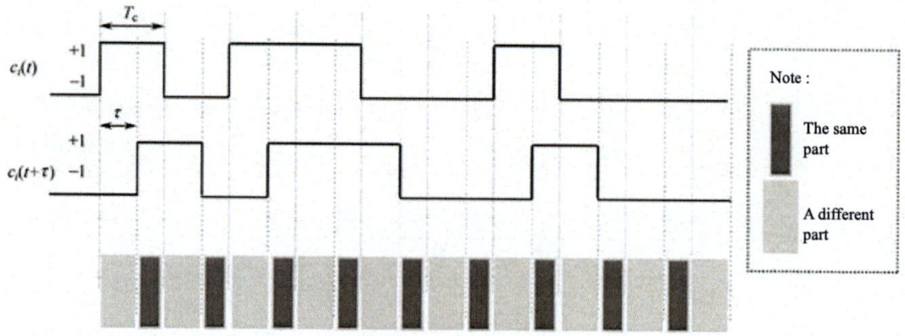

Fig. 3.9 Calculation of the C/A code autocorrelation function

moves to the left. The dark shade of the figure represents the same part of the two waveform signals, while the lighter shade stands for the difference between them. It is obvious that the product of the same part is 1, and the product of the different parts is -1, so the result of the final integration is the difference between the cumulative area of all the same parts and that of all the different parts.

When $\tau = 0$, then $c_i(t)$ and $c_i\ (t + \tau)$ are perfectly aligned, then

$$\begin{aligned} R_{i,i}(0) &= \frac{1}{T}\int_0^T c_i(t)c_i(t)\mathrm{d}t \\ &= \frac{1}{T}\int_0^T c_i^2(t)\mathrm{d}t \\ &= 1 \end{aligned} \tag{3.14}$$

Obviously, $R_{i,i}\ (\tau)$ reaches its maximum value at this time.

When $\tau \neq 0$, we should consider the case where τ is an integer multiple of T_c first, that is, $\tau = kTc$, where k is a non-zero integer, then

$$\begin{aligned} R_{i,i}(kT_c) &= \frac{1}{T}\int_0^T c_i(t)c_i(t + kT_c)dt \\ &= \frac{T_c}{T}\sum_{n=1}^{1023} c_i(n)c_i(n + k) \\ &= \frac{1}{N}\sum_{i=1}^{1023} c_i(n)c_i(n + k) \end{aligned} \tag{3.15}$$

where $c_i(n)$ is the discrete value generated by the C/A code generator described in Sect. 3.1.2, which is 0 or 1 in the digital logic circuit. Since the modulo-two addition of the digital circuit corresponds to the digital multiplication here, we need to convert the number 0 in the discrete code value to -1 to achieve the direct multiplication of numbers here.

It can be verified in Eq. (3.6) that there will only be three different values of $R_{i,i}(\tau)$ if τ is a non-zero integer multiple of T_c [4].

$$R_{i,i}(kT_c) = \left\{\frac{-1}{N}, \frac{-\beta(n)}{N}, \frac{\beta(n) - 2}{N}\right\} \tag{3.16}$$

where $\beta(n) = 1 + 2^{(n+2)/2}$, in which $\lfloor x \rfloor$ refers to the largest integer that does not exceed x. For the GPS C/A code, n = 10, then $\beta(n) = 65$, so, $R_{i.i}(kT_c) = \left\{\frac{-1}{1023}, \frac{-65}{1023}, \frac{63}{1023}\right\}$, where $k = 1, \ldots, 1022$. Considering $\tau = 0$ equals $k = 0$, we can conclude that only four finite values of the autocorrelation function of the C/A code are available when the time is offset by an integer multiple of T_c, which are $\left\{1, \frac{-1}{1023}, \frac{-65}{1023}, \frac{63}{1023}\right\}$.

This conclusion is true for all 37 GPS PRN codes. Figure 3.10 shows the autocorrelation function calculated by taking the C/A code of PRN6 as an example. The unit of its abscissa is T_c, namely a chip, while the unit of the ordinate is the value of the autocorrelation function, which is not normalized here. Figure 3.10a shows the autocorrelation function values for τ in the (0, T) interval, and Fig. 3.10b shows that autocorrelation function for τ in the (0, 5T) interval. Its periodicity can be clearly seen from this part. Apart from the maximum of 1023, the autocorrelation values at other code phase delays are limited to −1, −65, and 63.

The autocorrelation function for a white noise signal $n(t)$ is

$$R_n(\tau) = \begin{cases} 1 & \text{When } \tau = 0 \\ 0 & \text{Others} \end{cases} \tag{3.17}$$

The physical meaning of Eq. (3.17) is that the value of $n(t)$ at the current moment has no correlation with the signal value at any other moment, from which it can be directly inferred that we cannot estimate or predict the value of $n(t + \tau)$ at any other moment from $n(t)$ at the current moment. According to the signal and system theory, the power spectrum and the autocorrelation function are a pair of FFT (Fast Fourier Transform) pairs, and a flat noise spectrum similar to the white spectrum can be obtained through the FFT spectrum analysis of the function of Eq. (3.17), which is also how the "white noise" got its name.

Comparing the values of the autocorrelation function of Eq. (3.17) and Fig. 3.10, we see that the autocorrelation function of the C/A code is somewhat similar to the autocorrelation function of white noise. However, it is obviously not an autocorrelation function of true white noise. There is a clear difference between the two: the autocorrelation function of the C/A code is a periodic function, while that of the white

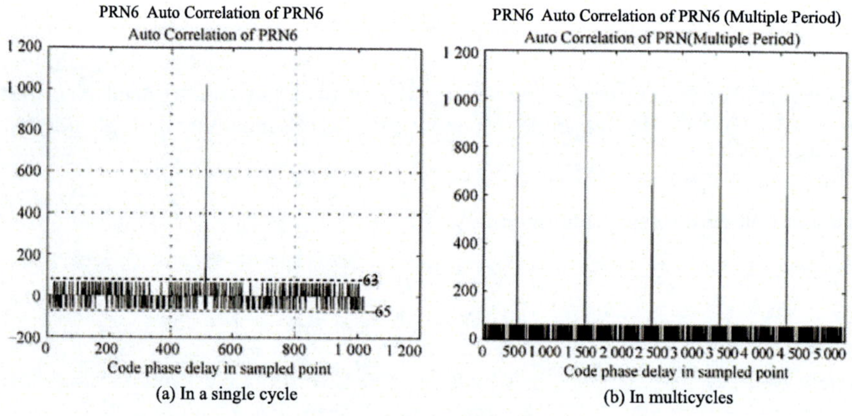

Fig. 3.10 The autocorrelation function of the GPS PRN6 C/A code

noise is not. Besides, when $\tau \neq 0$, the autocorrelation function value of the C/A code cannot always be 0, which is different from the real white noise. Therefore, we refer to the C/A code and the similar Gold code as "pseudo-random" codes, indicating that the autocorrelation function of this code is similar to a real random code but is not one.

The situation is somewhat complicated when $\tau \neq 0$ and is continuously variable. Firstly, we should consider the case where τ varies continuously between (0, T_c). According to the previous analysis, when τ is 0 in Fig. 3.9, the value of $R_{i,i}(\tau)$ is maximized since the values of $c_i(t + \tau)$ and $c_i(t)$ are 100% identical. As τ gradually increases, a difference between $c_i(t + \tau)$ and $c_i(t)$ begins to arise, while the length of the same part between them decrease and it is linearly related to τ. When τ increases to T_c, the value of $R_{i,i}(\tau)$ resumes it in $\tau = kT_c$ (when $k = 1$). It can be seen from the overall process that the ratio of the same part of $c_i(t + \tau)$ and $c_i(t)$ to the total period is linear with τ/T_c, in which the value of $R_{i,i}(\tau)$ varies between 1 and $R_{i,i}(T_c)$. When τ is continuously changing in [kT_c, (k + 1) T_c], a similar analysis can be made. At this time, the value of $R_{i,i}(\tau)$ varies linearly between $R_{i,i}(kT_c)$ and $R_{i,i}((k + 1)T_c)$, and the change is also linear with τ/T_c.

The above is a qualitative analysis of the trend of the function value of $R_{i,i}(\tau)$ in the case where τ is continuously variable, which will be demonstrated mathematically below. Without loss of generality, consider that when $kT_c < \tau < (k + 1)T_c$ and k is an integer, then $\Delta \tau = \tau - k T_c$. We can split the integral in Eq. (3.12) into two parts. The first part is the $\Delta\tau$ part of *ci*(*t*) and $c_i(t + \tau)$ in each chip, and the second part is the $(T_c - \Delta\tau)$ part of *ci*(*t*) and $c_i(t + \tau)$ in each chip, then

$$
\begin{aligned}
R_{i,\,i}(\tau) &= \frac{1}{T}\int_0^T c_i(t)c_i(t+\tau)\mathrm{d}t \\
&= \frac{1}{T}\sum_{n=0}^{1022}\int_{nT_c}^{(n+1)T_c} c_i(t)c_i(t+\tau)\mathrm{d}t \\
&= \frac{1}{T}\sum_{n=0}^{1022}\left[\int_{nT_c}^{nT_c+\Delta\tau} c_i(t)c_i(t+\tau)\mathrm{d}t + \int_{nT_c+\Delta\tau}^{(n+1)T_c} c_i(t)c_i(t+\tau)\mathrm{d}t\right] \\
&= \frac{1}{T}\sum_{n=0}^{1022}[c_i(n)c_i(n+k)\Delta\tau + c_i(n)c_i(n+k+1)(1-\Delta\tau)] \\
&= R_{i,i}(kT_c)\left(\frac{\Delta\tau}{T_c}\right) + R_{i,i}[(k+1)T_c]\left(\frac{T_c-\Delta\tau}{T_c}\right)
\end{aligned}
\tag{3.18}
$$

The results of Eq. (3.15) are used in the final step of Eq. (3.18), from which we can see that when $kT_c < \tau < (k + 1)T_c$, the value of $R_{i,i}(\tau)$ varies linearly between $R_{i,i}(kT_c)$ and $R_{i,i}((k + 1)T_c)$, which takes the form of the connection between $R_{i,i}(kT_c)$ and $R_{i,i}((k + 1)T_c)$.

Figure 3.11 shows the calculation result of the autocorrelation function of the PRN9 C/A code sampled at 16.368 MHz. In the Fig. 3.11a is a partial detail diagram near the maximum value of the autocorrelation function, and Fig. 3.11b is that of the autocorrelation function after a non-zero delay. As the sample rate is 16.368 MHz, there are 16 sample points in each chip, so the minimum unit of delay variation of the autocorrelation function is 1/16 T_c. As can be seen from Fig. 3.11, when [kT_c, (k + 1) T_c] is continuously changing, the value of $R_{i,i}(\tau)$ also changes linearly between $R_{i,i}(kT_c)$ and $R_{i,i}[(k+1)T_c]$ in the form of a saw-tooth wave, which thereby confirms the above analysis. It is also worth noting that the autocorrelation function values in Fig. 3.11 are not normalized.

The width of T_c is marked in Fig. 3.11. The autocorrelation function drops to a lower value immediately after a delay of one T_c from the peak, which is one

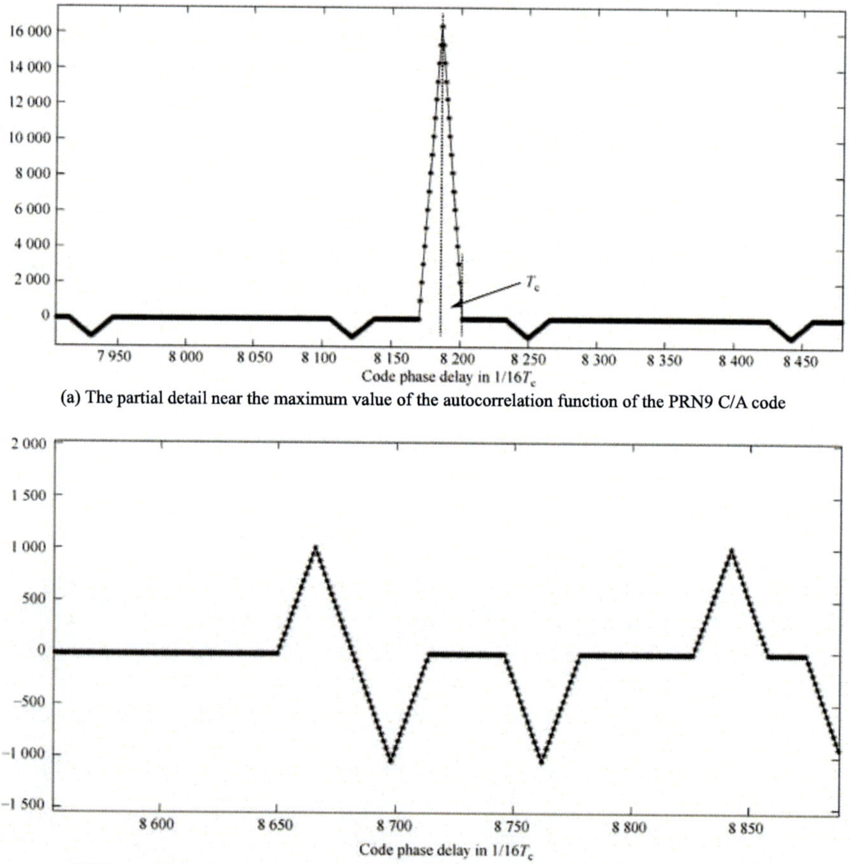

(a) The partial detail near the maximum value of the autocorrelation function of the PRN9 C/A code

(b) The partial detail near a non-zero delay of the autocorrelation function of the PRN9 C/A code

Fig. 3.11 Partial details of the GPS PRN9 autocorrelation function at the maximum value and a certain delay of $\left\{\frac{-1}{1023}, \frac{-65}{1023}, \frac{63}{1023}\right\}$. This feature causes the autocorrelation function of the C/A code to present a sharp peak. This is good for ranging applications, as the ranging accuracy directly relates to the sharpness of the autocorrelation peak: the sharper the autocorrelation peak, the easier the measurement of the arrival time of the pseudo-random code will be. As can be seen from Fig. 3.11, the smaller the T_c, the sharper the autocorrelation peak. For the GPS C/A code, $T_c \approx 0.977\,517\ \mu s$, and multiplied by the speed of light, it can be converted into distance. The corresponding distance of a C/A chip is approximately 300 m. When it can be guaranteed by the receiver that the measurement accuracy of the code phase of the C/A code is within 0.1 chip, the distance error can be restricted within 30 m. Modern GPS receivers can have even higher range accuracy when the signal from satellite has strong signal strength, for instance, there is no obstacle between satellite and receiver antenna.

We refer to peaks that are different from the maximum autocorrelation peak as secondary correlation peaks or side lobes. The largest autocorrelation peak is called the main lobe. The greater the ratio of the main lobe to the side lobe, the lower the possibility that the side lobe will be mistaken for the main lobe in actual processing. Mistaking the side lobes for the main lobe will have catastrophic consequences, as the distance error will deviate by hundreds or even tens of thousands of meters. To understand this, readers only need to review the correspondence between the distance and the C/A chip width. If N chips are staggered, the distance error will be 300 Nm. The ratio of the maximum side lobes to the main lobes of the GPS C/A code is

$$\frac{\max(\text{side lobes})}{\text{main lobes}} = \frac{65}{1023} \approx -23.94\ \text{dB} \tag{3.19}$$

The linear value is converted to decibels in Eq. (3.19), indicating that the maximum autocorrelation peak of the C/A code is at least 24 dB higher than the side-lobe peak. Therefore, when the signal quality is good, it is still unlikely that the side lobes will be mistaken for the main correlation peak.

Good autocorrelation of C/A codes is the basis for baseband signal processing, including signal acquisition, pseudo-code tracking, and pseudo-range measurement formation. It also plays a crucial role in suppressing multipath signals. This will be elaborated in subsequent chapters.

Unlike the autocorrelation function, the cross-correlation function measures the similarity of a signal and other signals after a certain period of time on the time axis. It is defined as a time-averaged cross-correlation function whose mathematical expression is

$$R_{i,j}(\tau) = \frac{1}{T}\int_0^T c_i(t)c_j(t+\tau)\mathrm{d}t, \quad \tau \in (-T/2, T/2) \tag{3.20}$$

The definition of each parameter in Eq. (3.20) is the same as that in Eq. (3.12). The only difference is that the subscript becomes (i, j), indicating the cross-correlation between the C/A codes of the i-th and j-th PRNs.

Because the signals transmitted by different GPS satellites share the same carrier frequency band, it is necessary to distinguish them by pseudo-code. In this case, the cross-correlation function between different C/A codes must be 0, which is

$$R_{i,j}(\tau) = \frac{1}{T}\int_0^T c_i(t)c_j(t+\tau)\mathrm{d}t = 0 \tag{3.21}$$

Equation (3.21) is the ideal case. If realized, it can lead to the conclusion that c_i (t) and c_j (t) are orthogonal. In fact, different C/A codes are only approximately orthogonal. PRN3 and PRN9 are taken as an example in Fig. 3.12. Compared with the autocorrelation function value field, the calculated value of the cross-correlation function, whose range is $\left\{\frac{-1}{1023}, \frac{-65}{1023}, \frac{63}{1023}\right\}$, only lacks the maximum value of the former.

The cross-correlation of C/A codes is very important for the design of modern receivers with high sensitivity. In places where strong and weak star signals coexist (such as cities, canyons, or partially occluded constructions) there is a Line-of-sight path between one or more satellites and receiver antennae, so the signal is strong, while the signals of other satellites are weak due to occlusion. Therefore, it is very likely to capture the cross-correlation peak of a strong star due to the coexistence of strong and weak signals. According to the above analysis, the ratio of the autocorrelation peak and max cross-correlation peak is

$$\frac{\text{autocorrelation peak}}{\max\left(\text{cross} - \text{correlation peak}\right)} = \frac{1023}{65} \approx 24\,\text{dB}$$

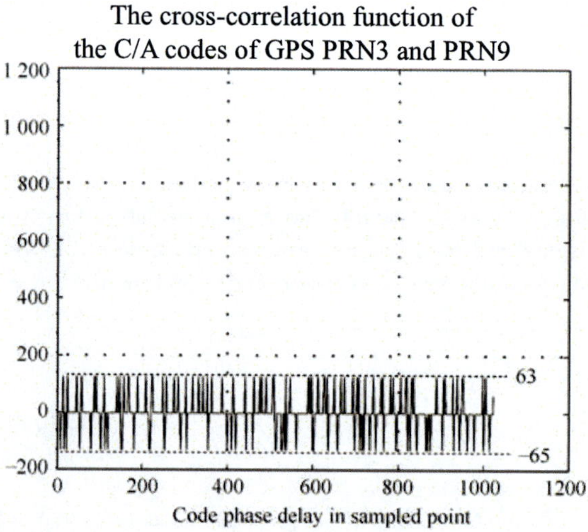

Fig. 3.12 The cross-correlation function of the C/A codes of GPS PRN3 and PRN9

Suppose that when the CN0 of a strong star is β dBHz, if the capture sensitivity of the receiver is higher than ($\beta - 24$) dBHz and the weak star's signal strength is lower than ($\beta - 24$) dBHz, it is possible for the receiver to capture the cross-correlation peak of the strong star. For example, in open-sky situations, the strength of GPS satellite signals is often higher than 45 dBHz. On such occasions, if the capture sensitivity of the receiver is higher than 21 dBHz, there is a risk of accidentally catching the cross-correlation peak of the strong star.

3.2 The BeiDou Signal

BDS is expected to achieve global coverage by 2020, prior to which it will only provide regional service covering China and the Asia Pacific region. At the time of writing this book, the receiver users of the BeiDou II can only receive the navigation signals defined in Ref. [5], so the navigation signals described in subsequent chapters all belong to the regional BDS system.

3.2.1 BDS Signal Structure

BeiDou satellites transmit navigation signals in the B1, B2, and B3 frequency bands. Since the navigation signals on the B3 frequency band are authorized and are not open to the public, the navigation signals introduced here fall into the B1 and B2 bands. The nominal carrier frequency of the B1 signal is 1561.098 MHz, and that of the B2 signal is 1207.140 MHz. Their mathematical expressions are

$$\begin{aligned} s_{\mathrm{B1}}(t) &= \sqrt{2P_{\mathrm{B1I}}}D_{\mathrm{B1I}}(t)c_{\mathrm{B1I}}(t)\cos[\omega_{\mathrm{B1}}t + \theta_{\mathrm{B1I}}] \\ &\quad + \sqrt{2P_{\mathrm{B1Q}}}D_{\mathrm{B1Q}}(t)c_{\mathrm{B1Q}}(t)\cos[\omega_{\mathrm{B1}}t + \theta_{\mathrm{B1Q}}] \end{aligned} \tag{3.22}$$

$$\begin{aligned} s_{\mathrm{B2}}(t) &= \sqrt{2P_{\mathrm{B2I}}}D_{\mathrm{B2I}}(t)c_{\mathrm{B2I}}(t)\cos[\omega_{\mathrm{B2}}t + \theta_{\mathrm{B2I}}] \\ &\quad + \sqrt{2P_{\mathrm{B2Q}}}D_{\mathrm{B2Q}}(t)c_{\mathrm{B2Q}}(t)\cos[\omega_{\mathrm{B2}}t + \theta_{\mathrm{B2Q}}] \end{aligned} \tag{3.23}$$

where P is the carrier power, c is the pseudo-random code, D is the navigation telegram bit, and θ is the initial phase of the carrier. The subscripts of these quantities are B1I, B1Q, B2I, and B2Q, indicating the I and Q paths of the B1 and B2 frequencies respectively. ω_{B1} and ω_{B2} are the carrier angular frequencies of B1 and B2 signals.

It can be seen from Eqs. (3.22) and (3.23) that the navigation signals on the B1 and B2 frequencies are modulated thorough the QPSK method. However, by 1st January 2013, the Chinese government had only disclosed the structure of the B1I and B2I signals. The structure of the B1Q and B2Q signals has not been publicly announced, so only the former will be described in this book. It is worth mentioning that although the Chinese government has not disclosed official documents for the B1Q, B2Q, and B3 signal formats, engineers and researchers all over the world have analyzed the characteristics of these undisclosed signals, such as pseudo-random code generation, carrier frequency, and signal bandwidth [6–10], to which readers can refer for a better understanding. However, the conclusions in these documents are not officially recognized, or only refer to signals in the intermediate test phase of the BDS plan that have not yet been determined. Therefore, they are for reference only.

Since the B1Q and B2Q signals are not yet available, the QPSK signal represented by Eqs. (3.22) and (3.23) can be regarded as BPSK signals, so its structure is similar to that of GPS L1 C/A code signals. The conclusion can be easily understood by comparing the first term of Eqs. (3.22) and (3.23) with that of Eq. (3.1). Because of the similarity between the BDS signal and the C/A code signal of the GPS L1 frequency, the processing method of GPS signals can be easily applied to BDS signals.

The B1 and B2 carrier signals are generated by atomic clocks on BeiDou satellites, but the reference frequency of these clocks has not yet been confirmed in public documents. The division among the carrier frequency, the pseudo-random code frequency, and the navigation message frequency has not been described in authoritative public literature. Therefore, it is impossible to draw a schematic diagram of the mechanism of the B1 and B2 signals similar to Fig. 3.1.

Since BeiDou satellites and GPS satellites operate together in space, they are inevitably affected by the effects of special and general relativity. However, unlike GPS, the BeiDou space constellation includes geostationary orbit satellites (GEO), tilted geosynchronous orbit satellites (IGSO), and medium-orbit satellites (MEO). Therefore, the situation of on-board atomic clocks on the BeiDou satellite being affected by the relativistic effect is more complicated than that of GPS satellites. Scholars in China and elsewhere have already conducted research and analysis on this issue. Table 3.4 shows a comparison of the effects of narrow and general relativity on GPS satellites and BeiDou GEO/IGSO/MEO satellites.

The orbital altitude and average velocity of the GPS satellites and BeiDou satellites are shown in Table 3.4. Because the special theory of relativity is related to the velocity of the satellite, the general relativity effect is related to the gravitational field strength of the satellite, which is mainly determined by the orbital height of the satellite.

Table 3.4 Comparison of the effects of relativity on GPS and BDS satellites

	GPS satellites	BeiDou GEO	BeiDou IGSO	BeiDou MEO
Orbital altitude (km)	20,715	35,786	35,786	21,528
Average speed (km/s)	3.835	3.075	3.075	3.683
Time dilation (μs/day)	−7.07	−4.54	−4.54	−6.52
Time gravitation (μs/day)	46.07	51.12	51.12	47.17
Synthetic effect (μs/day)	38.99	46.58	46.58	40.64

The time expansion term in the table results from the influence of special relativity, while the time gravitation term is influenced by general relativity. Their units are both microseconds/days, i.e. the on-board atomic clock is observed in the time span of one day, and the difference between the time of the clock and the standard time is recorded in microseconds. It can be seen from Table 3.4 that the clocks on the GEO/IGSO satellites and MEO satellites are different due to the relativistic effect. The orbital height of the GEO/IGSO satellites is greater, resulting in a larger value for time gravitation, while the average satellite speed is slightly lower than the GPS satellites, so the time expansion is smaller. The combined effect means that the on-board atomic clock becomes 46.6 μs faster per day. The orbital height of the MEO satellites is similar to that of the GPS satellites, as is their speed. Therefore, the final effect is that the on-board atomic clock of the MEO satellites is about 40.6 μs faster per day, which is similar to that of the GPS satellites, at 38.99 μs per day.

According to the results of Table 3.4, the on-board atomic clocks of the GEO/IGSO satellites are adjusted by -5.39×10^{-10} Hz in frequency, and those of the MEO satellites are adjusted by -4.28×10^{-10} Hz, the negative sign indicating the slowing of the clock. It should be noted that, like the adjustment of the on-board atomic clocks on GPS satellites, what is adjusted here is only the reference frequency of the atomic clock. When the satellite is operating in various positions, a position-related clock correction amount needs to be provided, which is included in the navigation message of BeiDou satellites.

The pseudo-random codes of the B1I and B2I signals have a code rate of 2.046 MHz and a code length of 2046 chips, so their code periods are both 1 ms. The ratio between the carrier frequency and the pseudo-code rate of B1 and B2 is as follows:

$$f_{\mathrm{B1}} = 1561.098\ \mathrm{MHz} = 763 f_0 \tag{3.24}$$

$$f_{\mathrm{B2}} = 1207.140\ \mathrm{MHz} = 590 f_0 \tag{3.25}$$

Here, $f_0 = 2.046$ MHz, which represents the rate of the BeiDou pseudo-random code. Therefore, there are 763 B1 carrier cycles or 590 B2 carrier cycles within a P1I or B2I pseudo-code chip.

Unlike GPS satellites, BeiDou satellites have two types of navigation messages: D1 and D2.

The D1 navigation message is modulated on the B1I and B2I signals of the MEO/IGSO satellites, and the D2 navigation message is modulated on the B1I and B2I signals of the GEO satellites. The rate of the D1 message is 50 sps, and is modulated by an NH code with a rate of 1.000 bps, while the D2 message has a rate of 500 sps and does not have an NH code. The navigation message of the BeiDou satellites is similar to that of GPS satellites, including satellite clock error correction parameters, satellite ephemeris and almanac data, ionospheric parameters, tropospheric parameters, UTC time parameters, and satellite operating conditions. The signal transmission time of the satellite is also provided. The D2 navigation message

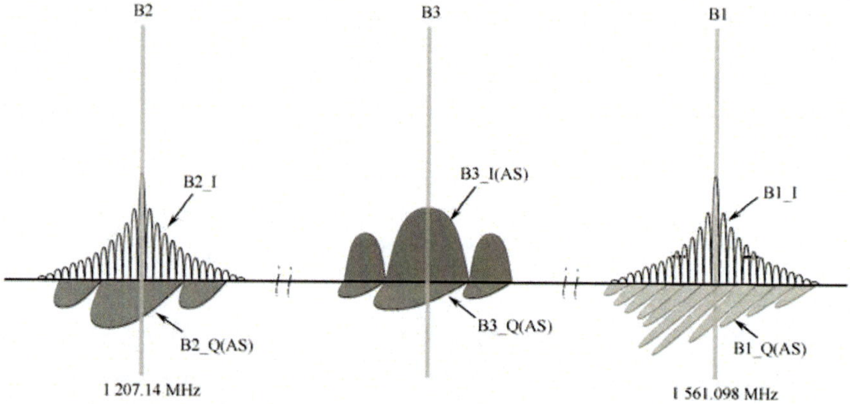

Fig. 3.13 Signal spectrum of the regional BeiDou service system

also broadcasts the integrity and differential information of the BeiDou system, as well as the ionospheric information of the grid point.

Figure 3.13 shows the spectrum of the navigation signal broadcasted by the BeiDou Regional Service System. Since the B3 frequency is not disclosed, it is not marked in the figure. The B1Q, B2Q, and B3I_Q signals are authorized signals.

The GPS signals and BeiDou navigation signals are compared in Table 3.5. The GPS signals include L1 C/A, P(Y), L1C, L2C and L5 signals, not the latest M-code. Meanwhile for the BeiDou signals, only the I signal of the B1 and B2 frequencies are presented in the table, as the B3 frequency is authorized, and the B1Q and B2Q are not yet disclosed.

3.2.2 *BDS Pseudo-Random Code Generator*

Similar to the GPS C/A code, the pseudo-random code modulated on the BDS B1I and B2I signals is also a Gold code, but its number of series is 11, which means that two 11th m-sequences are added to generate a balanced Gold code whose period is 211−1 = 2047 chips. However, the final pseudo-random code is generated through artificially truncating one chip, so the BeiDou pseudo-random code is a truncated code whose generating polynomial is

$$P_{G1} = 1 + X + X^7 + X^8 + X^9 + X^{10} + X^{11} \tag{3.26}$$

$$P_{G2} = 1 + X + X^2 + X^3 + X^4 + X^5 + X^8 + X^9 + X^{11} \tag{3.27}$$

Table 3.5 Comparison of GPS and BDS navigation signals

Item of comparison	GPS signals	BDS navigation signals
Satellite type	MEO	GEO/IGSO/MEO
Working frequency	L1: 1575.42 MHz L2: 1227.6 MHz L5: 1176.45 MHz	B1: 1561.098 MHz B2: 1207.14 MHz B3: Unknown
Modulation mode	L1 C/A: BPSK L1C: AltBOC P(Y): BPSK L2C: BPSK, Time−sharing modulation L5: QPSK	B1-I/B1-Q: QPSK B2-I/B2-Q: QPSK B3: Unknown
Pseudo−code rate	L1 C/A: 1.023 Mcps P(Y): 10.23 Mcps L2C: 511.5 kcps L5: 10.23 Mcps	I channel: 2.046 Mcps(B1I,B2I) Q channel: Unknown
Chip rate of the navigation message	L1 C/A: 50 bps P(Y): 50 pbs L2C: 25 bps L5: 50 bps	GEO 's' B1I and B2I: D2 code, 500 bps MEO/IGSO 's' B1I and B2I: D1 code, 50 bps, NH 1 kbps Satellite Q channel: Unknown
Multi-access mode	CDMA	CDMA
RF polarization	RHCP	RHCP
Satellite constellation	An elliptical orbit with a semi−major axis of 26,560 km, a eccentricity of 0.002–0.01, 6 orbital planes, and a total of 30–32 available satellites	MEO: An elliptical orbit with a semi−major axis of 27,906 km and an eccentricity of approximately 0.01, distributed over three orbital planes, currently [*] has four satellites IGSO: An elliptical orbit with a semi−major axis of approximately 42,164 km, distributed over three orbital planes, currently [*] has 5 satellites GEO: An elliptical orbit with a semi−major axis of approximately 42,164 km above the Equator, currently [*] has 5 satellites

[*] As of December 2012

The schematic diagram of the BeiDou pseudo-random code is included in the official BDS document, *BeiDou Navigation Satellite System Signal In-Space Interface Control Document V2.0* (hereinafter referred to as BeiDou ICD) as shown in Fig. 3.14.

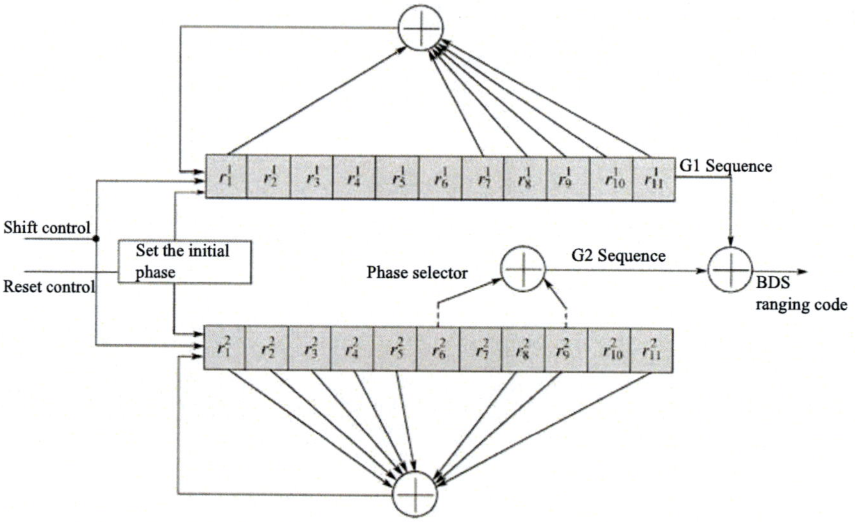

Fig. 3.14 Generation of the BDS pseudo-code with a tap selector

The method adopted in Fig. 3.14 is to set the phase taps of the G2 sequence, in which different PRN numbers correspond to different tap selector configurations. The frequency of the shift clock is 2.046 MHz. For the BeiDou satellite, the frequency is obtained by dividing the reference frequency of the atomic clock; for the user receiver, it comes from the numerically controlled oscillator (NCO) output controlled by the local pseudo-code tracking loop. The initial phase of the G1 and G2 register banks is not all "1"s, but is set to

$$\text{The initial phase of G1} = \left[0\ 1\ 0\ 1\ 0\ 1\ 0\ 1\ 0\ 1\ 0\right]$$
$$\text{The initial phase of G2} = \left[0\ 1\ 0\ 1\ 0\ 1\ 0\ 1\ 0\ 1\ 0\right]$$

BDS is different from GPS in the initial phase configuration, as the initial phase of the C/A code generator is all "1"s. Another difference is that unlike the GPS C/A code, the BeiDou pseudo-random code is intentionally truncated by 1 bit, which means that the G1 and G2 register sets are forced to be cleared when the working clock counts to 2046, then reset to the initial phase, and then start anew. This operation can also be understood as truncating the last bit from the 2047 bits generated from the balanced Gold code. It is because of the truncation that the BeiDou pseudo-code is no longer a balanced Gold code, and thus has different properties from the GPS C/A code.

According to BeiDou ICD2.0, the BeiDou G1and G2 sequences are generated in a very similar manner to the GPS C/A code. The processing is the same except that the number of register groups and the tap methods are different. Specifically, the update process of the G1 and G2 register sets is as follows.

$$
\begin{cases}
G1 \text{ register sets:} \\
F_1 = r_1^1 \oplus r_7^1 \oplus r_8^1 \oplus r_9^1 \oplus r_{10}^1 \oplus r_{11}^1 \\
r_1^1 = F_1 \\
r_i^1 = r_{i-1}^1, \quad i = 2, \ldots, 11 \\
\text{When CLK} = 2046, \text{reset and set the intial phase of the G1 register} \\
\text{G2 register sets:} \\
F_2 = r_1^2 \oplus r_2^2 \oplus r_3^2 \oplus r_4^2 \oplus r_5^2 \oplus r_8^2 \oplus r_9^2 \oplus r_{11}^2 \\
r_1^2 = F_2 \\
r_i^2 = r_{i=1}^2, \quad i = 2, \ldots, 11 \\
\text{When CLK} = 2046, \text{reset and set the intial phase of the G2 register}
\end{cases}
\tag{3.28}
$$

The pseudo-random output code is the modulo-two sum of the last G1 register and the contents of several G2 registers, with the first 2046 bits being taken, which is

$$
C_{\mathrm{BDS}}(k) = r_{11}^1(k) \oplus \left[r_{s1}^2(k) \oplus r_{s2}^2(k)\right], \quad k = 1, \ldots, 2046 \tag{3.29}
$$

where the subscripts s_1 and s_2 in Eq. (3.29) are the tap positions of the G2 linear shift register, and the generated BeiDou pseudo-code can be altered through changing the value of s_1 and s_2.

Similar to the GPS C/A code, the BeiDou pseudo-random code can also be generated by setting the initial phase of the G2 linear shift register group and the delay phase of the G2 sequence. The schematic diagram of the two methods is shown in Figs. 3.15 and 3.16, in which the former indicates the principle of the delayed phase method, and the latter explains the method of setting the initial phase. By comparing

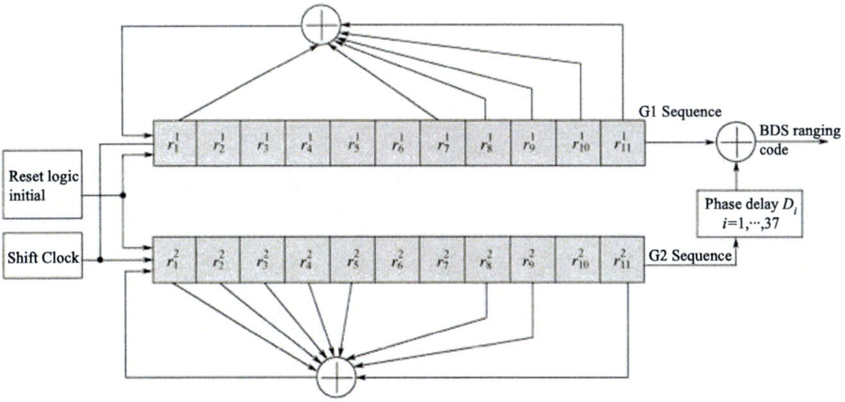

Fig. 3.15 Changing the BDS pseudo-random code through adjusting the phase delay of the G2 sequence

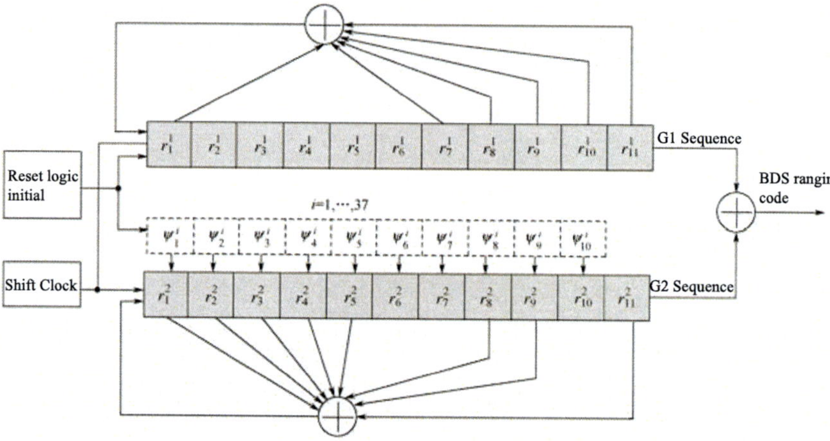

Fig. 3.16 Generation of the BDS pseudo-random code through changing the initial phase of the G2 register

it with Figs. 3.7 and 3.8, readers will see that the two methods are similar in form to the corresponding methods of generating GPS C/A codes.

The initial G1 and G2 phases in the delay phase method shown in Fig. 3.15 are both [010101010], and neither sequences are truncated. This means their length is 2047 chips. The phase of the G2 sequence is delayed by a value of D_i, in which i ranges from 1 to 37, corresponding to the 37 BeiDou ranging codes. Then, the new sequence obtained from the modulo-two addition of the G1 sequence and G2 sequences is truncated and the first 2046 chips are taken. The overall process can be written as

$$C_{\mathrm{BDS}}(k) = r_{11}^1(k) \oplus r_{11}^2(k + D_i), \quad k = 1, \ldots, 2046 \tag{3.30}$$

In the method of setting the initial phase of the G2 register as shown in Fig. 3.16, the initial phase and update steps of the G1 register are the same as those in the former two methods. The difference resides in the steps of setting the initial phase of the G2 register at reset.

$$\begin{cases} \text{G2 register sets:} \\ \text{Initialize:}\left[r_i^2\right] = \left[\Psi_i^j\right], \quad i = 1, \ldots, 11 \\ F = r_1^2 \oplus r_2^2 \oplus r_3^2 \oplus r_4^2 \oplus r_5^2 \oplus r_8^2 \oplus r_9^2 \oplus r_{11}^2 \\ r_1^2 = F \\ r_i^2 = r_{i-1}^2, \quad i = 2, \ldots, 11 \end{cases} \tag{3.31}$$

The different settings of $\Psi_1{}^i \ldots \Psi_{11}{}^i$ in Eq. (3.31) generate different BeiDou pseudo-random codes. The final BeiDou pseudo-code sequence is the modulo-two sum of the last bits of the G1 and G2 register sets, in which the first 2046 bits are taken, i.e.

$$C_{\mathrm{BDS}}(k) = r_{11}^{1}(k) \oplus r_{11}^{2}(k), \quad k = 1, \cdots, 2046 \tag{3.32}$$

There are 37 pseudo-random codes defined in BeiDou's ICD, of which the first five are assigned to the GEO satellites and the last 32 are assigned to the IGSO/MEO satellites. Table 3.6 shows the tap configuration, phase delay, and initial phase configuration (including hexadecimal and binary numbers) of all BeiDou pseudo-random codes. The reader can select the appropriate generation method based on real situation.

3.2.3 Autocorrelation and Cross-Correlation of BDS Pseudo-Random Codes

There are 37 pseudo-random codes defined in the BeiDou ICD, five of which are assigned to GEO satellites and the remaining thirty-two are assigned to IGSO and MEO satellites. Since the BeiDou pseudo-random code is a truncated Gold code, strictly speaking, the truncated pseudo-code is no longer a balanced Gold code, as the number of "1"s in nearly half of the 37 BeiDou pseudo-codes is two more than the number of "0"s, while the number of "1"s and "0"s in other BeiDou pseudo-codes is precisely equal. Besides, the "1"s in all GPS C/A codes are always one more than the number of "0"s, from which it can be concluded that the balance of the BeiDou pseudo-random code is not extremely skewed.

An important difference between the BeiDou pseudo-code and the GPS C/A code is the value distribution of their autocorrelation functions and cross-correlation functions. The definition of the autocorrelation function of the BeiDou pseudo-code is the same as Eq. (3.12), which can be written as

$$R_{i,i}(\tau) = \frac{1}{T}\int_{0}^{T} c_i(t)c_i(t+\tau)\mathrm{d}t, \quad \tau \in (-T/2, T/2) \tag{3.33}$$

where all the subscripts are defined the same as those of Eq. (3.12) but are replaced with the corresponding terms of the BeiDou pseudo-code.

The autocorrelation function of the BeiDou pseudo-code is still a periodic function whose period is the same as the BeiDou pseudo-code period, which is 1 ms. When the time delay $\tau = 0$, the value of $R_{i,i}(\tau)$ reaches the maximum, but when $\tau \neq 0$ and τ is the integer multiple of the chip displacement, the autocorrelation function of $R_{i,i}(\tau)$ will no longer conform to the theoretical value of the balanced Gold code, which means more than three different values will be taken.

Table 3.6 Summary of the three methods for generating BDS pseudo-random codes

Pseudo-code number	Satellite type	Setting of the tap selector ($s1$, $s2$)	Phase delay *Di* (Unit: chips)	Initial phase of G2[a]	
				(In hexadecimal notation)	(In hexadecimal notation)
1	GEO	1⊕3	713	0x187	00110000111
2	GEO	1⊕4	1582	0x639	11000111001
3	GEO	1⊕5	1415	0x1E6	00111100110
4	GEO	1⊕6	1551	0x609	11000001001
5	GEO	1⊕8	582	0x605	11000000101
6	MEO/IGSO	1⊕9	772	0x1F8	00111111000
7	MEO/IGSO	1⊕10	1312	0x606	11000000110
8	MEO/IGSO	1⊕11	1044	0x1F9	00111111001
9	MEO/IGSO	2⊕7	1550	0x704	11100000100
10	MEO/IGSO	3⊕4	360	0x7BE	11110111110
11	MEO/IGSO	3⊕5	711	0x061	00001100001
12	MEO/IGSO	3⊕6	1580	0x78E	11110001110
13	MEO/IGSO	3⊕8	1549	0x782	11110000010
14	MEO/IGSO	3⊕9	1104	0×07F	00001111111
15	MEO/IGSO	3⊕10	580	0x781	11110000001
16	MEO/IGSO	3⊕11	770	0x07E	00001111110
17	MEO/IGSO	4⊕5	359	0x7DF	11111011111
18	MEO/IGSO	4⊕6	710	0x030	00000110000
19	MEO/IGSO	4⊕8	1412	0x03C	00000111100
20	MEO/IGSO	4⊕9	1548	0x7C1	11111000001
21	MEO/IGSO	4⊕10	1103	0x03F	00000111111
22	MEO/IGSO	4⊕11	579	0x7C0	11111000000
23	MEO/IGSO	5⊕6	358	0x7EF	11111101111
24	MEO/IGSO	5⊕8	1578	0x7E3	11111100011
25	MEO/IGSO	5⊕9	1411	0x01E	00000011110
26	MEO/IGSO	5⊕10	1547	0x7E0	11111100000
27	MEO/IGSO	5⊕11	1102	0x01F	00000011111
28	MEO/IGSO	6⊕8	708	0x00C	00000001100
29	MEO/IGSO	6⊕9	1577	0x7F1	11111110001
30	MEO/IGSO	6⊕10	1410	0x00F	00000001111
31	MEO/IGSO	6⊕11	1546	0x7F0	11111110000
32	MEO/IGSO	8⊕9	355	0x7FD	11111111101
33	MEO/IGSO	8⊕10	706	0x003	00000000011
34	MEO/IGSO	8⊕11	1575	0x7FC	11111111100

(continued)

Table 3.6 (continued)

Pseudo-code number	Satellite type	Setting of the tap selector ($s1$, $s2$)	Phase delay *Di* (Unit: chips)	Initial phase of G2[a]	
				(In hexadecimal notation)	(In hexadecimal notation)
35	MEO/IGSO	9⊕10	354	0x7FE	11111111110
36	MEO/IGSO	9⊕11	705	0x001	00000000001
37	MEO/IGSO	10⊕11	353	0x7FF	11111111111

[a]LSB Corresponds to Ψ_1^i, MSB Corresponds to Ψ_{11}^i

$$R_{ij}(kT_c) \neq \left\{ \frac{-1}{N} \quad \frac{-\beta(n)}{N} \quad \frac{\beta(n)-2}{N} \right\} \tag{3.34}$$

where the definition of k, T_c, N, and $\beta(n)$ is the same as in Eq. (3.16).

The value of the autocorrelation function of the BeiDou PRN code is distributed between the maximum and the minimum value. If the value range is represented by a histogram, a continuous distribution can be seen, which is different from the three-valued characteristic of the GPS C/A code.

Computer simulation results show that the autocorrelation function values of different BeiDou PRN codes are also different, and their range is between [(−170, 162), 2046]. When $\tau = 0$, R_{ij} (kT_c) reaches the maximum value of 2046, and when $\tau \neq 0$ but is limited to an integer multiple of the chip, the autocorrelation function value is between (−170, 162).

The BeiDou PRN37 pseudo-code is taken as an example in Fig. 3.17, in which the distribution and value range of its autocorrelation function is shown. The left half of the figure is the value of the autocorrelation function, in which the abscissa represents the delay amount in units of chips, and the ordinate is the unnormalized autocorrelation function. It is clear that when the delay amount is 0, the autocorrelation value will reach 2046 (its maximum). The right half of the graph shows the range

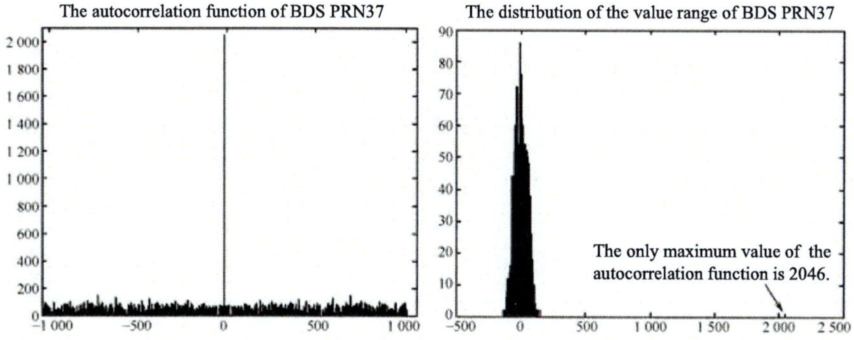

Fig. 3.17 The autocorrelation function (left) and the distribution of the value range (right) of BDS PRN37

distribution, in which the unnormalized autocorrelation value is on the abscissa, and the number of autocorrelation results when the delay is within the range of [0, 2046] chips is on the ordinate. When $\tau \neq 0$ and is limited to an integer multiple of the chip, the autocorrelation function of PRN37 has a value range between (−138, 150) and is continuously distributed within the range.

The autocorrelation function range of different BeiDou pseudo-codes varies. The autocorrelation function values of all 37 BeiDou pseudo-codes can be obtained by a computer simulation program, summarized in Table 3.7.

The definition of the cross-correlation function of the BeiDou PRN code is the same as that of Eq. (3.20), which is

$$R_{i,j}(\tau) = \frac{1}{T}\int_{0}^{T} c_i(t)c_j(t+\tau)\mathrm{d}t, \quad \tau \in (-T/2, T/2) \tag{3.35}$$

The subscripts in the above function need to be replaced by the corresponding items of the BeiDou pseudo-code.

Like that of the autocorrelation function, the value range of the cross-correlation functions of the BeiDou PRN codes also changes with the pseudo-codes. Suppose that there are M PRN codes, the number of PRN pairs will be ${C_M}^2$. When the value

Table 3.7 Range of the autocorrelation function (non-normalization) of BDS PRN1–PRN34

PRN	Range	PRN	Range
1	[(−122, 130), 2046]	2	[(−114, 122), 2046]
3	[(−126, 146), 2046]	4	[(−126, 110), 2046]
5	[(−118, 138), 2046]	6	[(−126, 126), 2046]
7	[(−146, 130), 2046]	8	[(−146, 138), 2046]
9	[(−142, 126), 2046]	10	[(−126, 138), 2046]
11	[(−170, 138), 2046]	12	[(−126, 122), 2046]
13	[(−146, 138), 2046]	14	[(−118, 130), 2046]
15	[(−130, 130), 2046]	16	[(−118, 134), 2046]
17	[(−138, 150), 2046]	18	[(−150, 142), 2046]
19	[(−122, 138), 2046]	20	[(−142, 114), 2046]
21	[(−134, 134), 2046]	22	[(−162, 162), 2046]
23	[(−162, 114), 2046]	24	[(−130, 134), 2046]
25	[(−142, 122), 2046]	26	[(−126, 126), 2046]
27	[(−138, 146), 2046]	28	[(−138, 114), 2046]
29	[(−106, 138), 2046]	30	[(−122, 134), 2046]
31	[(−158, 110), 2046]	32	[(−146, 118), 2046]
33	[(−134, 122), 2046]	34	[(−150, 126), 2046]
35	[(−134, 130), 2046]	36	[(−130, 138), 2046]
37	[(−138, 150), 2046]		

of M is large (such as M > 10), the value of ${C_M}^2$ will also be large. Therefore, it is not possible to enumerate the values of the cross-correlation functions of all BeiDou pseudo-code combinations. It can be understood from computer simulation that among all 37 BeiDou pseudo-codes, the value range of the cross-correlation functions between PRN1 and PRN37 falls within (−210, 202), and the minimum value (−210) occurs between PRN35 and PRN12, while the maximum value (202) occurs between PRN24 and PRN36. Taking BeiDou PRN1 and PRN10 as examples, the values of their cross-correlation functions and the range distribution are generated by computer programs. The result is shown in Fig. 3.18. The left half is the value of the autocorrelation functions, and the right half is the distribution of the value range of the functions. It can be seen that the distribution of the cross-correlation function and the autocorrelation function are similar, except that there is no maximum value of 2046 of the former.

It should be noted that the above analysis is based on the fact that there is no Doppler shift between two PRN codes. The result would be more complicated if there was. In the case of zero Doppler shift, the ratio of the autocorrelation peak and cross-correlation peak of the GPS C/A code is 20lg10 (1023/65) = 23.93 dB, while and the ratio between the two functions of the BeiDou PRN codes is 20lg10. (2046/210) = 19.77 dB. So, in the worst case, the cross-correlation suppression among different BeiDou PRN signals is about 4 dB weaker than GPS. Therefore, it is necessary to process the cross-correlation results in the acquisition of weak BeiDou signals.

When τ is not the delay of the whole digital film, the conclusion of Eq. (3.18) is still applicable to the BeiDou pseudo-code. The analysis process is the same as that of the GPS C/A code, and readers can analyze it themselves.

Further analysis of the characteristics of truncated Gold codes can be found in Refs. [11, 12].

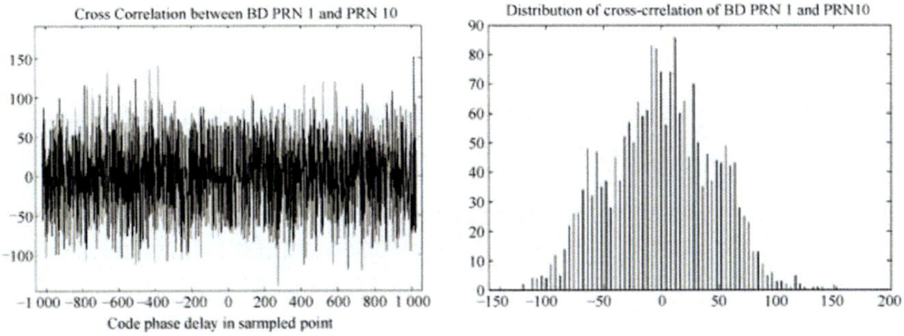

Fig. 3.18 The cross-correlation function (left) and the value range distribution (right) of BDS PRN1 and PRN10

3.3 Navigation Messages

This section will explain the navigation messages broadcasted by GPS and BeiDou satellites. The contents of this section refers to Refs. [1, 5]. GPS mainly covers the C/A code modulated navigation message of L1, while BeiDou covers the D1 code and D2 code of GEO/IGSO/MEO, i.e. the B1I signal and the navigation message modulated on the B2I signal. Due to the limitations of space, the navigation messages for the newly added L2C, L5, and L1C signals in GPS modernization are not mentioned here. Readers can refer to Refs. [1–3] for more technical details. In this section, the BeiDou navigation message refers specifically to the navigation message of the BeiDou regional coverage system. Since the navigation message of the BeiDou global system has not been disclosed, it is not discussed here. The content and structure of the future BeiDou system navigation message are likely to be changed significantly compared with the existing navigation message.

3.3.1 The GPS Navigation Message

The navigation message modulated on the GPS signal is the data information broadcast by the GPS satellite, that is, $D(t)$ in Eqs. (3.1) and (3.2), and the data rate is 50 bps/sps. Here, the meaning of bps is bits per second, and the concept of sps is the symbol rate, i.e. the rate of the symbol sequence generated after the original data bits are encoded by some channel. The channel coding can be CRC, interleaving, or convolution, and its general purposes are to improve the correction and error detection ability, and to increase anti-fading or anti-interference ability. The data of the GPS L1 frequency C/A code signal is not channel coded, so bps and sps are the same here. The length of each data bit in the GPS signal is 20 ms. The GPS receiver outputs one data bit every 20 ms after entering stable signal tracking, which determines that the maximum coherent integration time cannot be achieved in more than 20 ms during signal acquisition and tracking processing. At the same time, it can be seen that the data rate modulated in the GPS signal is relatively slow compared to other modern high-speed communication systems. This setting is designed to provide sufficient spreading gain for weak satellite signals, to ensure a sufficiently low error rate and a reasonable baseband signal processing performance.

Navigation messages play an important role in the overall GPS signal composition, and each satellite continuously transmits its own unique navigation message. In a nutshell, the role of navigation messages is as follows:

- To provide the signal's transmission time and the satellite's clock correction, which, together with the tracking state of the tracking loop, will provide the satellite's precise launch time;
- To provide satellite ephemeris data, and allow the receiver to obtain the exact position of the satellite based on the signal's launch time and ephemeris data.

- To provide additional information about the satellite, such as health status, whether the Anti-Spoofing Technology (AS) is turned on, and satellite configuration information (Block II/II-A/II-R).
- To offer ionospheric and tropospheric delay correction parameters that will provide the amount of correction for observation during positioning, thereby improving positioning accuracy.
- To provide almanac data for current satellites and other satellites. The almanac data can be used to calculate the general position coordinates of the satellite. The receiver uses this data to calculate the approximate orientation of the satellite for obtaining a signal quickly.

We already know that GPS satellites, as the known points of dynamic positioning, can determine their position given the precise signal transmission time when satellite orbital parameters are known. Therefore, obtaining an accurate signal transmission time is the key to GPS receiver positioning. However, strictly speaking, the precise signal transmission time is provided by both the navigation message and the tracking loop. The data bit length of the navigation message is 20 ms, which also determines that the navigation message can only provide a transmission time accurate to 20 ms. More accurate information must be provided by the tracking loop, especially the pseudo-code tracking loop.

Based on a bottom-up concept similar to that in software engineering, navigation messages can be divided into five different structural levels. The most basic structure is a bit length of 20 ms; the higher-level structure is a word, consisting of 30 data bits; the third-level structure is a sub-frame, consisting of 10 words, i.e. one sub-frame contains 300 data bits; the fourth-level structure is the page or main frame, consisting of five sub-frames; the fifth-level structure is a periodic navigation message composed of 25 pages. Each satellite continuously transmits a periodic navigation message consisting of 25 pages. The length of time for each level of the navigation message is shown in Table 3.8.

Based on the above analysis, the structure of the navigation message is demonstrated in Fig. 3.19. It can be seen that the GPS navigation message sends a complete navigation message structure every 12.5 min, including 25 main frames, 125 sub-frames, 3750 words, or 75,000 bits.

The first word of each sub-frame gives the TLM code, and its Chinese name can be translated into telemetry code. The telemetry code starts with an 8-bit leading character, which is 10001011. The leading character is mainly used as the search synchronization character. The ninth to 22nd bits of the telemetry word are reserved for privileged users, and the 23–30 bit is the check code. How to apply this telemetry code to the sub-frame synchronization in the demodulated message is described in detail in subsequent chapters. The content of the telemetry word is shown in Fig. 3.20.

Table 3.8 The length of time for each component of the navigation message

Component	Data bits	Code	Sub-frame	Main frame (page)	25 main frames
Length of time	20 ms	600 ms	6 s	30 s	12.5 min

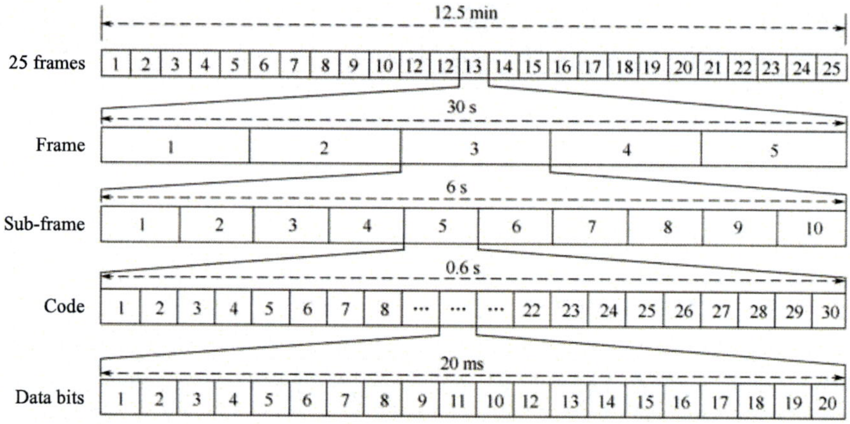

Fig. 3.19 The components of the GPS navigation message and their length

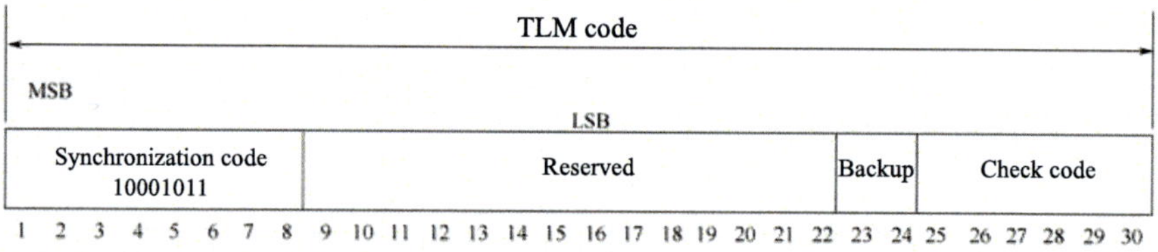

Fig. 3.20 Bit content of the telemetry code

The second word in the telegram sub-frame provides the timestamp mentioned in the previous section, and is called a HOW word (crossover word or conversion word in Chinese). In order to clearly understand how to obtain the launch time of the current signal from the crossover word, we must first review the concept of GPS time. Having read Sect. 1.3.4, we already know that GPS time treats continuous time as a period of time in weeks, resetting every Saturday at midnight going into Sunday, and then accumulating until the beginning of the next week. There are 604,800 s in a week. The timestamp in the GPS signal is for GPS time in seconds. There is a special word in the navigation message to describe this timestamp, which is TOW (Time Of Week). The tracking loop of the receiver can know the transmission time of the current signal according to the timestamp after being able to demodulate the navigation message. As mentioned above, the content of the navigation message can only be accurate to 20 ms. Here, we do not discuss how to increase the accuracy of the launch time to the level required for positioning. Since each sub-frame lasts for 6 s, a total of 100,800 (ie 604,800/6) sub-frames can be sent in one week, and the GPS week count is incremented when the sub-frame counts to 100,799. The bit content of a crossover word is shown in Fig. 3.21.

The first 17 bits of the crossover word are the intra-week time represented by the Z-number, which is broadcast in high-order first and low-order next. Here, the Z-number

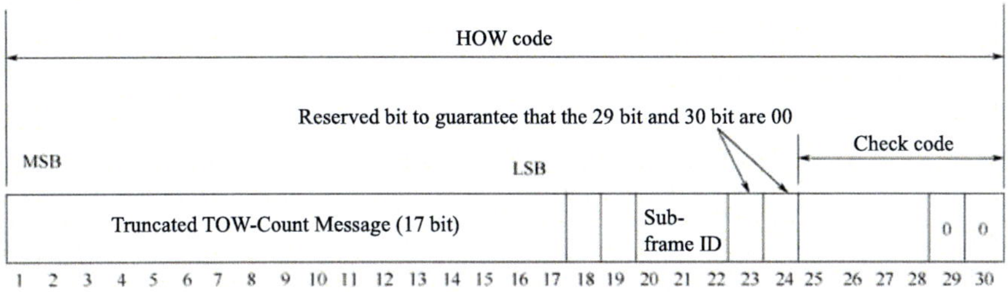

Fig. 3.21 Bit content of a crossover word

is a manually defined GPST timing unit with a length of 1.5 s, which is actually the period length of the X1 sequence of the GPS signal generation. Obviously, one sub-frame takes a total of four Z-numbers in time. The time of a week (604,800 s) will have a total of 403,200 Z-numbers, which means that 19 bits are required in binary, while the HOW in the navigation message only reserves 17 bits to record the Z-numbers. This seems like a contradiction. In fact, the Z-number recorded in the navigation message is a "truncated" Z-number, which is the result of discarding the last two bits of the Z-number and retaining only the high 17 bit. The reason for this is easy to understand: every 1 increase in the "truncated" Z-number means that the GPST is increased by 6 s (4 Z-numbers), and 6 s is the length of a sub-frame. So the "truncated" Z-number in the HOW can be thought of as a counter for the sub-frame during the week. The maximum number that can be recorded by a 17-bit binary number is 131,072, and the maximum value of the truncated Z-number within a week is 100,800, so 17 bits are used here to indicate the count overflow problem without a limited word length.

It should be noted that the Z-number provided by the second word of the current sub-frame is the end of the current sub-frame and the intra-week count of the start time of the next sub-frame, instead of the intra-week count of the current sub-frame start time. Therefore, after decoding the 17-bit truncated Z-number of this sub-frame by decoding in the receiver, you must multiply this value by 6 and subtract 4.8 s to get the start time of the next word in the current sub-frame. The reason for multiplying by 6 has been explained earlier. The 4.8 s is because the HOW is in the second word of the current sub-frame, and the next word (the third word) of this sub-frame has a time difference of 4.8 s from the beginning of the next sub-frame. Figure 3.22 clearly shows this, and it must be noted in the receiver programming, otherwise you will get the wrong signal transmission time.

After obtaining the Z-number in the HOW, the GPS time of any subsequent time can be obtained according to the bit count of the navigation message. In this regard, the details of the process are illustrated by the following example. For example, the truncated Z-number of the receiver demodulation to the current sub-frame is 1000. From the above analysis, it can be understood that the GPS time from the third word of the sub-frame is $1000 \times 6 - 4.8 = 5995.2$ s, then at any time in the future, such as at the 10th bit of the fourth word, the receiver can know that the current transmission time is:

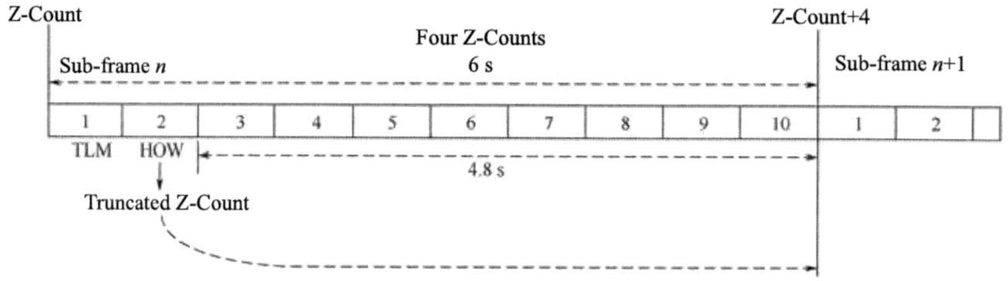

Fig. 3.22 The relationship between the Z-count of the current sub-frame and the time count in a week

$$\text{GPST} = 1000 \times 6 - 4.8 + 40 \times 0.02 = 5996.0\,\text{s}$$

The "40" in the formula is the 10th bit of the fourth word, that is, the bit number from the starting bit number of the third word; 0.02 s is the length of one bit. It must be pointed out that the GPST calculated by the above formula is unstable. This is because the smallest unit of the navigation message is a bit, and the length of one bit is 20 ms, so the GPST obtained from the navigation message has an ambiguity of 20 ms. That is to say, the portion of the transmission time with an accuracy higher than 20 ms cannot be obtained by the navigation message. The transmission time obtained in the above formula cannot be directly used for positioning because it will bring about an error of thousands of kilometers. In order to solve this 20 ms ambiguity, more precise information must be obtained from the state parameters of the pseudo-code tracking loop. This problem will be solved after the receiver implements the tracking of the signal.

So far, we have explained how to use the truncated Z-number to calculate the rough GPST of signal transmission. It should be noted that although the Z-number is provided by the HOW in the second word of each sub-frame, we cannot ignore the effect of the TLM in the first word. This is because the TLM in the first word provides the leading character, and the subsequent navigation message is correctly demodulated only after synchronization with the leading character. Therefore, for receiver to obtain the HOW, it must first demodulate and correctly identify the TLM. In fact, in the internal software of the receiver, the first step in demodulating the navigation message is to find the correct TLM character.

The first two words of each sub-frame of the GPS navigation message are telemetry words and crossover words, but the subsequent content is different. They can basically be divided into two categories: directly related to the positioning solution, or related to other satellites. The former is set to calculate the exact position of the satellite, mainly its orbit parameters, its health status, the positional accuracy URA, the clock correction, and other parameters. The latter includes the almanac data of all satellites, the ionospheric delay correction parameters, UTC time parameters, and information on the health status of all satellites. The data directly related to the positioning solution is repeated in the first three sub-frames of each sub-frame, and each main frame is repeatedly broadcasted, i.e. every 30 s, so that the receiver can analyze the information necessary for calculating positioning in the navigation message data of up to 30 s. The data related to other satellites is dispersed in the fourth and fifth sub-frames of different main frames. All of the broadcasts need 12.5 min, i.e. the length of 25 main frames.

The telegrams directly related to the positioning calculation are mainly in the first, second, and third sub-frames, and generally include the following data:

(1) First sub-frame

- o Week-Number (WN): takes up the 61st to 70th bits, the meaning of which has been explained above. Since WN was reset once in 1999, the receivers after that need to add 1024 when reading the number of weeks to get the current correct GPS week.
- o User-Range-Accuracy (URA): takes up the 73rd–76th bits. This parameter gives the use of the satellite's data to achieve positioning, and the resulting user position is 1σ statistical error estimates. This parameter ranges from 0 to 15. The smaller the value, the more precise it is.
- o Satellite health status: takes up the 77th–82nd bits. This parameter demonstrates the health status of the satellite's launching signal. The upper part of the parameter (MSB) indicates the general condition of the signal: if MSB $=$ 0, it means that the navigation message data is normal; if MSB $=$ 1, it means that the navigation message data is abnormal, and the lower five bits indicate different abnormal conditions, including if the signal power is slightly weaker than the normal value, if there is a certain component data loss (P component or C/A component), or if the satellite is turned off. In general, the satellite data is only used when MSB $=$ 0.
- o Issue of Data, Clock (IDOC): takes up the 83rd and 84th bits and the 211th–218th bits, among which the former is the upper two bits and the latter is the lower eight bits. A change in this parameter means that the satellite correction parameters have been updated and the receiver needs to be ready to update its local satellite parameters.
- o Estimated group delay differentia: takes up the 197th–204th bits and is generally shown in T_{GD}. This is a parameter that is used to compensate for the group delay effect of the satellite clock.
- o Satellite clock correction factor: a total of four correction factors. t_{oc} is based on the 219th–234th bits, a_{f0} is based on the 27th–292nd bits, a_{f1} is based on the 249th–264th bits, and a_{f2} is based on the 241st–248th bits. These correction factors are used in the calculation of satellite positioning.

The bit content of the first sub-frame is shown in Fig. 3.23.

To explain more clearly, the data bits contained in all 10 words in a sub-frame are regarded as consecutive 300 bits. For example, the first word takes up the first to 30th bits, and the second word takes up the 31st–60th bits. It can be reasoned that the 10th word takes up the 271st–300th bits. The reason for this is that some parameters take up different bit fields of multiple words. The following sub-frame contents are also explained in this way.

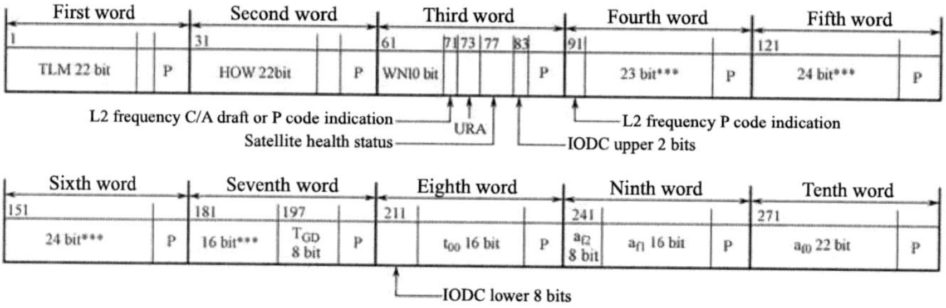

Fig. 3.23 Bit content of the first sub-frame of the GPS navigation message

(2) Second sub-frame

- IODE (Issue of Data, Ephemeris): takes up the 61st–68th bits of sub-frame 2. In fact, the 271st–278th bits of sub-frame 3 also transmit IODE data. Normally, the two IODEs should be equal to the lower eight bits of the IODC in sub-frame 1. Any time that the three pieces of data (the IODE of sub-frame 2 and sub-frame 3, and the lower eight bits of IODC of sub-frame 1) are not equal, it means that the ephemeris data has changed, and the receiver should pay attention to receiving new ephemeris data. Together, IODE and IODC provide users with a very convenient way to check the validity of ephemeris data.
- *Crs* (Amplitude of the sine harmonic correction term to the orbit radius): takes up the 69th–84th bits. The unit is m and the scaling factor is 2^{-5}.
- Δn (Mean motion difference from computed value at reference time): takes up the 91st–106th bits. The unit is πrad/s and the scaling factor 2^{-43}. This is calculated by using the satellite's average angular velocity and using the formula $\sqrt{\mu/a^3}$ to obtain the difference between the values, where μ is the universal gravitational constant of the Earth for GPS satellites, and a is the semi-major axis of the satellite's elliptical orbit. From this relationship, the average angular velocity after satellite's correction is $n = \sqrt{\mu/a^3} + \Delta n$.
- $M0$ (Mean anomaly at reference time): The upper eight bits take up the 107th–114th bits, and the lower 24 bits take up the 121st–144th bits. The unit is πrad and the scaling factor is 2^{-31}.
- *Cuc* (Amplitude of the cosine harmonic correction to the argument of latitude: takes up the 151st–166th bits. The unit is rad and the scaling factor is 2^{-29}.
- e_s (Elliptical eccentricity of the satellite orbit): The upper eight bits take up the 167th–174th bits, and the lower 24 bits take up the 181st to the 204th bits. The scaling factor is 2^{-33}.
- C_{us} (Amplitude of the sine harmonic correction to the argument of latitude): takes up the 211st–226th bits. The unit is rad and the scaling factor is 2^{-29}.

- $\sqrt{a}$ (The square root of the semi-major axis of the satellite): The upper eight bits take up the 227th–234th bits, and the lower 24 bits take up the 241st–264th bits. The unit is $m^{1/2}$ and the scaling factor is 2^{-19}.
- t_{oe} (Ephemeris reference time): takes up the 271st–286th bits. The unit is second and the scaling factor is 2^4.

The bit content of the second sub-frame is shown in Figure 3.24

(3) Third sub-frame

- C_{ic} (Amplitude of the cosine harmonic correction term to the angle of inclination): takes up the 61st–76th bits of the third sub-frame. The unit is rad and the scaling factor is 2^{-29}.
- Ω_e (Longitude of the ascending node of the satellite's orbit): the upper eight bits take up the 77th–84th bits of the third sub-frame, and the lower 24 bits take up the 91st–114th bits. The unit is πrad and the scaling factor is 2^{-31}.
- C_{is} (Amplitude of the sine harmonic correction term to the angle of inclination): takes up the 121st–136th bits of the third sub-frame. The unit is rad and the scaling factor is 2^{-29}.
- i_0 (Inclination angle at the reference time): the upper eight bits take up the 137th–144th bits of the third sub-frame, and the lower 24 bits take up the 151st–174th bits. The unit is πrad and the scaling factor is 2^{-31}.
- C_{rc} (Amplitude of the cosine harmonic correction term to the orbit radius): takes up the 181st–196th bits of the third sub-frame. The unit is meters and the scaling factor is 2^{-5}.
- ω (Argument of perigee): The upper eight bits take up the 197th–204th bits of the third sub-frame, and the lower 24 bits take up the 211th–234th bits. The unit is πrad and the scaling factor is 2^{-31}.
- Ω (Rate of right ascension): takes up the 241st–264th bits of the third sub-frame. The unit is πrad/s and the scaling factor 2^{-43}.
- IODE: takes up the 271st–278th bits of the third sub-frame (meaning explained earlier).
- IDOT: Inclination angle at the reference time: takes up the 279th–292nd bits of the third sub-frame. The unit is πrad/s and the scaling factor is 2^{-43}.

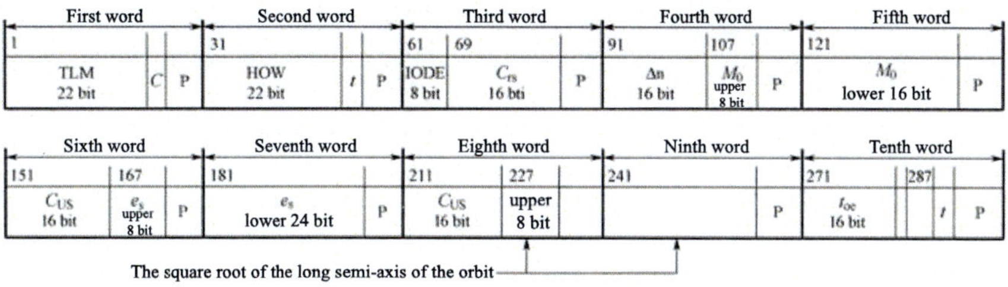

Fig. 3.24 Bit content of the second sub-frame of the GPS navigation message

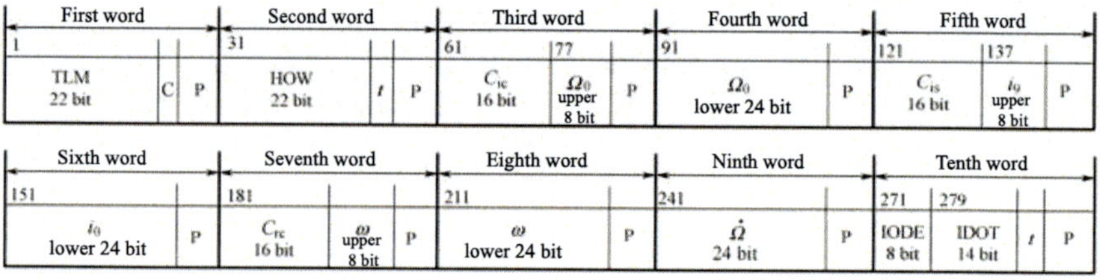

Fig. 3.25 Bit content of the third sub-frame of the GPS navigation message

The bit content of the third sub-frame is shown in Fig. 3.25

Data related to other satellites includes almanac data, ionospheric delay parameters, UTC time parameters, AS marks, and health indicators. These are scattered in the fourth and fifth of the 25 main frames, so the decoding must be treated differently for the primary frame number corresponding to the current sub-frame. Table 3.9 lists the contents of the fourth and fifth sub-frames of each main frame.

Data related to other satellites is not the information that is necessary for the receiver to achieve positioning, but the overall performance of the receiver is closely related to these parameters. The parameters of the almanac data are based on the Kepler satellite model, which can basically correspond to the ephemeris data, but without the disturbance correction term in the ephemeris data. The almanac data can generally be considered as a subset or a "lite" version of the ephemeris data. The purpose of the almanac data is to calculate the position of the satellite. However, since the accuracy of the almanac is lower than that of the ephemeris, the calculated satellite position error is large. Therefore, it is not used to directly implement the

Table 3.9 Bit content of the fourth and fifth sub-frames of the GPS navigation message

Number of sub-frame	Main frame Number	Description
Fourth sub-frame	1, 6, 11, 12, 16, 19, 20, 21, 22, 23, 24	Reserved
	2, 3, 4, 5, 7, 8, 9, 10	Almanac parameters of satellites 25-32
	13	NMCT
	14, 15	Reserved by the system
	17	Special information
	18	Ionospheric delay correction parameters and UTC time parameters
	25	A-S marks and health indicators of satellites 25–32
Fifth sub-Frame	1–24	Almanac parameters of satellite 1–24
	25	Almanac parameters of satellite 1–24

positioning function, but is used to search at the receiver. When the satellite is used, its orientation is determined roughly in order to shorten the search time. The almanac data can occupy fewer bits during the transmission process. As can be seen above, a set of ephemeris data needs to be transmitted in three sub-frames, while a set of almanac data can be transmitted by only one sub-frame. At the same time, the accuracy of the almanac data is low, so the time validity period that can be tolerated is much longer than for the ephemeris data. The ephemeris data generally has a validity period of 2–4 h, while the almanac data has a validity period of several months, so receiver can store it in non-volatile memory after demodulating a valid set of almanac data. This information can then be used to improve the speed of searching for and capturing signals.

There are eight ionospheric parameters, namely $[\alpha_1, \alpha_2, \alpha_3, \alpha_4, \beta_1, \beta_2, \beta_3, \beta_4]$. They are based on the Klobuchar model and can be used to obtain pseudo-range without differential correction or dual-frequency observation. The observation is corrected to improve the positioning accuracy. The UTC parameter gives the cumulative leap second correction between GPST and UTC time, and other necessary clock correction parameters that can be used to obtain an accurate UTC time based on the current GPST. The satellite's health indicator is used to detect the availability of satellite-borne signals and to avoid the use of satellites that are not functioning properly. Due to the limitations of space, a detailed description of these parameters is not offered here. Interested readers can view them in Ref. [1].

3.3.2 The BDS Navigation Message

There are two kinds of navigation messages for BeiDou satellites, namely D1 and D2. The D1 navigation message rate is 50 bps and is modulated with a 1 kHz secondary code (NH code). The D1 navigation message is broadcast on the B1I and B2I signals of the BeiDou IGSO and MEO satellites. The D2 navigation message rate is 500 bps, and the D2 navigation message is broadcast on the B1I signal and the B2I signal of the BeiDou GEO satellite. The content of the messages on D1 and D2 includes information directly related to the positioning calculation, as well as information related to other satellites. The D2 message also includes enhanced service information, such as difference and integrity information about the BeiDou system and the ionosphere of the grid point.

The BeiDou navigation message was coded before transmission, and includes bit interleaving and BCH coding. The main purpose of bit interleaving is to discretize continuous errors, thereby improving the ability to resist continuous errors. The processing steps of bit interleaving can be seen from the following operational formulas.

Input bit stream:

[b_0, b_1, b_2, b_3, b_4, b_5, b_6, b_7, b_8, b_9, b_{10}, b_{11}, b_{12}, b_{13}, b_{14}, b_{15}, b_{16}, b_{17}, b_{18}, b_{19}, b_{20}, b_{21}].

After interleaving:

[b_0, b_2, b_4, b_6, b_8, b_{10}, b_{12}, b_{14}, b_{16}, b_{18}, b_{20}] + [b_1, b_3, b_5, b_7, b_9, b_{11}, b_{13}, b_{15}, b_{17}, b_{19}, b_{21}].

The input bit stream in the above equation is serially processed at every other bit to obtain two parallel sequences; one is an odd-decimated sequence, and the other is an even-decimated sequence, and the number of bits per sequence is eleven. The selected sequence length is 11 bits because the check mode of the BeiDou navigation message is BCH (15, 11, 1) coding. That is, the code length is 15 bits, the information bit is 11 bits, and the error correction ability is one bit. Therefore, before performing BCH coding, it is first necessary to perform bit grouping on the input data to obtain an information code group of 11 bits in length. Each of the two 11-bit information code groups is BCH-encoded to obtain two 15-bit code groups, and then the two code groups are subjected to one bit and serial processing to obtain a new 30-bit to-be-sent code group.

At the receiver, the process is just the opposite. First, the demodulated data is deinterleaved. In fact, the processing format is the same as the encoding. It is a 1-bit serial-to-parallel conversion, except that this time, two sets of 15-bit input code groups are obtained every 30 bits. The two input code groups perform BCH decoding to obtain an 11-bit information code group and a 4-bit check code word. They then convert the two sets of information code groups and check code words into a 2two-bit information bit and the 8-bit check bit, which is a word in the BeiDou navigation message. The overall process of interleaving, deinterleaving, and BCH encoding and decoding is shown in Fig. 3.26. The upper part is the schematic diagram of the encoding process, and the lower part is the schematic diagram of the decoding process.

The generator polynomial of the BeiDou BCH code is $g(x) = x^4 + x + 1$, which determines the coding block diagram of BCH (15, 11, 1), as shown in Fig. 3.27.

In the figure, the initial state of the shift register is 0, the switch K1 is connected, and K2 is disconnected. As the 11 information bits of the shift clock pass through the gate, the content is changed by K1 driving the shift register, and the tap setting constitutes g(x). In the divider circuit, when all 11 information bits have been shifted, the 4-bit check code word is reserved among [r_1, r_2, r_3, r_4]. Then, K1 is disconnected, K2 remains connected, and a 4-parity bit is output so that it constitutes a 15-bit BCH code with the previous 11-bit information. Then, K1 is connected and K2 disconnected. The shift clock drives the next cycle and repeats.

The block diagram for BeiDou BCH (15, 11, 1) is shown in Fig. 3.28. The 4-bit shift register and feedback tap settings form the g(x) divider circuit, and the initial state of the shift register is still all zeros. The input stream is also split into two paths, one through the divider to generate a 4-bit error correction code, and the other in a 15-bit error correction buffer. When the shift clock completes 15 beats, among it the

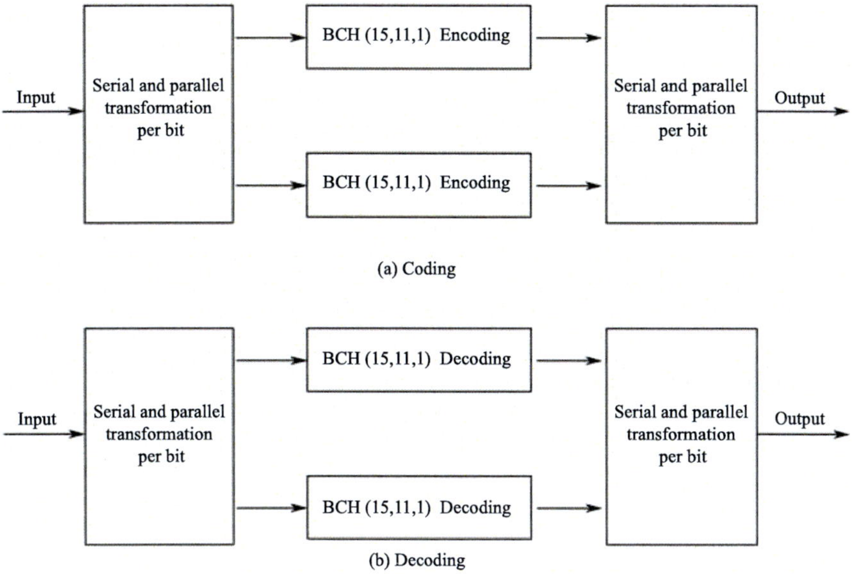

Fig. 3.26 Coding and decoding of the BDS navigation message

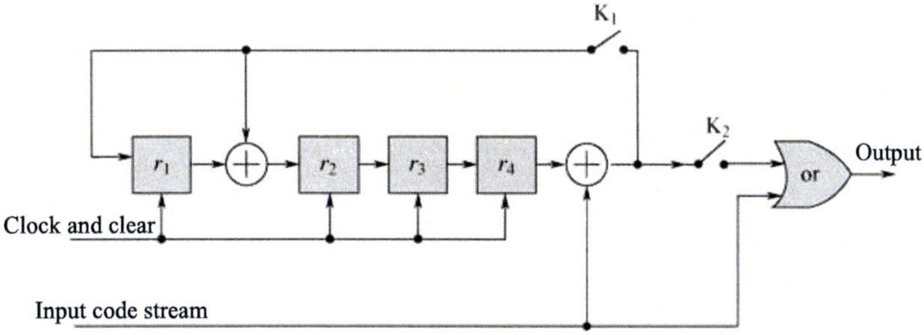

Fig. 3.27 Coding block diagram for BeiDou BCH (15, 11, 1)

[r_1, r_2, r_3, r_4] is a 4-bit error correction code. The error bit information is obtained through the BCH error correction code figure, and then the information in the error correction buffer is performed. The error correction here is obtained by using the error correction code. The error correction here is performed by using the 15-bit error correction signal obtained by the error correction code table and the buffered input code stream. The contents of the error correction code table are shown in Table 3.10, where [r_1, r_2, r_3, r_4] is all 0. This means that there is no error, otherwise the error bit in the input code stream can be corrected by finding the corresponding error correction signal according to its content. It is worth noting that the above error

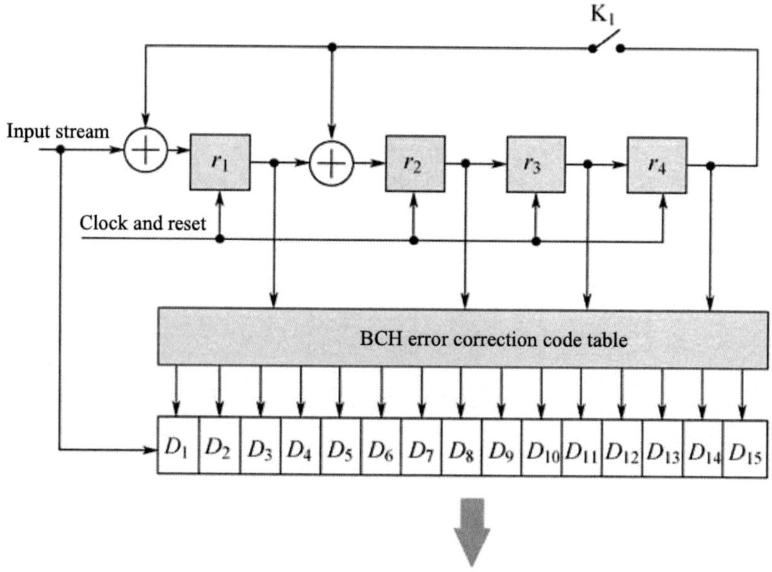

Fig. 3.28 The decoding process of BCH (15, 11, 1)

Table 3.10 BCH error correction codes

[*r*1, *r*2, *r*3, *r*4]	15 bit error correction signal	[*r*1, *r*2, *r*3, *r*4]	15 bit error correction signal
0000	000000000000000	0001	000000000000001
0010	000000000000010	0011	000000000010000
0100	000000000000100	0101	000000100000000
0110	000000000100000	0111	000010000000000
1000	000000000001000	1001	100000000000000
1010	000001000000000	1011	000000010000000
1100	000000001000000	1101	010000000000000
1110	000100000000000	1111	001000000000000

correction mechanism only works when one bit of data is in error. If two or more error bits are presented, the above mechanism cannot correct the error.

Similar to the structure of GPS navigation messages, the structure of BeiDou navigation messages also includes superframes, main frames, sub-frames, words, and bits. The concept of superframes here is a complete set of navigation messages. Unlike the GPS navigation message, the BeiDou navigation message is divided into D1 and D2 navigation messages, and the structure of the two is slightly different.

The superframe of the D1 navigation message lasts 12 min and contains 24 main frames. One main frame lasts 30 s and contains five sub-frames. One sub-frame lasts six seconds and contains 10 words. One word lasts 0.6 s, including 30 bits. Therefore, a D1 superframe contains 36,000 bits.

The superframe of the D2 navigation message lasts six minutes and contains 120 main frames. One main frame lasts three seconds and contains five sub-frames. One sub-frame lasts 0.6 s, and contains 10 words. One word lasts 0.06 s, and includes 30 bits. Since the bit rate of the D2 navigation message is 500 bps, the number of bits transmitted by the D2 navigation message is 10 times that of the D1 navigation message in the same time. A D2 superframe contains 180,000 bits.

The hierarchical structure of the D1 navigation message and D2 navigation message and the time relationship between them are shown in Figs. 3.29 and 3.30 respectively. The format of the check code of the first word and the subsequent word of each sub-frame is different; the check code of the first word is four bits, and the check code of the subsequent two to nine words is eight bits. The difference between two check codes is presented in the chart.

The content of the navigation message of the D1 code is very similar to that of the GPS navigation message. Sub-frames 1, 2, and 3 transmit data directly related to the positioning calculation, while sub-frames 4 and 5 transmit data related to other satellites. Sub-frames 1, 2, and 3 are broadcasted every 30 s, which ensures that the receiver can demodulate the ephemeris and other data necessary for positioning calculation within up to 30 s, while the contents of sub-frames 4 and 5 include satellite-related data that varies according to the main frame number (or page number). A complete navigation message requires 24 main frames, i.e. 12 min, to complete. Since each main frame is repeated, the message content of sub-frames 1, 2, and 3 has no main frame information; the message content of sub-frames 4 and 5 has a 7-bit main frame, and the receiver is in the demodulation sub-frame. The 4 and 5 of the message needs to be processed according to the main frame number. Figure 3.31a shows the arrangement of the content of the D1 navigation message.

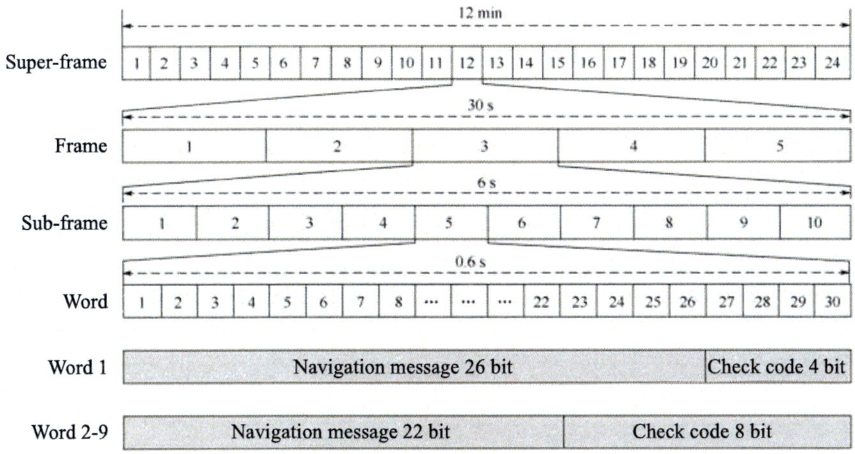

Fig. 3.29 Organizational structure and time length of the BDS D1 code navigation message

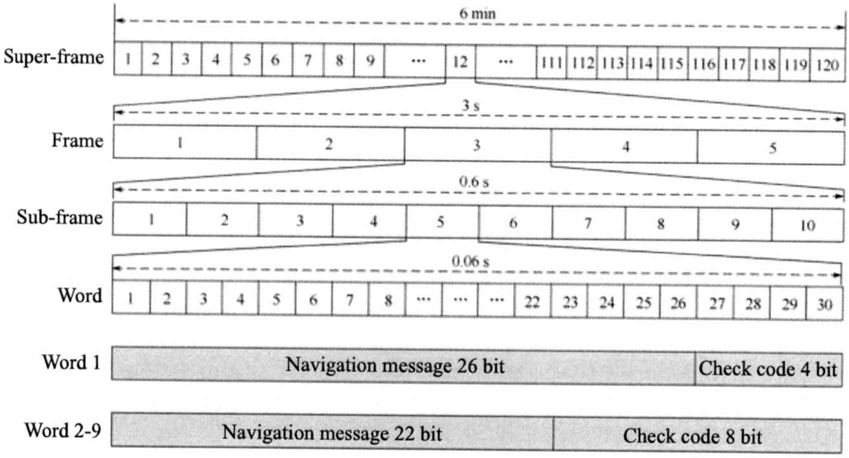

Fig. 3.30 Organizational structure and time length of the BDS D2 code navigation message

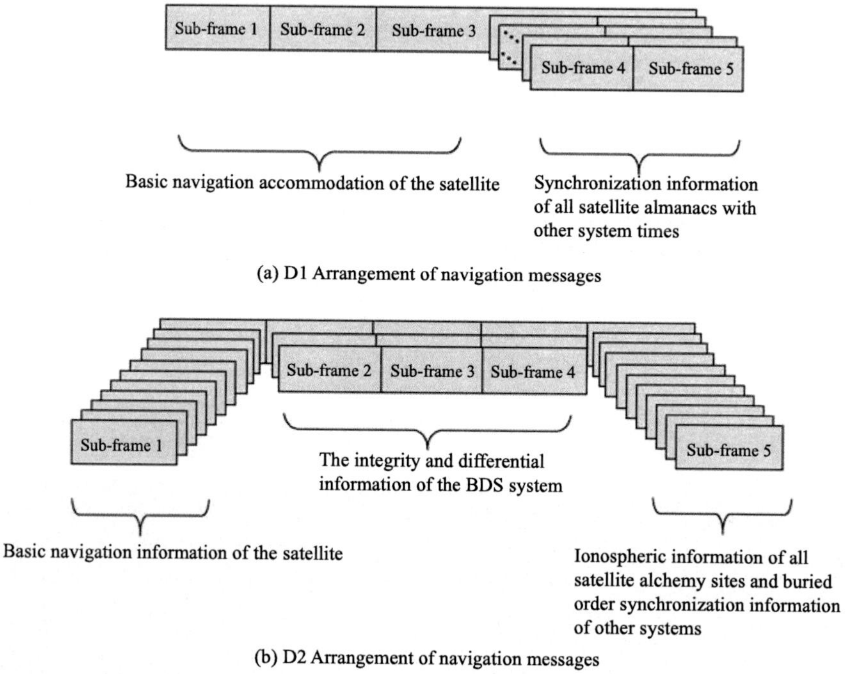

Fig. 3.31 Arrangement of the content of the D1 and D2 navigation messages

The content of the navigation message of the D2 code is quite different from that of the GPS navigation message. First, the superframe of the D2 code contains 120 main frames, while the superframe of the GPS has only 25 main frames, and the number of main frames is more than that of GPS. Secondly, the main frame duration of the D2 code is three seconds, which is only 1/10 of the duration of the GPS main frame. This is because the data rate of the D2 code is 10 times the data rate of GPS; the main difference is the navigation message content. In the arrangement, the D2 navigation message not only contains data directly related to the positioning calculation and data related to other satellites, but also includes the integrity and differential information of the BeiDou satellite system. The data directly related to the positioning calculation is transmitted in sub-frame 1, and 10 primary frames are transmitted. The data related to other satellites is transmitted in sub-frame 5, and 120 main frames are transmitted. The BeiDou satellite system is intact. The completeness and difference information are mainly in sub-frames 2, 3, and 4, and are transmitted in six main frames. Figure 3.31b shows the arrangement of the content of the D2 navigation message.

Similar to the GPS navigation message, the sub-frame of the BeiDou navigation message also contains the synchronization code and the time information within the week. The first 11 bits of the first word of each sub-frame are the synchronization code, and the content is "1110001010". Before the receiver starts demodulating the navigation message, it must first confirm the search for the synchronization code, meaning that the sub-frame synchronization has been completed. The sync code is followed by four bits of reserved information, followed by a 3-bit sub-frame number and the upper eight bits of the SOW, as shown in Fig. 3.32.

SOW means the second-count in the week. This is because BeiDou is also represented by the number of weeks and the number of seconds in the week, so the SOW information here is similar to the TOW information of GPS. BeiDou's SOW has a total of 20 digits, and a maximum range of 1,048,576 s. This is greater than 604,800 s, so no truncation is required. The lower 12 of the SOW is transmitted in the upper 12 bits of the second word of each sub-frame, and any sub-frame of the D1 and D2 navigation messages is set as such, as shown in Fig. 3.32.

It should be noted that the second minute corresponding to the SOW in the BeiDou sub-frame refers to the time corresponding to the rising edge of the first pulse of the synchronization code of the sub-frame, which is slightly different from the meaning of the TOW information of GPS.

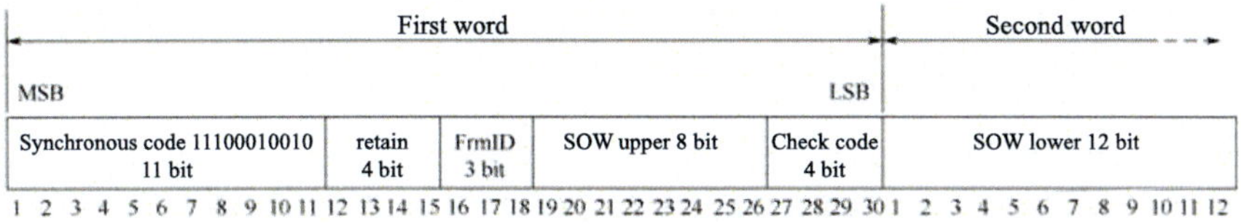

Fig. 3.32 Synchronization code and SOW information of the BDS navigation message

The BeiDou Week Number (WN) information is transmitted in 13 bits, which draws on the experience of GPS messages, because the BeiDou WN needs to accumulate 8192 weeks. This will cause overflow clearing only after 157.1 years. Compared with 19.6 years in the GPS navigation message, the overflow resetting period has been greatly increased. Considering the life cycle of existing electronic products, a receiver that is designed according to the existing BeiDou signal format does not have to consider the problem of resetting the overflow of the BeiDou week during the service period.

The most important data directly related to the positioning solution is the satellite's ephemeris data, which is transmitted by sub-frames 1, 2, and 3 in the D1 navigation message, and by sub-frame 1 in 10 main frames in the D2 navigation message. The definitions of ephemeris data in D1 and D2 navigation messages are the same. Of course, the two are different in the arrangement of their respective sub-frame contents. Table 3.11 shows the definition of the parameters of the BeiDou satellite calendar, the number of bits, the scale factor, and the unit. Such information and the ephemeris parameters are applicable to the BeiDou GEO/IGSO/MEO satellite.

Table 3.11 Ephemeris parameters of the BeiDou satellite

Parameter	Description	No. of bits	Scale factor	Effective range	Unit
t_{oe}	Ephemeris reference time	17	2^{3}	604,792	s
$\sqrt{A}$	Square root of semi-major axis	32	2^{-19}	8192	$m^{1/2}$
E	Eccentricity	32	2^{-33}	0.5	
ω	Argument of perigee	32[a]	2^{-31}	± 1	π
Δ_n	Mean motion difference between computed value and average motion rate	16[a]	2^{-43}	$\pm 3.73 \times 10^{-9}$	π/s
M_0	Mean anomaly at reference time	32[a]	2^{-31}	± 1	π
Ω_0	Longitude of ascending node of orbital plane at weekly epoch	32[a]	2^{-31}	± 1	π
$\dot{\Omega}$	Rate of right ascension	24[a]	2^{-43}	$\pm 9.54 \times 10^{-7}$	π/s
i_0	Rate of inclination angle	32[a]	2^{-31}	± 1	π
IDOT	Inclination angle at reference time	14[a]	2^{-43}	$\pm 9.31 \times 10^{-10}$	π/s
C_{uc}	Amplitude of cosine harmonic correction to the argument of latitude	18[a]	2^{-31}	$\pm 6.10 \times 10^{-5}$	Radian
C_{us}	Amplitude of sine harmonic correction to the argument of latitude	18[a]	2^{-31}	$\pm 6.10 \times 10^{-5}$	Radian

(continued)

Table 3.11 (continued)

Parameter	Description	No. of bits	Scale factor	Effective range	Unit
C_{rc}	Amplitude of cosine harmonic correction term to the orbit radius	18[a]	2^{-6}	±2 048	m
C_{rs}	Amplitude of sine harmonic correction term to the orbit radius	18[a]	2^{-6}	±2048	m
C_{ic}	Amplitude of cosine harmonic correction term to the angle of inclination	18[a]	2^{-31}	$\pm 6.10 \times 10^{-5}$	Radian
C_{is}	Amplitude of sine harmonic correction term to the angle of inclination	18[a]	2^{-31}	$\pm 6.10 \times 10^{-5}$	Radian

[a]is the binary complement. Its top digit is the sign bit

The arrangement of the content of sub-frames 1, 2, and 3 in the D1 navigation message is shown in Fig. 3.33. The arrangement of the content of sub-frame 1 (10 pages) in the D2 navigation message is shown in Fig. 3.34. The receiver mainly demodulates the ephemeris parameters according to Fig. 3.33. The information given in Fig. 3.34 and Table 3.10 is not described in detail here.

The contents of sub-frames 4 and 5 in D1 navigation message are shown in Table 3.12.

Unlike the GPS navigation message, the BeiDou system moves the ionospheric delay parameters to sub-frames 1, 2, and 3, and repeats every 30 s, so it can be obtained when demodulating the BeiDou satellite ephemeris parameters. The ionospheric parameters allow ionospheric correction to be used immediately when the solution is obtained.

Table 3.13 shows the arrangement of the content of sub-frames 2, 3, 4, and 5 in the D2 navigation message. Sub-frame 2 and sub-frame 3 transmit the BeiDou system integrity and difference information and satellite identification, regional user distance accuracy (RURA), and user differential Distance Error (UDRE) and equivalent clock correction Δt. Sub-frame 4 is reserved for use by the system. Sub-frame 5 transmits all satellite almanac data, grid point ionospheric information, BeiDou time and other satellite time system parameters, and UTC time parameters. For a more detailed description and instructions on how to use this information, see Ref. [5].

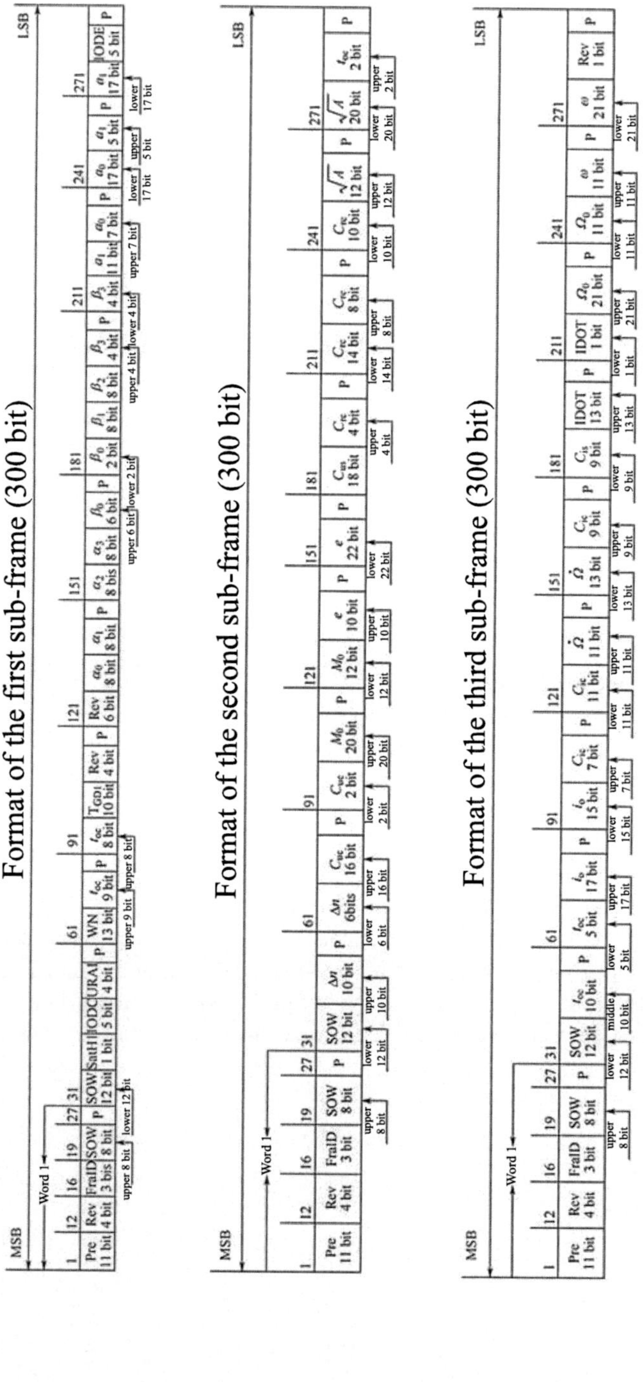

Fig. 3.33 Format of the first, second, and third sub-frame of the BeiDou D1 navigation message

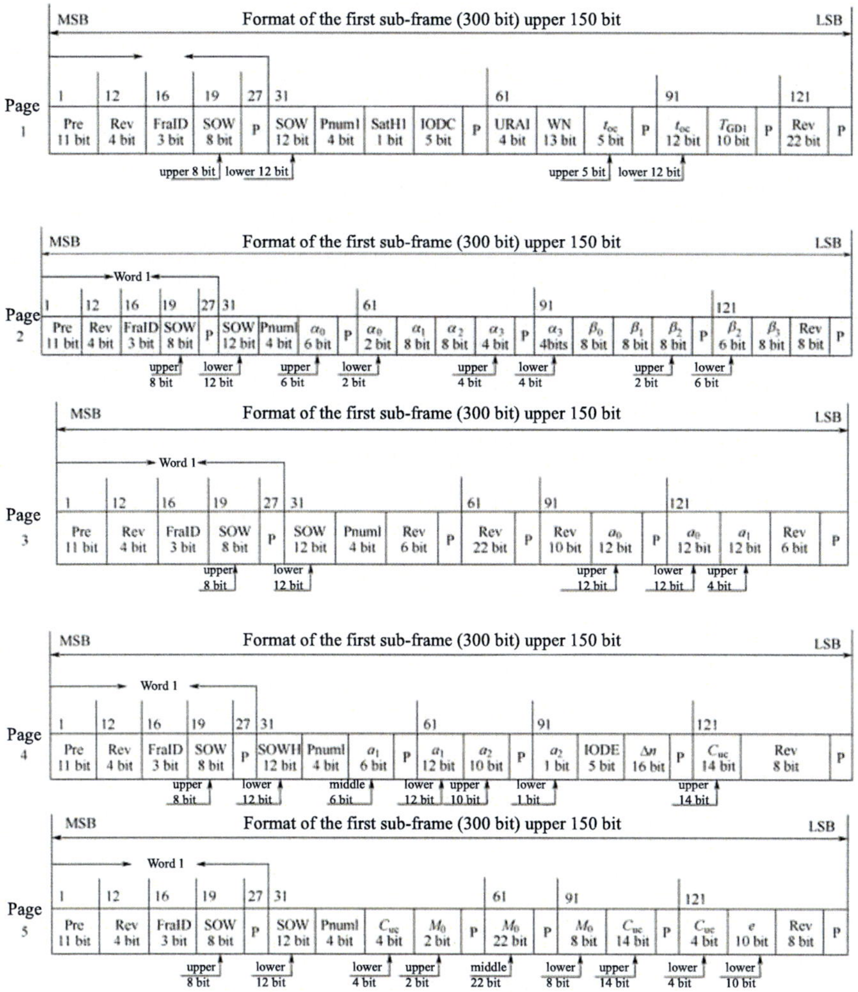

Fig. 3.34 Format of the first sub-frame of the BeiDou D2 navigation message[1]

3.4 Time Relationships Among Various Satellite Signals

In the previous sections of this chapter, we learned about the basic structure and main features of the navigation signals transmitted by a single GPS and BeiDou satellite. We already know that each satellite constantly broadcasts its own navigation message, and that the signal is used every once in a while. A timestamp is added to the navigation message. For GPS signals, this time interval is six seconds; for BeiDou, the time interval of the D1 signal is six seconds, and the time interval of the D2 signal

[1]Figs. 3.33 and 3.34 are extracted from Ref. [10].

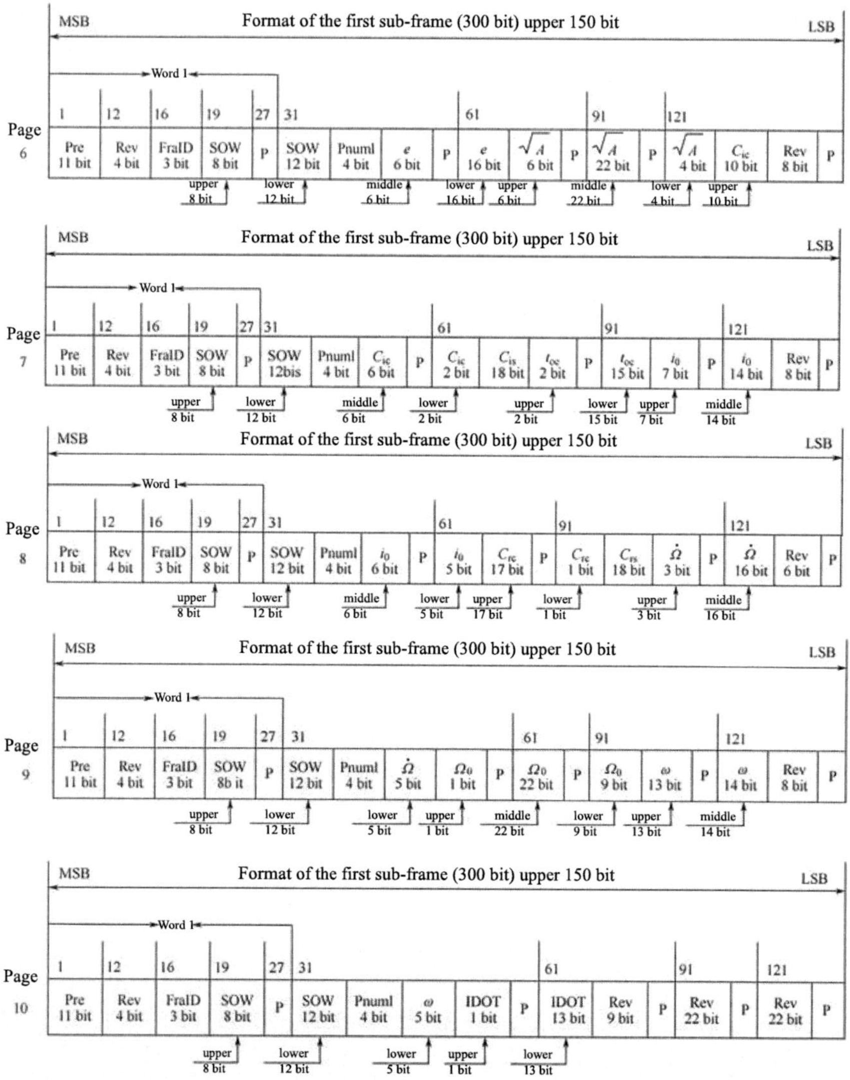

Fig. 3.34 (continued)

is 0.6 s. The data structure of the navigation message is skillfully organized so that once the receiver has reliably kept tracking of the signal, the signal transmission time at any point can be derived from the timestamp. However, what is often overlooked by beginners is that the signals transmitted by all GPS satellites in space are strictly synchronized in time, as are the signals transmitted by all BeiDou satellites. These are the most fundamental qualities that allow GPS and BeiDou satellite signals to

Table 3.12 Content of the fourth and fifth sub-frames in the BDS D1 navigation message[2]

Sub-frame	Main frame	Contents
Fifth sub-frame	1–6	Almanac parameters of satellites 25–30
	7	Health indicators of satellites 1–19
	8	Health indicators of satellites 20–30, week numbers of the almanac parameters
	9	BDS time and other time system parameters
	10	BDS time and UTC time parameters
	11–24	Reserved for the system
Fourth sub-frame	1–24	Almanac parameters of satellites 1–24

Table 3.13 The content of the second, third, fourth, and fifth sub-frames in the BDS D2 navigation message[3]

Sub-frame	Main frame	Description
Second	1–6	Integrity and difference information of the BDS system
Third	1–6	Regional user distance accuracy (RURA) and clock correction parameters
Fourth	1–6	Reserved for the system
Fifth	1–13 61–73	Grid point ionospheric information
	35	Health indicators of satellites 1–19
	36	Health indicators of the BDS satellites 20 to 30, and the number of weeks of the almanac data
	37–60	Almanac parameters of satellites 1–24
	95–100	Almanac parameters of satellites 25–30
	101	BDS time and other time system parameters
	102	BDS time and UTC time parameters
	14–34 74–94 103–120	Reserved for the system

achieve positioning. In this sense, the satellite navigation system is not so much a positioning system as a strict time synchronization system.

First, the atomic clocks on each satellite remain strictly synchronized. One of the most basic characteristics of the atomic clock is its extremely high frequency stability. In addition, the ground control station constantly monitors the state of the atomic clock on the satellite and corrects the time error. Even if there is a deviation between the atomic clock and GPST and BDT, the ground station sends corrections at any time to maintain strict synchronization of the clocks on the satellite. In fact,

[2]The BeiDou system is in a period of rapid adjustment. This content may change in the future. Please update it according to BeiDou's official website at any time.

[3]See Footnote 1.

the synchronization deviation of the atomic clocks on different GPS satellites is controlled within five to 10 ns.

Under the premise that the atomic clocks on the satellites are synchronized well, the navigation signals transmitted by different GPS and BeiDou satellites have the basic conditions for maintaining synchronization. Taking GPS satellites as an example, GPS time is divided into 100,800 time slots in a week. Each time period is six seconds—exactly the same length as one sub-frame signal. Between midnight on Saturday and Sunday morning, all GPS satellites start transmitting the first sub-frame of the week based on the time provided by their atomic clock, then the second sub-frame and so on, until the following Saturday and week thereafter. The point of alternating is in the morning. Therefore, from the starting point of the navigation message, the navigation signals transmitted by all GPS satellites at the same GPS time have the same timestamp. The same conclusion can be drawn from a similar analysis of the BeiDou satellite.

Another point that is easily overlooked by beginners is that the pseudo-code phase in the signals transmitted by different satellites is also synchronized. Although the pseudo-codes of different satellite modulations are different, at the same time (GPST or BDT), all pseudo-code phases are the same when the navigation signals of different satellites leave the transmitting antenna of the satellite.

The pseudo-code phase synchronization here is only for the same satellite navigation system. The pseudo-code phases of one GPS satellite and another GPS satellite are synchronized, as are the pseudo-code phases of one BeiDou satellite and another BeiDou satellite. However, the pseudo-code phases of a GPS satellite and a BeiDou satellite are not necessarily synchronized because of the inherent systemic bias between GPST and BDT. However, this strict phase synchronization relationship no longer exists when they are received by the receiver antenna. This is explained in Fig. 3.35.

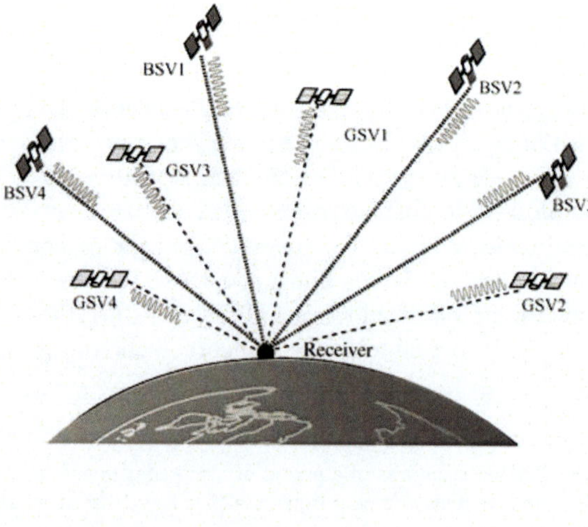

Fig. 3.35 Navigation signals from different satellites have different path delays

Figure 3.35 shows the situation of four GPS satellites and four BeiDou satellites. Four satellites were chosen because this is the minimum required to achieve three-dimensional positioning. Of course, if a joint solution is used, a total of eight satellite signals are available here. Since each satellite is located at a different location in space, the distance from the receiver antenna is different. It is assumed that the four GPS satellites are γ_{Gi}, $i = 1, 2, 3, 4$ from the receiver antenna, and four BeiDou satellite distance receivers. The antennae are γ_{Bi}, $i = 1, 2, 3, 4$ respectively, so the transmission time of the navigation signal from the satellite to the receiver is:

$$\begin{aligned} \text{GPS satellite: } \tau_{Gi} &= \gamma_{Gi}/c, i = 1, 2, 3, 4 \\ \text{BD satellite: } \tau_{Gi} &= \gamma_{Bi}/c, i = 1, 2, 3, 4 \end{aligned} \tag{3.36}$$

c is the speed of light. τ_{Gi} and τ_{Bi} are different here.

If at time t, the receiver antenna receives signals from the eight satellites, then the received signal is actually transmitted eight times before the current time. These eight moments are respectively:

$$\begin{aligned} t_{GSV,i} &= t - \tau_{Gi} \\ t_{BSV,i} &= t - \tau_{Bi} \end{aligned} \tag{3.37}$$

The subscripts BSV and GSV represent the BeiDou satellite and the GPS satellite respectively: $i = 1, 2, 3, 4$.

Since τ_i is different, $t_{SV,i}$ will also become jagged, so the navigation signals of different satellites at the antenna end of the receiver no longer maintain a strict time synchronization relationship, but show a difference between satellite signals. The phase relationship reflects the user's location information. At the same time, only when all satellite signals are strictly synchronized at the transmitting end does the seemingly chaotic signal phase at the receiving end make sense. Otherwise, if the phase of the signal is already randomly distributed at the transmitting end, then the phase at the receiving end has no meaning at all. The task of the receiver is to obtain the user's position information using a certain algorithm based on these seemingly chaotic signal phases ($t_{SV,i}$). This process is shown in Fig. 3.36.

Figure 3.36a shows the situation when four GPS satellites and four BeiDou satellites transmit signals. The upper part of Fig. 3.36a is the signal transmitted by the GPS satellite, and the lower part is the signal transmitted by the BeiDou satellite. Both the GPS and BeiDou satellites transmit the sub-frame 1–5 signal. The BeiDou IGSO/MEO satellite signal (D1 code) is used here for convenience. The sub-frames of the four GPS satellites are time-synchronized. The rising edge of sub-frame 1 of the GPS satellite is represented by Z-number (n), and the sub-frames of the four BeiDou satellites are also time-synchronized. The rising edge timing of sub-frame 1 of the satellite is represented by SOW(n). The figure shows a systematic time deviation between the GPS satellite signal and the rising edge of the sub-frame of the BeiDou satellite signal. The figure is represented by T. This time deviation is the systematic deviation between GPST and BDT. It is estimated as a system state

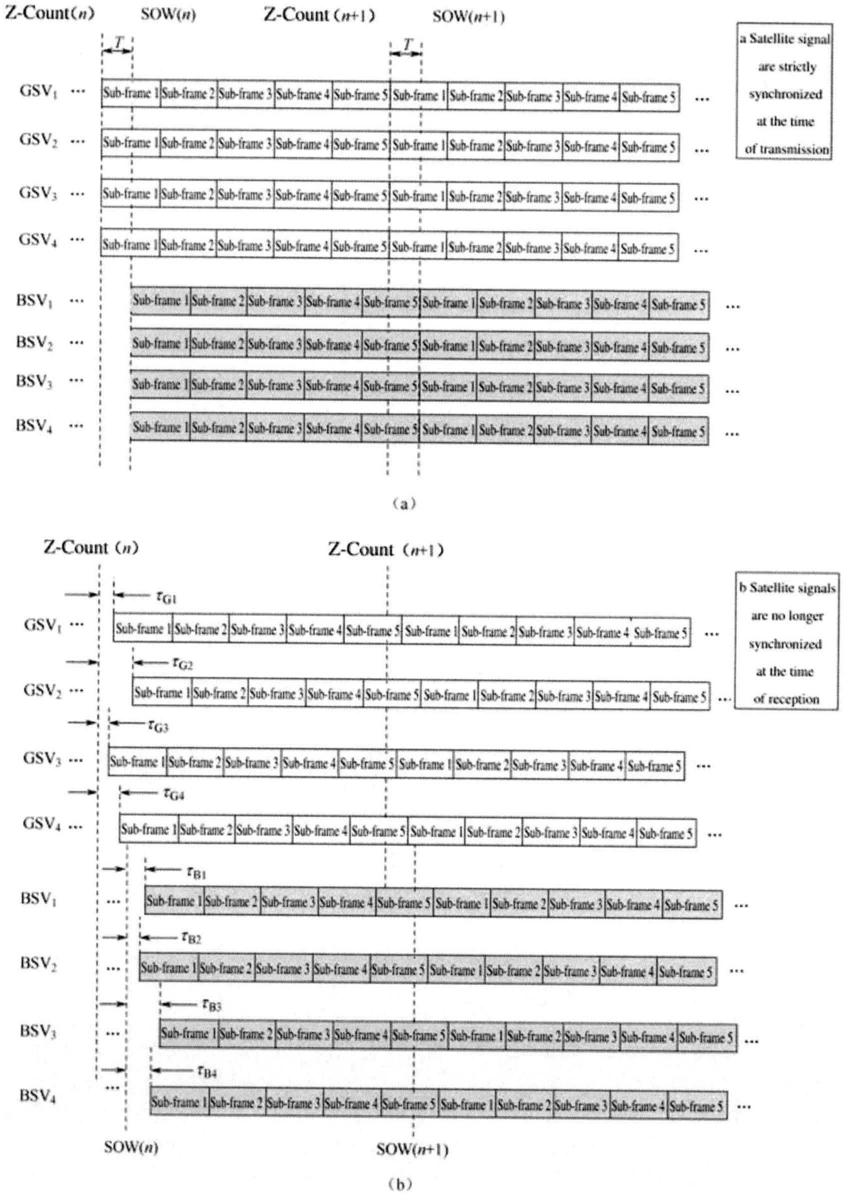

Fig. 3.36 Time relationships among different GPS and BDS satellite signals at the time of transmission and reception

quantity, which will be explained in detail later. The pseudo-code phase of the signal is not shown in the figure, but within each data bit, the pseudo-random code phases in different satellite signals of the same satellite system are synchronized.

Figure 3.36b shows the receiver antenna receiving information. The dotted line in the lower half of Fig. 3.36b indicates that the Z-number (n) and SOW(n) moments are aligned with the corresponding moments of the upper half. This can be understood as observing the signal transmission timing and signal reception from the same time frame. It can be seen that the sub-frame synchronization relationship among the four GPS satellites and the four BeiDou satellite signals has been disturbed at the time of signal reception. Some satellites have more postponement and others have less. A signal with a long postponement indicates that the satellite is further away from the receiver antenna, and vice versa.

The strict synchronization relationship between different satellite signals is the premise through which satellite navigation systems can achieve positioning. Figure 3.36 shows that only the signals from different satellites maintain a strict synchronization relationship at the time of transmission, and the transmission delay τ_i of different satellite signals becomes meaningful. Different transmission delays and the distance between the satellite antenna and the receiver antenna have a strict linear correlation. This is also a fundamental element of the premise of pseudo-range observation and the working principle of GPS and BeiDou receivers. Only by grasping this premise can we understand the physical meaning behind the pseudo-range equation.

References

1. Navstar GPS Space Segment/Navigation User Interfaces. IS-GPS-200G, 5 Sept 2012
2. Navstar GPS Space Segment/User Segment L5 Interfaces. IS-GPS-705, Rev A, 8 June 2010
3. Navstar GPS Space Segment/User Segment L1C Interfaces. IS-GPS-800, Rev A, 8 June 2010
4. Gold R (1968) Maximum recursive sequences with 3-valued recursive cross correlation functions. IEEE Trans Inf Theor 14:154–156
5. China Satellite Navigation System Management Office. Beidou Satellite Navigation System Space Signal Interface Control Document (Open Service Signal) Version 2.0. Dec 2013
6. Gibbons G (2008) Compass in the rearview mirror. Inside GNSS, Feb 2008. pp 62–63
7. Grelier T, Dantepal J, Delatour A, Ghion A, Ries L (2007) Initial Observations and Analysis of Compass MEO Satellite Signals. Inside GNSS, May/June 2007. pp 39–43
8. Grelier T, Dantepal J, Ries L, Delatour A, Issler J-L, Avila-Rodriguez JA, Wallner S, Hein GW (2007) Compass signal structure and first measurements, ION GNSS 2007, Sept 2007, Fort Worth, TX
9. Xingxin G, Chen A, Lo S, De Lorenzo D, Enge P (2007) The compass MEO satellite codes. Inside GNSS, July/Aug 2007. pp 36–43
10. Xingxin G, Chen A, Lo S, De Lorenzo D, Enge P (2008) Compass-M1 broadcast codes and their application to acquisition and tracking. In: Proceedings of the ION National Technical Meeting 2008, San Diego
11. Haixia W, Jiaoyan S (2008) An analysis of the characteristics of truncated balanced gold codes. Inf Technol 32(7):47–51
12. Huang J, Shi Z, Bao Z (2006) A statistical analysis of truncated balanced gold codes. Syst Eng Electron Technol 28(5)
13. Ashby N (2003) Relativity in the global positioning system. Living Rev Relativity 6

14. Kouba J (2002) Relativistic time transformation in GPS. GPS Solutions 5(4)
15. Da D, Yang Y (2005) Theoretical study on relativistic effect error of Beidou navigation satellite clock. In: National time and frequency symposium, Xi'an, 1 Nov 2005
16. Tsui JB-Y (2008) Fundamentals of global positioning receivers: a software approach, 2nd edn. Wiley
17. Kaplan ED (1996) Understanding GPS principles and applications. Artech House Publishers
18. Misra P, Enge P (2008) Global positioning system—sSignal, measurement and performance (Second Edition). Electronic Industry Press, Beijing
19. China Satellite Navigation System Management Office. Beidou Satellite Navigation System Open Service Performance Specification Version 1.0. Dec 2013
20. China Satellite Navigation System Management Office. Beidou Satellite Navigation System Development Report V2.2. Dec 2013
21. Gold R. Optimal binary sequences for spread spectrum multiplexing. IEEE Trans Inf Theor IT-13:619–621
22. Holmes J (1990) Coherent spread spectrum systems. Krieger Publishing Company, Malabar, Florida
23. Spiker J (1996) GPS signal structure and theoretical performance. In: Parkinson B, Spiker J, Axelrad P, Enge P (eds) Global positioning system: theory and applications, vol I
24. Baojun F (2007) Applications of relativity theory in modern navigation. National Defense Industry Press, Beijing

Chapter 4
Capture and Tracking of the Signal

The GPS and BDS satellite signals are down-converted from the radio frequency to the intermediate frequency through the RF front-end. In order to demodulate the navigation message, it must undergo a series of signal processing operations such as carrier tracking, pseudo-code tracking, bit synchronization, and sub-frame synchronization. In some literature, carrier tracking and pseudo-code tracking are called carrier synchronization and pseudo-code synchronization respectively. They are the same process but with different names. In this process, signal acquisition can be seen as the first step in signal processing within a GPS receiver. Only after signal acquisition is completed is it possible to start signal tracking. Signal tracking includes carrier tracking and pseudo-code tracking, performing carrier tracking and pseudo-code tracking respectively. After completion of carrier and pseudo-code stripping, the signal is the satellite data "bit", which is the navigation message. Then, signal demodulation, extracting pseudo-range observations, extracting carrier phase and Doppler observations, and other baseband signal processing can be performed.

All code division multiple access (CDMA) systems have signal acquisition problems, because multiple signals share the same carrier frequency. They only distinguish each other by orthogonal coding, so the signal existing in the received signal must first be judged. Further, the approximate carrier frequency and pseudo-code phase of the signal are searched. BDS and GPS systems are no exceptions as typical code division multiple access systems. Moreover, the high-speed movement of space satellites makes signal acquisition more complicated and difficult. The direct reason for this is the high-speed change of the relative distance between the satellite and the receiver. The Doppler shift caused by the carrier frequency changes drastically. "Time To First Fix (TTFF)" measures the time it takes for the receiver to power up to achieve effective positioning. Generally, without auxiliary information, the GPS receiver spends most of its time on signal acquisition in all TTFF sessions. Hence, increasing the speed of signal acquisition is significant for shortening TTFF.

Y. Lu, *BDS/GPS Dual-Mode Software Receiver*, Navigation: Science and Technology 10, https://doi.org/10.1007/978-981-16-1075-2_4

After the GPS and BDS receivers have completed the signal acquisition, they have a general estimate of the carrier frequency and pseudo-code phase of the signal. The term "general" is used here to describe the result of signal acquisition compared to the outcome of the tracking loop. In general, based on the result of the signal acquisition, the accuracy of the estimation of the carrier frequency is in the range of a few hundred hertz to several tens of hertz, and the estimation accuracy of the pseudo-code phase is within ± 0.5 chips. This accuracy is not sufficient to achieve demodulation of the navigation message data, since the demodulated data must generally be entered after going into a stable phase locked state, where phase locking refers to both the carrier phase and the pseudo-code phase that is being kept locked. At the same time, with the relative motion of the satellite and the receiver, the carrier frequency and pseudo-code phase of the signal received by the antenna can still change in the moment. What's more, the clock drift and random jitter of the receiver's local clock can also affect the locking of the acquired signal. Therefore, if the carrier tracking loop and the pseudo-code tracking loop are not continuously dynamically adjusted, the acquired signal will quickly become unlocked. Essentially, signal tracking is a dynamic adjustment strategy for loop parameters, which is taken to achieve stable tracking of the signal.

In this chapter, we first lay out the basic concepts of signal acquisition and why the receiver should perform it. The purpose of signal acquisition is explained. Then, several commonly used methods of signal acquisition are detailed, including hardware correlators, matched filters, and software FFT based signal acquisition methods. Two parallel acquisition schemes are introduced for the requirements of high sensitivity receivers: the combination of matched filters and FFT, and the fast acquisition method using phase compensation and synchronous data blocks, carrying out performance analysis and computational analysis of the two schemes. Based on the signal acquisition, the basic principle, theoretical model, and performance analysis of the signal tracking loop are explained in detail. The linear phase-locked loop, frequency-locked loop, phase detector, and discriminator are introduced, as well as the GPS and BDS receivers. In accordance with the characteristics of the hardware, the implementation method of the Costas loop will be explained. Furthermore, we process the actual data for different tracking loops, and analyze and compare the results.

4.1 Signal Acquisition

4.1.1 The Concept of Signal Acquisition

In CDMA systems, signals transmitted by different sources are distinguished by different pseudo-random codes. In general, CDMA systems can share the same carrier frequency and time. Signal acquisition problems exist in all CDMA systems. There are several reasons for this.

Reason 1: Since all sources of the CDMA system share the same carrier frequency and channel time, signals from all possible sources are inevitably mixed together at the antenna of the receiver. When the receiver is power on, it has no idea which signal source the current antenna is receiving from. This problem is especially significant in satellite navigation receiver systems because the receiver can only receive satellite signals that fall within the Line-of-Sight of the antenna. It is impossible for the receiver to receive when the satellite is running to the back of the Earth. Only after knowing the source of the currently received signals can the receiver track and decode them. In this sense, signal acquisition is a step that must be completed by GPS and BDS receivers before subsequent signal processing.

Reason 2: The introduction of pseudo-random codes broadens the spectrum of the signal, and accordingly the power of the signal can be reduced to a very low level. In the case of GPS and BDS signals, because the distance between the satellite and the ground exceeds 20,000 km, a huge path loss occurs. In layman's terms, the received signal level is often much lower than the background noise level. That is, the signal is "annihilated" by noise. In this case, weak signals must be extracted from the noise by signal capture.

Reason 3: According to the characteristics of CDMA signals, strong autocorrelation of pseudo-random codes must be used to achieve signal tracking and data demodulation. However, the premise is that the correct pseudo-random code phase must be found before the strong autocorrelation function can be used, and the randomness of the receiver power-on period determines the phase randomness of the received signal. Therefore, the pseudo-random code phase of a signal must be informed by signal acquisition. In fact, this complements the first reason: only after the correct pseudo-code phase is found and the spike of the autocorrelation function is obtained can the pseudo-code modulated signal contained in the received signal be known.

For GPS and BDS signals, there is a particular reason why signal acquisition becomes more necessary and more complex, namely the Doppler effect.

Considering the motion of satellites, we first take GPS satellites as an example to arrive at some general conclusions, and then apply these conclusions to the BDS satellite. As described in the previous section, the GPS satellite is a medium-orbiting satellite that moves at a constant time in space. We can roughly estimate the speed of the satellite's movement. The orbit of the satellite is near-circular. Here we approximate it as a circle without introducing too great an error (after all, the eccentricity of the GPS satellite orbit is only about 0.01), so the satellite trajectory can be considered as a circle with a radius of 26,560 km. The cycle of the satellite is about 11 h and 58 min, or 43,080 s, so we can estimate its average angular velocity:

$$\omega = 2\pi/43{,}080 \approx 1.4585 \times 10^{-4}\ \text{rad/s}$$

It should be pointed out here that, strictly speaking, the motion of the satellite is not in uniform velocity. According to Kepler's second law, the satellite is the fastest in the perigee, but its orbit is near-circular, the difference is not very great. Then the estimated average satellite speed is:

$$V_s = r \cdot \omega = 2.656 \times 10^7 \times 1.4585 \times 10^{-4} \approx 3.87 \tag{4.1}$$

Due to the Doppler effect, this high-speed motion will inevitably cause a Doppler shift in the signal obtained by the receiver. Simple estimation shows that the frequency shift caused by the signal of the L1 carrier frequency can reach

$$\delta f_{L_1} = \frac{\overline{v}}{c} f_{L_1} \approx 20.3 \text{ kHz} \tag{4.2}$$

In fact, the Doppler shift is only related to the radial velocity component of the relative motion. Assuming that the receiver is in a stationary state on the Earth's surface, a Doppler shift of up to 20 kHz must be generated by the satellite facing the Earth's surface where the receiver located. In general, the satellite always travels around the Earth, so the radial velocity component of its relative motion cannot reach 3.87 km/s relative to a certain point on the Earth's surface. Figure 4.1 offers a quantitative analysis.

The position of the GPS satellite in Fig. 4.1 is denoted by P, the position of the receiver is denoted by A, the flight path of the satellite is indicated by the dotted line circle, and the point O is the center of mass of the Earth, where the satellite's orbit is approximated as circular, and the center of the circle is at point O. The radius of the Earth (that is, the distance of AO is R_e) and the distance of the satellite's orbit from point O (that is, PO) is represented by R_s. Apparently $R_e \approx 6378$ km, and $R_s \approx 26{,}560$ km. Suppose that the angle between the satellite and the receiver's connection AP and OP is θ, and the radial speed between the receiver and the user when the receiver is at rest is

$$V_d = V_s \sin(\theta) \tag{4.3}$$

Applying the cosine theorem and the sine theorem to the triangle AOP, there is

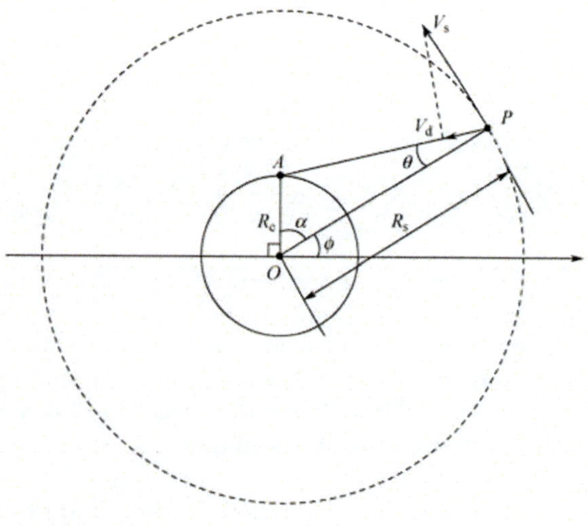

Fig. 4.1 The relationship between the Doppler shift and the observation angle of GPS satellites

$$\frac{R_e}{\sin(\theta)} = \frac{|AP|}{\sin(\alpha)} \tag{4.4}$$

$$|AP| = \sqrt{R_e^2 + R_s^2 - 2R_e R_s \cos(\alpha)} \tag{4.5}$$

Substituting Eq. (4.5) into Eq. (4.4) yields an expression of $\sin(\theta)$:

$$\sin(\theta) = \frac{R_e \sin(\alpha)}{\sqrt{R_e^2 + R_s^2 - 2R_e R_s \cos(\alpha)}} \tag{4.6}$$

Then the equation of the radial velocity V_d is:

$$V_d = \frac{V_s R_e \sin(\alpha)}{\sqrt{R_e^2 + R_s^2 - 2R_e R_s \cos(\alpha)}} \tag{4.7}$$

In Eq. (4.7), V_s, R_s and R_e can all be regarded as constants. Only the angle of α changes with the position of the satellite. What we care about is when V_d can obtain the maximum value, and what that maximum value is. Then, taking the derivative of α for Eq. (4.7) and letting it be 0, we get

$$\frac{dV_d}{d\alpha} = \frac{V_s R_e[\cos(\alpha)(R_e^2 + R_s^2) - R_e R_s \cos^2(\alpha) - R_e R_s]}{[R_e^2 + R_s^2 - 2R_e R_s \cos(\alpha)]^{3/2}} = 0 \tag{4.8}$$

Solving Eq. (4.8) we see that when $\cos(\alpha) = \frac{R_e}{R_s}$, V_d takes the maximum value $\frac{V_s R_e}{R_s}$. If you look more carefully at Fig. 4.1, *AP* and *OA* are at right angles at this point. That is, the satellite is at the moment when the horizon rises from point A. The maximum value of V_d at this time:

$$V_{dM} = \frac{V_s R_e}{R_s} \approx \frac{3870 \times 6378}{26{,}560} \approx 929.3 \text{ m/s} \tag{4.9}$$

Tsui makes a similar analysis of this problem in Ref. [1], concluding that the maximum radial velocity component of the relative motion between the receiver and the satellite on the Earth's surface is about 929 m/s. This outcome is consistent with the outcome of (4.9). The resulting maximum Doppler shift is

$$f_{dM} = V_{dM}\frac{f_{L1}}{c} \approx \frac{929.3 \times 1575.42 \times 10^6}{3 \times 10^8} \approx 4880.3 \text{ Hz} \tag{4.10}$$

Here, c is the speed of light and f_{L1} is the carrier frequency of L1. The above analysis concludes that the maximum Doppler shift that GPS satellite signals can produce for stationary receivers on the Earth's surface often does not meet the above assumptions. First of all, the receiver cannot be in a static state. Regardless of the motion dynamics of the receiver carrier itself, an object that is stationary on the surface of the Earth is always in motion as the Earth rotates. Secondly, only the receiver antenna position is in the orbit plane of the satellite in the above analysis, otherwise the velocity in the radial direction between the satellite and the receiver is not simply $V_s \sin(\theta)$, but rather the multiplicative sinusoidal component of the angle between the vector of the satellite receiver and the satellite's orbital plane. However, Eq. (4.10) can still be considered as a good reference value for the upper limit of the Doppler shift obtained by receivers on the Earth's surface, because the flight speed of satellites is the main factor of the Doppler shift in general civilian applications.

The above analysis can also be applied to the L2 frequency, but since the carrier frequency value of the L2 frequency is not as high as L1, the Doppler effect produced by the same radial velocity is not as obvious as the L1 frequency, so it is omitted here. Readers can analyze it by themselves.

The situation for the BDS satellite is slightly more complicated. On the one hand, the BDS carrier frequency is different from that of GPS. On the other hand, because the BDS space constellation contains three kinds of satellites (GEO/IGSO/MEO), the orbital height of the various types of satellites is different. The flight speed and flight range are different, which results in different Doppler shift results for different types of BDS satellites, meaning that they need to be analyzed separately.

Figure 4.2 shows the projection of the trajectory of BDS GEO/IGSO/MEO satellites on the Earth's surface. The GEO satellite uses PRN1 as an example, the IGSO satellite uses PRN6 as an example, and the MEO satellite uses PRN11 as an example. The figure only shows one cycle. The satellite trajectory is 12 h and 55 min for the MEO satellite and 24 h for the GEO/IGSO satellite. The Doppler shift for MEO satellites can be similarly analyzed using GPS satellites, but GEO and IGSO satellites and MEOs are different. If the analysis is carried out according to the method for GPS satellites, then the orbital radius of GEO is 42,100 km, and the average speed is

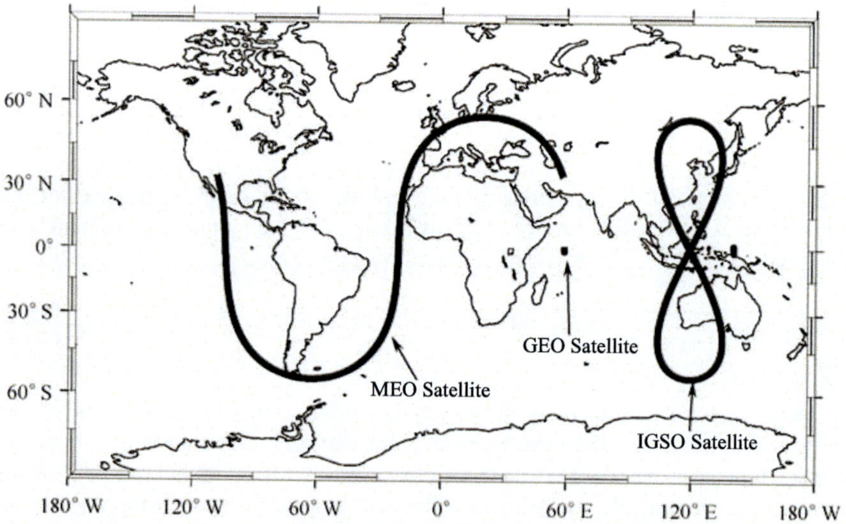

Fig. 4.2 The satellite track of the BDS space constellation GEO/IGSO/MEO

$$V_s = r \cdot \omega = 4.21 \times 10^7 \times 2\pi/(24 \times 3600) \approx 3.07\,\text{km/s}$$

However, the objects on the surface of the Earth also move with the Earth's rotation. The GEO satellites and the objects on the Earth's surface remain relatively static. Therefore, calculating the maximum Doppler shift of GEO for stationary receivers on the Earth's surface according to Eq. (4.9) is obviously unreasonable. The situation for the IGSO satellite is more complicated, as can be seen from its flight path, so it is even more impossible to calculate the maximum Doppler shift by Eq. (4.9).

In practice, the maximum satellite operating speed in a cycle can be calculated based on the GEO/IGSO satellites' orbital parameters broadcast by the BDS satellite, such as ephemeris data and almanac data. The calculated speed is the flight speed in the ECEF coordinating system. Then, the Doppler shift of the receiver is calculated relative to the Earth's surface. According to the satellite's actual orbital parameters, the maximum speed of the GEO satellite can be calculated to be between 80 and 90 m/s, and the maximum flight speed of the IGSO satellite is about 2800 m/s. Considering when the satellite appears in the horizon direction of the receiver, the maximum Doppler shift is generated, so there is

$$\text{GEO 卫星：}\ V_{\text{dM}} = \frac{V_{\text{s,GEO}} R_{\text{e}}}{R_{\text{s,GEO}}} \approx \frac{90 \times 6378}{42\,100} \approx 13.6\ \text{m/s} \tag{4.11}$$

$$\text{IGSO 卫星: } V_{\text{dM}} = \frac{V_{\text{s,IGSO}} R_{\text{e}}}{R_{\text{s,IGSO}}} \approx \frac{2800 \times 6378}{42100} \approx 424.2\ \text{m/s} \tag{4.12}$$

42,100 km is the orbital radius of the GEO and IGSO satellites. The maximum Doppler shifts of the BDS GEO and IGSO calculated at the B1 frequency according to Eqs. (4.11) and (4.12) are

$$f_{\text{dM,GEO}} = V_{\text{dM,GEO}} \frac{f_{\text{B1}}}{c} \approx \frac{13.6 \times 1561.098 \times 10^6}{3 \times 10^8} \approx 70.8\,\text{Hz} \tag{4.13}$$

$$f_{\text{dM,IGSO}} = V_{\text{dM,IGSO}} \frac{f_{\text{B1}}}{c} \approx \frac{424.2 \times 1561.098 \times 10^6}{3 \times 10^8} \approx 2207.4\,\text{Hz} \tag{4.14}$$

The BDS MEO satellite can be calculated using Eqs. (4.1) and (4.9).

$$V_{\text{s,MEO}} = r_{\text{MEO}} \cdot \omega_{\text{MEO}} = 2.791 \times 10^7 \times 1.3531 \times 10^{-4} \approx 3.78\,\text{km/s} \tag{4.15}$$

$$V_{\text{dM,MEO}} = \frac{V_{\text{s,MEO}} R_{\text{e}}}{R_{\text{s,MEO}}} \approx \frac{3780 \times 6378}{27{,}900} \approx 864.1\ \text{m/s} \tag{4.16}$$

$$f_{\text{dM,MEO}} = V_{\text{dM,MEO}} \frac{f_{\text{B1}}}{c} \approx \frac{864.1 \times 1561.098 \times 10^6}{3 \times 10^8} \approx 4496.5\,\text{Hz} \tag{4.17}$$

From the above analysis, it can be seen that for the BDS receiver on the Earth's surface, the maximum Doppler shifts produced by the GEO/IGSO/MEO satellites are approximately 70.8 Hz, 2207.4 Hz, and 4496.5 Hz respectively.

In addition to the Doppler shift due to the high-speed motion of the satellite, the deviation of the receiver's own RF clock crystal also shifts the received signal carrier frequency to a theoretical value. Theoretical calculations show that a crystal deviation of 1 ppm can cause a carrier frequency deviation of 1.57 kHz on the L1 carrier frequency and a carrier frequency deviation of 1.56 kHz at the B1 frequency. Therefore, in the design of the RF front-end of the receiver, the quality of the crystal's frequency stability is critical, as it not only affects the performance of the subsequent tracking loop, but also determines the search range of the signal capture. In the absence of a priori information, the carrier frequency search range of the satellite signal is determined by the Doppler shift and the crystal stability. The crystal oscillators used in the RF front-end of modern GPS receivers are generally temperature compensated crystal oscillators (TCXO), which have a frequency stability of ± 1 ppm or a higher crystal stability. If a ± 1 ppm crystal is used, the Doppler frequency search range for GPS and BDS GEO/IGSO/MEO satellites needs to add ± 1.57 kHz (GPS satellite) and ± 1.56 kHz (BDS satellite) in Eqs. (4.10), (4.13), (4.14), and (4.17).

Considering these factors comprehensively, it can be considered that the GPS and BDS signals finally obtained by the receiver antenna not only have an ambiguity in the pseudo-code phase, but also have a certain amount of ambiguity in the carrier frequency. In order to achieve stable tracking of the signal, it is necessary to solve the problem of pseudo-code phase ambiguity and carrier frequency ambiguity at the same time.

Signal capture in GPS and BDS receivers can be considered as a three-dimensional search. The first dimension is from the direction of the PRN code, the second dimension is the direction from the pseudo-code phase, and the third dimension is the direction from the Doppler shift.

From the direction of the PRN code, if there is no auxiliary information at the beginning of the receiver power-on, and there is no knowledge of the current zenith satellite constellation distribution, the number of PRN codes of GPS satellites that may exist in the signal received from the antenna is thirty-two. All possible BDS satellites have a PRN code of 37 (this range can be narrowed down to 14 BDS satellites before the BDS Global System is completed in 2020), and every possible PRN must be tried one by one. If there is other auxiliary information to reduce the search volume, the search time can be greatly shortened. For example, Warm-Start or Hot-Start involves the use of past almanac data or ephemeris data and local time saved by the receiver. In cases where the local approximate position is known, the current elevation of satellites can be roughly derived. The distribution of the GPS and BDS satellite constellations greatly limits the search space of the PRN code, thus shortening the time required for the search process.

From the search of the pseudo-code phase direction, it is first necessary to generate a local pseudo-code. This local pseudo-code is correlated with the input signal by setting different local pseudo-code phases. By using the strong autocorrelation of the pseudo-code, a strong correlation value can be generated only when the pseudo-code of the local code and the signal are aligned, and once the correlation value of a certain pseudo-code phase exceeds a predetermined threshold. It can be considered that the correct local pseudo-code phase is found. As can be seen in Sects. 3.1.3 and 3.2.3, the position of the autocorrelation spike is very sensitive to the pseudo-code phase. If the pseudo-code phase difference exceeds one chip, the spike will be quickly lost. Thus, once the high correlation peak appears, it can be considered that the difference between the pseudo-code phase of the input signal and the phase of the local pseudo-code is already within one chip.

The search from the Doppler shift direction is obtained by multiplying and integrating the value of the local carrier and adjusting the local carrier. If the carrier frequency of the local carrier and the input signal are very close, the input signal is obtained by coherent integration. The high frequency carrier component is removed, and the difference frequency component (i.e. the component close to zero frequency) is accumulated. Since the possible carrier frequency values of the input signal are not known in advance, it is necessary to try by setting different local carrier values. The selection of the frequency step is based on a compromise between search sensitivity and search efficiency: for a certain frequency search interval, the small frequency step increases the number of frequency bins to be searched, thereby directly increasing the coverage of all frequency search range. With the amount of work required, the result is increased search time, but the benefit is to increase the sensitivity of the search. The relationship between sensitivity and frequency step size is further explained in conjunction with the mathematical expressions in Sect. 4.1.2.

Figure 4.3 shows a two-dimensional search diagram for a given GPS signal with a PRN. The ordinate on the left represents the search range of the pseudo-code phase, and the abscissa represents the search range of the Doppler shift. The search range

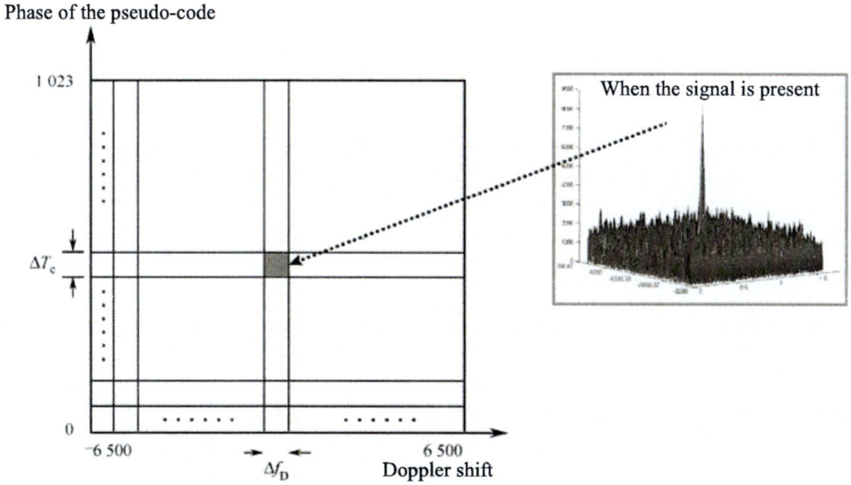

Fig. 4.3 Two-dimensional search graph for signal capture

of the pseudo-code phase is 1023 chips. Of course, if the BDS signal is searched, the search range of the pseudo-code phase should be 2046 chips, and the search step value of the pseudo-code phase in the figure is ΔT_c. The search range of the Doppler shift is set to [−6500 Hz, 6500 Hz], which can be considered as [(±5.0 kHz) + (±1.5 kHz)], where the two values of 5.0 and 1.5 kHz are analyzed before. The search step value of the Doppler shift in the figure is Δf_D. According to the previous analysis, the search step value ΔT_c of the pseudo-code phase should be smaller than a pseudo-code chip. The smaller the ΔT_c, the more refined the search result will be. However, this will cause increased computational complexity. The Doppler frequency shift search step Δf_D is related to the length of the coherent integration time. A simple conclusion is that the longer the coherent integration time, the smaller the Doppler search step value, otherwise the correlation peak is easily missed. When the local pseudo-code phase and Doppler shift are aligned with the pseudo-code phase of the input signal and the Doppler shift, the correlation peak shown on the right side of Fig. 4.3 will appear for the shape and statistical properties of the correlation peak. This analysis will be expanded in subsequent chapters.

In actual signal capture, the search of the pseudo-code phase and the search of the carrier frequency are actually done at the same time. This is because, if only the pseudo-code phase search is completed and the carrier component still exists, then the signal obtained by multiplying the input signal and the local pseudo-code after completion of the pseudo-code stripping is still a high frequency with respect to the integration time. The signal and correlation operation must pass through the integrator. The integration of the high frequency signal does not result in a very high peak. Conversely, if only the carrier frequency search is completed and the pseudo-code component still exists, then the input signal multiplied by the local carrier will become a low frequency component, but the existence of the pseudo-code means that the signal will still be a spread spectrum signal, so high correlation peaks will not appear after passing through the integrator. Only after the simultaneous completion of the pseudo-code stripping and carrier stripping will the obtained low-frequency continuous wave signal have a relatively high correlation peak through the integrator. Therefore, in the actual signal capture process, the local pseudo-code phase and the local carrier frequency are set at the same time. The correlation operation is then performed, and finally the correlation result is checked to determine whether signal capture has been realized. That is to say, carrier stripping and pseudo-code stripping are both indispensable.

4.1.2 *Signal Capture Based on the Time Domain Correlator*

The basic structure of the time domain correlator commonly used in GPS/BDS receivers is shown in Fig. 4.4.

The “time domain” here is relative to the FFT-based search algorithm that will be introduced later, because the FFT algorithm itself involves frequency domain and time domain conversion, and the correlator here only operates on the time domain.

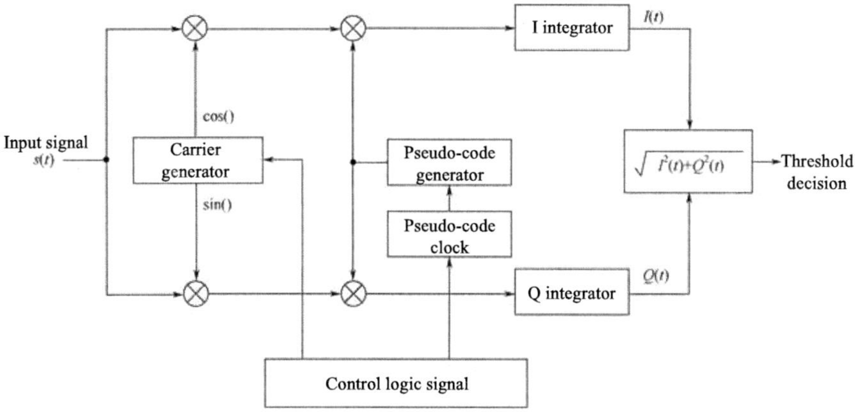

Fig. 4.4 The basic structure of the time domain correlator

The time domain correlator here can process GPS signals as well as BDS signals, and only needs to adjust the pseudo-code generator and pseudo-code clock to adapt to GPS and BDS signals. For convenience, if no special explanation is given, the subsequent explanation will be based on the GPS signal.

The time domain correlator includes a local carrier generator, a local pseudo-code generator, I-channel and Q-channel way multipliers, an integrator, and corresponding control circuits. The input signal is the ADC-sampled IF digital signal output from the RF front-end. It is first multiplied by the sin() and cos() components of the local carrier to obtain the I-channel and Q-channel components, and then correlated with the local pseudo-code at a certain pseudo-code phase respectively. The correlation operation is performed at the code phase, and finally the integration result is given by the integrator. The time of the integrator is an integer multiple of 1 ms, that is, an integer multiple of the C/A code period. The control circuit controls the frequency of the local carrier. At a fixed carrier frequency, the phase of the local pseudo-code is slid. The phase sliding range is from 1 to 1023 chips, and the step value of each pseudo-code sliding can be set to be less than or equal to the value of one chip. Generally half a chip is a good choice—a reasonable compromise between the required computation and the correlation peak's capture fineness. For each carrier frequency and pseudo-code phase, the I/Q correlator outputs the correlation result. If the correlation operation of all 1023 chips is completed at the current carrier frequency value and the spike exceeding the threshold has not been obtained, the current carrier frequency is changed, and then all 1023 pseudo-code phase searches are repeated. The steps described here are two-dimensional searches for a PRN code. When all the current possible carrier frequencies and pseudo-code phases are completed, the relevant peaks that do not meet the requirements are still present, indicating either that the currently received signal does not include the PRN pseudo-code signal, or that the signal corresponding to the PRN code is too weak. Therefore, the control logic needs to consider changing the current PRN code and re-searching all of the possible pseudo-code phases and Doppler shifts.

Mathematical formulas can be used to illustrate the process of signal acquisition and the relationship between capture sensitivity and efficiency.

The input signal is the IF sample of the RF output. Its mathematical expression is as follows.

$$s_{IF}(t) = \sqrt{2P_s}C(t-\tau)D(t-\tau)\cos[\omega_{IF}t + \phi(t)] + n(t) \tag{4.18}$$

In this equation, P_s is the signal power; $C(t)$ is the C/A code with a value of ± 1; τ represents the time delay brought about during transmission; $D(t)$ is the navigation message bit, as described above, of which the bit rate is 50 bps; ω_{IF} is the intermediate frequency carrier frequency, whose value is determined by the RF circuit and Doppler shift; $\varphi(t)$ is the initial carrier phase; $n(t)$ is white noise, and its power spectral density is considered constant, with $N_0/2$ as an indicator.

The local carrier generator outputs two signals, which are expressed respectively as

$$I_{ca}(t) = \sqrt{2}\cos[(\omega_{IF} + \Delta\hat{\omega})t + \hat{\phi}_0] \tag{4.19}$$

$$Q_{ca}(t) = \sqrt{2}\sin[(\omega_{IF} + \Delta\hat{\omega})t + \hat{\phi}_0] \tag{4.20}$$

The frequency of the local carrier is $((\omega_{IF} + \Delta\hat{\omega}))$, which is different from the carrier frequency of the input signal by $\Delta\hat{\omega}$, where $\Delta\hat{\omega}$ is unknown. One of the purposes of signal capture is to adjust the local carrier frequency so that $\Delta\hat{\omega}$ is as small as possible. Different frequency wells correspond to different $\Delta\hat{\omega}$. When the capture is achieved, the frequency difference is small, and when the receiver starts to stabilize the tracking signal, the frequency difference can be considered as close to zero.

The I/Q signals of the local carrier are respectively sent to the mixer and multiplied by the input signal. The results are

$$I(t) = \sqrt{2P_s}C(t-\tau)D(t-\tau)\cos[\omega_{IF}t + \phi(t)] \times \sqrt{2}\cos[(\omega_{IF} + \Delta\hat{\omega})t + \hat{\phi}_0] \tag{4.21}$$

$$Q(t) = \sqrt{2P_s}C(t-\tau)D(t-\tau)\cos[\omega_{IF}t + \phi(t)] \times \sqrt{2}\sin[(\omega_{IF} + \Delta\hat{\omega})t + \hat{\phi}_0] \tag{4.22}$$

Since there is an integrator behind the multiplier, it can be regarded as a low-pass filter, so the high-frequency components in the above equation can be ignored. The following simplified results are obtained:

$$I(t) = \sqrt{P_s}C(t-\tau)D(t-\tau)\cos[\phi(t) - \Delta\hat{\omega}t - \hat{\phi}_0] \tag{4.23}$$

$$Q(t) = \sqrt{P_s}C(t-\tau)D(t-\tau)\sin[\phi(t) - \Delta\hat{\omega}t - \hat{\phi}_0] \tag{4.24}$$

It should be noted that Eqs. (4.23) and (4.24) only represent the signal components of the mixer output. The noise components are listed separately. They are as follows:

$$n_{\mathrm{I}}(t) = n(t) \times \sqrt{2}\cos[\hat{\phi}(t)] \tag{4.25}$$

$$n_{\mathrm{Q}}(t) = n(t) \times \sqrt{2}\sin[\hat{\phi}(t)] \tag{4.26}$$

Its mean value $E(n_{\mathrm{I}}) = E(n_{\mathrm{Q}}) = 0$, and power spectral density $\mathrm{PSD}(n_{\mathrm{I}}) = \mathrm{PSD}(n_{\mathrm{Q}}) = N_0/2$. The noise amount of Eqs. (4.25) and (4.26) has a phase rotation compared with the original noise amount, but the power spectral density has not changed. For this conclusion, we only consider the case where white noise and $e^{\mathrm{j}\hat{\phi}(t)}$ are multiplied in the frequency domain to correspond to the translation of the spectrum, but the spectrum shape and distribution do not change at all.

Assuming that the local pseudo-code generated by the local pseudo-code generator can be expressed as $C(t-\hat{\tau})$, which is compared with Eq. (4.18), then the deviation of the pseudo-code phase of the local pseudo-code and the signal is $\Delta\tau = \tau - \hat{\tau}$.

Assuming that the integration time is T_{I}, the output of the I-channel and Q-channel integrators are

$$\begin{aligned}\overline{I}_{\mathrm{P}} &= \int_0^{T_{\mathrm{I}}} \sqrt{P_s}C(t-\tau)C(t-\hat{\tau})D(t-\tau)\cos[\Delta\hat{\omega}t + \delta\phi_0]\mathrm{d}t + \overline{N}_{\mathrm{I}} \\ &= D\sqrt{P_s}\int_0^{T_{\mathrm{I}}} C(t-\tau)C(t-\hat{\tau})\cos[\Delta\hat{\omega}t + \delta\phi_0]\mathrm{d}t + \overline{N}_{\mathrm{I}}\end{aligned} \tag{4.27}$$

$$\begin{aligned}\overline{Q}_{\mathrm{P}} &= \int_0^{T_{\mathrm{I}}} \sqrt{P_s}C(t-\tau)C(t-\hat{\tau})D(t-\tau)\sin[\Delta\hat{\omega}t + \delta\phi_0]\mathrm{d}t + \overline{N}_{\mathrm{Q}} \\ &= D\sqrt{P_s}\int_0^{T_{\mathrm{I}}} C(t-\tau)C(t-\hat{\tau})\sin[\Delta\hat{\omega}t + \delta\phi_0]\mathrm{d}t + \overline{N}_{\mathrm{Q}}\end{aligned} \tag{4.28}$$

In the above two equations, $\delta\phi_0 = \hat{\phi}_0 - \phi(t)$; T_{I} is the integration time, as described above—generally the C/A code period, which is an integer multiple of 1 ms; the bit rate of $D(t)$ is 50 bps, so $D(t)$ remains unchanged within 20 ms. Generally, $T_s <$ 20 ms is selected, so $D(t)$ can be considered constant within the integration time, and can be moved outside the integral number.

It is now possible to explain how to acquire the signal by adjusting the local pseudo-code phase and the local carrier frequency according to Eqs. (4.27) and

(4.28). In Eqs. (4.27) and (4.28), there are two variables $\hat{\tau}$ and $\Delta\hat{\omega}$. They are difficult to analyze, so we can fix one variable first and analyze the influence of the change of the other variable on the integrator result.

First, assuming that the local carrier frequency and the carrier frequency of $s(t)$ are exactly the same, that is, $\Delta\hat{\omega} = 0$, then Eqs. (4.27) and (4.28) become

$$\begin{aligned} \overline{I}_{\mathrm{P}} &= D\sqrt{P_{\mathrm{s}}}\cos[\delta\phi_0]\int_0^{T_{\mathrm{I}}} C(t-\tau)C(t-\hat{\tau})\mathrm{d}t + \overline{N}_{\mathrm{I}} \\ &= D\sqrt{P_{\mathrm{s}}}\cos[\delta\phi_0]R(\Delta\tau)T_{\mathrm{I}} + \overline{N}_{\mathrm{I}} \end{aligned} \tag{4.29}$$

$$\begin{aligned} \overline{Q}_{\mathrm{P}} &= D\sqrt{P_{\mathrm{s}}}\sin[\delta\phi_0]\int_0^{T_{\mathrm{I}}} C(t-\tau)C(t-\hat{\tau})\mathrm{d}t + \overline{N}_{\mathrm{Q}} \\ &= D\sqrt{P_{\mathrm{s}}}\sin[\delta\phi_0]R(\Delta\tau)T_{\mathrm{I}} + \overline{N}_{\mathrm{Q}} \end{aligned} \tag{4.30}$$

$R(\Delta\tau)$ in Eqs. (4.29) and (4.30) is a pseudo-code autocorrelation function as defined by Eq. (3.12). According to the analysis of the autocorrelation property of the PRN code, this result indicates that when the phase of the sliding local pseudo-code is consistent with the pseudo-code phase of the input signal, the integrator output reaches a maximum value; when the phase difference is greater than one C/A chip, the integrator output becomes a noise-like output.

Now suppose that the phase difference between the local pseudo-code and the input signal pseudo-code is 0, i.e. $\Delta\tau = 0$, we can find the effect of the change of the local carrier frequency on the output of the integrator. At this time, Eqs. (4.27) and (4.28) become

$$\begin{aligned} \overline{I}_{\mathrm{P}} &= D\sqrt{P_{\mathrm{s}}}\int_0^{T_{\mathrm{I}}} C^2(t-\tau)\cos[\Delta\hat{\omega}t + \delta\phi_0]\mathrm{d}t + \overline{N}_{\mathrm{I}} \\ &= D\sqrt{P_{\mathrm{s}}}T_{\mathrm{I}}\mathrm{sinc}\left(\frac{\Delta\hat{\omega}T_{\mathrm{I}}}{2}\right)\cos\left[\frac{\Delta\hat{\omega}}{2}T_{\mathrm{I}} + \delta\phi_0\right] + \overline{N}_{\mathrm{I}} \end{aligned} \tag{4.31}$$

$$\begin{aligned} \overline{Q}_{\mathrm{P}} &= D\sqrt{P_{\mathrm{s}}}\int_0^{T_{\mathrm{I}}} C^2(t-\tau)\sin[\Delta\hat{\omega}t + \delta\phi_0]\mathrm{d}t + \overline{N}_{\mathrm{Q}} \\ &= D\sqrt{P_{\mathrm{s}}}T_{\mathrm{I}}\mathrm{sinc}\left(\frac{\Delta\hat{\omega}T_{\mathrm{I}}}{2}\right)\sin\left[\frac{\Delta\hat{\omega}}{2}T_{\mathrm{I}} + \delta\phi_0\right] + \overline{N}_{\mathrm{Q}} \end{aligned} \tag{4.32}$$

If s_{IP} and s_{QP} are the signal parts of $\overline{I}_P$ and $\overline{Q}_P$ respectively, then the signal power represented by Eqs. (4.31) and (4.32) is

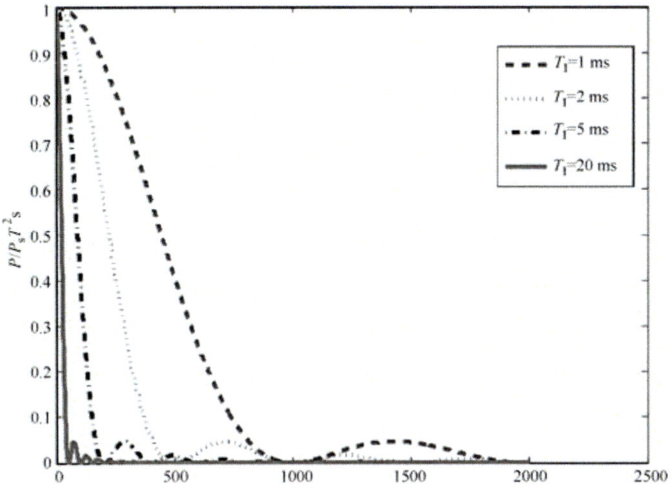

Fig. 4.5 The relationship between the peak and frequency error of normalized correlator under different coherent integration time lengths

$$\overline{P} = s_{\mathrm{IP}}^2 + s_{\mathrm{QP}}^2 = P_s T_{\mathrm{I}}^2 \mathrm{sinc}^2\left(\Delta\hat{\omega} T_{\mathrm{I}}/2\right) \tag{4.33}$$

The function sinc(x) behaves near the zero point: when $x \to 0$, then sinc(x) $\to 1$; the maximum value of the sinc(x) function is 1. It can then be concluded that when $\Delta\hat{\omega} \to 0$, the integrator outputs the maximum signal power. This is shown in Fig. 4.5.

Figure 4.5 shows the relationship between the result of normalizing the peak power value of the correlator and the frequency error under different coherent integration times. The figure shows four different integration times: 1, 2, 5, and 20 ms. In the figure, the abscissa is the carrier frequency error in Hertz, and the ordinate is the normalized signal power value of the output, which is the ratio of the power output of the correlator to the maximum power. The figure shows that for the integration time of 1 ms, the normalized power decreases as $\Delta\omega$ increases in the main lobe. When $\Delta\omega = 1000$ Hz, there is no signal power output; the maximum value of the first side lobes is around $\Delta\omega = 1430$ Hz, but the normalized side lobes have a maximum value of 0.047 and a decibel of -13 dB, indicating that the first side lobes and the main lobe peaks differ by about 13 dB. For the integration time of 20 ms, the corresponding frequency error is reduced to 50 Hz when the output power is close to zero for the first time, and the maximum value of the first side lobes is about $\Delta\omega = 72$ Hz. At this time, the normalized first side lobes are still about 13 dB away from the main lobe peak. The effect of the frequency step size of the search on the sensitivity can be understood easily from this figure. The longer the integration time, the faster the curve in the graph decays, which means that the signal power outputted by the integrator is more sensitive to the error of the carrier frequency. Therefore, the larger the frequency step, the larger the possible frequency error and the weaker the signal component of the correlator output, so the easier it is to miss the signal.

Figure 4.5 also shows that the longer the integration time, the narrower the width of the main lobe, which means that the same carrier frequency error is more likely to lose signal power in the case of long-term integration. Therefore, the longer the integration time T_I, the smaller the step size of the carrier frequency search Δf_D should be. In practice, the step size setting of the carrier frequency search must be considered in conjunction with the integration time. In general, $\Delta f_D = 1/T_I$ can be set. Of course, it can be set smaller, but this means that the number of frequency wells required to cover the same Doppler range is increased, thereby increasing the calculation.

From Eq. (4.33) we can also see the effect of the length of the integration time on the output of the integrator. The longer the T_I, the larger the value of $T_I \text{sinc}(\Delta\hat{\omega}T_I/2)$, which means that the stronger the signal outputted by the integrator, the worse the tolerance to the carrier frequency error, and the greater the possibility that a data bit jump occurs in the longer integration time. If the bit jump occurs during the integration time, the coherent integration result will be offset. The shorter the T_I, the output of the integrator, the weaker the rate is. However, T_I can be no shorter than one C/A code period, so 1 ms is the shortest integration time.

The correlation peaks in Fig. 4.5 have been normalized so that the effect of the change in integration time length on the correlation peak intensity of the integrator output is not seen. Figure 4.6 shows the relationship between the output peak and frequency deviation of the non-normalized correlator. The integration time is 1, 2, and 5 ms. The maximum correlation peak obtained by the integration time of 20 ms is too high. The details cannot be clearly displayed in one figure, so the correlation peak of the 20 ms integration time is not given in Fig. 4.6. It can be seen from the

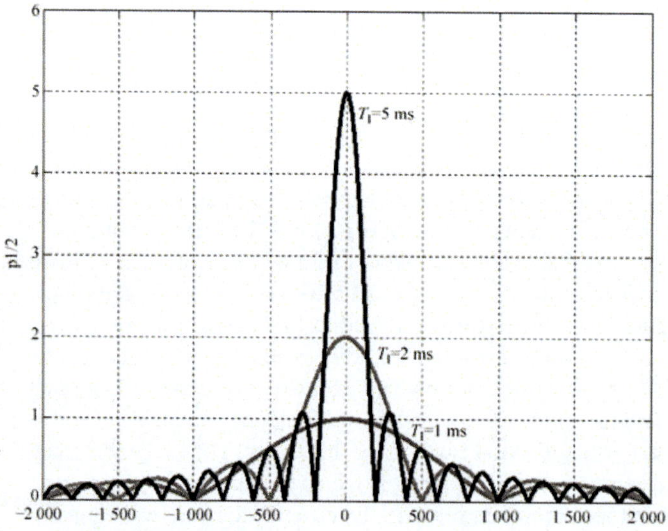

Fig. 4.6 The relationship between the peak and frequency error of non-normalized correlators with different coherent integration times

figure that the longer the coherent integration time, the larger the peak value of the correlator output; the main lobe width is narrowed, i.e. the frequency selectivity is stronger. This can be seen as two results of increasing the length of the coherent integration time. On the one hand, it is possible to obtain a stronger correlation peak, which is advantageous for finding a weaker signal; on the other hand, the tolerance to the carrier frequency error deteriorates, and the frequency search is performed. The step size must be shortened, otherwise it is prone to missed detection.

The term "coherent integration" can be more clearly explained by combining (4.31) and (4.32). The concept of coherence comes from the coherent wave in physics, which is defined as two waves with the same frequency, the same phase, and the same direction of propagation. What is the relationship between the coherent integration in the receiver and the coherent wave in physics? There is no strict uniform definition, but coherent integration means that the continuity of the carrier phase is maintained during the integration process. It is only necessary to observe the integration process of the I-channel and Q-channel for a longer period of time. The integral interval of Eqs. (4.31) and (4.32) is $[0, T_\mathrm{I}]$, and the change integration interval is $[T_\mathrm{I}, 2T_\mathrm{I}]$.

$$\begin{aligned}\overline{I}_{\mathrm{P},1} &= D\sqrt{P_\mathrm{s}}\int_{T_\mathrm{I}}^{2T_\mathrm{I}} C^2(t-\tau)\cos[\Delta\hat{\omega}t+\delta\phi_0]\mathrm{d}t+\overline{N}_\mathrm{I} \\ &= D\sqrt{P_\mathrm{s}}T_\mathrm{I}\mathrm{sinc}\left(\frac{\Delta\hat{\omega}T_\mathrm{I}}{2}\right)\cos\left[\frac{3\Delta\hat{\omega}}{2}T_\mathrm{I}+\delta\phi_0\right]+\overline{N}_\mathrm{I}\end{aligned} \tag{4.34}$$

$$\begin{aligned}\overline{Q}_{\mathrm{P},1} &= D\sqrt{P_\mathrm{s}}\int_{T_\mathrm{I}}^{2T_\mathrm{I}} C^2(t-\tau)\sin[\Delta\hat{\omega}t+\delta\phi_0]\mathrm{d}t+\overline{N}_\mathrm{Q} \\ &= D\sqrt{P_\mathrm{s}}T_\mathrm{I}\mathrm{sinc}\left(\frac{\Delta\hat{\omega}T_\mathrm{I}}{2}\right)\sin\left[\frac{3\Delta\hat{\omega}}{2}T_\mathrm{I}+\delta\phi_0\right]+\overline{N}_\mathrm{Q}\end{aligned} \tag{4.35}$$

If Eqs. (4.34) and (4.35) are written in the plural form, then

$$\overline{I}_{\mathrm{P},1}+\mathrm{i}\overline{Q}_{\mathrm{P},1} = D\sqrt{P_\mathrm{s}}T_\mathrm{I}\mathrm{sinc}\left(\frac{\Delta\hat{\omega}T_\mathrm{I}}{2}\right)e^{\mathrm{j}\left[\frac{3\Delta\hat{\omega}}{2}T_\mathrm{I}+\delta\phi_0\right]}+\overline{N}_{\mathrm{I,Q}}(T_\mathrm{I}) \tag{4.36}$$

If the integral interval is changed to $[2T_\mathrm{I}, 3T_\mathrm{I}]$, then

$$\overline{I}_{\mathrm{P},2}+\mathrm{i}\overline{Q}_{\mathrm{P},2} = D\sqrt{P_\mathrm{s}}T_\mathrm{I}\mathrm{sinc}\left(\frac{\Delta\hat{\omega}T_\mathrm{I}}{2}\right)e^{\mathrm{j}\left[\frac{5\Delta\hat{\omega}}{2}T_\mathrm{I}+\delta\phi_0\right]}+\overline{N}_{\mathrm{I,Q}}(2T_\mathrm{I}) \tag{4.37}$$

By analogy, the integration interval can be continuously changed, and a result similar to Eqs. (4.36) and (4.37) can be derived. The carrier phase in the equation changes with time at the angular frequency $\Delta\hat{\omega}$. It can be seen from Eqs. (4.36) and (4.37) that the I-channel and Q-channel integrals maintain the continuity of the carrier phase, but only if the modulo operation of Eq. (4.33) is performed.

$$\left|\overline{I}_{\mathrm{P},1} + \mathrm{i}\overline{Q}_{\mathrm{P},1}\right|^2 = P_{\mathrm{s}} D^2 T_{\mathrm{I}}^2 \mathrm{sinc}^2\left(\frac{\Delta\hat{\omega} T_{\mathrm{I}}}{2}\right),$$

$$\left|\overline{I}_{\mathrm{P},2} + \mathrm{i}\overline{Q}_{\mathrm{P},2}\right|^2 = P_{\mathrm{s}} D^2 T_{\mathrm{I}}^2 \mathrm{sinc}^2\left(\frac{\Delta\hat{\omega} T_{\mathrm{I}}}{2}\right)$$

It can be seen that the continuity of the carrier phase no longer exists, leaving only the amplitude information, as shown in Fig. 4.1. It should be noted that in order to keep the expression concise, the noise term does not appear in the above equation.

In cases where the local carrier phase and the carrier phase of the input signal are strictly synchronized, the real part of the signal $\overline{I}_{\mathrm{P},k}$ contains all the signal energy, and the imaginary part $\overline{Q}_{\mathrm{P},k}$ is 0 or only the noise term, which is not required at this time. For the modulo operation, only the $\overline{I}_{\mathrm{P},k}$ threshold decision can be made. However, in cases where the frequency difference or phase difference cannot be strictly synchronized, the modulo operation is necessary. At this time, the single-channel $\overline{I}_{\mathrm{P},k}$ or $\overline{Q}_{\mathrm{P},k}$ will rotate with the carrier phase difference.

Therefore, the signal energy value cannot be directly obtained from the modulo operation, and the carrier signal phase value is removed. The measured signal energy value is independent of the carrier phase difference, so the threshold decision can be stably performed.

In the left half of Fig. 4.7, only $\overline{I}_{\mathrm{P},k} + i\overline{Q}_{\mathrm{P},k}$, k = 0, 1, … in the complex plane; the right half is the I-channel and Q-channel results of the integral in the complex plane after the modulo operation. If only coherent integration is performed, then $\overline{I}_{\mathrm{P},k} + i\overline{Q}_{\mathrm{P},k}$ rotates at a certain angle, where the angle represents its carrier phase. This means that the phase information can be clearly seen, and the result after the modulo operation no longer has phase information.

So far, we have analyzed the influence of the pseudo-code phase error and carrier frequency error on the signal power outputted by the correlator. It can also be seen that the process of signal acquisition is to constantly adjust the local carrier frequency and pseudo-code phase until the process of the pseudo-code phase difference and carrier frequency difference are both zero. Of course, an error of 0 is an ideal state. In fact, as long as the threshold is less than a certain value, the integrator outputs a

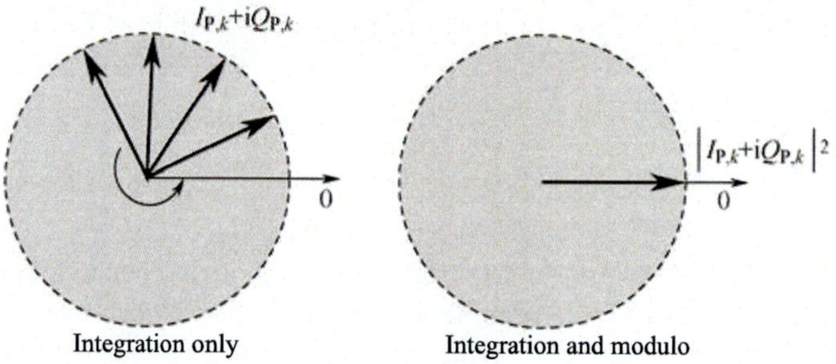

Fig. 4.7 Coherent integration and carrier phase continuity

strong correlation result when the signal is present and the signal strength is sufficient, thereby transferring control to the tracking loop.

Although the process by which the time domain correlator acquires the signal is not difficult to understand, it reveals the principle of signal acquisition. In particular, Figs. 4.5 and 4.6 illustrate the integration time length and frequency error versus the correlation peak. The effects of several signal acquisition methods in subsequent chapters are specific, but some are different in specific implementations. The mathematical principles behind them and the final conclusions are similar to the time domain correlators, which can be understood in this section.

The shortcomings of the time domain correlator are also obvious. It can be seen from the structure that for a combination of carrier frequency and pseudo-code phase, the time for completing one integration is fixed, that is, all possible pseudo-code phases are completed in the series. Combined with carrier frequency, the time of each integral operation is determined by the coherent integration time T_I. Assuming that there are N frequency points and M pseudo-code phases to search, the total time overhead required is $N \times M \times T_I$, and it can be seen that all frequency and phase combinations are completed. The search time is also fixed, which limits the speed of the search, especially if the acquisition time of the weak signal takes too long. Therefore, the acquisition engine of modern receivers often no longer uses the time domain correlator for capture. One solution that can improve search speed is to increase the number of correlators, and use multiple correlators to process in parallel at the same time. This is to reduce the time overhead by increasing the use of hardware resources to increase the search speed. Parallel correlators can also be added to achieve serial hybrid search processing [2, 3], but their degree of parallelism is limited. At present, another improved method is to store the digital samples of the intermediate frequency data, and then use the high-speed logic circuit to process the data with a very high clock, that is, a buffer-playback mode. The processing speed of this mode is determined by the clock speed. The high-speed clock can speed up the acquisition, but increase the complexity of the circuit, while the power consumption of the high-speed logic circuit is quite large.

4.1.3 Signal Acquisition Based on the Matched Filter

According to the signal system theory, the matched filter is a linear filter with the largest ratio of the instantaneous power of the signal to the average noise of the noise under white noise conditions. It is an optimal filter. The transfer function of the matched filter is the conjugate of the signal spectrum, so the matched filter must exactly match the spectrum of the input signal. Once the input signal changes, the original matched filter must change its transfer function to accommodate the new signal spectrum. In this sense, the matched filter can also be considered as related reception. The meaning of the two is the same, except that the matching filter emphasizes the representation in the frequency domain, and the related concept focuses on the expression of the time domain.

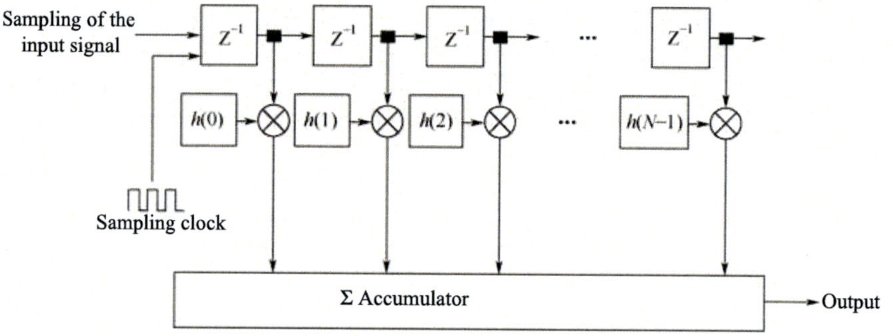

Fig. 4.8 Block diagram of baseband signal capture using matched filters

Therefore, it is easy to imagine the use of matched filters for GPS and BDS signal acquisition. The principle of capture using a matched filter is shown in Fig. 4.8. Considering that modern matched filters are often implemented on digital signals, discrete time domain representations are used here. The input sampled signal is subjected to the necessary down-conversion and down-sampling processing to obtain an intermediate frequency or baseband signal, which is then driven by the sampling clock and associated with the matched filter. Each drive clock produces a correlation result output. $h(0)$, $h(1)$, ..., $h(N)$ are matched filter coefficients. According to the matched filter theory, they should be arranged in reverse order of pseudo-code values, i.e. $C(N-1)$, $C(N-2)$, ..., $C(0)$, where C is the pseudo-code value and N is the number of sample points in a pseudo-code period. If the sampling clock is set to 1.023 MHz, with one sampling point in each pseudo-code chip, then $N = 1023$, and the output value of the matched filter of each sampling clock corresponds to the correlation result of a pseudo-code phase, and all 1023 pseudo-code phase correlation operations can be completed with 1 ms or 1023 sample clocks.

Assuming that the sample sequence of the input signal is represented by $s(n)$, the output of the matched filter can be expressed as

$$\mathrm{MF}_{\mathrm{out}}(i) = \sum_{k=0}^{N-1} s(i-k)h(k) = \sum_{k=0}^{N-1} s(i-k)C(N-1-k) \tag{4.38}$$

Equation (4.38) shows that the output of the matched filter is actually the correlation value between the input signal and the local pseudo-code signal. In physics it is the same as the time-domain correlator's sliding correlation method in Sect. 4.1.2, except that this method adjusts the phase of the local pseudo-code every time the coherent integration is performed. The phase of the local pseudo-code of the matched filter method is fixed, the phase of the input signal is sliding, and the output is at its maximum when the phases are identical.

Note: The block diagram shown in Fig. 4.8 is based on the baseband signal. If the input signal is an IF signal, the parameters of the matched filter must take the IF carrier component into account. In this case, the matched filter coefficients need

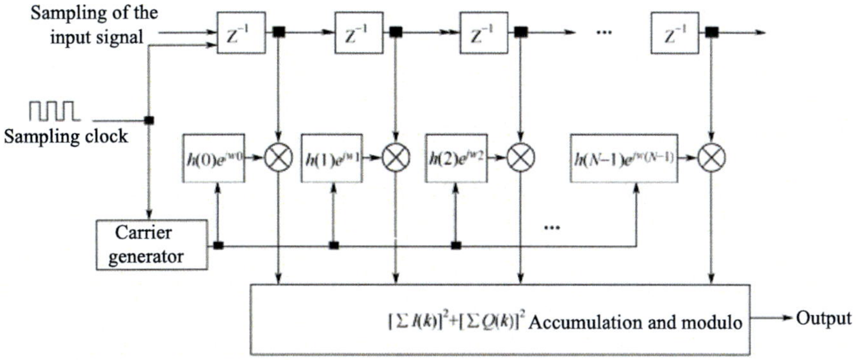

Fig. 4.9 Block diagram of IF signal capture using matched filters

two sets, corresponding to the I-channel and Q-channel components. Figure 4.9 is a block diagram of the matched filter capture when inputting IF sampled data. At this time, a carrier generator is used to generate the I-channel and Q-channel components of the Doppler shift signal. The signal is multiplied by the pseudo-code value as a matched filter. As the coefficient, the corresponding accumulator link increases the modulo operation. Each time a different Doppler shift value is set, the coefficients of the matched filter need to be changed, and the accumulated modulo output of each sample clock matched filter corresponds to the correlation output of different pseudo-code phases at the carrier frequency. When the input signal and the local pseudo-code are in phase, and the intermediate frequency carrier frequency is the same as the frequency of the local carrier generator, the accumulation and modulo unit outputs the maximum correlation peak.

Although the output of the matched filter and the physical meaning of the output of the time domain correlator are the same, in terms of implementation form, the matched filter can output the relevant result under the current pseudo-code phase when each sampling clock arrives. In this way, the correlation result of all pseudo-code phases can be completed within one pseudo-code period. In this sense, the matched filter is equivalent to 1023 parallel time domain correlators, realizing the pseudo-code domain (time domain) parallel search. In the dimension of the carrier Doppler frequency, serial processing is still needed, so parallel search in the carrier's Doppler shift dimension is not implemented.

The above analysis is for the GPS signal. For the BDS signal, the coefficient of the matched filter must correspond to the BDS pseudo-code. Assuming that the sampling clock is 2.046 MHz, one sampling point within each pseudo-code chip, then $N = 2046$. In this case, the chip result of the correlation result of the matched filter output is 1 pseudo-code chip. If the sampling clock is changed to 4.092 MHz, with two sampling points in each pseudo-code chip, $N = 4092$ at this time. The chip precision of the correlation result outputted by the matched filter is half pseudo-code chip. Chip precision increases, meaning that the computation needs to be doubled.

4.1.4 Signal Capture Based on FFT

The basic idea of the scheme is based on the relationship between DFT and signal convolution. The convolution of the real-time domain corresponds to the multiplication of the frequency domain, and the multiplication of the time domain corresponds to the convolution of the frequency domain.

$$s(t) \otimes h(t) \Leftrightarrow S(f) \cdot H(f) \tag{4.39}$$

$$S(f) \otimes H(f) \Leftrightarrow s(t) \cdot h(t) \tag{4.40}$$

$s(t)$ and $h(t)$ in Eqs. (4.39) and (4.40) correspond to two time-domain signals respectively. $S(f)$ and $H(f)$ correspond to their Fourier transform respectively, and $\otimes$ denotes a convolution operation.

If the result of multiplying the intermediate frequency signal $s(t)$ of the ADC by the in-phase component and the quadrature component of the local carrier generator is regarded as two intermediate variables $s_{\mathrm{I}}(t)$ and $s_{\mathrm{Q}}(t)$, that is:

$$s_{\mathrm{I}}(t) = s(t)\sqrt{2}\cos[(\omega_{\mathrm{IF}} + \Delta\hat{\omega})t] \tag{4.41}$$

$$s_{\mathrm{Q}}(t) = s(t)\sqrt{2}\sin[(\omega_{\mathrm{IF}} + \Delta\hat{\omega})t] \tag{4.42}$$

The output of the local pseudo-code generator is recorded as a local signal $s_{\mathrm{L}}(t) = C(t)$, and the correlation results of the integrator outputs of I-channel and Q-channel are

$$I(\tau) = \int_0^{T_{\mathrm{I}}} s_{\mathrm{I}}(t)s_{\mathrm{L}}(t+\tau)\mathrm{d}t \tag{4.43}$$

$$Q(\tau) = \int_0^{T_{\mathrm{I}}} s_{\mathrm{Q}}(t)s_{\mathrm{L}}(t+\tau)\mathrm{d}t \tag{4.44}$$

The convolution expression for any two time domain signals $x(t)$ and $y(t)$ is

$$Z(t) = \int_0^{T_{\mathrm{I}}} x(\tau)y(t-\tau)\mathrm{d}\tau \tag{4.45}$$

Comparing Eqs. (4.43), (4.44), and (4.45), it can be seen that the signal outputted by Eq. (4.45) and the integrator are very similar in mathematical expression. The only difference is the symbol of the function argument τ.

If we perform a Fourier transform on Eq. (4.43), we obtain

$$\begin{aligned}\mathrm{DFT}\{I\} &= \int_0^{T_I}\left[\int_0^{T_I} s_I(t)s_L(t+\tau)\mathrm{d}t\right]e^{-j2\pi f\tau}\mathrm{d}\tau \\ &= \int_0^{T_I} s_I(t)\left[\int_0^{T_I} s_L(t+\tau)e^{-j2\pi f(t+\tau)}\mathrm{d}\tau\right]e^{j2\pi ft}\mathrm{d}t \\ &= S_L(f)\int_0^{T_I} s_I(t)e^{j2\pi ft}\mathrm{d}t \\ &= S_L(f)S_I^*(f)\end{aligned} \tag{4.46}$$

Here, $S_L(f)$ and $S_I(f)$ are the Fourier transforms of $s_L(t)$ and $s_I(t)$ respectively, and $S_I^*(f)$ is the complex conjugate of $S_I(f)$. The same treatment of Eq. (4.44) can lead to a similar conclusion, that is

$$\mathrm{DFT}\{Q\} = S_L(f)S_Q^*(f) \tag{4.47}$$

Equations (4.46) and (4.47) can be rewritten as

$$I(\tau) = \mathrm{IDFT}\left[\mathrm{DFT}(s_L)\cdot\mathrm{DFT}^*(s_L)\right]$$

$$Q(\tau) = \mathrm{IDFT}\left[\mathrm{DFT}(s_L)\cdot\mathrm{DFT}^*\left(s_Q\right)\right]$$

The DFT* in the above equation represents the conjugate operation of the Fourier transform. Therefore, that the integral I-channel and Q-channel operations in the time domain correlator can be performed by applying DFT to the input signal and the local signal, and then multiplying the conjugate. The result is obtained by an IDFT inverse transform. Since the DFT and IDFT of the sequence have fast algorithms, the processing speed can be greatly improved compared with the conventional operation.

Most of the actual engineering practice deals with discrete sequences, that is

$$s_I(t)\text{ discretization} \Rightarrow s_I(0), s_I(1), \ldots, s_I(N-1)$$

$$s_Q(t)\text{ discretization} \Rightarrow s_Q(0), s_Q(1), \ldots, s_Q(N-1)$$

$$s_L(t)\text{ discretization} \Rightarrow s_L(0), s_L(1), \ldots, s_L(N-1)$$

The corresponding $I(\tau)$ and $Q(\tau)$ are discretized into $I(k)$ and $Q(k)$, $k = 0, 1, \ldots, N-1$

$$I(k) = \sum_{i=0}^{N-1} s_{\mathrm{L}}(i) s_{\mathrm{I}}(\lceil i + k \rceil_N) \tag{4.48}$$

$$Q(k) = \sum_{i=0}^{N-1} s_{\mathrm{L}}(i) s_{\mathrm{Q}}(\lceil i + k \rceil_N) \tag{4.49}$$

The operation $[\cdot]_{\mathrm{N}}$ in the above equation is to take the remainder of N, that is $\lceil i + k \rceil_{\mathrm{N}} = \mathrm{mod}(i + k, N)$.

The correlation operation here is of the cyclic kind, not linear; this is similar to the relationship between circular convolution and linear convolution. Because of this cyclic correlation, the length of the input and output sequences here is N. If it is linearly related, the length of the input sequence is N and the length of the output sequence will be $2N - 1$.

It can be seen from Eqs. (4.48) and (4.49) that the conventional method is used to recalculate N times the multiplication and addition operations every time k is changed, and traversing the complete part k requires a total of N^2 multiplication operations; while using FFT, a total of three times FFT and N-time multiplication and addition operation is required. If $N\mathrm{lb}(N)$ is assumed for the multiplication and addition operation required for one FFT, the total multiplication and addition operation amount is $[3\mathrm{lb}(N) + 1]N$, which can reduce the operation by an order of magnitude. In practice, the FFT of the local pseudo-code signal can be calculated and stored in advance, so that a part of the calculation can be reduced in real-time capture.

Using FFT to implement signal capture, the actual signal processing flow is briefly described below:

Step 1: Multiplying the input intermediate frequency signal and the in-phase and quadrature signals outputted by the local carrier generator, and obtaining a complex signal $I_{\mathrm{L}} + \mathrm{j}Q_{\mathrm{L}}$ of the baseband through the low-pass filter;
Step 2: Performing an FFT on the complex signal obtained in the first step;
Step 3: Performing FFT on the pseudo-code signal outputted by the local pseudo-code generator and taking a conjugate;
Step 4: Multiplying the results of the second and third steps and making the product into an IFFT transform;
Step 5: Modeling the IFFT result of the fourth step and applying a threshold to the result. If the appearance of the spike is strong enough, it indicates that signal capture has been achieved. The position corresponding to the peak corresponds to the pseudo-code phase. At this time, the frequency value of the local carrier is the carrier frequency where the signal is located. If there is not a strong enough spike, the local carrier is reset. The frequency of the device is the value of the next frequency well, repeating Step 1 to Step four.

Figure 4.10 is a schematic diagram of the signal processing flow for sampling this method.

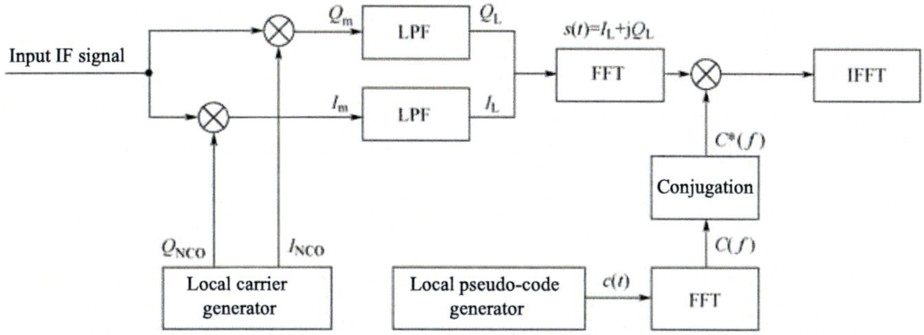

Fig. 4.10 Principle of signal capture based on the FFT algorithm

The above FFT method can obtain the correlation result of all N pseudo-code phases after completing one process (two FFTs and one IFFT), so parallel processing of the pseudo-code domain is realized, and the search of the Doppler shift dimension is still Serial, that is, each time the frequency of the local carrier generator is changed, a new process needs to be performed, namely two FFTs, one IFFT, and N times multiply and add operations.

The sampling rate of the points participating in the FFT operation determines the pseudo-code precision of the correlation peak result. The sampling frequency can be measured by the number of sampling points in the unit chip. This processing can better adapt to the different chip rates of GPS and BDS signals. For example, if the number of sampling points in a unit chip is one, the chip precision of the correlation peak obtained by the FFT is one chip; if the number of sampling points in the unit chip is two, the chip precision of the correlation peak is half. So, for a more accurate chip precision, it is necessary to increase the sample rate. At the same time, the operation will increase.

The length of the signal sequence involved in the FFT operation determines the Doppler shift accuracy of the result. If the signal length is T_I, the Doppler frequency resolution of the correlation peak is $1/T_I$ Hz, so it is determined according to the signal length when setting the frequency step value. The basic principle is that the step value is less than or equal to $1/T_I$ Hz. This is consistent with the conclusion in Fig. 4.6 of the time domain correlator. The signal length of the FFT operation is equivalent to the integral length of the time domain correlator, so the longer the signal length, the sharper the correlation peak shape, the higher the peak value and the frequency, and the more sensitive the error. Figure 4.11 shows the results of FFT capture of a GPS signal. Figure 4.11a is for processing 1 ms data, Fig. 4.11b is for 2 ms data, and Fig. 4.11c shows the processing of the 5 ms data. The frequency step values were distributed at 1 kHz, 500 Hz, and 200 Hz in the three cases, and the shape and peak value of the correlation peak verified the previous analysis. It should be noted that the length of the signal sequence is limited by the data bit hopping and cannot exceed the length of one data bit. Therefore, the frequency accuracy cannot be improved by increasing the length of the signal sequence without limitation.

We will now discuss another signal capture scheme using the FFT method.

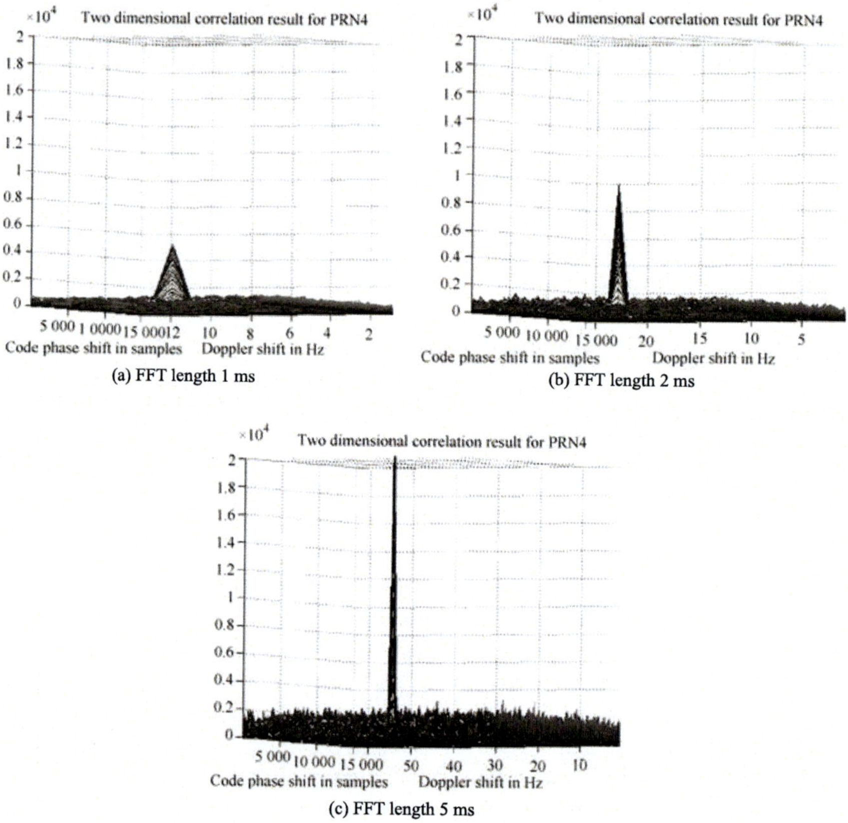

(a) FFT length 1 ms

(b) FFT length 2 ms

(c) FFT length 5 ms

Fig. 4.11 FFT capture results for GPS signals of different time lengths

Considering the intermediate frequency signal shown in Eq. (4.18), we multiply the local pseudo-code signal $C\ (t-\hat{\tau})$ and the intermediate frequency signal to obtain

$$y(t)=\sqrt{2P_s}C(t-\tau)C(t-\hat{\tau})\cos[\omega_{\mathrm{IF}}t+\phi(t)]+n'(t) \tag{4.50}$$

Since the signal length of the general coherent processing is less than 20 ms, the GPS navigation telegram can be regarded as a constant, so $D(t)$ in Eq. (4.18) is removed. $n'(t)$ is the result of the multiplication of the original noise term and $C(t-\hat{\tau})$, because $C(t-\hat{\tau})=\pm1$, and $C(t-\hat{\tau})$ is the balance code, so the mean and variance of the noise $n'(t)$ are the same as $n(t)$. When the pseudo-code phase difference $\Delta\tau=\tau-\hat{\tau}$ is 0, Eq. (4.50) becomes

$$y(t)=\sqrt{2P_s}\cos[\omega_{\mathrm{IF}}t+\phi(t)]+n'(t) \tag{4.51}$$

It can be seen that $y(t)$ becomes a mixed signal of a continuous wave signal and additive noise, because $n'(t)$ has flat spectral characteristics of white noise, and the continuous wave has line spectrum characteristics. Therefore, in Eq. (4.51), a spectral spike in the continuous wave component can be observed by performing a DFT transform. The principle of the overall process is shown in Fig. 4.12.

The block diagram based on this idea is shown in Fig. 4.13.

The form in Fig. 4.13 is very similar to the matched filter method. The IF input signal is driven by the sampling clock to slide a pseudo-code phase each time. After each sliding phase, the input signal is multiplied by the local pseudo-code $C(t)$. However, the multiplied result is not the cumulative output given by the accumulator, but the peak of the spectrum after DFT. When the maximum peak exceeds a certain threshold, it means that a pseudo-code is captured. The pseudo-code phase at this time indicates the pseudo-code phase information of the input signal at this time, and the position in the FFT sequence where the spectral peak is located indicates the magnitude of the Doppler shift amount.

This method differs from the method in Fig. 4.10 in that it performs pseudo-code stripping by serially sliding the pseudo-code phase and multiplying it, so it is serially processed in the pseudo-code phase dimension. However, through an FFT operation,

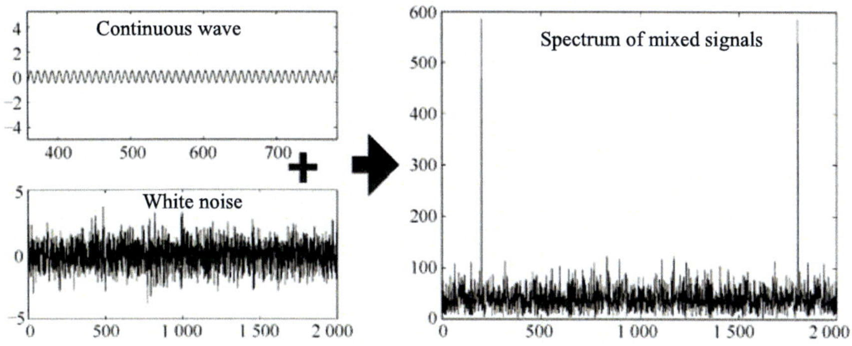

Fig. 4.12 Mixed signal of a continuous wave and white noise, and its spectrum

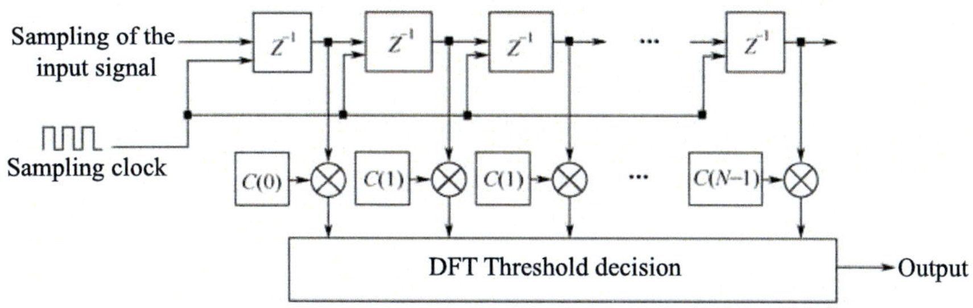

Fig. 4.13 Multiplying the input signal and local pseudo-code to perform DFT processing

it is possible to achieve signal search in all Doppler shift coverage, so it can be considered parallel in the dimension of the Doppler shift.

The GPS signal is taken as an example to analyze the computation required for the method. Since the GPS C/A code period is 1023 chips, the input signal has a time length of 1 ms, taking $N = 1023$. In this case, the input IF signal sampling frequency needs to be reduced to 1.023 MHz, as shown in Fig. 4.10. The sampling clock f_s is 1.023 MHz, and the local pseudo-code signal is also 1023 samples. Each sampling clock is equivalent to sliding the pseudo-code phase of the input signal by a pseudo-code chip and then multiplying it by the local pseudo-code sample value. As a result, 1023 samples are obtained, and the FFT of the sample sequence after multiplication can obtain the correlation result of all Doppler shifts in the case of the pseudo-code phase. A total of 1023×1023 points of FFT is required, and the calculation is striking. If N is changed to 2046, the pseudo-code phase accuracy becomes half of a pseudo-code chip. However, the operation is larger at this time, so the method is not widely used in practice.

The above analysis is based on the GPS signal. If the signal is for BDS, the calculation is larger, as the pseudo-random code of the BDS signal is 2046 chips, which is twice the pseudo-code period of the GPS signal. Therefore, if the same chip precision is used, the calculation is four times the calculation for processing GPS signals.

Due to the above situation, because the spectral resolution of the N-point FFT is f_s/N, the Doppler shift range covered by the N-point FFT is $f_s/2$ (considering the Nyquist frequency). In the above example, $f_s = 1.023$ MHz, the Doppler shift coverage is 515.5 kHz, and the Doppler shift range of relative motion between satellites and users mostly falls within the range of [−10 kHz, +10 kHz], which can be seen directly to N. Point-by-phase multiplication of points to perform FFT operations results in a lot of invalid results, and is a waste of computational effort. Of course, it is also possible to use the chirp Z-transform (CZT transform) to search only for the frequency range in which the Doppler shift may exist [4].

Although this FFT capture method is of little significance in practice, it shows theoretically that parallel processing can also be performed in the dimension of a Doppler shift. If the two-dimensional parallel processing of the pseudo-code phase dimension and the Doppler shift dimension can be realized, the speed of signal acquisition will be greatly accelerated.

4.1.5 Signal Acquisition Combing Short-Time Correlation Matched Filter and FFT

If we compare Fig. 4.9 with Fig. 4.13, we can see that the two are very similar in form. The only difference is that in the processing of the latter step, the matched filter scheme in Fig. 4.9 accumulates and outputs the result of multiplying the matched filter coefficients with the input signal samples, and the FFT scheme in Fig. 4.13

performs DFT processing and modulo decision. The scheme in Fig. 4.9 implements parallel processing of pseudo-code phase dimensions, while the scheme in Fig. 4.13 implements parallel processing of Doppler shift dimensions.

Analyzing the scheme in Fig. 4.9, the biggest problem is that the accumulation of the last step of the matched filter results in the loss of the frequency information of the continuous wave sequence after the pseudo-code stripping, so the carrier signal must be added to the local signal to achieve pseudo-code stripping. At the same time, carrier stripping is performed. The scheme in Fig. 4.13 is implemented without any accumulation during carrier stripping, and spectral analysis of all multiplied samples must be performed by FFT to find possible Doppler shifts, which leads to invalid frequencies and coverage, and unnecessary computational waste. Therefore, it can be considered as combining the two to realize two-dimensional parallel processing [5, 6]. The basic idea is to segment and accumulate the sequence after pseudo-code stripping, so as to ensure the smooth completion of this stripping. At the same time, a valid Doppler frequency range is preserved in the sequence of numbers that guarantees the summation of segmentation.

The signal processing block diagram of this new scheme is shown in Fig. 4.14.

The scheme divides the result of multiplying the matched filter coefficients and the input data into L segments, as shown in Fig. 4.14. Assuming that the length of each segment of data is T_{p}, the total coherent integration time is LT_{p}. As a result of accumulating the multiplication of each time segment, a total of L accumulated sums is obtained. The L accumulated sums are sent to the DFT unit, and the capture of signals can be completed by taking the modulus of the results and making threshold judgments. This scheme splits the matched filter in Fig. 4.8 into L shorter matched filters, so it is called a short-time correlation matched filter. This is the simple process of the

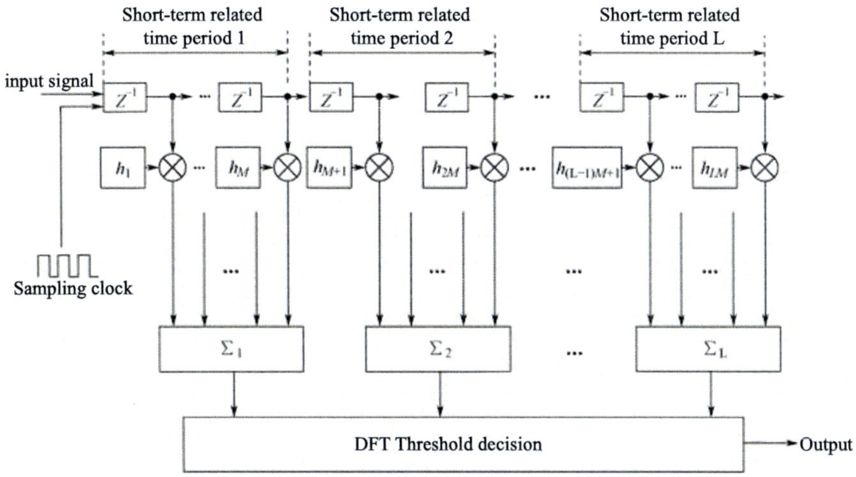

Fig. 4.14 Capture scheme with short-term correlation matched filter and FFT

short-term correlation matched filter and FFT acquisition scheme. The following is a mathematical analysis, which is based on the GPS signal.

Consider a GPS IF data sample sequence of 1 ms in length, which can be denoted as

$$s_{\mathrm{IF}}(mT_s) = \sqrt{2P_s}C(mT_s - \tau)\cos[\omega_{\mathrm{IF}} mT_s + \phi_0] + n(mT_s), \quad m = [0, N-1] \tag{4.52}$$

In Eq. (4.52), T_s is the sampling interval, C is the pseudo-random code, ω_{IF} is the intermediate frequency carrier frequency, φ_0 is the initial phase, and $n(mT_s)$ is the Gaussian white noise, fitting the $N(0, \sigma^2)$ distribution. The value of m is 0 to $N-1$, indicating that the number of samples in 1 ms is N. Since there is no navigation telegram jump in a pseudo-random code period, the navigation message bit item is ignored here.

The length of each short-term correlation integral is M samples. Here, T_p is used to indicate the duration of the short-term correlation, so $T_p = MT_s$. Then, the short-term correlated accumulated output of the ith segment is:

$$I_i = \sum_{k=iM+1}^{(i+1)M} s_{\mathrm{IF}}(kT_s) \cdot \cos(\omega_L \cdot kT_s) \cdot c(kT_s) \tag{4.53}$$

$$Q_i = \sum_{k=iM+1}^{(i+1)M} s_{\mathrm{IF}}(kT_s) \cdot \sin(\omega_L \cdot kT_s) \cdot c(kT_s) \tag{4.54}$$

In Eqs. (4.53) and (4.54), ω_L is the local carrier frequency and $c(kT_s)$ is the local pseudo-random code. After simplification, the result is

$$I_i = \frac{1}{2}\sqrt{2P_s}R(\tau) \cdot \frac{\sin\left(\frac{\Delta\omega}{2}T_p\right)}{\sin\left(\frac{\Delta\omega}{2}T_s\right)} \cdot \cos[i\Delta\omega T_p + \varphi] + N_I(i) \tag{4.55}$$

$$Q_i = -\frac{1}{2}\sqrt{2P_s}R(\tau) \cdot \frac{\sin\left(\frac{\Delta\omega}{2}T_p\right)}{\sin\left(\frac{\Delta\omega}{2}T_s\right)} \cdot \sin[i\Delta\omega T_p + \varphi] + N_Q(i) \tag{4.56}$$

In Eqs. (4.55) and (4.56), $\Delta\omega = \omega_{\mathrm{IF}} - \omega_L$ is the carrier frequency difference between the input signal and the local signal; φ is the fixed phase difference, and its value is

$$\varphi = \frac{1}{2}\Delta\omega(T_P - T_s) + \phi_0$$

$R(\tau)$ is a normalized short-term correlation function, defined as follows:

$$R_L(\tau) = \frac{1}{M} \cdot \sum_{k=iM+1}^{(i+1)M} C(mT_s - \tau) \cdot c(mT_s) \tag{4.57}$$

The N_I and N_Q in Eqs. (4.55) and (4.56) are still Gaussian white noise, fitting the $N(0, M\sigma^2)$ distribution.

When the phase of the local pseudo-random code and the pseudo-code phase of the input signal are the same, $\tau \to 0$; when the local carrier frequency and the input signal carrier frequency are the same, $\Delta\omega \to 0$, and the signal energy term of the short-time correlation integral is

$$\lim_{\substack{\tau\to 0\\ \Delta\omega\to 0}} \sqrt{I_i^2 + Q_i^2} = \sqrt{\frac{P_s}{2}\frac{T_P}{T_s}} = M\sqrt{\frac{P_s}{2}} \tag{4.58}$$

The short-term correlation score is only 1/*L* of the total correlation integral compared with all related points, which is also in line with intuitive thinking, because the number of samples participating in the operation is only 1/*L* of the total number of samples.

If $\tau = 0$ is fixed and $\Delta\omega$ is changed, the curve of Eq. (4.58) is similar to that of Figs. 4.5 and 4.6. The longer the short integration time, the higher the signal correlation peak but the sharper the shape, indicating a stronger frequency selection. Conversely, the lower the signal correlation peak of Fig. 4.6 but the wider the shape, the weaker the frequency selectivity.

The short-term correlation result is a sequence of numbers $\{I_n - jQ_n, n = 1, \ldots L\}$ with a sampling interval of T_p. The number of samples in all periods is *L*. Performing FFT transformation on the sequence, the result is

$$F(\omega_k) = \sum_{n=0}^{L-1} (I_n - jQ_n) e^{-j\omega_k n T_p} \tag{4.59}$$

In Eq. (4.59), $\omega_k = \frac{2\pi k f_p}{L}$, $k = 0, \ldots, L-1, f_p = \frac{1}{T_p}$.

The signal component of $(I_n - jQ_n)$ is as follows:

$$\text{Signal}\{I_n - jQ_n\} = \frac{1}{2}\sqrt{2P_s}R(\tau) \cdot \frac{\sin\left(\frac{\Delta\omega}{2}T_p\right)}{\sin\left(\frac{\Delta\omega}{2}T_s\right)} \cdot e^{j(n\Delta\omega T_p + \varphi)} \tag{4.60}$$

Here, in order to make the equation more concise, we only consider the signal component of $(I_n - jQ_n)$, then substitute Eq. (4.60) into Eq. (4.59) and simplify it to obtain the real and imaginary parts of $F(\omega_k)$.

$$\text{Re}[F(\omega_k)] = \frac{1}{2}\sqrt{2P_s}R(\tau) \cdot \frac{\sin(\frac{\Delta\omega}{2}T_p)}{\sin(\frac{\Delta\omega}{2}T_s)} \cdot \frac{\sin(\frac{\Delta\omega - \omega_k}{2}LT_p)}{\sin(\frac{\Delta\omega - \omega_k}{2}T_p)} \cdot \cos(\Phi_k) \tag{4.61}$$

$$\mathrm{Im}[F(\omega_k)] = \frac{1}{2}\sqrt{2P_s}R(\tau)\cdot\frac{\sin\left(\frac{\Delta\omega}{2}T_p\right)}{\sin\left(\frac{\Delta\omega}{2}T_s\right)}\cdot\frac{\sin\left(\frac{\Delta\omega-\omega_k}{2}LT_p\right)}{\sin\left(\frac{\Delta\omega-\omega_k}{2}T_p\right)}\cdot\sin(\Phi_k) \tag{4.62}$$

In the equation, $\Phi_k = (L-1)(\Delta\omega - \omega_k)T_p + \varphi$.

With the beat of the sampling clock, the input signal is equivalent to sliding the pseudo-code phase; each time one sampling point is slipped, L short-term correlation matching filters output L short-term correlation results. When the phases of the local pseudo-code and the pseudo-code of the input signal are aligned, $R(\tau) = 1$, and the normalized signal power component is

$$|F(\omega_k)|_{\mathrm{norm}} = \left|\frac{\sin\left(\frac{\Delta\omega}{2}T_p\right)}{\sin\left(\frac{\Delta\omega}{2}T_s\right)}\cdot\frac{\sin\left(\frac{\Delta\omega-\omega_k}{2}LT_p\right)}{\sin\left(\frac{\Delta\omega-\omega_k}{2}T_p\right)}\right| \tag{4.63}$$

The first term in Eq. (4.63) is provided by short-term correlation, and the second term is provided by FFT processing, which is analyzed below.

When $\Delta\omega \to 0$, the first term reaches the maximum value:

$$\max_{\Delta\omega}\left\{\left|\frac{\sin\left(\frac{\Delta\omega}{2}T_p\right)}{\sin\left(\frac{\Delta\omega}{2}T_s\right)}\right|\right\} = \frac{T_p}{T_s} = M$$

When $\omega_k \to \Delta\omega$, the second term reaches the maximum value:

$$\max_{\omega_k}\left\{\left|\frac{\sin\left(\frac{\Delta\omega-\omega_k}{2}LT_p\right)}{\sin\left(\frac{\Delta\omega-\omega_k}{2}T_p\right)}\right|\right\} = \frac{LT_p}{T_p} = L$$

Therefore, when $\Delta\omega = 0$, $\omega_k = 0$, the maximum value of Eq. (4.63) is obtained.

$$\max_{\Delta\omega,\omega_k}|F(\omega_k)|_{\mathrm{norm}} = ML \tag{4.64}$$

FFT processing can only obtain the results of L frequency points, in other words, ω_k can only take a finite discrete value. Therefore, only when the local carrier and the input signal carrier frequency are equal (at this time the 0 processing results of FFT processing) will the short-term correlation result reach the maximum value. The processing result is the same as the processing gain for the time domain correlation processing of all input signals, both of which are ML, wherein the short-time correlation provides M times the gain, and the FFT provides L times the gain. The matched filter implements a parallel search of the pseudo-code phase dimension, while the FFT implements a parallel search of the Doppler shift dimension. Although this method provides the same processing gain as the conventional method, the two-dimensional parallel search in the frequency domain and the time domain is realized, and the processing speed of the conventional method is greatly improved.

The short-term correlation matched filter and FFT scheme can be seen as an improvement to the FFT scheme shown in Fig. 4.13. Meanwhile, Fig. 4.13 can be seen as a special case of the short-term correlation matched filter and FFT scheme, where L is the full sample sequence. As for the number of samples, the length of the data block for each short-term correlation operation is M = 1. There is no short-term correlation operation in the scheme shown in Fig. 4.13, which results in the number of samples participating in the FFT operation being the number of samples in the total integration length. However, in this way, the Doppler frequency coverage is very wide. It covers a lot of invalid areas, and at the same time leads to a huge computation. By combining short-term correlation and FFT, the number of samples participating in the FFT operation can be greatly compressed. Calculations can be reduced by controlling the Doppler frequency coverage area within a reasonable range, making full use of the time domain parallel of the matched filter and the frequency domain parallel of the FFT.

Assuming that the total IF data length is T ms, where T is selected as an integer greater than or equal to 1, and is equally divided into L segments, the length of each segment is $T_{\mathrm{p}} = T/L$ ms. Since the number of points participating in the FFT operation is L, it is generally taken that L is a power of two, or that L is converted to a power of two by zero padding. The frequency resolution of the FFT operation is inversely proportional to T, which is $1000/T$ Hz, so the frequency coverage of the FFT operation is $\mathrm{L}\cdot 1000/T$ Hz. The more segments, the larger the frequency coverage, and the more Doppler shift results can be obtained by one FFT.

Taking the data length of 1 ms as an example, we can consider two cases, the first of which is a segmentation configuration of $L = 8$, and the second is a segmentation configuration of $L = 16$. In both cases, the frequency resolution is 1 kHz. In the first case, the 8-point FFT can cover the 8 kHz Doppler shift searching range; in the second case, the 16 kHz searching range can be covered. Although the more segments, the wider the Doppler shift range of the FFT coverage, it does not mean that the number of segments can be increased without limit to achieve better results, because the first term in Eq. (4.63) also affects the final processing gain. When the number of segments is larger, it allows for a wider range of Doppler frequency coverage, but those FFT results that deviate from the true Doppler frequency are attenuated by the effects of short-term correlation envelopes.

Figures 4.15 and 4.16 show cases where 1 ms of data is divided into $L = 8$ segments and $L = 16$ segments respectively. The dotted line shows the short-term correlation integral envelope, that is, the function envelope of the first term in Eq. (4.63). The solid line inside the dashed line is the result of each frequency point of the FFT operation. The FFT points in the two figures are 8 and 16 respectively, so there are 8 solid peaks and 16 solid peaks each. The final processing gain is determined by the combination of the short-term correlation integral envelope and the FFT operation. It can be seen from the figure that when $\Delta\omega \to 0$, the first result of the FFT (i.e. the first solid line peak) gets the maximum value. However, as $\Delta\omega$ increases, the other results of the FFT are attenuated, and although the last result of the FFT can cover the Doppler frequencies of 8 kHz/16 kHz, its attenuation has exceeded 10 dB.

Figures 4.15 and 4.16 also show how segmentation has no effect on frequency resolution and only affects frequency coverage. Although the frequency coverage in Figs. 4.15 and 4.16 is 8 and 16 kHz, the frequency spacing between adjacent solid

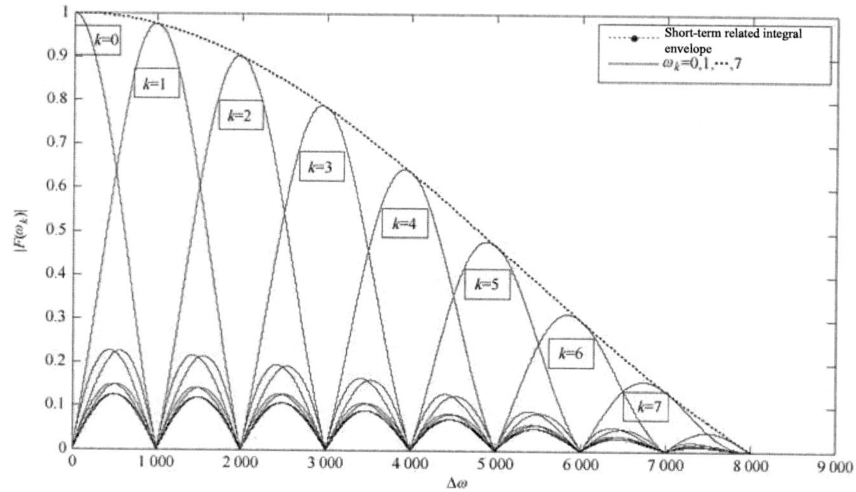

Fig. 4.15 Processing gain of short-term correlation + FFT, from 1 ms IF data divided into 8 segments

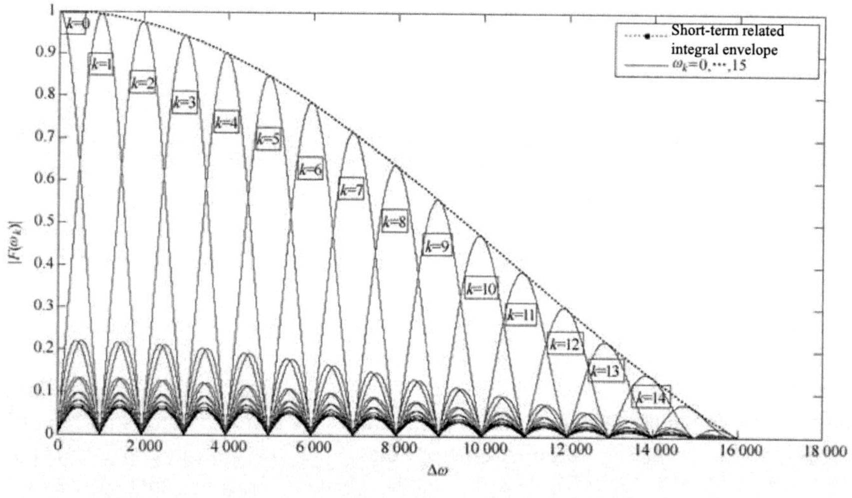

Fig. 4.16 Processing gain of short-term correlation + FFT, from 1 ms IF data divided into 16 segments

peaks is 1 kHz, which is determined by the length of the sample sequence involved in the operation. In this example, the data length is 1 ms, so the interval between the peaks is 1 kHz. If the data length is 2 ms, the frequency interval between the solid peaks becomes 500 Hz. Therefore, the frequency resolution can be increased by lengthening the sequence, and the Doppler frequency coverage can be improved by increasing the number of L segments.

In practical applications, the total data length involved in the short-term correlation matching filter and FFT processing cannot exceed the maximum length allowed by the data bit hopping, otherwise the problem of correlation peak cancellation will occur. For the GPS signal, since the GPS data bit length is 20 ms, the maximum data length is less than 20 ms. In terms of the BDS signal, the 1 kHz *NH* code is modulated on the D1 code, so the maximized length when the data does not jump is only 1 ms. This limits the processing performance of the method to the BDS signal to some extent.

4.1.6 Signal Capture Based on Data Blocking and Frequency Compensation

Consider the data of length L ms, where $L \geq 1$ is an integer. Suppose that there is a total of N sample points in 1 ms, then N is obviously determined by the sampling frequency. Then, there is a total of NL sample points in L ms. We divide these sample points into L data blocks. Each data block contains exactly 1 ms of data samples, as shown in Fig. 4.17.

Generally, signals will contain signals from multiple satellites. Here, we first analyze the signal of one satellite. The discrete data sample $s_k(n)$ obtained by the satellite with a PRN of k can be expressed as

$$s_k(n) = D_k(nt_s)C_k(nt_s)e^{j2\pi f_k nt_s} + v_k(nt_s) \tag{4.65}$$

Here ts is the sampling period, $Nts = 1$ ms. $n \in [0, NL - 1]$, D_k is the navigation message data bit, and C_k is the pseudo-random code. f_k is the signal carrier frequency. Due to the Doppler shift, the carrier frequency of each satellite is different. The value is between the theoretical intermediate frequency plus [−5 kHz, +5 kHz]. This search range needs to be adjusted based on different crystal oscillators. v_k (nts) is Gaussian white noise that fits the $N(0, \sigma^2)$ distribution. Since the carrier amplitude in Eq. (4.65) is 1, the noise power is actually normalized here:

$$\sigma^2 = \frac{N_0 B_w}{2A^2}$$

where A is the actual received carrier signal amplitude; B_w is the signal bandwidth of the RF front-end; and $N_0/2$ is the noise bilateral power spectral density.

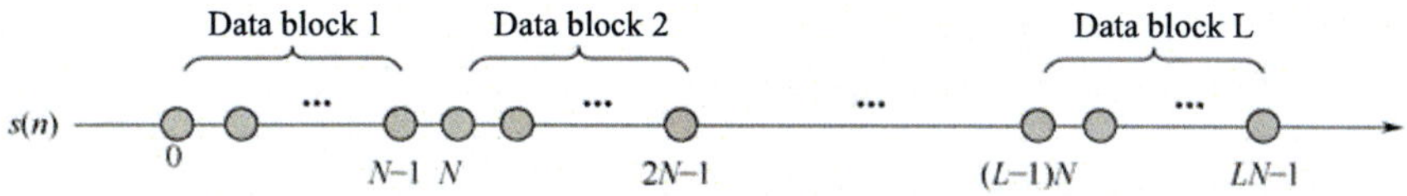

Fig. 4.17 *L* ms sample data is divided into blocks, with each block including 1 ms of data

To omit the noise term in Eq. (4.65), consider only the signal portion. If the length of L is less than the telegram bit period, then D_k can be regarded as a constant, and the pseudo-code signal $C_k(i)$ is a periodic function. For GPS and BDS signals, the period is 1 ms, which is represented by discrete fields:

$$C_k(it_s) = C_k[(i+N)t_s], \quad i = 0, 1, \ldots, (L-1)N-1$$

There are three components in the signal part of $s_k(n)$: $\{D_k, C_k, e^{j2\pi fkt}\}$, where the components D_k and C_k are unchanged at the same position of different data blocks. So, if these corresponding samples of the same position of the block in L data are superimposed, then

$$\begin{aligned}\sum_{i=0}^{L-1} s_k(n+iN) &= D_k(nt_s)C_k(nt_s)e^{j2\pi f_k nt_s}\sum_{i=0}^{L-1} e^{j2\pi f_k iNt_s} + \sum_{i=0}^{L-1} v_k(nt_s + iNt_s) \\ &= S\{s_k(n)\}G(L, f_k) + \sum_{i=0}^{L-1} v_k(nt_s + iNt_s)\end{aligned} \tag{4.66}$$

In Eq. (4.66), $S\{s_k(n)\}$ is the signal component in $s_k(n)$, and $G(L, f_k)$ is a new gain function.

$$G(L, f_k) = \sum_{i=0}^{L-1} e^{j2\pi f_k iNt_s}, \quad n = 0, 1, \ldots, N-1 \tag{4.67}$$

Through the block superposition described above, the L segment data is merged into a piece of data, and the accumulated signal component is equivalent to the original signal multiplied by the function $G(L, f_k)$. The properties of $G(L, f_k)$ are shown as follows:

$$G(L, f_k) = \sum_{i=0}^{L-1} e^{j2\pi f_k iNt_s} \Rightarrow |G(L, f_k)| \le \sum_{i=0}^{L-1} |e^{j2\pi f_k iNt_s}| = L \tag{4.68}$$

Equation (4.68) shows that the maximum value that $G(L, f_k)$ can take is L, which is obtained when $e^{j2\pi fkiNts} = 1$. When $G(L, f_k)$ takes the maximum value, it means that the original signal is amplified L times. The value of $G(L, f_k)$ is related to f_k. Direct block accumulation of the input signal does not guarantee that $G(L, f_k)$ is the maximum value. A carrier adjustment signal with a frequency of Δf_k can be generated locally. The signal is a continuous wave signal, which can be expressed as $p_k(n) = e^{j2\pi \Delta fknts}$. Multiplying $p_k(n)$ and the input signal $s_k(n)$ to obtain a new signal $y_k(n)$ after the carrier frequency adjustment,

$$y_k(n) = D_k(nt_s)C_k(nt_s)e^{j2\pi(f_k+\Delta f_k)nt_s} + v_k(nt_s)e^{j2\pi\Delta f_k nt_s} \tag{4.69}$$

The introduction of $p_k(n)$ only changes the carrier frequency of the original signal, and has no effect on D_k and C_k. The noise component in Eq. (4.69) is a certain phase of rotation on the original noise component, but the power spectral density does not change. Here, a new white noise signal is denoted as $vk'(t)$.

Now, we accumulate a similar block of $y_k(n)$,

$$Y_k(n) = \sum_{i=0}^{L-1} y_k(n + iN) = S\{y_k(n)\}G(L, f_k') + \sum_{i=0}^{L-1} v_k'(n + iN) \tag{4.70}$$

In this equation, the definitions of $f_k' = f_k + \Delta f_k$, $G(L, f_k')$ are the same as they are in Eq. (4.67).

Analyzing $G(L, f_k')$ shows that when $f_k' N t_s$ is an integer, then $e^{j2\pi f k' i N ts} = 1$, $\forall i \in [0, L-1]$. At this time, the L term's accumulation term in $G(L, f_k')$ is 1, which accumulates and gets the maximum value. Since $N \cdot t_s = 1$ ms, the condition that $f_k' N t_s$ is an integer is as follows:

$$\mathrm{mod}(f_k', 1\ \mathrm{kHz}) = 0 \tag{4.71}$$

Here, mod(x, y) means that x takes a modulo operation on y. The meaning of Eq. (4.71) is that f_k' is an integer multiple of 1 kHz. When this condition is satisfied, $G(L, f_k')$ achieves the maximum value L, so

$$\begin{aligned} Y_k(n) &= S\{y_k(n)\} \sum_{i=0}^{L-1} e^{j2\pi i[\text{certain integer}]} + \sum_{i=0}^{L-1} v_k'(n + iN) \\ &= S\{y_k(n)\} \cdot L + \sum_{i=0}^{L-1} v_k'(n + iN) \end{aligned} \tag{4.72}$$

That is, the signal strength in the accumulated data block is enhanced.

In fact, Eq. (4.72) is still possible for the original f_k. For example, after the Doppler shift, the signal carrier frequency is exactly an integer multiple of 1 kHz, so the original signal can be directly accumulated to increase the signal strength.

It is only because the effect of the Doppler frequency is uncontrollable that we cannot guarantee that the received signal carrier frequency satisfies Eq. (4.71). The introduction of the carrier adjustment signal $p_k(n)$ allows us to control the value of f_k' to satisfy Eq. (4.71), which can increase the signal strength after carrier adjustment by L times.

Regardless of complicated mathematical equations and analysis from the intuitive perspective, the original signal cannot be directly accumulated because the starting carrier phase of each 1 ms data block is not synchronized. The introduction of $p_k(n)$ means that the starting carrier phase of the adjusted signal per data blocks is the same, and does not affect the data bits and pseudo-code bits in the signal structure. In this sense, $p_k(n)$ can be regarded as the carrier phase adjustment sequence. In the data sequence after phase adjustment, the data bits of the signal remain unchanged, the pseudo-code bits remain at a periodicity of 1 ms, and the initial carrier phase of multiple data blocks is also synchronized. Thus, accumulating multiple data blocks leads to a linear accumulation of signal strengths, the magnitude of which increases in signal strength and is proportional to the number of data blocks participating in the accumulation. The entire process can be shown as Fig. 4.18.

For convenience of description, we refer to Δf_k which can hold Eq. (4.71) as 'optimized' Δf_k. Below are the characteristics of the optimized Δf_k.

1. The optimized Δf_k is only valid for satellite signals with a PRN of *k*.

The reason for this conclusion is obvious. Because the Doppler shifts of various satellite signals are different, the result of the carrier frequency being 1 kHz is different, so the optimized Δf_k of different satellites must be searched. Since the result of the carrier frequency's 1 kHz remainder determines Δf_k, the next property of Δf_k can be derived.

2. The optimized Δf_k is cyclical, and it is the reciprocal of the data block period.

In the above analysis, if Δf_k is optimized for the signal of satellite *k*, then $(\Delta f_k + n \times 1\ \text{kHz})$ is also optimized for the signal of satellite *k*, where *n* is an arbitrary integer. Further analysis shows that the period is inversely proportional to the length of the unit data block. If the length of the unit data block is 1 ms, the period is 1 kHz; if the bit block length is 2 ms, then the period becomes 500 Hz, and so on.

3. Accumulating the data after phase compensation of the optimized Δf_k can increase the signal-to-noise ratio by 10 lg*L* decibels.

Accumulating the *L* data block increases the signal strength by a factor of *L* and increases the signal power by a factor of L^2. For the noise term, assuming that the noise in the original signal is Gaussian white noise, its distribution is $v_k'(n + iN) \sim N(0, \sigma^2)$, and the accumulated noise is $\sum_{i=0}^{L-1} v_k'(n + iN) \sim \mathcal{N}(0, L\sigma^2)$, in which the power increase of the noise is *L* times. However, the speed of the noise increase

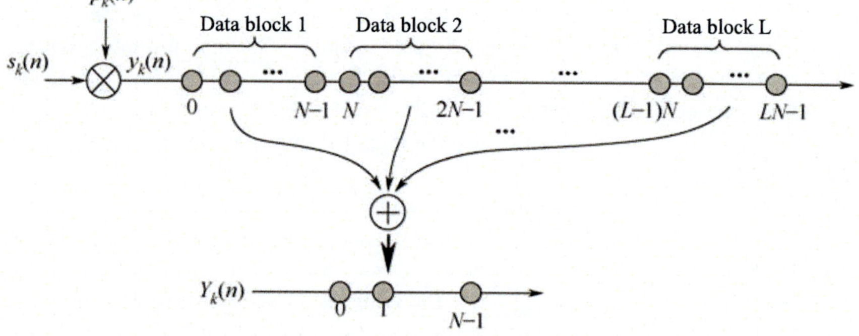

Fig. 4.18 Increased signal strength by data block accumulation after carrier phase adjustment

cannot keep up with the speed of the signal power increase. When $L = 10$, the signal-to-noise ratio will increase by 10 dB; when $L = 20$, the signal-to-noise ratio will increase by 13 dB, but L cannot exceed the period of the navigation message bit transition.

There is no shortcut to find the optimized Δf_{k}. It can be searched according to the above second characteristic. The search range is [0, 1 kHz]. Compared to the original Doppler shift range, the search range of Δf_{k} is much smaller. The step frequency of the search can be determined according to the characteristic of $G(L, \Delta f_{\mathrm{k}})$. Figure 4.19 is the $G(L, \Delta f_{\mathrm{k}})$ for the various L and Δf_{k} codomains.

As can be seen in Fig. 4.19, the larger L is, the narrower the main lobe of $G(L, \Delta f_{\mathrm{k}})$. This is consistent with Fig. 4.6 because L corresponds to the length of the data involved in the accumulation, which is equivalent to the integral length in Fig. 4.6. In general, the width of the first zero-crossing point of $G(L, \Delta f_{\mathrm{k}})$ can be selected as the search step value of Δf_{k}. When $L = 5$, the search step of Δf_{k} can be set to 200 Hz. When $L = 10$, the search step is set to 100 Hz, so the maximized sensitivity loss due to frequency error is about 2 dB.

After determining the frequency step value, a set of Δf_{k} is generated. Each Δf_{k} can obtain a carrier phase compensation and the accumulated data Y_{kI}, (n) as shown in Eq. (4.72), where the subscript l corresponds the number of Δf_{k}. The remaining question is how to find the optimized Δf_{k} from these I 1 ms data blocks.

The first processing method is to perform a squaring operation on the l data blocks. After squaring, the signal components in Y_{kl}, (n) become continuous wave signals, and the FFT operation is performed to find the peak value of the spectral

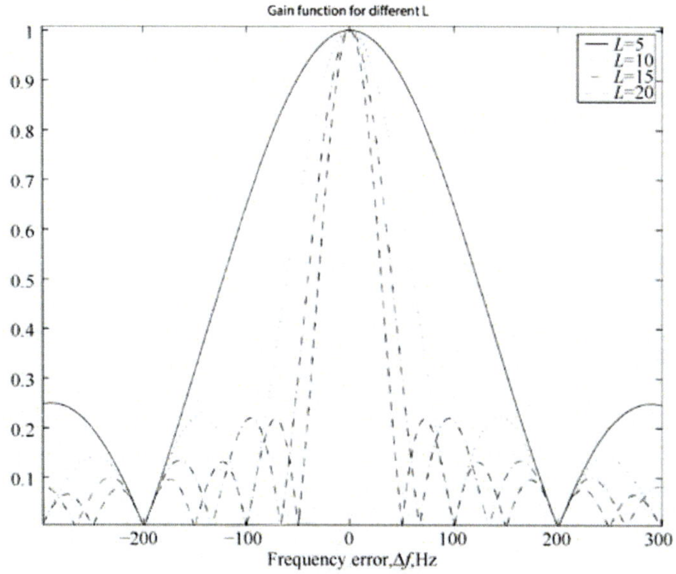

Fig. 4.19 Normalized $G(L, \Delta f_{\mathrm{k}})$ for different L values

components. The frequency position corresponding to the peak and the pseudo-code phase position are analyzed to know the number and the pseudo-code phase of PRN contained in the signal. Based on this idea, squaring both sides of Eq. (4.72) will give:

$$Y_k^2(n) = D_k(n)^2 C_k(n)^2 e^{j2\pi 2(f_k+\Delta f_k)nt_s} \cdot L^2 + V(n)^2 + 2LV(n)D_k(n)C_k(n)e^{j2\pi(f_k+\Delta f_k)nt_s} \tag{4.73}$$

In Eq. (4.73), $V(n) = \sum_{i=0}^{L-1} v_k'(n+iN)$ is the noise term after the L block is accumulated. Since $v_k'(n+iN)$ are spaced from each other in 1 ms, and can be considered as independent of each other, then $V(n) \sim N(0, L\sigma^2)$. To simplify this analysis, it can be written as

$$\eta_n(L) = V(n)^2 + 2LV(n)D_k(n)C_k(n)e^{j2\pi(f_k+\Delta f_k)nt_s} \tag{4.74}$$

Performing a theoretical analysis on $\eta_n(L)$, we can obtain

$$E[\eta_n(L)] = L\sigma^2 \tag{4.75}$$

$$\mathrm{var}[\eta_n(L)] = 2L^2\sigma^4 + 4L^3\sigma^2 \tag{4.76}$$

The signal-to-noise ratio after square operation can be obtained:

$$S/N_{平方} = \frac{L^2}{2\sigma^4}\left(\frac{1}{1+2L/\sigma^2}\right) \tag{4.77}$$

In Eq. (4.77), when $2L/\sigma^2 >> 1$,

$$S/N_{平方} \approx \frac{L}{4\sigma^2}$$

The signal spectrum obtained by the square operation is relatively obvious. When $2L/\sigma^2 << 1$,

$$S/N_{平方} \approx \frac{L^2}{2\sigma^4}$$

The above equation approaches 0 in the case of $2L/\sigma^2 << 1$, so a significant signal spectrum cannot be obtained by the square operation at this time. From these theoretical analyses, it can be seen that for a strong signal, the signal spectrum can be detected by a squaring operation. The squaring operation of the weak signal makes the detection of the signal spectrum worse. The length limitation of the data bit jump makes it impossible to satisfy the condition $2L/\sigma^2 >> 1$ by simply increasing L.

Therefore, the capture effect of this method on weak signals is not obvious. Note that σ^2 in the above analysis is the normalized noise power in the IF sampled data. For GPS and BDS, there is often $\sigma^2 > 1$.

Another method is to perform the FFT capture shown in Fig. 4.10 for$l$ data blocks. The larger the l is, the larger the number of FFTs and IFFTs required, so significant computation must be performed.

In order to achieve the accumulation of data blocks in the above process, the optimized Δf_{k} must be searched. The search process attempts to cover the [0, 1 kHz] range by serially trying different Δf_{k}. Inspired by the characteristics of the carrier phase of the sampled signal at the same position of the data block, the FFT algorithm can be used to search for Δf_{k}, as shown in Fig. 4.20.

Figure 4.20 rearranges the data in Fig. 4.17 and divides it into a two-dimensional matrix. Each thick frame in the figure represents an IF sampling sample. The arrow between the sampling points indicates the direction of the data stream. Looking from the horizontal direction of the matrix, the data is divided into L horizontal blocks; from the vertical, the data is divided into N vertical blocks. Each horizontal block represents 1 ms of sample data, and each vertical block has L data. The time interval between the data is 1 ms.

Firstly, considering the ith vertical block, where $i = 0, 1, \ldots, N-1$, which contains L data, respectively

$$[s(i), s(N+i), \ldots s((L-1)N+i)]$$

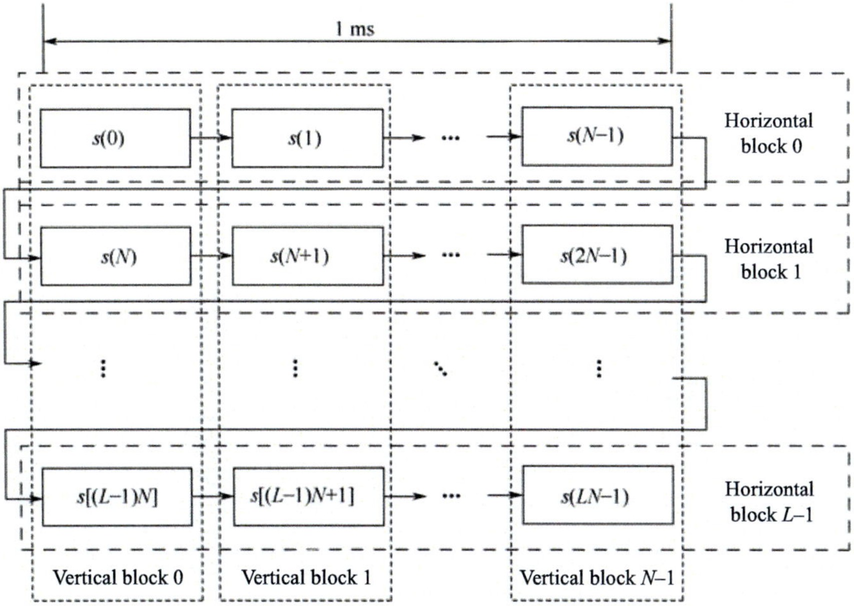

Fig. 4.20 Rearranging IF data samples by horizontal and vertical blocks

Regardless of the noise term first, the mathematical expression of each sample in the vertical block is

$$\begin{cases} s(i) = C_k D_k e^{j2\pi f_k i T_s + \phi_0} \\ s(N+i) = C_k D_k e^{j2\pi f_k (i+N)T_s + \phi_0} = s(i)e^{j2\pi f_k N T_s} \\ \vdots \\ s[(L-1)N+i] = C_k D_k e^{j2\pi f_k (i+(L-1)N)T_s + \phi_0} = s(i)e^{j2\pi f_k (L-1)NT_s} \end{cases} \tag{4.78}$$

In Eq. (4.78), the meanings of D_k, C_k, *and* f_k are the same as in Eq. (4.65).

Since $NTs = 1$ ms, the sampling frequency of the sampled data in the vertical block is 1 kHz. If the carrier frequency ω_0 is divided into integer multiples and remainders of 1 kHz,

$$f_k = F \cdot 1\ \text{kHz} + \Delta f_k \tag{4.79}$$

Here, F is an integer, $0 < \Delta f_k < 1$ kHz, so the sampling point in the *i*th vertical block can be simplified to $s(i)$, $s(i)e^{j2\pi\Delta f_k NT_s}, \ldots, s(i)e^{j2\pi\Delta f_k (L-1)NT_s}$.

Performing FFT transformation on these L sample points, according to the theory of FFT transformation, the L results are

$$Y(i,m) = \sum_{n=0}^{L-1} s(i)e^{j\Delta\omega_k nNT_s} e^{-j\omega_m nNT_s} \tag{4.80}$$

In Eq. (4.80), $\omega_m = \frac{2\pi m f_p}{L}$, $m = 0, 1, \ldots, L-1$, where $f_p = 1$ kHz. $\Delta\omega_k = 2\pi\Delta f_k$, in the latter analysis the subscript k is removed for simplicity. After simplification, Eq. (4.80) can be written as:

$$Y(i,m) = s(i)\frac{1 - e^{j(\Delta\omega - \omega_m)LNT_s}}{1 - e^{j(\Delta\omega - \omega_m)NT_s}} \tag{4.81}$$

Replace the data of the *i*th vertical block with $Y(i, k)$ to obtain a new data block matrix as shown in Fig. 4.21.

At this point, the data in the *m*th horizontal block can be written as:

$$s(0)A(m), s(1)A(m), \ldots, s(N-1)A(m), m = 0, 1, \ldots, L-1$$

In this equation, $A(m) = \frac{1 - e^{j(\Delta\omega - \omega_m)LNT_s}}{1 - e^{j(\Delta\omega - \omega_m)NT_s}}$, Considering the sampling moment of the *m*th horizontal block data, $A(m)$ is a public term that is irrelevant to the sampling time.

After FFT and rearrangement of the vertical block data, the data sampling interval in the horizontal block is T_s, and the number of sampling points is N, which includes a periodic pseudo-random code signal. If we apply the FFT capture method in Fig. 4.10

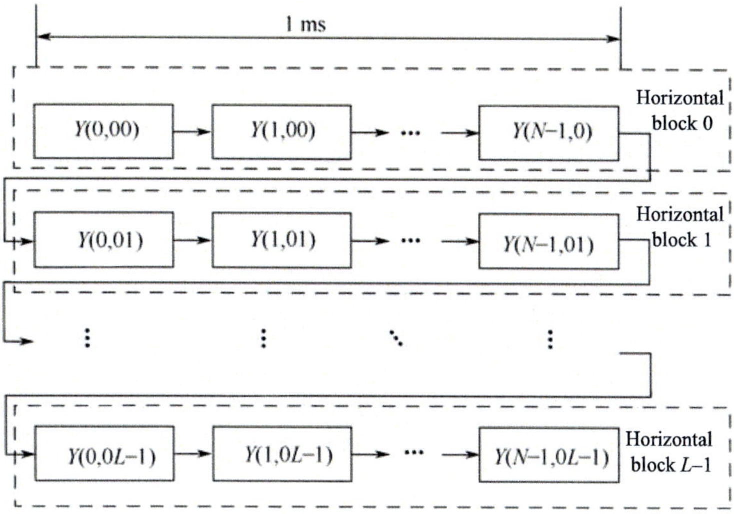

Fig. 4.21 Vertical block data and data block matrix after FFT

to N, we can obtain the information of the pseudo-code phase. The whole process is shown in Fig. 4.22.

In Fig. 4.22, all horizontal block data is multiplied by the FFT of the local pseudo-code + carrier mixed signal by FFT, and then the IFFT transform is used to obtain the correlation result of all pseudo-code phases. A detailed analysis of the overall process is offered in Sect. 4.1.4. The only difference is that the IF data participating in the correlation operation has a common factor $A(m)$. After simplification, the model of $A(m)$ is

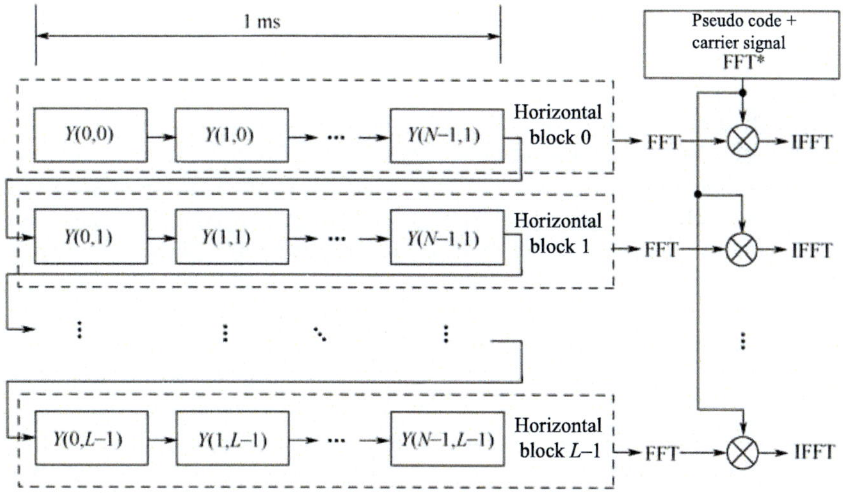

Fig. 4.22 FFT code phase parallel capture for horizontal block data

$$|A(m)| = \left| \frac{\sin\left[\frac{\Delta\omega - \omega_m}{2} LNT_s\right]}{\sin\left[\frac{\Delta\omega - \omega_m}{2} NT_s\right]} \right| \tag{4.82}$$

The maximum value of Eq. (4.82) is L, which is achieved when $\Delta\omega = \omega_m$. In this sense, $A(m)$ can be regarded as a frequency selection factor, except that the part below f_k Hz is selected below 1 kHz. The 1 kHz portion of f_k, which is integer F in Eq. (4.79), is determined by setting the local carrier frequency by FFT capture of the horizontal block data.

The local carrier frequency setting interval in Fig. 4.22 is fixed at 1 kHz, and 10 frequency points are required to cover the frequency range of [−5 kHz, +5 kHz]. Corresponding to each frequency point f_i, $i = 0, 1, \ldots, 9$, a set of mixed signal data of the local pseudo-code and carrier can be obtained. After the FFT of this data, the multiplication of the FFT of L horizontal blocks, and the performing of IFFT, an $L \times N$ matrix is finally obtained. The longitudinal coordinate of the matrix $(0, 1, \ldots, N - 1)$ corresponds to the pseudo-code phase information, and the horizontal coordinate $(0, 1, \ldots, L - 1)$ corresponds to the carrier's Doppler frequency information, where the Doppler frequency value corresponding to the mth row of data is

$$f(m) = f_i \cdot 1\ \text{kHz} + \omega_m, \quad m=0, 1, \ldots, L - 1 \tag{4.83}$$

The above equation shows that the search resolution of the Doppler frequency is determined by ω_m. From the definition of ω_m above, the frequency granularity of ω_m is 1 kHz divided by L, so the larger the L is, the higher the Doppler frequency resolution will be. In practice, in order to consider the requirements of the fast FFT algorithm, L is generally a power of two. However, the total duration of the data cannot exceed the boundary of the data bit transition. Therefore, L can be 2, 4, 8, or 16. If $L = 4$, the frequency resolution is 250 Hz; if $L = 8$, the frequency resolution is 125 Hz, and so on.

In the actual design, the FFT operation of the mixed signal of the local pseudo-code + carrier can be completed ahead of time and stored in the RAM. The FFT capture is directly read from the RAM to participate in the operation so the calculation can be reduced. The price is an increase in RAM capacity overhead. In this case, the operation of each horizontal block data requires two FFT operations—one for the FFT operation of the horizontal block data, and the other for the IFFT operation on the multiplied data. For an $L \times N$ input data block, if the Doppler frequency range is [−5 kHz, +5 kHz], a total of $2L \times 10$ number of N-point FFT operations are required, considering that the FFT operation on the vertical block data needs N number of L-points FFT operation to be performed. The total FFT operation is

$$C_{\text{Block}} = 10 \times 2L\, \text{number of N-point FFT} + N\, \text{number of L-point FFT}$$

If the FFT capture is performed directly on the $L \times N$ input data, the scheme of pre-storing the FFT result of the local pseudo-code and the carrier mixed signal is still adopted, and the number of frequency points required to achieve the same

Doppler frequency resolution is $10 \times L$. However, each frequency point still needs to complete two FFTs, but at this time, the number of data points of the FFT is $L \times N$, and the total FFT operation is

$$C_{\text{Normal}} = \text{FFT of } 20 \times L \text{ number of } LN\text{-point}$$

If the calculation amount of the n-point FFT is $n\text{lb}(n)$, the total operation amount of the block processing is

$$C_{\text{Block}} = 20 \cdot L \cdot N \cdot \text{lb}(N) + N \cdot L \cdot \text{lb}(L) \tag{4.84}$$

The total amount of computation for regular FFT processing is

$$\begin{aligned} C_{\text{Normal}} &= 20 \cdot L \cdot LN\text{lb}(LN) \\ &= 20 \cdot L^2 \cdot N \cdot \text{lb}(N) + 20 \cdot L^2 \cdot N \cdot \text{lb}(L) \end{aligned} \tag{4.85}$$

When $L > 1$, it is obvious that the value of Eq. (4.84) is smaller than Eq. (4.85). Figure 4.23 shows the comparison of the operations of the two methods corresponding to several different L lengths when $N = 1024$. The scheme of block processing and then FFT capture achieves fast capture by making full use of the FFT fast algorithm to reduce computation.

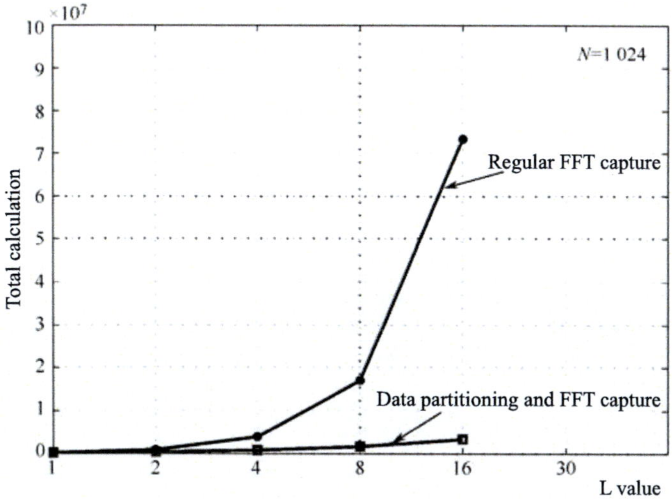

Fig. 4.23 Comparison of computation between time block processing when $L = 1, 2, 4, 8, 16$, and conventional FFT

4.1.7 Threshold Setting for Signal Capture

The previous sections explained several different methods of signal capturing. Although the specific implementation details of these methods vary, the mathematical principles are not the same. Ultimately, the threshold is created to determine whether satellite signals exist. Also, the carrier frequency of the signal and an estimate of the pseudo-code phase is determined. Some basic knowledge of coherent integration and threshold decisions has been described in Sect. 4.1.2. This section provides a detailed theoretical analysis of the setting of threshold decisions.

According to the analysis in Sect. 4.1.2, when the pseudo-code phase difference between the local pseudo-code and the input signal is $\Delta\tau$ and the carrier frequency difference is $\Delta\hat{\omega}$, the coherent integration result of the I-channel and Q-channel integrated outputs is

$$\overline{I}_P = D\sqrt{P_s}T_I \text{sinc}\left(\frac{\Delta\hat{\omega}T_I}{2}\right)R(\Delta\tau)\cos\left[\frac{\Delta\hat{\omega}}{2}T_I + \delta\phi_0\right] + \overline{N}_I \tag{4.86}$$

$$\overline{Q}_P = D\sqrt{P_s}T_I \text{sinc}\left(\frac{\Delta\hat{\omega}T_I}{2}\right)R(\Delta\tau)\sin\left[\frac{\Delta\hat{\omega}}{2}T_I + \delta\phi_0\right] + \overline{N}_Q \tag{4.87}$$

The definitions of the parameters in Eqs. (4.86) and (4.87) are the same as in Eqs. (4.31) and (4.32). $\overline{N}_I$ and $\overline{N}_Q$ are Gaussian white noises, fitting the $N(0, \sigma^2)$ distribution. If $A = \sqrt{P_s}T_I$ is defined, then when the pseudo-code phase and the carrier frequency are strictly aligned, the signal to noise ratio is

$$SN = \frac{A^2}{\sigma^2} \tag{4.88}$$

Note: The signal-to-noise ratio here is not the signal-to-noise ratio of the IF data, but the baseband signal-to-noise ratio after carrier stripping and pseudo-code stripping.

In cases where the carrier frequency error is not 0, since there is phase rotation, it is impossible to make a threshold decision only by the value of $\overline{I}_P$. The reason for this has been explained in Sect. 4.1.2. In this case, by using the pair (4.86) and (4.87) to perform the modulo operation to obtain the decision threshold:

$$\begin{aligned}\left|\overline{I}_P + i\overline{Q}_P\right| &= \sqrt{I_P^2 + Q_P^2} \\ &= A\left|\text{sinc}\left(\frac{\Delta\hat{\omega}T_I}{2}\right)\right|R(\Delta\tau) + N_R\end{aligned} \tag{4.89}$$

Equation (4.89) indicates a random variable whose distribution is Ricean.

Ricean distribution is named after the American scientist Stephen O. Rice, who analyzed the statistical properties of this random variable in 1945 [7]. The most common description of Ricean distribution is a certain amplitude of sine wave.

The envelope probability density function distribution of the narrowband Gaussian process can be expressed by the following formula

$$p(z|A, \sigma) = \frac{z}{\sigma^2} e^{-\frac{(z^2+A^2)}{2\sigma^2}} \mathrm{I}_0\left[\frac{Az}{\sigma^2}\right] \tag{4.90}$$

In this equation, A is the amplitude of the sine wave; σ^2 is the variance of the narrow-band Gaussian process; $\mathrm{I}_0(x)$ is the first-order zero-order modified Bessel function. Figure 4.24 is a probability density function for the Ricean distribution, which gives a plot of the probability density function for six different A values. It can be seen that the probability density function curve of the Ricean distribution is determined by A and σ^2, where A determines the approximate position of the center of the curve, and σ 2 determines the degree of concentration of the curve.

When capturing a signal, there are two situations: no signal exists, or only a noise component exists. This case is denoted by H_0 hypothesis. Another situation is that there is a signal, and the superposition of the signal and the noise is similar to a sine wave plus a narrow-band Gaussian random process, so it conforms to the definition of the above-mentioned Les distribution. This situation is denoted by the H_1 hypothesis.

In the situation of the H_0 hypothesis, A = 0, at which time Eq. (4.90) degenerates into Rayleigh distribution, where the probability density function is determined only by σ^2:

$$p(z) = \frac{z}{\sigma^2} e^{-\frac{z^2}{2\sigma^2}} \tag{4.91}$$

So we can consider the Rayleigh distribution as a special case of Rice distribution.

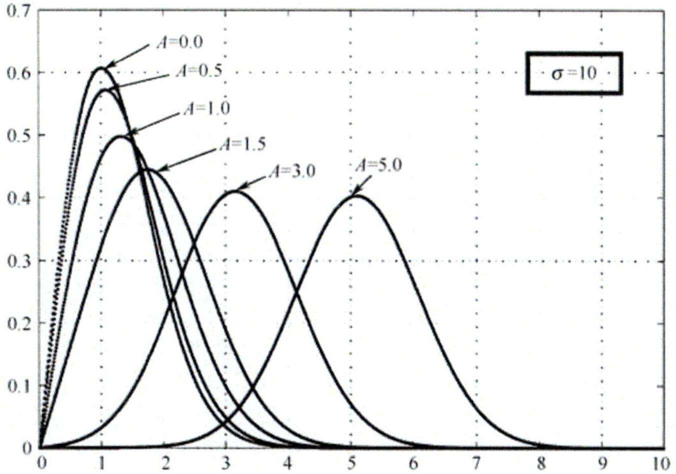

Fig. 4.24 PDF function for Rice distribution

With the knowledge of the probability density distribution of H_0 and H_1, the threshold decision can be regarded as the judgment of the random variable $\sqrt{I_p^2 + Q_p^2}$ according to the preset threshold value V_t: when $\sqrt{I_p^2 + Q_p^2}$ exceeds the threshold value, the signal is presented as a statement when $\sqrt{I_p^2 + Q_p^2}$ does not exceed the threshold.

So, under the assumption of H_0 and H_1, there are four cases:

1. Under the assumption of H_0, $\sqrt{I_p^2 + Q_p^2}$ exceeds the threshold.
2. Under the assumption of H_1, $\sqrt{I_p^2 + Q_p^2}$ does not exceed the threshold.
3. Under the assumption of H_0, $\sqrt{I_p^2 + Q_p^2}$ does not exceed the threshold.
4. Under the assumption of H_1, $\sqrt{I_p^2 + Q_p^2}$ exceeds the threshold.

Figure 4.25 vividly illustrates the physical meaning of the four decision results by using a graphical combination of probability density distribution functions. The area of the shaded part under the probability density function curve indicates the probability of occurrence of the corresponding event. The (a) event is called a false alarm event, indicating that the signal is incorrectly considered to exist in the absence of a signal, and its probability is represented by P_{fa}. The (b) event is called a missed detection event, indicating that there is a signal present. It is erroneously considered that no signal exists, and its probability is represented by P_m. The (c) event is called a correct discarding event, indicating that in the absence of a signal, it is correctly considered that no signal exists, and its probability is represented by P_T. The (d) event is called detection. An event indicates that a signal is correctly considered to exist in the presence of a signal, and its probability is represented by P_D. Obviously,

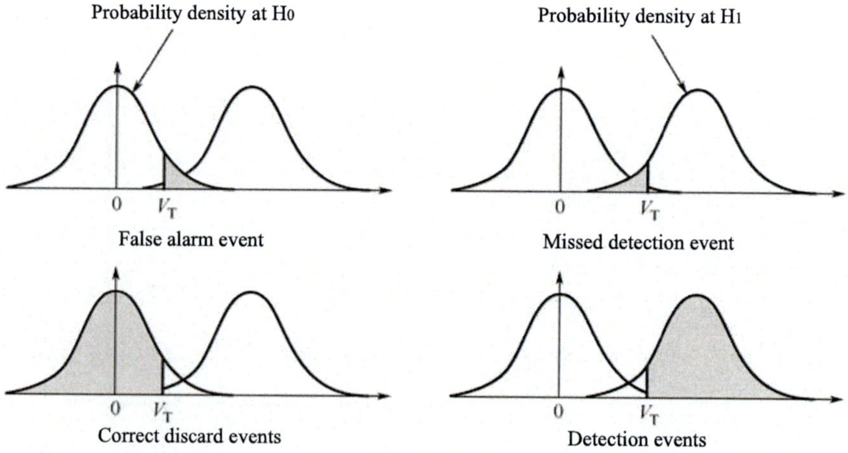

Fig. 4.25 Four cases of threshold decisions

$P_D + P_m = 1, P_{fa} + P_T = 1$, and it is generally desirable that P_{fa} and P_m are as small as possible, and that the larger the P_D and P_T, the better it will be.

In the receiver design, the false alarm event and the detection event are more important than the missed detection event and the correct discarding event, because the false alarm event often means that the subsequent traction and tracking process must be continued for the non-existent signal, and the hardware overhead of these processes and the time overhead is overwhelmed. False alarm events must be avoided as much as possible. The success of the detection event is directly related to the success rate of the receiver to the signal capture process, and is also directly related to the performance indicators such as the receiver TTFF. So, false alarm probability and detection probability are generally two key factors that must be considered when a threshold is set.

According to the calculation of the shaded area in Fig. 4.25, the expressions of P_{fa} and P_D can be given:

$$P_{fa} = \int_{V_T}^{+\infty} p(z|H_0)\,dz \tag{4.92}$$

$$P_D = \int_{V_T}^{+\infty} p(z|H_1)dz \tag{4.93}$$

In order to avoid square loss caused by non-coherent integration, only a single decision is analyzed here. For a single test decision, the decision input is $\sqrt{I_p^2 + Q_p^2}$, then given P_{fa} and σ^2, can determine the threshold V_T according to Eqs. (4.92) and (4.93).

$$\begin{aligned} P_{fa} &= \int_{V_T}^{+\infty} \frac{z}{\sigma^2} e^{-\frac{z^2}{2\sigma^2}}\,dz \\ &= e^{-\frac{V_T^2}{2\sigma^2}} \end{aligned} \tag{4.94}$$

It can be seen from the above equation that V_T can be uniquely determined at this time, that is,

$$V_T = \sigma\sqrt{-2\ln(P_{fa})} \tag{4.95}$$

After determining the V_T, the P_D can be determined according to the functional equation of $p(z|H_1)$. It can be seen that V_T, P_{fa} and P_D are related to each other when (A, σ^2) is fixed. Generally, one parameter cannot be determined independently, and then the other two parameters are determined independently. The choice of V_T values is expected to make the P_D as close as possible to 1, and P_{fa} as close as possible to

0. It is not possible to achieve this at the same time. The actual V_T value must be a compromise between these two.

In the actual GPS and BDS signal capture process, σ^2 may be determined according to the distribution of the IF sampling data, the quantization level, the IF sampling rate, the integration time length, and the proportional coefficient of the local pseudo-code carrier signal, or a noise correlation may be independently allocated. The channel estimates σ^2 in real time, and then determines A according to the lower limit CN0 of the capture sensitivity and the length of the coherent integration time, so that $p(z|H_0)$ and $p(z|H_1)$ can be determined. Then the V_T value can be set according to the above analysis. The V_T value is closely related to the details of the specific capture scheme implementation. For example, if you change the quantization bit width, the coherent integration length, or the IF sampling rate, you will need to re-estimate A and σ^2 to reset the V_T value correctly.

It should be noted that the discussion here is about a single decision on the envelope. There are many kinds of decision strategies in practice. Reference [8] explains a method for searching all possible $(\Delta\hat{\omega}, \Delta\tau)$ combinations and then selecting the highest value of $\sqrt{I_p^2 + Q_p^2}$ for decision. Reference [9] explains the Tong search method and the decision of M among N. The main purpose of these methods is to improve the completeness and robustness of signal capture. The selection of threshold values must be based on the probability distribution function of specific decision events. See the chapter on signal capture for more details.

4.1.8 Coherent Integration and Non-coherent Integration

Coherent integration is a very common but also critical operation in GPS and BDS receivers. Coherent integration is required in both signal capture and signal tracking processing. The coherent integration described in Sect. 4.1.2 is characterized by maintaining the carrier phase at the continuity of time. This section will explain the role of coherent integration and the difference between coherent and non-coherent integration.

Since a modern satellite receiver performs capture and tracking processing on the digital signal after the ADC sample is quantized, the discrete domain signal is analyzed here. Considering that the RF front-end bandwidth is B_w, the noise bilateral power spectral density is $N_0/2$, and the sampling rate is T_s, the signal sampled by the ADC can be expressed as

$$s_{IF}(mT_s) = \sqrt{2P_s}C(mT_s - \tau)\cos[\omega_{IF}mT_s + \phi_0] + n(mT_s), \quad m = [0, 1, \ldots] \tag{4.96}$$

In this equation, P_s is the carrier power; C is the pseudo-random code; ω_{IF} is the intermediate frequency carrier frequency; φ_0 is the initial phase; and $n(mT_s)$ is the Gaussian white noise, obeying the $N(0, \sigma^2)$ distribution, where $\sigma^2 = B_N$. Considering that the coherent integration length is less than the data bit hopping period, the data bit entries are omitted here.

If the local pseudo-code phase, the local carrier frequency, and the τ and ω_{IF} in Eq. (4.96) are identical (that is, the pseudo-code and the carrier are completely synchronized) and carrier stripping and pseudo-code stripping can be realized at this time, then the I-channel and Q-channel accumulators are passed. After the low-pass filtering characteristic (represented by the integrator in the continuous time domain, represented by the accumulator), Eq. (4.96) will be simplified to

$$\begin{aligned} s_{\rm I}(mT_{\rm s}) &= \sqrt{P_{\rm s}}\cos(\Phi) + n_{\rm I}(mT_{\rm s}) \\ s_{\rm Q}(mT_{\rm s}) &= \sqrt{P_{\rm s}}\sin(\Phi) + n_{\rm Q}(mT_{\rm s}) \end{aligned} \tag{4.97}$$

In this equation, Φ is the carrier phase difference with a fixed angle.

Considering that the coherent integration length is 1 ms and the number of sampling points is M, then the sum of the I-channel and Q-channel is

$$\begin{aligned} {\rm Sum}_{\rm I} &= \sqrt{P_{\rm s}}M\cos(\Phi) + \sum_{m=0}^{M-1} n_{\rm I}(mT_{\rm s}) \\ {\rm Sum}_{\rm Q} &= \sqrt{P_{\rm s}}M\sin(\Phi) + \sum_{m=0}^{M-1} n_{\rm Q}(mT_{\rm s}) \end{aligned} \tag{4.98}$$

In Eq. (4.98), the signal power is $M^2P_{\rm s}$, and the noise term variance is $M\sigma^2$, so the accumulated signal-to-noise ratio is

$$SN = \frac{M^2P_{\rm s}}{M\sigma^2} = M\frac{P_{\rm s}}{\sigma^2} \tag{4.99}$$

The accumulated signal-to-noise ratio is M times before the accumulation. Here, the boost of the signal-to-noise ratio is called the coherent gain, written in decibels:

$$\text{Coherent gain} = 10\lg(M)\,\text{dB}$$

It can be seen that the coherent gain is determined by M. When $M = 2046$, the coherent gain is $10\lg(2046) \approx 33$ dB. Since M is determined by the IF sampling rate, can it be improved by increasing the IF sampling rate? The conclusion can be drawn directly from the expression of the coherent gain, but in cases where the bandwidth of the digital sampling signal is outputted by the ADC, simply increasing the IF sampling rate means that the noise term in Eq. (4.96) is no longer "white noise", that is, it is no longer irrelevant in time, but exhibits certain correlation characteristics. Hence, the noise term variance in Eq. (4.98) is no longer $M\sigma^2$, but $f(M)\sigma^2$, where $f(M)$ is a value between M and M^2, depending on the correlation of the noise term in time. In this case, the coherent gain is no longer 10lg (M), but less than 10lg (M).

If you increase the bandwidth of the RF front-end, you can improve the sampling rate of the ADC while still ensuring the white noise characteristics of the noise term. It seems that we can increase the coherent gain by increasing M. However, it is necessary to carefully consider that while increasing the bandwidth of the RF front-end, the noise variance σ^2 in Eq. (4.96) is also increased; the coefficient of the increase of σ^2 and B_w are linear, and the sampling rate and Bw is also generally linear, that is, the multiple of M increase and the multiple of σ^2 increase are synchronous. Hence, the coherent integral gain cannot be increased by simply increasing the bandwidth of the RF front-end.

The above is an analysis of the characteristics of the coherent gain by discretely sampled signals and the statistical properties of the noise samples. The following angles are analyzed from the perspective of noise power spectral density and signal equivalent bandwidth.

We know that in order to ensure the loss of useful signal components as much as possible while suppressing out-of-band noise and interference, the RF front-end bandwidth B_w of the receiver can be considered as the bandwidth of the satellite navigation signal. For GPS signals, $B_w = 2.046$ MHz, and for BDS signals, $B_w = 4.092$ MHz.

After the coherent integration, assuming that the coherent integration time length is T, the equivalent signal bandwidth becomes $1/T$. The reason is that the signal has become a narrowband signal after pseudo-code stripping; the signal component is not affected, and the power of the noise component then becomes N_0/T. The smaller the equivalent signal bandwidth, the smaller the noise power, which means that the higher the signal-to-noise ratio, the multiple of the signal-to-noise ratio is the ratio of the RF bandwidth to the baseband signal bandwidth. For example, when $T = 1$ ms, the noise bandwidth becomes 1 kHz, and for the GPS signal, the signal-to-noise ratio is improved to

$$\frac{2.046 \text{ MHz}}{1 \text{ kHz}} = 2046 \approx 33 \text{ dB}$$

The same result can be obtained from an analysis of the signal bandwidth and the noise power spectral density. The same analysis can be performed on the BDS signal.

By summarizing the above analytical process, it can be concluded that if the coherent integration time is T, the coherent gain can be expressed as

$$\text{Gain} = 10\,\text{lg}\left(\frac{B_w}{B_b}\right) \text{ (dB)} \tag{4.100}$$

In Eq. (4.100), $B_b = 1/T$, and T is the coherent integration time length. In addition to the coherent integration time length, the actual coherent gain is also affected by many factors such as whether the pseudo-code phase is aligned, whether the carrier frequency is consistent, the ADC quantization noise, and the RF front-end noise figure.

Although (4.100) is simple, it is very useful in practice. Taking the GPS signal as an example, when the antenna equivalent noise temperature is 290 K, the noise power spectral density is about −174 dBm/Hz. If the input signal power is −130 dBm, and the RF front-end bandwidth is 2.046 MHz, the RF signal noise ratio is

$$-130\,\text{dBm} - (-174 + 63)\,\text{dBm} = -19\,\text{dB}.$$

According to Ref. [1], the baseband signal-to-noise ratio must be above 14 dB for reliable signal capture and tracking, so a coherent gain of 35 dB must be provided, as seen in Eq. (4.100). A coherent integration of $T = 2$ ms, which means that 2 ms are required to ensure sufficient coherent gain for good signal capture and tracking results.

According to the above analysis, the coherent gain can be increased by lengthening the coherent integration time, thereby improving the signal-to-noise ratio of the baseband signal. However, the influence of the bit jump of the navigation message cannot increase the coherent integration time without limitation. In this case, non-coherent integration can be adopted. Integrating (or accumulating) is the way to improve the signal-to-noise ratio.

The biggest difference between non-coherent integration and coherent integration is that non-coherent integration no longer maintains carrier phase continuity (as explained in Sect. 4.1.2) where the squared loss due to non-coherent integration is quantified.

The definition of squared loss is not the same among various research scholars. The definition in Ref. [10] is used here, that is

$$\text{Square loss} = \frac{\text{SNR}(\sqrt{I_\text{p}^2 + Q_\text{p}^2})}{\text{SNR}(\overline{I}_\text{p} + \text{i}\overline{Q}_\text{p})} \tag{4.101}$$

The meaning of Eq. (4.101) is the ratio of the signal-to-noise ratio after modulo $\overline{I}_\text{p} + \text{i}\overline{Q}_\text{p}$ to the signal-to-noise ratio of $\overline{I}_\text{p} + \text{i}\overline{Q}_\text{p}$. Reference [1] analyzes the non-coherent integration loss from the perspective of the given detection probability and false alarm probability. Reference [10] analyzes the square loss from the perspective of the probability distribution density function. The conclusions are not consistent, but only the basis of the problem and the starting point are different, and they are all reasonable under their respective contexts [11]. The following analysis will be based on the theory in Ref. [10], because from the intuitive physical meaning, this theory is more realistic.

Non-coherent integration does not accumulate $\overline{I}_\text{p} + \text{i}\overline{Q}_\text{p}$ of the I-channel and Q-channel integrator outputs, but accumulates its mode $\sqrt{I_\text{p}^2 + Q_\text{p}^2}$. The statistical characteristics of the random variables corresponding to the modulo operation are analyzed in Sect. 4.1.7. According to the analysis results in Sect. 4.1.7, the probability distribution of $\sqrt{I_\text{p}^2 + Q_\text{p}^2}$ is divided into two cases: (1) H_0 assumes a Rayleigh distribution; (2) H_1 assumes a Rice distribution. From the visual analysis, the modulo operation includes the square, add, and square root operations of the I-channel and Q-channel integration results. A new random variable is generated in the overall process, and the new random variable has the following differences compared with the coherent integration result:

1. H_1 assumes that the magnitude of the signal correlation peak has changed.
2. H_0 and H_1 assume that the mean of $\sqrt{I_p^2 + Q_p^2}$ is no longer 0, and the mean value of noise contained in the coherent integration result $\overline{I}_p + i\overline{Q}_p$ is zero.
3. H_0 assumes that the variance of $\sqrt{I_p^2 + Q_p^2}$ has changed.

Figure 4.26 shows the result of coherent integration of an actually acquired GPS signal. The integration time is 1 ms. The picture on the left shows the result of $\overline{I}_p+i\overline{Q}_p$, and the picture on the right shows the result of $\sqrt{I_p^2 + Q_p^2}$. In the left picture, there are actually two curves corresponding to the I integral and the Q integral. Although the two curves are not entangled with each other due to the limitation of the display ratio, it is obvious that the noise average is 0. The picture on the right is $\sqrt{I_p^2 + Q_p^2}$, which is the result of squaring, adding, and taking the square root of the two curves on the left. The whole curve is positive. It can be seen that the noise mean is no longer 0, but higher than 0.

First, we assume that the signal amplitude of the I-channel and Q-channel results of the coherent integrated output is A_c, and the respective noise powers of the two channels are σ_c^2, where the subscript c indicates the coherent integral, considering that the noise components of the I-channel and Q-channel channels are mutually orthogonal. Therefore, the signal to noise ratio is

$$\mathrm{SNR}(\overline{I}_p + i\overline{Q}_p) = \frac{A_c^2}{2\sigma_c^2} \tag{4.102}$$

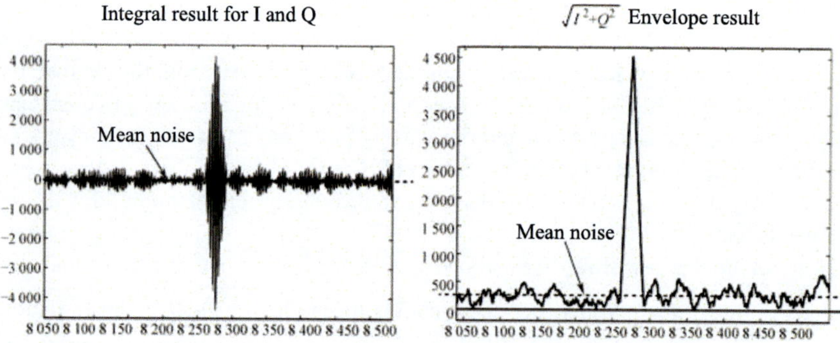

Fig. 4.26 Comparison of noise mean value between $\overline{I}_p + i\overline{Q}_p$ (left) and $\sqrt{I_p^2 + Q_p^2}$ (right)

Let's look at the case of non-coherent integration. The first step in non-coherent integration is to perform the modulo operation on $\overline{I}_{\mathrm{p}} + \mathrm{i}\overline{Q}_{\mathrm{p}}$.

Considering the H0 hypothesis, $\sqrt{I_{\mathrm{p}}^2 + Q_{\mathrm{p}}^2}$ is Rayleigh distribution, and its probability density function is shown by Eq. (4.91). At this time, the noise mean and variance are as shown in Eqs. (4.103) and (4.104) respectively [10].

$$\mu_{\mathrm{n0}} = \sigma_{\mathrm{c}}\sqrt{\frac{\pi}{2}} \tag{4.103}$$

$$\sigma_{\mathrm{n0}}^2 = \frac{4-\pi}{2}\sigma_{\mathrm{c}}^2 \tag{4.104}$$

In the case of H_1 hypothesis, $\sqrt{I_p^2 + Q_p^2}$ is Rice distribution, and its probability density function is shown by Eq. (4.90). At this point, the mean noise value [10] is

$$\mu_{\mathrm{n1}} = \sigma_{\mathrm{c}}\sqrt{\frac{\pi}{2}}e^{-\frac{A_{\mathrm{c}}^2}{4\sigma_{\mathrm{c}}^2}}\left[\left(1+\frac{A_{\mathrm{c}}^2}{2\sigma_{\mathrm{c}}^2}\right)\mathrm{I}_0\left(\frac{A_{\mathrm{c}}^2}{4\sigma_{\mathrm{c}}^2}\right) + \frac{A_{\mathrm{c}}^2}{2\sigma_{\mathrm{c}}^2}I_1\left(\frac{A_{\mathrm{c}}^2}{4\sigma_{\mathrm{c}}^2}\right)\right] \tag{4.105}$$

In Eq. (4.105), $\mathrm{I}_0(\cdot)$ and $\mathrm{I}_1(\cdot)$ are distributed as first-order zero-order and first-order modified Bessel functions.

If $\gamma = A_{\mathrm{c}}^2/2\sigma_{\mathrm{c}}^2$ is the signal-to-noise ratio of the coherent integration, and Eq. (4.105) can be simplified to

$$\mu_{\mathrm{n1}}(\gamma, \sigma_{\mathrm{c}}) = \sigma_{\mathrm{c}}\sqrt{\frac{\pi}{2}}e^{-\frac{\gamma}{2}}\left[(1+\gamma)\mathrm{I}_0\left(\frac{\gamma}{2}\right) + \gamma I_1\left(\frac{\gamma}{2}\right)\right] \tag{4.106}$$

The mean of $\sqrt{I_{\mathrm{p}}^2 + Q_{\mathrm{p}}^2}$ is a function of γ and σ. The meaning of Eq. (4.106) is the mean value of $\sqrt{I_{\mathrm{p}}^2 + Q_{\mathrm{p}}^2}$ after the modulo operation when the signal is present. In the right graph in Fig. 4.12, μ actually includes two parts: one part is the peak due to the signal autocorrelation peak; the other is the increase of the autocorrelation peak due to the noise mean no longer being zero. This extra boost is the component represented by Eq. (4.103). When calculating the signal power, the part of μ_{n1} that is increased due to the non-zero noise mean value should be subtracted, so the signal-to-noise ratio at this time is

$$\begin{aligned}\mathrm{SNR}\left(\sqrt{I_{\mathrm{p}}^2 + Q_{\mathrm{p}}^2}\right) &= \frac{(\mu_{\mathrm{n1}} - \mu_{\mathrm{n0}})^2}{\sigma_{\mathrm{n0}}^2} \\ &= \frac{\pi}{4-\pi}\left[e^{-\frac{\gamma}{2}}\left[(1+\gamma)I_0\left(\frac{\gamma}{2}\right) + \gamma I_1\left(\frac{\gamma}{2}\right)\right] - 1\right]^2\end{aligned} \tag{4.107}$$

Combining Eqs. (4.102) and (4.107), according to Eq. (4.101), the square loss at this time is

$$\text{Square loss} = \frac{\pi}{(4-\pi)\gamma}\left[e^{-\frac{\gamma}{2}}\left[(1+\gamma)I_0\left(\frac{\gamma}{2}\right)+\gamma I_1\left(\frac{\gamma}{2}\right)\right]-1\right]^2 \quad (4.108)$$

Figure 4.27 shows the plot of the squared loss and the coherent integrated signal-to-noise ratio γ plotted according to Eq. (4.108). The horizontal axis represents the range of γ, the unit is dB, and the vertical axis is the corresponding squared loss. The unit is also dB. In the figure, γ varies from −20 to 30 dB, and the corresponding squared loss is −20.4 to +6.4 dB.

It may seem puzzling that the squared loss becomes positive, because it means that the modulo operation is not "loss" but "enhancement" for the signal-to-noise ratio. However, the rigorous mathematical derivation from the above does yield such a result, and the actual Monte Carlo simulation data results confirm this [10]. In fact, it may be easier to understand from the perspective of experience and intuitiveness. In the case of high signal-to-noise ratio, the variance of the error is made smaller by the modulo operation, which can be calculated through Eq. (4.104). The biggest change in error is that the mean is no longer 0, which is equivalent to adding a positive DC component, as shown by the result of Eq. (4.103). However, at the same time, the squared results of the I-channel and Q-channel also increase the signal amplitude. When the signal-to-noise ratio is high, the signal amplitude increases faster than the noise average value, and the result is that the output signal-to-noise ratio is greater than the input signal-to-noise ratio according to Eq. (4.107).

As can be seen in Fig. 4.27, when the coherent integrated signal-to-noise ratio is greater than 3.4 dB, the squared loss begins to be greater than 0. Therefore, before the non-coherent integration, if the signal-to-noise ratio is greater than 3.4 dB by a certain time correlation integral, it results in an additional signal to noise ratio increase. If the signal-to-noise ratio of the correlation is low, then the modulo operation will bring a non-negligible weakening to the signal-to-noise ratio. For example,

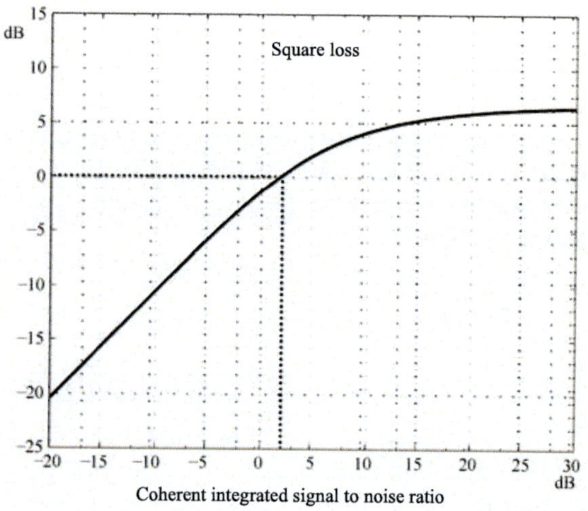

Fig. 4.27 Coherent integral signal-to-noise ratio γ and squared loss

when $\gamma = -20$ dB, the signal-to-noise ratio after modulo reduction will be reduced by about 20.4 dB, which is a result that needs to be avoided.

The above analysis shows the influence of the modulo operation on the signal-to-noise ratio of the coherent integration result. After that, several non-coherent accumulations are needed, assuming that the number of non-coherent accumulations is L. Because the results of L coherent integration modulo ($\sqrt{I_{p,k}^2 + Q_{p,k}^2}$, k = 0, 1, …, L), the random variables contained can be considered as time-independent. Then, the signal amplitudes are increased by L times, and the noise root mean square is only increased by $\sqrt{L}$, so it can be considered that the improvement of the signal-to-noise ratio caused by L-order non-coherent accumulation is

$$\text{Non-coherent accumulating gain} = 10\lg(L)\ (\text{dB})$$

If the coherent integration time length is T, the non-coherent accumulation time is L, and the RF front-end bandwidth is B_w. Combing Eq. (4.100) for the correlation integral gain, and considering the above conclusion about the square loss and the non-coherent accumulation gain, the total gain is

$$\text{Gain} = 10\lg\left(\frac{B_w}{B_b}\right) + 10\lg(L) - \textit{square loss}\ (\text{dB}) \tag{4.109}$$

In Eq. (4.109), $B_b = 1/T$, and the square loss is calculated by Eq. (4.108). The calculation of the squared loss is related to the signal-to-noise ratio of the coherent integration result, so it is difficult to give a certain value at the beginning of the system design. Different signal strengths and different correlation integration times will result in different square losses. In general, the coherent integration time should be delayed as much as possible to ensure that there is no danger of data bit jump. This can ensure a large coherent integrated signal-to-noise ratio, and a smaller square loss can be obtained. However, when the coherent integration time is extended, it also causes the Doppler shift search space to increase, raising the amount of computation. Therefore, the coherent integration time length and the non-coherent accumulation number need to be carefully considered, and a balance point between performance and computation amount must be selected.

4.2 Signal Tracking

After the acquisition of the signal, a rough estimate of the carrier frequency and pseudo-code phase of the signal will be conducted in GPS and BDS receivers. The word "rough" is used here to describe the acquisition of a signal relative to the tracking loop. In general, based on the acquisition of the signal, the estimated accuracy of the carrier frequency is several tens to several hundreds of hertz, and the estimation accuracy of the pseudo-code phase is within ±0.5 chips. However, the accuracy cannot guarantee the stable demodulation of the navigation message data, as the demodulation of data must generally be conducted in a stable tracking state, that is, the carrier frequency difference is 0, the carrier phase difference is close to 0, and the pseudo-code phase difference is within 0.01 ~ 0.1 chips. With the relative motion of the satellite and the receiver, the carrier frequency and pseudo-code phase of the signal received by the antenna also changes accordingly. Even more challenging is the fact that the clock drift and random jitter of the local clock on the receiver will also affect the locking of the captured signal. Hence, without the continuous dynamic adjustment of the carrier NCO and the pseudo-code NCO, the captured signal will quickly lose lock. Signal tracking is by its very nature a dynamic adjustment strategy for loop parameters to achieve stable tracking of signal carriers and pseudo-codes.

Signal tracking has two purposes: one is to track the carrier components in satellite navigation signals, and the other is to track the pseudo-code components. Therefore, there must be two tightly coupled tracking loops inside the receiver. Simultaneous analysis of such loops is so difficult that they are analyzed separately in this chapter, which means that one loop is assumed to be stably locked when the other is under analysis. This kind of processing not only facilitates theoretical analysis, but also accords with the actual working state of the receiver. In practical applications, when the receiver is in a stable working state, the two loops must be stably locked.

After the synchronization of the carrier, the pseudo-codes, and the bits is achieved, the in-phase integrator of the tracking loop will generate the navigation telegram bits. The navigation signal can be demodulated when the signal-to-noise ratio is sufficiently high, and the pseudo-range, carrier phase, and Doppler observation can also be extracted. The information will be sent to the subsequent navigation solving unit for PVT solution. Therefore, the signal tracking link plays a crucial role in the signal processing of the receiver, and the performance of the signal tracking function will directly affect the overall performance index and user experience of the receiver.

Before we begin to analyze the tracking loops inside the BDS and GPS receivers, we must have a deep understanding of how the basic phase-locked loop works, which serves as the basis for the subsequent theoretical analysis. So, let us begin with the basic phase-locked loop.

4.2.1 The Basic Phase-Locked Loop

The basic phase-locked loop is shown in Fig. 4.28.

It can be seen that the basic phase-locked loop is actually a closed-loop feedback control system, including a phase detector, voltage-controlled oscillator, and loop filter. In Fig. 4.28, $v_i(t)$ is the input signal. Since the phase-locked loop is designed to lock the phase of the input signal, the actual input is the phase of $v_i(t)$, as shown by $\theta_i(t)$. The phase of the local signal generated by the device is written as $\theta_1(t)$, so the phase difference outputted by the phase detector is $\hat{\theta} = \theta_i(t) - \theta_l(t)$. The high frequency components of the phase difference are filtered out by the loop filter, and the value after the procedure—which is also the output of the phase-locked

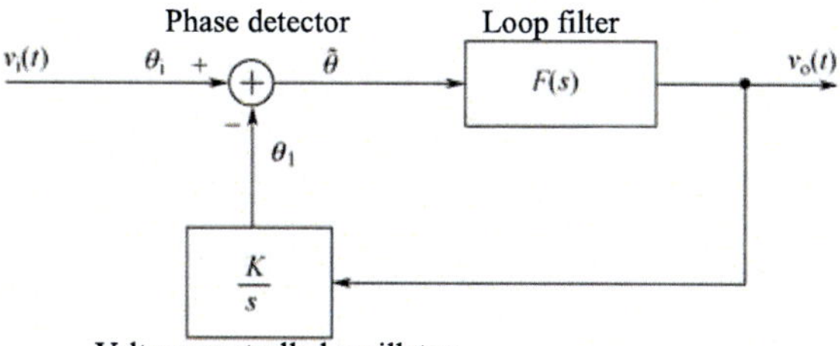

Fig. 4.28 Block diagram of the basic phase-locked loop function

loop, written as $v_o(t)$—is used to control the voltage-controlled oscillator. For the convenience of successive analysis, the model shown in the figure is the frequency domain model, and the one used in reality is the time domain model.

The output frequency of the voltage-controlled oscillator is proportional to the input. Assuming that $u(t)$ is the input of the voltage controlled oscillator, then

$$f_{\text{vco}}(t) = ku(t) \tag{4.110}$$

where $f_{\text{vco}}(t)$ is the frequency of the output signal of the voltage-controlled oscillator, and k is the voltage-controlled gain of the oscillator in radians/volts.

Since the phase is the integral of the frequency, that of the output signal can be expressed as

$$\theta_{\text{vco}}(t) = \int_0^t f_{\text{vco}}(\tau)\mathrm{d}\tau \tag{4.111}$$

Taking the Laplace transform of Eq. (4.111) and substituting Eq. (4.110) in it, the frequency domain model of the voltage-controlled oscillator can be obtained.

$$\theta_{\text{vco}}(s) = \frac{k}{s}U(s) \tag{4.112}$$

Here, $U(s)$ is the Laplace transform of $u(t)$, i.e. $U(s) = L[u(t)]$, where $L[\cdot]$ is a Laplacian. Looking back at Fig. 4.28, where the input to the voltage-controlled oscillator is the output of the loop filter,

$$\begin{aligned} v_o(s) &= \hat{\theta}(s)F(s) \\ &= [\theta_i(s) - \theta_l(s)]F(s) \end{aligned} \tag{4.113}$$

According to Eq. (4.112), we get

$$\theta_l(s) = \frac{k}{s}[\theta_i(s) - \theta_l(s)]F(s) \tag{4.114}$$

Making a further transformation of the equation, then

$$\theta_1(s) = \frac{kF(s)}{s + kF(s)}\theta_i(s) \tag{4.115}$$

According to Eq. (4.115), the phase difference $\hat{\theta}(s)$ can be written as

$$\hat{\theta}(s) = \frac{s}{s + kF(s)}\theta_i(s) \tag{4.116}$$

So, the Loop transfer function $H(s)$ and error transfer function $E(s)$ can be written as

$$H(s) = \frac{\theta_1(s)}{\theta_i(s)} = \frac{kF(s)}{s + kF(s)} \tag{4.117}$$

$$E(s) = \frac{\hat{\theta}(s)}{\theta_i(s)} = \frac{s}{s + kF(s)} \tag{4.118}$$

It is easy to work out that the relation of $H(s)$ and $E(s)$ is

$$E(s) = 1 - H(s) \tag{4.119}$$

It can be seen from the above analysis that the transfer function of the phase-locked loop is determined by the loop filter. Therefore, we can say that the performance of the phase-locked loop is largely related to the loop filter. After the fixation of the form of $F(s)$, the transfer function $H(s)$ can be obtained using Eq. (4.117), and the corresponding equivalent noise bandwidth is thereto defined as

$$B_n = \int_0^\infty |H(\mathrm{j}2\pi f)|^2 \mathrm{d}f \tag{4.120}$$

The physical meaning of the equivalent noise bandwidth can be considered as a linear system with white noise passing through the transfer function $H(s)$. If the power spectral density of the single sideband noise is N_0, the noise power outputted by the system is

$$P_{N_0} = \int_0^\infty |H(\mathrm{j}2\pi f)|^2 N_0 \mathrm{d}f = B_n N_0$$

It can be seen that P_{N_0} is similar to the result of the above-mentioned white noise passing through a linear system with a bandwidth of B_n. In this sense, Eq. (4.120)

Table 4.1 The noise bandwidth formula of first-order, second-order, and third-order loops [12]

Loop order	$H(s)$	B_n
1	$\frac{a_0}{b_0+b_1 s}$	$\frac{a_0^2}{4b_0 b_1}$
2	$\frac{a_0+a_1 s}{b_0+b_1 s+b_2 s^2}$	$\frac{a_0^2 b_2+a_1^2 b_0}{4b_0 b_1 b_2}$
3	$\frac{a_0+a_1 s+a_2 s^2}{b_0+b_1 s+b_2 s^2+b_3 s^3}$	$\frac{a_2^2 b_0 b_1+(a_1^2-2a_0 a_2)b_0 b_3+a_0^2 b_2 b_3}{4b_0 b_3(b_1 b_2-b_0 b_3)}$

defines the equivalent noise bandwidth corresponding to the closed-loop transfer function $H(s)$.

B_n will be used later to analyze the loop thermal noise characteristics. The error transfer function $E(s)$ obtained from Eq. (4.118) is used to analyze the response of the loop to dynamic stresses, specifically, to analyze the steady-state phase error of the input excitation under dynamic stress according to Laplace's final value theorem.

For the closed-loop transfer function of the n-order loop, the transfer function of the phase-locked loop is generally written as

$$H(s) = \frac{a_0 + a_1 s + a_2 s^2 + \cdots + a_{n-1} s^n}{b_0 + b_1 s + b_2 s^2 + \cdots + b_{n-1} s^n} \tag{4.121}$$

The order of the phase-locked loop is determined by the highest power of the s operator contained in the denominator of $H(s)$. Substituting Eq. (4.121) into Eq. (4.120), the noise bandwidth of different loops can be obtained. The loops commonly used in GPS and BDS receivers include the first-order loop, second-order loop, and third-order loop. Lindsey offers a table for B_n, as shown in Table 4.1, which helps us to calculate the value of B_n corresponding to $H(s)$ of the first-, second-, and third-order loops.

The highest power of the s operator contained in the denominator of $H(s)$ is actually the number of poles of H(s). According to the knowledge of the signal and the system, the integrator of the time domain corresponds to the 1/s operator of the Laplacian domain. Thus, the number of poles of $H(s)$ is the same as the number of integrators in the loop. It can therefore be inferred that the number of phase-locked loops is determined by the number of integrators in the closed loop, which will be demonstrated in subsequent analysis of the first-, second-, and third-order loops. Crucially, the voltage-controlled oscillator is an indispensable device in the phase-locked loop, and an integrator is already installed in it.

Comprehensive analysis of the phase-locked loop is a difficult and cumbersome task [13]. The phase-locked loop in the receiver includes the carrier loop and the pseudo-code loop. Engineers are usually concerned about the transient response and steady-state response of the loop. Below is a brief introduction to these two features.

The three common inputs for phase-locked loops are shown in Fig. 4.29, distributed from left to right as phase steps, frequency steps, and frequency ramps. The phase step input adds a value of θ_0 to the original phase, while the frequency step input adds a value of $\Delta\omega t$ to the original phase. This means that frequency

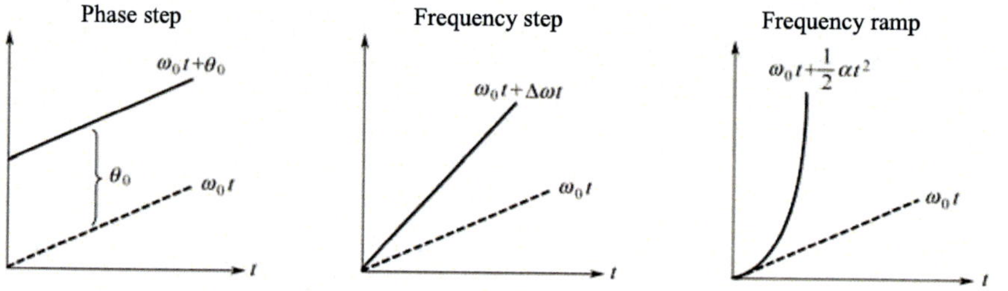

Fig. 4.29 Three common phase inputs for a phase-locked loop

discontinuity occurs, and the frequency ramp input adds an acceleration component α of the frequency to the original one.

Since the carrier phase in the satellite receiver corresponds to the distance between the receiver and the satellite, and the carrier frequency corresponds to the velocity between the receiver and the satellite, then the three phase inputs in Fig. 4.29 correspond to the mutation in distance, speed, and acceleration. If the distance vector between the satellite and the receiver is $\boldsymbol{R}$, then the three phase inputs in the figure correspond to $\Delta\boldsymbol{R}$, $\Delta\dot{\boldsymbol{R}}$ and $\Delta\ddot{\boldsymbol{R}}$.

Suppose the input quantity of the phase-locked loop is $u(t)$ in the time domain, and is written as $u(s)$ after being converted to the s field, then the output can be obtained through Eq. (4.117).

$$\theta_1(s) = H(s)u(s)$$

Transforming the above expression into the time domain, then

$$\theta_1(t) = L^{-1}[H(s)u(s)] \tag{4.122}$$

Then, according to the result of Eq. (4.122), the transient response analysis of the output of the phase-locked loop can be performed. The transient response of a phase-locked loop primarily measures the response speed and performance of changes in the loop input. For example, when a mutation in the frequency or phase of the input signal occurs, it is a case of whether the loop can keep up with the change and remain locked as quickly as possible, or whether the loop lock will fail due to its inability to keep up with the pace of change.

Laplace's final value theorem can be directly applied to the analysis of the steady-state response, i.e.

$$\hat{\theta}_{1,\infty} = \lim_{s\to 0} sH(s)u(s) \tag{4.123}$$

Unlike the transient response, the steady-state response measures the ability of the phase-locked loop to track the input signal after entering a stable tracking state, or whether there is a stable deviation. The local phase output and phase error quantities in the steady state can be obtained by applying the final value theorem to the $H(s)$ and $E(s)$ functions of the loop. Both transient and steady-state responses require some input excitation, but the focus is different.

We will now analyze some common loop filters.

1. First-Order Loop

The loop filter corresponding to the first-order loop is an all-pass filter, and there is no reactance component. Its filter function is

$$F(s) = 1$$

Bringing the above formula into Eqs. (4.117) and (4.118), we can get the $H(s)$ and $E(s)$ of the first-order loop.

$$H(s) = \frac{k}{s+k}, \quad E(s) = \frac{s}{s+k}$$

Comparing them with the parameters of the first-order loop in Fig. 4.1, we can get that $a_0 = k$, $b_1 = 1$, $b_0 = k$. According to the calculation formula of B_n in the table, the noise bandwidth of the first-order loop is

$$B_n = \frac{k}{4} \tag{4.124}$$

(1) Phase Step Input

When the input signal is a phase step signal, $\theta_\mathrm{i}(t) = \theta_0 u(t)$ [where $u(t)$ is a unit step function, the same below], and its Laplace transform can be written as $\theta_\mathrm{i}(s) = \frac{\theta_0}{s}$, then according to final value theorem, the steady-state phase difference is

$$\hat{\theta}_{e,\infty} = \lim_{s\to 0} sE(s)\theta_\mathrm{i}(s) = \lim_{s\to 0} \frac{\theta_0 s}{s+k} = 0$$

(2) Frequency Step Input

When the input signal is a frequency step signal, $\theta_\mathrm{i}(t) = \Delta\omega t u(t)$, and its Laplace transform is $\theta_\mathrm{i}(s) = \frac{\Delta\omega}{s^2}$, then its steady-state phase difference is

$$\hat{\theta}_{e,\infty} = \lim_{s\to 0} sE(s)\theta_\mathrm{i}(s) = \lim_{s\to 0} \frac{\Delta\omega}{s+k} = \frac{\Delta\omega}{k} \tag{4.125}$$

That is, its steady-state phase difference is not zero.

(3) Frequency Ramp Input

When the input signal is ramped up by frequency, $\theta_\mathrm{i}(t) = \frac{1}{2}\alpha t^2$ and its Laplace transform is $\theta_\mathrm{i}(s) = \frac{\alpha}{s^3}$, then the steady-state phase difference is

$$\hat{\theta}_{e,\infty} = \lim_{s \to 0} sE(s)\theta_{\mathrm{i}}(s) = \lim_{s \to 0} \frac{\alpha}{s(s+k)} = \infty \tag{4.126}$$

That is, the steady-state phase difference is infinite, which means that the loop is out of lock.

According to the above analysis, the first-order loop cannot track the phase input signals of frequency step and frequency ramp up particularly well. In fact, the satellite signal received by the GPS receiver is affected by the Doppler effect and the dynamics of the local user. There will always be frequency steps, more or less. With the aid of external information, frequency steps and higher-order frequency variations can be deducted, and the first-order loop can be used for tracking. However, in the absence of external assistance, the first-order loop will generally not be used in the tracking loop in BDS and GPS receivers.

2. Second-Order Loop

There are many kinds of loop filters that can perform the second-order loop, such as a resistive volume divider, an active proportional integrator, and a passive proportional integrator. Here, we take the active proportional integrator as an example. Readers can perform a similar analysis for other filters.

For an active proportional integrator, $F(s) = \frac{1+s\tau_1}{s\tau_2}$, then $H(s)$ and $E(s)$ can be written as

$$H(s) = \frac{\frac{k\tau_1}{\tau_2}s + \frac{k}{\tau_2}}{s^2 + \frac{k\tau_1}{\tau_2}s + \frac{k}{\tau_2}}, \quad E(s) = \frac{s^2}{s^2 + \frac{k\tau_1}{\tau_2}s + \frac{k}{\tau_2}}$$

In circuit and control theory analysis, writing the denominator in the above two formulae into a normalized expression tends to make subsequent analysis easier, i.e.

$$\left(s^2 + \frac{k\tau_1}{\tau_2}s + \frac{k}{\tau_2}\right) \triangleq (s^2 + 2\zeta\omega_{\mathrm{n}}s + \omega_{\mathrm{n}}^2) \tag{4.127}$$

where, ω_{n} and ζ are defined as follows:

$$\omega_{\mathrm{n}} \triangleq \sqrt{\frac{k}{\tau_2}}, \quad \zeta \triangleq \frac{1}{2}\omega_{\mathrm{n}}\tau_1 \tag{4.128}$$

In the above formula, ω_{n} represents the natural frequency of the PLL, the physical meaning of which is that for the change of $\theta_{\mathrm{i}}(t)$, the output of the PLL, $\theta_1(t)$, will generate a transient response, whose expression is similar to the damped oscillation. Since the angular frequency of the damped oscillation is ω_{n}, the corresponding damping coefficient is ζ. According to the control theory, when ζ is very small, the transient response needs a large overshoot to reach the steady state, which means that it is underdamped. When ζ is large, because the system is too damped, although there is no overshoot, it will still take a long time for the system to reach a steady state, which means it is over-damped. Therefore, generally speaking, the critical optimization value $\zeta = 0.707$ will be selected. With the definition of ω_{n} in Eq. (4.128), $H(s)$ can be rewritten as:

$$H(s) = \frac{2\zeta\omega_{\mathrm{n}}s + \omega_{\mathrm{n}}^2}{s^2 + 2\zeta\omega_{\mathrm{n}}s + \omega_{\mathrm{n}}^2}$$

According to Fig. 4.1,

$$a_1 = 2\zeta\omega_{\mathrm{n}}, \quad a_0 = \omega_{\mathrm{n}}^2$$
$$b_2 = 1, \quad b_1 = 2\zeta\omega_n, \quad b_0 = \omega_n^2$$

Therefore, the equivalent noise bandwidth of the second-order loop can be obtained:

$$B_n = \frac{1}{8}\omega_n\left(4\zeta + \frac{1}{\zeta}\right) \tag{4.129}$$

Let us consider the steady-state phase difference of the second-order loop for the three input phases.

(1) Phase Step Input

When the input signal is a phase step signal, $\theta_{\mathrm{i}}(t) = \theta_0 u(t)$, $\theta_{\mathrm{i}}(s) = \frac{\theta_0}{s}$, then according to final value theorem, the steady-state phase difference is

$$\hat{\theta}_{e,\infty} = \lim_{s\to 0} sE(s)\theta_{\mathrm{i}}(s) = \lim_{s\to 0} \frac{\theta_0 s^2}{s^2 + 2\zeta\omega_{\mathrm{n}}s + \omega_{\mathrm{n}}^2} = 0 \tag{4.130}$$

(2) Frequency Step Input

When the input signal is a frequency step signal, $\theta_{\mathrm{i}}(t) = \Delta\omega t u(t)$, $\theta_{\mathrm{i}}(s) = \frac{\Delta\omega}{s^2}$, then its steady-state phase difference is

$$\hat{\theta}_{e,\infty} = \lim_{s\to 0} sE(s)\theta_{\mathrm{i}}(s) = \lim_{s\to 0} \frac{\Delta\omega s}{s^2 + 2\zeta\omega_{\mathrm{n}}s + \omega_{\mathrm{n}}^2} = 0 \tag{4.131}$$

(3) Frequency Ramp Input

When the input signal is ramped up by frequency, $\theta_{\mathrm{i}}(t) = \frac{1}{2}\alpha t^2$, $\theta_{\mathrm{i}}(s) = \frac{\alpha}{s^3}$ then the steady-state phase difference is

$$\hat{\theta}_{e,\infty} = \lim_{s\to 0} sE(s)\theta_{\mathrm{i}}(s) = \lim_{s\to 0} \frac{\alpha}{s^2 + 2\zeta\omega_{\mathrm{n}}s + \omega_{\mathrm{n}}^2)} = \frac{\alpha}{\omega_{\mathrm{n}}^2} \tag{4.132}$$

That is, the steady-state phase difference is not zero.

From this we can see that the second-order loop can track the phase step and frequency step signal well, but not the frequency ramp signal. However, it is still able to handle the satellite signals received by BDS and GPS receivers most of the time. At the same time, the second-order loop is also ideal in terms of complexity and stability, so the second-order loop is the tracking loop that is commonly used in civilian receivers. However, the second-order loop is not suitable for highly dynamic motion applications because it cannot handle the unbiased tracking of the frequency ramping input signal.

3. Third-Order Loop

The loop filter of the third-order loop commonly used in BDS and GPS receivers is often expressed in the following form [9]:

$$F(s) = \tau_1 + \frac{\tau_2}{s} + \frac{\tau_3}{s^2}$$

So, H(s) and E(s) can be written as

$$H(s) = \frac{k\tau_1 s^2 + k\tau_2 s + k\tau_3}{s^3 + k\tau_1 s^2 + k\tau_2 s + k\tau_3}, \quad E(s) = \frac{s^3}{s^3 + k\tau_1 s^2 + k\tau_2 s + k\tau_3}$$

Comparing them to Fig. 4.1, then

$$a_2 = k\tau_1, \ \ a_1 = k\tau_2, \ \ a_0 = k\tau_3$$
$$b_3 = 1, \ \ b_2 = k\tau_1, \ \ b_1 = k\tau_2, \ \ b_0 = k\tau_3$$

Therefore, the equivalent noise bandwidth of the third-order loop can be derived:

$$B_n = \frac{k\tau_1}{4} + \frac{k\tau_2^2}{4(k\tau_1\tau_2 - \tau_3)} \tag{4.133}$$

(1) Phase Step Input

When the input signal is a phase step signal, $\theta_i(t) = \theta_0 u(t)$, $\theta_i(s) = \frac{\theta_0}{s}$, then the steady-state phase difference is

$$\hat{\theta}_{e,\infty} = \lim_{s\to 0} sE(s)\theta_i(s) = \lim_{s\to 0} \frac{\theta_0 s^3}{s^2 + k\tau_1 s^2 + k\tau_2 s + k\tau_3} = 0 \tag{4.134}$$

(2) Frequency Step Input

When the input signal is a frequency step signal, $\theta_i(t) = \Delta\omega t u(t)$, $\theta_i(s) = \frac{\Delta\omega}{s^2}$, then its steady-state phase difference is

$$\hat{\theta}_{e,\infty} = \lim_{s\to 0} sE(s)\theta_i(s) = \lim_{s\to 0} \frac{\Delta\omega s^2}{s^2 + k\tau_1 s^2 + k\tau_2 s + k\tau_3} = 0 \tag{4.135}$$

(3) Frequency Ramp Input

When the input signal is ramped up by frequency, $\theta_{\mathrm{i}}(t) = \frac{1}{2}\alpha t^2, \theta_{\mathrm{i}}(s) = \frac{\alpha}{s^3}$, then the steady-state phase difference is

$$\hat{\theta}_{e,\infty} = \lim_{s\to 0} sE(s)\theta_{\mathrm{i}}(s) = \lim_{s\to 0} \frac{\alpha s}{s^2 + k\tau_1 s^2 + k\tau_2 s + k\tau_3} = 0 \tag{4.136}$$

It can be seen from the above analysis that the third-order loop can track the frequency ramping input signal without bias, so it is more suitable for the tracking loop design in highly dynamic situations. However, unlike the first- and second-order loops, the third-order loop has stability problems. The first- and second-order loops are unconditionally stable. According to the stability judgment conditions of Rouse-Hurwitz, the system is stable when the following conditions are met:

$$k\tau_1 > 0,\ k\tau_2 > 0,\ k\tau_3 > 0,\ \ k\tau_1 \cdot k\tau_2 > k\tau_3 \tag{4.137}$$

Equation (4.137) only shows a very loose stability constraint. In Ref. [9], E. D. Kaplan proposes that when $Bn \leq 18$ Hz, stability can be better guaranteed in the third-order loop.

Observing the $H(s)$ and $E(s)$ of the first-, second-, and third-order loops, it can be seen that the first-order loop has only one adjustment coefficient: k, and the second-order loop has two adjustment systems: ζ and ω_{n}—according to Eq. (4.127)—while the third-order loop has four adjustment coefficients: k, τ_1, τ_2 and τ_3. In fact, $k\tau_1$, $k\tau_2$, and $k\tau_3$ can be combined and regarded as three coefficients, so the third-order loop actually has only three adjustment coefficients. In the design of the real receiver tracking loop, since the NCO is implemented by digital logic or software programming, and the voltage control gain can be set by itself, we can simply set it to 1 and convert k to other coefficients. Then, the implementation of the loop filters of the first-, second-, and third-order loops can be indicated in Fig. 4.30.

The coefficients in Fig. 4.30 and those in the above formula correspond as shown in Table 4.2. The calculation formula of Bn is also given.

After implementing different loop filters according to Fig. 4.30, the filter coefficient(s) must be determined, which, for the first-order loop, means only ω_{n}. According to the relationship between B_n and ω_n in Table 4.2, ω_n can be determined according to the set equivalent noise bandwidth. For the second-order loop, it is necessary to determine the value of a_2 and ω_n. One of the principles is to select the damping coefficient of the second-order loop as the critical optimization value, i.e. $\zeta = 0.707$. Then, it can be proved that when $a_2 = 1.414$, $B_n = 0.53\ \omega_n$, which means that the value of ω_n can also be determined based on B_n, the set equivalent noise bandwidth. For the third-order loop, the value of a_3, b_3, and ω_n needs to be determined. The optimization values given in Ref. [9] are $a_3 = 1.1$, b_3 2.4. Substituting the values of a_3 and b_3 into B_n, then $B_n = 0.7845\omega_n$. ω_n can also be determined according to B_n, the set equivalent noise bandwidth.

Figure 4.30 shows that the order of the phase-locked loop corresponds to the number of integrators in the loop. The last integrator in Fig. 4.30 is provided by the

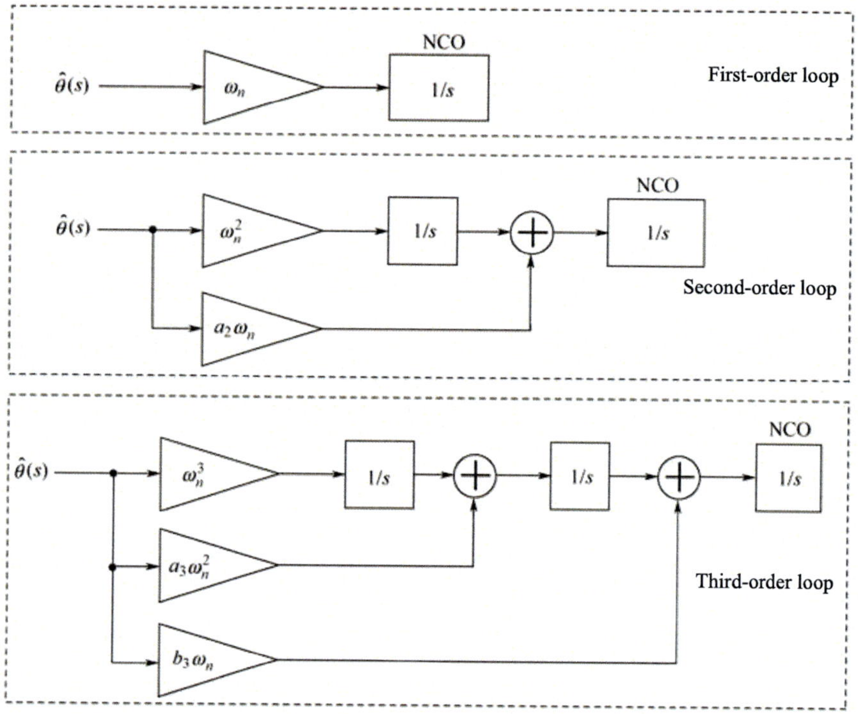

Fig. 4.30 The three-loop filter implementation method: first-order, second-order, and third-order loops

Table 4.2 Coefficient correspondence of the loop filter and the calculation formula of B_n

	Coefficient in Fig. 4.30	Coefficient in formula	B_n
First-order loop	ω_n	k	$\omega_n/4$
Second-order loop	$(\omega_n)^2$	k/τ_2	$\frac{(1+a_2^2)}{4a_2}\omega_n$
	$a_2\omega_n$	$k\tau_1/\tau_2$	
Third-order loop	$(\omega_n)^3$	$k\tau_3$	$\frac{(a_3b_3^2+a_3^2-b_3)}{4(a_3b_3-1)}\omega_n$
	$a_3(\omega_n)2$	$k\tau_2$	
	$b_3\omega_n$	$k\tau_1$	

NCO. In general, the loop filter of the N-order phase-locked loop contains $N-1$ integrators. If NCO is added, then there are N integrators. Similar to the form of a loop filter of the third-order loop, for the Nth order loop, it can be assumed that $F(s) = a_0 + a_1/s + \ldots + a_{N-1}/s^{N-1}$. Then, its closed-loop transfer function is

$$H(s) = \frac{a_0 s^{N-1} + a_1 s^{N-2} + \cdots + a_{N-1}}{s^N + a_0 s^{N-1} + a_1 s^{N-2} + \cdots + a_{N-1}} \tag{4.138}$$

and the error transfer function is

$$E(s) = \frac{s^N}{s^N + a_0 s^{N-1} + a_1 s^{N-2} + \cdots + a_{N-1}} \tag{4.139}$$

Defining the natural frequency of the loop $\omega_n = (a_{N-1})^{1/N}$, we can use the final value theorem to prove that the steady-state phase difference of the input dynamic stress is

$$\begin{aligned} \hat{\theta}_{e,\infty} &= \lim_{s \to 0} sE(s)\theta_i(s) = \lim_{s \to 0} \frac{s^{N+1}\theta_i(s)}{s^N + a_0 s^{N-1} + a_1 s^{N-2} + \cdots + a_{N-1}} \\ &= \lim_{s \to 0} \frac{s \cdot [s^N \theta_i(s)]}{\omega_n^N} = \lim_{t \to \infty} \frac{\mathrm{d}^N \theta_i(t)/\mathrm{d}t^N}{\omega_n^N} \end{aligned} \tag{4.140}$$

Since the physical meaning of $\theta_i(t)$ is the carrier phase in the input signal received by the receiver, and the phase is proportional to the distance, then Eq. (4.140) means that phase-locked loops of different orders can track dynamic stress inputs of different orders. The higher the dynamic, the higher the order that will be needed for the phase-locked loop. However, at the same time, the coefficient selection and stability problems of the high-order phase-locked loop will become more prominent. It can be seen from Eq. (4.140) that in the circumstance of stable tracking, the higher the natural frequency ω_n, the smaller the steady-state phase difference will be. Because B_n and ω_n are linear, the higher the value of ω_n, the larger the noise bandwidth will be. Therefore, generally speaking, although the steady-state phase difference can be reduced with larger ω_n, the increased value of B_n will also degrade the thermal noise performance of the loop.

4.2.2 *Thermal Noise Performance Analysis of the Linear Phase-Locked Loop*

Because the physical quantity to be tracked by the phase-locked loop is the phase of the input signal, noise analysis equals the analysis of the noise of the output phase. Assuming that the Gaussian white noise is superimposed on the input signal, it is obvious that the noise is not directly superimposed on the carrier phase of the signal. In fact, these noises are directly reflected in the amplitude of the signal. However, the presence of noise will affect the precision of the moment when the carrier wave passes the zero point. If the carrier phase is measured by a simple record of the moment, then the accuracy of the phase measurement will be affected, which will indirectly affect the stability of the carrier phase. This is the effect of additive noise on the carrier phase, which is shown in Fig. 4.31.

The upper left part of Fig. 4.31 is a pure sine wave signal $A\sin(\omega_0 \mathrm{t} + \varphi)$, and the upper right part is a Gaussian white noise, following the $N(0, \sigma 2)$ distribution pattern.

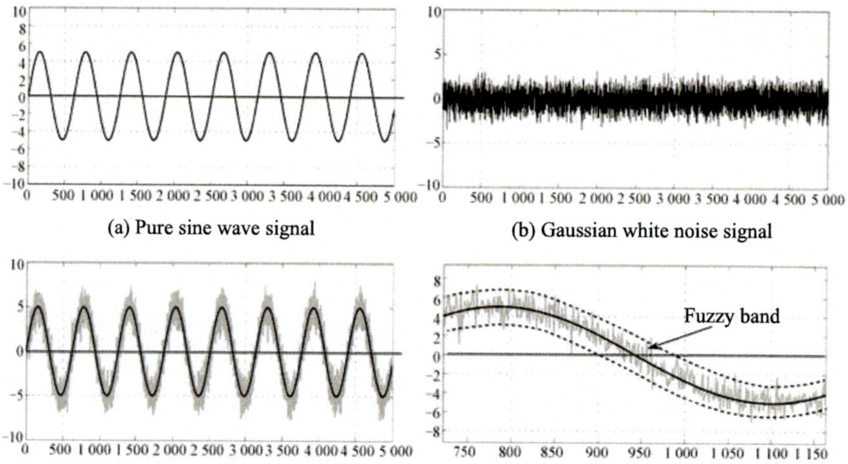

(a) Pure sine wave signal (b) Gaussian white noise signal

(c) Superimposed mixed signal (d) Partial detail of the mixed signal

Fig. 4.31 Effect of additive Gaussian white noise on the carrier phase

The lower left part is the mixed signal after the superposition of the two. There are many "burrs" on the sine wave signal, and the lower right part is the detailed figure of the mixed signal near the zero point. The middle solid black line is the original clean sine wave signal for comparison. The moment when the original sine wave signal crosses the zero point is fixed, but due to the superposition of noise, the moment when the mixed signal crossed the zero point changes randomly. It is no longer a certain moment, but a "fuzzy band" whose width indicates the magnitude of the phase jitter.

If Gaussian white noise $n(t) \sim N(0, \sigma^2)$ is superimposed on a certain amplitude sine wave signal $A\sin(\omega_0 t + \varphi)$, the mathematical expression of the mixed signal can be written as follows:

$$s(t) = A\sin(\omega_0 t + \varphi) + n(t) \tag{4.141}$$

After measuring the phase of $s(t)$, it can theoretically deduced that $\sigma_{\theta_n}^2$, the variance of the phase jitter due to the superposition of additive noise can be written as

$$\sigma_{\theta_n}^2 = \frac{1}{2\mathrm{SNR}} \tag{4.142}$$

Here, SNR is the signal-to-noise ratio of the mixed signal, i.e. $\mathrm{SNR} = P_s/P_n$, and P_s and P_n are the power of the signal and noise respectively.

Figure 4.32 is a block diagram of the phase-locked loop after noise superimposition. The white noise $n(t)$ is added to the signal after it comes out of the phase detector, while the noise in real cases is added to the input signal before the signal goes through the phase detector. Generally speaking, the most commonly used phase detector is a multiplier. The input signal is multiplied by the local carrier signal through the

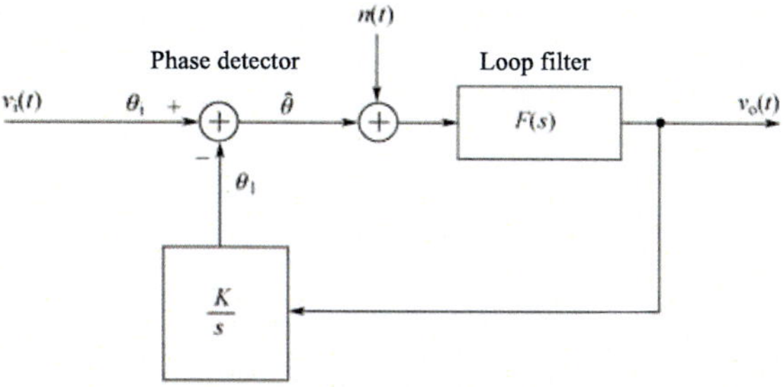

Fig. 4.32 Block diagram of a linear phase-locked loop taking additive noise into account

phase detector, so the influence of the phase detector on the noise only changes the noise phase, and the noise power spectrum and noise distribution stay unchanged. Therefore, the equivalence makes sense. The purpose of introducing it here is to make subsequent analysis easier.

Now, the phase difference signals $\hat{\theta}(t)$ and $n(t)$ are input into the voltage controlled oscillator. Following the method of analysis in Sect. 4.2.1, we can work out that

$$\begin{aligned} \theta_1(s) &= \theta_i(s) - \hat{\theta}(s) = \frac{kF(s)}{s}[\hat{\theta}(s) + n(s)] \quad \Rightarrow \\ \hat{\theta}(s) &= \frac{s}{s + KF(s)}\theta_i(s) - \frac{kF(s)}{s + kF(s)}n(s) \\ &= E(s)\theta_i(s) - H(s)n(s) \end{aligned} \tag{4.143}$$

Here, $n(s)$ is the Laplace transformation of $n(t)$.

Equation (4.143) is a very useful formula, from which it can be concluded that in cases when white noise exists in the phase-locked loop input, the phase noise of the output phase will be composed of two parts, one of which is the steady-state phase difference of the system caused by dynamic stress, and the other is the random phase difference caused by the input additive noise. The steady-state phase difference of the system has been discussed in detail in Sect. 4.2.1. It can be deduced from Eq. (4.140) that as long as there is high-order dynamic stress in the phase input signal of the phase-locked loop, the steady-state phase difference will always exist. Even if the input signal is a very clean sine wave, there will still be a steady-state phase difference. While the random phase difference is caused by noise. The last term in Eq. (4.143) is the phase error component generated by the input noise. By carefully observing the expression form in Eq. (4.143), we can see that the random phase difference is similar to the input noise $n(t)$ going through a linear system with a transfer function of $H(s)$. Assuming that the unilateral power spectral density of n(t) is N_0, the mean value of the noise signal outputted by the linear system is

$$\sigma_{H(s)}^2 = \int_0^{\infty} |H(j2\pi f)|^2 N_0 \mathrm{d}f \tag{4.144}$$

According to Eq. (4.142), when the input signal power of the phase-locked loop is P_i, the variance of the phase jitter caused by the random phase difference will be

$$\delta\theta_n^2 = \frac{\sigma_{H(s)}^2}{2P_i} = \frac{N_0}{2P_i}\int_0^{\infty} |H(j2\pi f)|^2 df$$

According to the equation above and the definition of the phase-locked loop equivalent noise bandwidth B_n, we can see that:

$$\delta\theta_n^2 = \frac{N_0 B_n}{2P_i} \tag{4.145}$$

It can be concluded from the above analysis that the system's steady-state phase difference is positively correlated with the error transfer function $E(s)$, and the stochastic phase difference caused by additive noise is positively correlated with the closed-loop system transfer function $H(s)$. Therefore, the random phase difference can be reduced through reducing the noise bandwidth B_n, but the steady-state phase difference will increase. Increasing the noise bandwidth B_n can help reduce the steady-state phase difference, but the random phase difference will deteriorate. The requirements for stochastic phase difference and steady-state phase difference are contradictory, and designers must consider the trade-off between the two.

The tracking loop in BDS and GPS receivers is much more complicated in implementation than in Fig. 4.32, and the analysis of the corresponding loop noise is also much more difficult. However, the fundamental structure of the loop after linearization is still the basic phase-locked structure. In addition to the influence of the random and steady-state phase difference analyzed above, the effect of clock jitter on the tracking loop is not negligible, for which the clock on the GPS receiver must have high stability. Due to the limitations of space, the influence of clock jitter will not be described in detail in this chapter. Interested readers can refer to Ref. [9] in Chap. 5 and Ref. [8] in Chap. 12.

The carrier and pseudo-code tracking loop in BDS and GPS receivers will be described in detail in the following two sections. Since the two loops are coupled to each other, their parameters affect each other's performance, so it is very difficult to analyze both loops simultaneously. Therefore, in this book, analysis of these two loops is carried out separately. It is assumed that the pseudo-code phase is aligned in the analysis of the carrier loop, and the carrier phase has been stably locked in the analysis of the pseudo-code loop. This will greatly simplify the process of analysis, and at the same time correspond to the stabilized status of the locked loop in reality.

4.2.3 Carrier Tracking Loop

In most cases, carrier synchronization in BDS and GPS receivers is achieved through the Costas loop, which was invented by American engineer John P. Costas in the 1950s, and is considered to have imposed "a profound impact on the field of modern digital communications" [14]. In this section, a detailed introduction will be given on the application of the Costas loop in the baseband signal processing of satellite navigation and positioning receivers.

1. The Basic Structure of the Costas Loop

Figure 4.33 shows the basic structure of the Costas loop. The main logic units include the multiplier, carrier NCO, loop filter, phase detector, and integral output unit (Integrate and Dump). The input signal is first pseudo-code stripped by a pseudo-code generator, and then multiplied by the in-phase and quadrature components outputted by the local carrier NCO. Then, the result is integrated and inverted by the in-phase and quadrature integrators to obtain the in-phase component $I(t)$ and the quadrature component $Q(t)$. The integration result will be sent to the phase detector to obtain the phase error signal, which will be sent as an input to the loop filter. The obtained error signal will return to the carrier NCO to help complete the tracking adjustment of the signal carrier. Since the in-phase and quadrature signals of the local carrier are involved, the Costas loop is also referred to as an in-phase orthogonal loop. In terms of signal flow, the Costas loop is similar to the basic phase-locked loop introduced earlier, a closed-loop feedback system. But, the generation of the phase error signal in the Costas loop is slightly more complicated than that in the basic phase-locked loop. The benefits of using phase inversion signals from in-phase and quadrature signals are shown below.

The pseudo-code generator in Fig. 4.33 is unique to BDS and GPS receivers. It is not essential for the simple synchronization of the carrier signal. However, in terms of the structural characteristics of BDS and GPS ranging signals, the pseudo-code

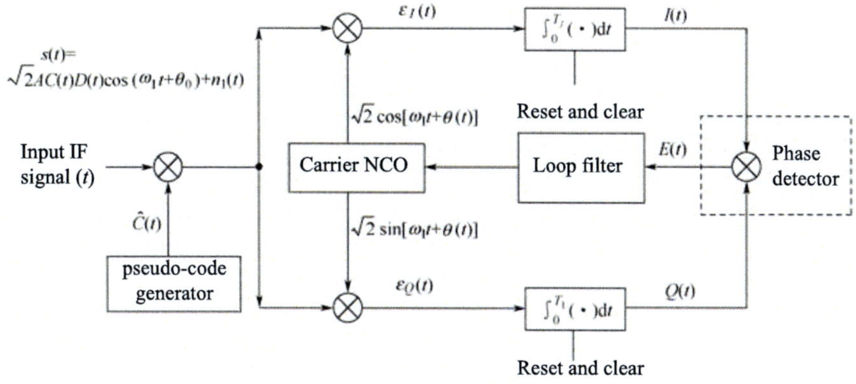

Fig. 4.33 Basic structure of the Costas loop

generator is indispensable because the only way to remove the pseudo-code signal is to allow the integral output unit to integrate over a length of 1 ms, otherwise the integration result will be similar to that of the noise due to the presence of the pseudo-code component. The pseudo-code generator in Fig. 4.33 must be adjusted according to the signal to be processed, which means that it must be set as the GPS pseudo-code generator when the GPS signal is under process, and as the BDS pseudo-code generator when the BDS signal is under process.

In Fig. 4.33 we assume that the input signal $s(t) = \sqrt{2}AC(t)D(t)\cos(\omega_{\mathrm{I}}t + \theta_0) + n_{\mathrm{I}}$. Here, A is the signal amplitude, $C(t)$ is the pseudo-random code, $D(t)$ is the data bit, ω_{I} is the carrier frequency, θ_0 is the carrier initial phase, and $n_{\mathrm{I}}(t)$ is the Gaussian white noise, whose unilateral power spectral density is assumed to be N_0. Noise terms are temporarily ignored in this section because the noise performance of the Costas loop will be specifically analyzed in the following chapters.

First, $s(t)$ is multiplied by the local pseudo-code $C(t)$ generated by the pseudo-code generator, where $\hat{C}(t) = C(t + \tau)$, and where τ represents the pseudo-code phase difference between the local pseudo-code and the input signal. When the pseudo-code tracking loop is stable, $\tau \approx 0$, so

$$C(t)\hat{C}(t) \approx C^2(t) = 1$$

Therefore, it can be considered that after the local pseudo-code is multiplied, it will be removed from the input signal, which means that the process of pseudo-code de-spreading is completed. The signal processed by the pseudo-code generator multiplier can be expressed as:

$$s(t)\hat{C}(t) \approx \sqrt{2}AD(t)\cos(\omega_I t + \theta_0) \tag{4.146}$$

$D(t)$ is the data bit. For the GPS and BDS D1 code signals, the data code period is 20 ms; for the BDS D2 code signal, the data period is 2 ms. The length of the integrator should generally not exceed the data code period. Otherwise, the data bit hopping will be included in the integration time, resulting in the offset of the partial integration result, thus affecting the result of the whole integration. Erroneous phase error signals will be generated when the error is severe. The integration time T_{I} is typically an integer multiple of 1 ms, such as 5, 10, and 20 ms. This is because the pseudo-code period adopted for both BDS and GPS signals is 1 ms. It should be noted here that when integration over a length of 1 ms is performed after the bit synchronization to ensure that the integration time does not cross the bit transition boundary, and for the BDS D1 code signal, peeling of the secondary code can occur.

Signals are output from the local carrier NCO in two channels; one is the in-phase channel $\sqrt{2}\cos[\omega_{\mathrm{I}}t + \theta(t)]$, and the other is the orthogonal path $\sqrt{2}\sin[\omega_{\mathrm{I}}t + \theta(t)]$. The output of the orthogonal channel is actually obtained by phase-shifting the output of the in-phase channel by 90°. Here, $\theta(t)$ is the difference between the local carrier phase and the input signal carrier frequency $\omega_{\mathrm{I}}t$, in which difference may only exists in the initial phase of the carrier, or the carrier frequency, so it is represented by a function of time: t.

The output signals of the multipliers of the in-phase and quadrature branches are

$$\varepsilon_{\mathrm{I}}(t) = AD(t)\cos(\omega_{\mathrm{I}}t + \theta_0)\cos[\omega_{\mathrm{I}}t + \theta(t)] \tag{4.147}$$

$$\varepsilon_{\mathrm{Q}}(t) = AD(t)\cos(\omega_{\mathrm{I}}t + \theta_0)\sin[\omega_{\mathrm{I}}t + \theta(t)] \tag{4.148}$$

To simplify the analysis, noise terms are omitted from Eqs. (4.147) and (4.148).$\varepsilon_{\mathrm{I}}(t)$ and $\varepsilon_{\mathrm{Q}}(t)$ are integrated by the integrator. With the results of Eqs. (4.31) and (4.32), the outputs of the in-phase integrator and the quadrature integrator can be obtained.

$$I(t) = AD(t)T_{\mathrm{I}}\mathrm{sinc}\left(\frac{\Delta\omega T_{\mathrm{I}}}{2}\right)\cos\left[\frac{\Delta\omega}{2}T_{\mathrm{I}} + \phi(t)\right] \tag{4.149}$$

$$Q(t) = AD(t)T_{\mathrm{I}}\mathrm{sinc}\left(\frac{\Delta\omega T_{\mathrm{I}}}{2}\right)\sin\left[\frac{\Delta\omega}{2}T_{\mathrm{I}} + \phi(t)\right] \tag{4.150}$$

In Eqs. (4.149) and (4.150), $\Delta\omega = \theta'(t)$ is the frequency difference between the local carrier signal and the input signal, and $\varphi(t)$ is defined as follows:

$$\phi(t) \triangleq \theta(t) - \theta_0 \tag{4.151}$$

It can be seen that $\varphi(t)$ is actually the phase difference between the input signal and the local carrier.

When entering the carrier tracking phase, the frequencies of the input signal and the local carrier signal are already very close, and are often much smaller than the bandwidth of the integrator, i.e. $\Delta\,\omega << 1/T_{\mathrm{I}}$, then $\Delta\,\omega T_{\mathrm{I}} \approx 0$, so Eqs. (4.149) and (4.150) can be approximated as

$$I(t) \approx AD(t)T_{\mathrm{I}}\cos[\phi(t)] \tag{4.152}$$

$$Q(t) \approx AD(t)T_{\mathrm{I}}\sin[\phi(t)] \tag{4.153}$$

A multiplier is adopted in the classic Costas loop to complete the phase discrimination, as shown in the dashed box in Fig. 4.33. However, the actual phase detector is not necessarily a multiplier. Several different phase detectors will be described in the following sections. When the phase detector function is performed by the multiplier, the output phase error signal $E(t)$ is

$$E(t) = I(t)Q(t) = \frac{1}{2}A^2D^2(t)T_{\mathrm{I}}^2\sin[2\phi(t)] \tag{4.154}$$

The value of $D(t)$ in Eq. (4.154) becomes 1 after squaring, and it can be seen that the multiplier acts as a phase detector so that the Costas loop is insensitive to the

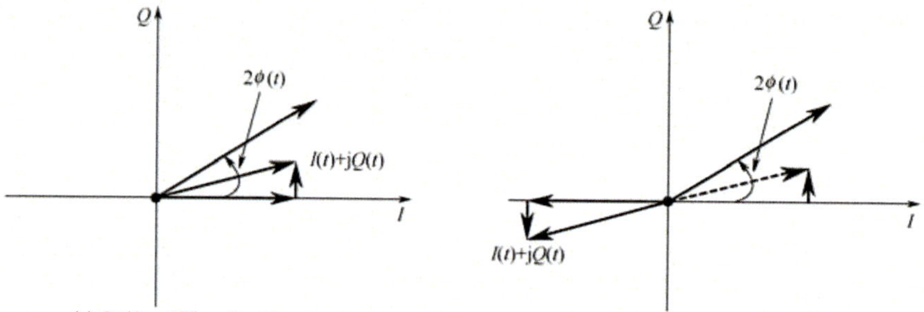

(a) $D(t) = 1$ The situation (b) $D(t) = -1$ The situation

Fig. 4.34 The Costas loop is insensitive to changes in data bits

+1 and −1 values of the data bits. After normalization, the phase error in Eq. (4.154) is $\sin[2\varphi(t)]$. When $\varphi(t)$ is small, $\sin[2\varphi(t)] \approx 2\varphi(t)$, and normalization here can be done by using the value of $\sqrt{I^2(t) + Q^2(t)}$ as the divisor.

Figure 4.34 shows the insensitivity of the Costas loop to data bit hopping. Figure 4.34a is the case when $D(t) = 1$ and Fig. 4.34b is when $D(t) = -1$. In both cases, the phase detector output is $2\varphi(t)$.

After most of the high frequency components are filtered out from the phase error signal outputted by the phase detector through the loop filter, the signal will function as the adjustment signal of the carrier NCO, bringing the local carrier frequency and the input signal frequency closer and closer, and eventually achieving the locking of the frequency and phase. Readers can refer to the analysis of the loop filter $F(s)$, the system transfer function $H(s)$, the error transfer function $E(s)$, and the dynamic stress steady-state phase difference of the basic phase-locked loop in Sect. 4.2.1 for further analysis of the entire closed loop.

When phase tracking is implemented, $2\varphi(t) \approx 0$, then according to Eqs. (4.149) and (4.150), data bits will be generated by the I-channel integrator. The Q-channel integrator does not have any signal components, only some noise signals, so the data bits can be read directly from the I-channel and then sent to subsequent units for navigation message demodulation.

2. Phase Detector and Discriminator

There are many phase detectors used in the Costas loop. Table 4.3 shows the four that are commonly used in BDS and GPS receivers, namely $I(t) \times Q(t)$, $\text{sign}[I(t)] \times Q(t)$, $Q(t)/I(t)$, and $\text{atan}[Q(t)/I(t)]$, where sign [·] represents an f h operation with a value of +1 or −1.

The expression of the phase difference of the four phase detector outputs in Table 4.3 is not difficult to understand. It can be easily obtained by substituting Eqs. (4.152) and (4.153). The noise term can be ignored during the derivation to simplify the analysis. It is worth mentioning that the range of the output phase difference of the four phase detectors is −90° to +90°. For $I(t) \times Q(t)$, it is obvious, because the period of $\sin[2\varphi(t)]$ is 180°. Yet for $\text{sign}[I(t)] \times Q(t)$, there might be some misunderstanding, because according to the table, the output phase difference

Table 4.3 Commonly used phase detectors in the Costas loop

Phase detector	Difference of output phases	Property
$Q(t) \times I(t)$	$\sin[2\phi(t)]$	The classic Costas ring phase detector has an approximately optimized phase discrimination characteristic in the case of low SNR; the phase slope will be greatly affected by the signal amplitude
$Q(t) \times \text{sign}[I(t)]$	$\sin\phi(t)$	In cases of high SNR, it has approximately optimal phase discrimination characteristics. Its phase discrimination will be greatly affected by the signal amplitude, but the calculation amount is small
$Q(t)/I(t)$	$\tan\phi(t)$	Under both high and low signal-to-noise ratios, it has near-optimal phase-detection characteristics. The phase-detecting slope will not be affected by the signal amplitude, and will diverge when the phase difference is ±90°
$\text{atan}[Q(t)/I(t)]$	$\phi(t)$	It is the inverse tangent function of the two quadrants, which has the best phase discrimination characteristics in cases of high SNR and low SNR. The phase slope will not be affected by the signal amplitude

is $\sin\varphi(t)$. Intuitively, the period should be 360°, and the phase discrimination range should be −180° to +180°. However, due to the value hopping of the product term of sign[$I(t)$], the phase difference of the output hops at ±90°, so that the phase discrimination range repeats the case between [−90°, 0°] at [90°, 180°] and the case between [0°, 90°] at [−180°, 90°]. The output range of the phase detector corresponding to $Q(t)/I(t)$ and atan[Q(t)/$I(t)$] is also relatively easy to understand, and readers can analyze it independently.

Figure 4.35 shows the comparison of the phase difference outputs of the four phase detectors. For better comparison, only the phase difference term is shown; the signal amplitude $AD(t)T_{\mathrm{I}}$ term is omitted, and the phase differences are all converted to angles. It can be seen from the figure that the linearity of atan[$Q(t)/I(t)$] is the best of the four, but the computation is intensive. The linearity of sign[$I(t)$] × $Q(t)$ follows that of atan[$Q(t)/I(t)$], and the computation is less intensive. Since division exists in $Q(t)/I(t)$ and atan[$Q(t)/I(t)$], there may be a problem of dividing by zero, but the concern of normalization is eliminated. The computation of $I(t) \times Q(t)$ and sign[$I(t)$] × $Q(t)$ is modest, but normalization needs to be conducted, otherwise the phase difference signal output may fluctuate with the change of signal amplitude.

In practice, in addition to tracking the carrier phase, the Costas loop can also be used to track the carrier frequency. The structure is basically the same as the loop structure shown in Fig. 4.33, except that the phase detector is replaced by a detector. The main function of the frequency discriminator is to detect the frequency difference between the input signal and the local carrier. The error signal used to control the NCO is therefore not the phase difference but rather the frequency difference. When

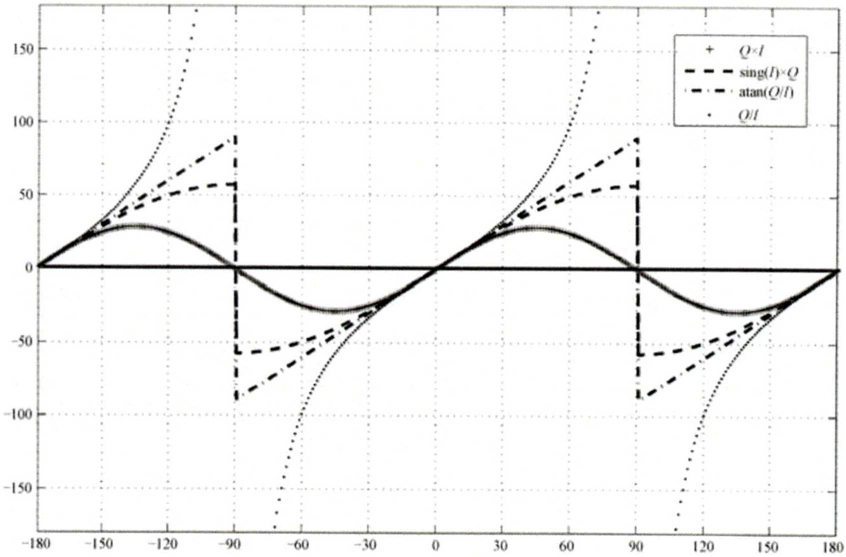

Fig. 4.35 Phase detection curves of the four most commonly used Costas loop phase detectors

the loop is stable, the frequency of the input carrier is locked. The Costas loop that tracks the input carrier frequency is also called FLL (Frequency Lock Loop), or the "Automatic Frequency Control" (AFC) loop. Since the frequency is the differentiation of the phase, the basic principle of a discriminator is to subtract the phase of the adjacent two frequencies to generate the estimated value of the frequency difference, that is,

$$\Delta f(kT_s) = \frac{\phi(kT_s) - \phi[(k-1)T_s]}{T_s} \tag{4.155}$$

In a digital discrete system, the calculation of Eq. (4.155) can be represented by Fig. 4.36.

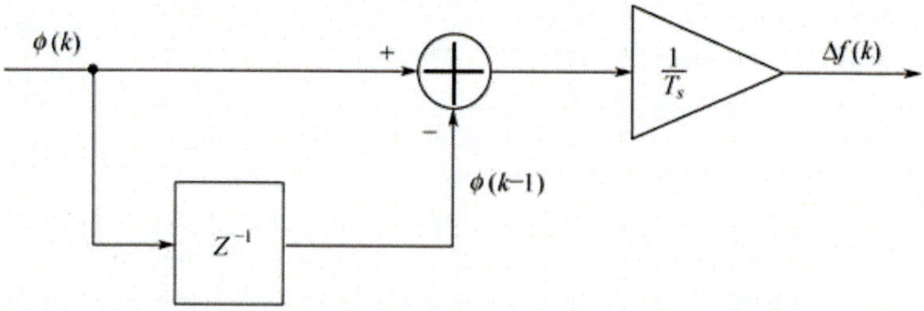

Fig. 4.36 The principle of calculating frequency difference in a discrete system

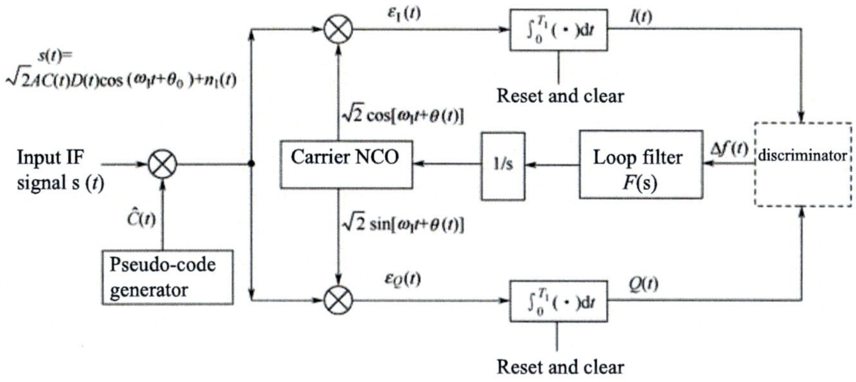

Fig. 4.37 Basic structure of the Costas FLL

Since the phase is the integral of the frequency, after the frequency difference of the output of the discriminator passes through the loop filter, a first-order integrator needs to be added before being sent to the phase NCO. Figure 4.37 shows the basic structure of the Costas frequency-locked loop. Readers can work out the difference between the frequency-locked loop and the phase-locked loop by comparing it with Fig. 4.33.

The frequency tracking loop is more stable than the phase loop, and the frequency capture range of the former is much larger than the latter. In this regard, the phase tracking loop tracks the phase difference intuitively, and the phase difference is the integral of the frequency difference, so the frequency difference must be close to 0 to ensure that the phase difference is within a certain range. Therefore, the phase tracking loop has higher requirements on the phase difference between the input signal and the local carrier. In cases where the phase difference is relatively large, the frequency loop is generally used to gradually reduce the frequency difference between the input signal and the local carrier to within a certain range, and then the phase loop can take over the loop control. In BDS and GPS receivers, since the data bit transition time is unknown after the signal acquisition, the frequency difference between the input signal and the local carrier is still relatively large (≈ 100 ~ 500 Hz), so the phase tracking loop cannot directly maintain the phase lock. This is when a frequency pull is needed. In this process, the frequency loop can be used to gradually reduce the frequency difference to within the capture range of the phase loop. Phase stabilization tracking is then achieved through the phase-locked loop.

Table 4.4 is a list of commonly used discriminator schemes, where the mathematical meanings of point multiplication and cross multiplication are as follows:

$$\textit{Point multiplication} = I_k I_{k+1} + Q_k Q_{k+1} = A^2 D^2(t) T_I^2 \cos[\phi(k+1) - \phi(k)] \tag{4.156}$$

Table 4.4 Several commonly used frequency discriminators in the Costas loop

Discriminator	Difference of output frequency	Frequency discrimination property
$\frac{\text{cross product}}{t_2-t_1}$	$\frac{\sin(\phi_2-\phi_1)}{t_2-t_1}$	It has near-optimized frequency discrimination characteristics at low SNR, and will be greatly affected by signal amplitude
$\frac{\text{cross product} \times \text{sign(dot product)}}{t_2-t_1}$	Segmented $\frac{\sin(\phi_2-\phi_1)}{t_2-t_1}$	It has near-optimized frequency discrimination characteristics at high SNR, and will be greatly affected by signal amplitude
$\frac{\text{atan2(cross product, dot product)}}{t_2-t_1}$	$\frac{\phi_2-\phi_1}{t_2-t_1}$	It has maximum likelihood estimation, and has optimized frequency discrimination characteristics in high SNR and low SNR. It will not be greatly affected by signal amplitude, but has a large amount of computation
$\frac{\text{cross product} \times \text{dot product}}{t_2-t_1}$	$\frac{\sin[2(\phi_2-\phi_1)]}{t_2-t_1}$	It has near-optimized frequency discrimination characteristics at high SNR, and will be greatly affected by signal amplitude

Note dot product $= I_1 \times I_2 + Q_1 \times Q_2$, cross product $= I_1 \times Q_2 - I_2 \times Q_1$; the subscripts 1 and 2 represent signal output of integrators at two adjacent times

$$\textit{Cross multiplication} = I_k Q_{k+1} - Q_k I_{k+1} = A^2 D^2(t) T_I^2 \sin[\phi(k+1) - \phi(k)] \tag{4.157}$$

The results of Eqs. (4.152) and (4.153) are used in the derivation of Eqs. (4.156) and (4.157), and noise terms are omitted.

From Eqs. (4.156) and (4.157), we can obtain the output frequency difference signals of the four discriminators in Table 4.4. Among the four discriminators, except for the atan2 four-quadrant arctangent discriminator, the other three are related to the amplitude of the input signal, so normalization is required. The normalization factor is $1/\sqrt{I^2(t) + Q^2(t)}$ or $1/[I^2(t) + Q^{2i}(t)]$. In addition to the normalization, it also needs to be divided by two integration time intervals, i.e. $(t_2 - t_1)$. No matter which type of discriminator, the I-channel and Q-channel integration results of the two adjacent frequencies are always needed. So, it is necessary to ensure that the two integration moments do not cross the edge of the data transition. Otherwise, as soon as the $D(t)$ of the two integration moments changes, $D^2(t)$ in Eqs. (4.156) and (4.157) will not make sense. Then, the discriminator will generate an incorrect frequency difference result, and continuing to use this frequency difference to correct the loop will lead to undesirable results. In this sense, the best time to use the cross-multiply and point-multiplier phase detectors is after the bit synchronization is completed.

Figure 4.38 shows the frequency-recognition curves of the four discriminators in the table. The figure shows that the atan2 four-quadrant arctangent discriminator

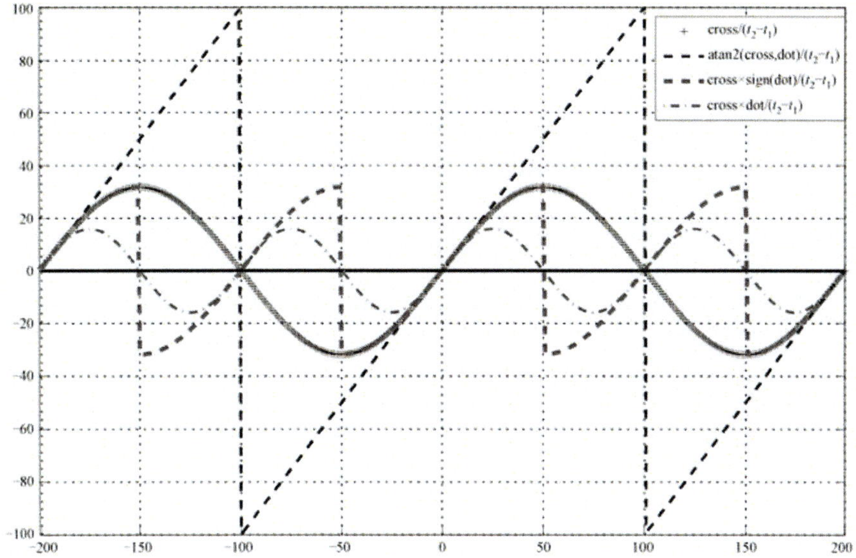

Fig. 4.38 Frequency discrimination characteristic curves of the four discriminators (with an integral interval of 5 ms)

has the best linearity, but the amount of computation required is also the largest. The frequency pulling range of the cross-multiplier is the same as the atan2 four-quadrant tangent discriminator; both are -100 to $+100$ Hz. The cross-multiply × sign (point multiplication) discriminator is the range of the segmented $\sin(\varphi_2 - \varphi_1)$ function whose frequency pulling range is only -50 to $+50$ Hz. The frequency pulling range is related to the integration time interval and is the reciprocal of the integration time interval. Therefore, if the integration time interval is increased to 10 ms, the frequency pulling range of each discriminator in Fig. 4.38 needs to be narrowed by half.

3. Thermal Noise Performance Analysis of the Costas Loop

The subsections above have analyzed the steady-state phase difference of the phase-locked loop with the presence of dynamic stress on the input signal. There are also other factors affecting the phase difference in reality, such as the clock phase jitter and the random phase difference caused by the Allan variance. Another more significant factor is the phase difference caused by thermal noise in the input signal. Therefore, the thermal noise performance of the Costas loop will be discussed this subsection.

The noise term is ignored in both of the above subsections in order to highlight the signal presence and simplify the analysis. However, in this section, the noise term will be restored, and will be assumed as Gaussian white noise. The object of the study is the PLL of the Costas loop; the phase detector type is $I(t) \times Q(t)$, and the coherent integration time length is assumed to be T_{I}.

First, we assume that the noise n_{I} in the input signal $s(t)$ in Sect. 4.2.3 is a narrow-band Gaussian random process, written as

$$n_{\rm I}(t) = \sqrt{2}n_{\rm c}\cos(\omega_{\rm I}t + \theta_0) + \sqrt{2}n_{\rm s}\sin(\omega_{\rm I}t + \theta_0) \tag{4.158}$$

Here, we assume that the unilateral noise power spectral density of n_c and n_s is N_0, and the distribution pattern is that of the Gaussian white noise. The mean is 0, and n_c and n_s are stationary Gaussian processes, which are mutually uncorrelated or statistically independent. It is reasonable to assume that the noise in the received satellite signal is a narrowband Gaussian random process because the bandwidth of the RF front-end is much smaller than the carrier frequency, and the spectrum of the noise is limited to a narrow frequency band centered on the carrier frequency.

The local carrier signal is multiplied by the input signal after the pseudo-code is stripped to obtain $\varepsilon_{\rm I}(t)$ and $\varepsilon_{\rm Q}(t)$, which can be written respectively as

$$\varepsilon_{\rm I}(t) = 2AD(t)\cos(\omega_{\rm I}t + \theta_0)\cos[\omega_{\rm I}t + \theta(t)] + \sqrt{2}n_{\rm I}(t)\cos[\omega_{\rm I}t + \theta(t)]$$

$$\varepsilon_{\rm Q}(t) = 2AD(t)\cos(\omega_{\rm I}t + \theta_0)\sin[\omega_{\rm I}t + \theta(t)] + \sqrt{2}n_{\rm I}(t)\sin[\omega_{\rm I}t + \theta(t)]$$

Substituting Eq. (4.158) into the above two equations, we can obtain the following equations after re-arranging and omitting the high frequency components.

$$\varepsilon_{\rm I}(t) = AD(t)\cos[\phi(t)] + n_{\rm c}\cos[\phi(t)] - n_{\rm s}\sin[\phi(t)] \tag{4.159}$$

$$\varepsilon_{\rm Q}(t) = AD(t)\sin[\phi(t)] + n_{\rm c}\sin[\phi(t)] + n_{\rm s}\cos[\phi(t)] \tag{4.160}$$

The local NCO will get the error signal to adjust the local carrier phase each time after the integrator completes the current integration, and the time of the integration is very short. It can therefore be considered that $\varphi(t)$ is constant during the integration operation, so the time variable t can be omitted from the following analysis. According to the above analysis, the data bit $D(t)$ remains unchanged during the integration operation, so the time variable t in $D(t)$ can also be omitted. Then, the signals outputted by the I-channel and Q-channel integrators can be simplified as

$$I(t) = ADT_{\rm I}\cos\phi + N_{\rm c}\cos\phi - N_{\rm s}\sin\phi$$

$$Q(t) = ADT_{\rm I}\sin\phi + N_{\rm c}\sin\phi + N_{\rm s}\cos\phi$$

where

$$N_{\rm c} = \int_0^{T_{\rm I}} n_{\rm c}{\rm d}t, \quad N_{\rm s} = \int_0^{T_{\rm I}} n_{\rm s}{\rm d}t$$

According to the properties of Gaussian white noise, N_c and N_s are still Gaussian distributions, the mean is still 0, and the variance becomes $\sigma_{N_c}^2 = \sigma_{N_s}^2 = N_0 T_{\rm I}$.

Since $D^2(t) = 1$, and according to the analysis above, the output of the $I(t) \times Q(t)$ phase detector can be obtained.

$$\begin{aligned} z(t) &= [ADT_\mathrm{I}\cos\phi + N_\mathrm{c}\cos\phi - N_\mathrm{s}\sin\phi][ADT_\mathrm{I}\sin\phi + N_\mathrm{c}\sin\phi + N_\mathrm{s}\cos\phi] \\ &= \underbrace{\frac{1}{2}A^2T_\mathrm{I}^2\sin 2\phi}_{\textit{signal term}} \\ &\quad \underbrace{+ (ADT_\mathrm{I}N_\mathrm{c} + \frac{N_\mathrm{c}^2}{2} - \frac{N_\mathrm{s}^2}{2})\sin 2\phi + (ADT_\mathrm{I}N_\mathrm{s} + N_\mathrm{c}N_\mathrm{s})\cos 2\phi}_{\text{noise term}} \end{aligned} \tag{4.161}$$

The signal term in Eq. (4.161) is the phase difference signal obtained by the phase detector, which is consistent with the result of Eq. (4.154). It can be seen that the signal strength is proportional to the input signal amplitude A^2 and to the integration time ${T_I}^2$. This is consistent with intuitive analysis. The longer the integration time is, the stronger the phase difference signal will be. However, it should be noted that the premise here is that $\varphi(t)$ is constant during the integration time. If $\varphi(t)$ contains the frequency difference, and the frequency difference is already greater than $1/T_\mathrm{I}$, then the conclusion will no longer stand. This has already been explained in the analysis of the frequency pull range of the discriminator.

Now, it is time to thoroughly analyze the noise term in Eq. (4.161). To simplify the analysis, we can set that

$$N(t) = \left(ADT_\mathrm{I}N_\mathrm{c} + \frac{N_\mathrm{c}^2}{2} - \frac{N_\mathrm{s}^2}{2}\right)\sin 2\phi(t) + (ADT_\mathrm{I}N_\mathrm{s} + N_\mathrm{c}N_\mathrm{s})\cos 2\phi(t)$$

It is easy to verify that $N(t)$ has a mean of 0, so its variance

$$\begin{aligned} \sigma_N^2 &= E[N^2(t)] \\ &= E\left[A^2T_\mathrm{I}^2N_\mathrm{c}^2 + \frac{N_\mathrm{c}^4}{4} + \frac{N_\mathrm{s}^4}{4} - \frac{N_\mathrm{s}^2N_\mathrm{c}^2}{2}\right]\sin^2 2\phi(t) \\ &\quad + E\left[A^2T_\mathrm{I}^2N_\mathrm{s}^2 + N_\mathrm{c}^2N_\mathrm{s}^2\right]\cos^2 2\phi(t) \end{aligned} \tag{4.162}$$

E[·] in the above formula means mathematical expectation.

Since N_c and N_s are still Gaussian distributed and their variance is N_0T_I, then Eq. (4.162) can be further simplified.

Since

$$E[N_\mathrm{c}^2] = E[N_\mathrm{s}^2] = N_0T_\mathrm{I} \tag{4.163}$$

$$E[N_\mathrm{c}^4] = E[N_\mathrm{s}^4] = 3E[N_\mathrm{s}^2] = 3N_0^2T_\mathrm{I}^2 \tag{4.164}$$

Substituting Eqs. (4.163) and (4.164) into Eq. (4.162), then

$$\sigma^2_{N(t)} = A^2T_{\mathrm{I}}^2N_0T_{\mathrm{I}} + N_0^2T_{\mathrm{I}}^2 \tag{4.165}$$

Since the noise term in the adjacent integration unit can be regarded as irrelevant, $N(t)$ appears to be "white" in time. So, the autocorrelation function of the "white" noise $N(t)$ can be written as

$$R_{N(t)}(\tau) = \begin{cases} \sigma^2_{N(t)}\left[1 - \frac{|\tau|}{T_I}\right], & when\ \tau| < T_I \\ 0, & when\ |\tau| > T_I \end{cases}$$

Applying the Fourier transformation to $\mathrm{R}_N(\tau)$, we can obtain the noise power spectral density of $N(t)$, which can be written as

$$S_N(f) = (A^2T_{\mathrm{I}}^3N_0 + N_0^2T_{\mathrm{I}}^2)T_s\left(\frac{\sin\pi fT_{\mathrm{I}}}{\pi fT_{\mathrm{I}}}\right)^2 \tag{4.166}$$

Figure 4.39a, b illustrate $R_{\mathrm{N}}(\tau)$ and $S_{\mathrm{N}}(f)$ respectively. The intersection of the dotted line and the frequency axis in Fig. 4.39b is the equivalent noise bandwidth B_{n} of the loop. B_{n} is given in the figure to illustrate that when $B_{\mathrm{n}} << 1/T_{\mathrm{I}}$, $S_{\mathrm{N}}(f)$ can be considered as approximately flat, which means that $N(t)$ can be regarded as the white noise within the loop noise bandwidth B_{n}, and its average power spectral density $N_0' \approx (A^2T_{\mathrm{I}}^4N_0 + N_0^2T_{\mathrm{I}}^3)$. This approximation can greatly simplify the successive analysis. In fact, the noise bandwidth of the carrier tracking loop in BDS and GPS receivers is generally within a few tens of hertz. If the integration time is 1 ms, then $1/T_{\mathrm{I}} = 1$ kHz, and the requirement of the bandwidth can be met. When the integration time increases, the noise bandwidth also needs to be reduced correspondingly so that the noise can still be regarded as white noise in the B_{n} range. Under the condition of $\mathrm{B} << 1/T_{\mathrm{I}}$, the closed-loop noise power can be written as $N_0'B_n$, so the conclusion of Sect. 4.2.2 can be used to obtain the variance of the random phase difference of the carrier loop.

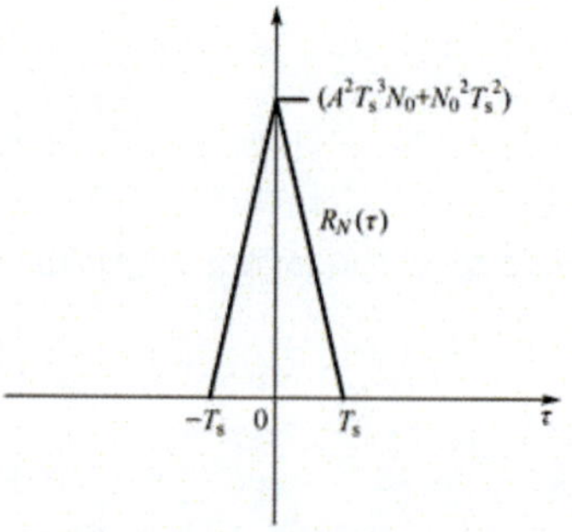

(a) $N(t)$ Autocorrelation function $R_N(t)$

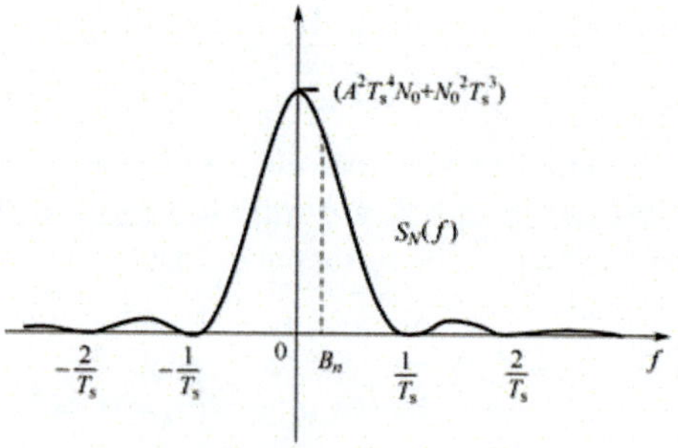

(b) $N(t)$ Noise power densitya $S_N(f)$

Fig. 4.39 Random variance of carrier loop

$$\overline{\sigma}_n^2 = \frac{1}{2\text{SNR}} = \frac{N_0' B_n}{2P_i}, \quad \left(\text{here,} \;\; P_i = \left[\frac{A^2 T_\text{I}^2}{2}\right]^2\right)$$
$$= \frac{2N_0 B_n}{A^2}\left(1 + \frac{N_0}{A^2 T_\text{I}}\right) \tag{4.167}$$

where, $P_i = \left[\frac{A^2 T_\text{I}^2}{2}\right]^2$ is obtained from the signal term of Eq. (4.161).

Since N_0 is the unilateral noise power spectral density of n_c and n_s, the carrier-to-noise ratio of the input signal CN_0 is defined follows,

$$\text{CN}_0 \triangleq \frac{A^2}{2N_0}$$

Substituting the definition of CN_0 into Eq. (4.167), we can then write $\overline{\theta_n^2}$ as

$$\overline{\sigma}_n^2 = \frac{B_n}{\text{CN}_0}\left[1 + \frac{1}{2\text{CN}_0 T_\text{I}}\right]$$
$$= \frac{B_n}{\text{CN}_0 S_L} \tag{4.168}$$

Equation (4.168) indicates the relationship between the Costas loop PLL phase difference and the input carrier-to-noise ratio, the equivalent noise bandwidth, and the coherent integration time, in which

$$S_L = \frac{1}{1 + \frac{1}{2\text{CN}_0 T_\text{I}}} \tag{4.169}$$

is referred to as the "square loss." This equation shows that the square loss has a close relationship with the input carrier-to-noise ratio: the stronger the carrier-to-noise ratio of the input signal, the smaller the square loss will be, and vice versa. At the same time, Eq. (4.169) also shows that the length of the integration time can offset the effect of the input carrier-to-noise ratio. For weaker signals, a longer integration time can help reduce the squared loss. So, the above derivation indicates that the root cause of the square loss is the multiplication of the noise term in the process of $I(t) \times Q(t)$, which leads to the higher moment of the noise term.

It can be deduced from Eq. (4.168) that the root mean square of the random phase difference caused by thermal noise is

$$\overline{\sigma}_n = \sqrt{\frac{B_n}{\text{CN}_0}\left[1 + \frac{1}{2\text{CN}_0 T_\text{I}}\right]} \tag{4.170}$$

where, the unit of $\overline{\sigma}_n$ is radians, which can be converted into the following two forms:

$$\overline{\sigma}_{n,D} = \frac{360}{2\pi}\sqrt{\frac{B_n}{\mathrm{CN_0}}\left[1+\frac{1}{2\mathrm{CN_0}T_\mathrm{I}}\right]} \quad (°) \tag{4.171}$$

$$\overline{\sigma}_{n,M} = \frac{\lambda}{2\pi}\sqrt{\frac{B_n}{\mathrm{CN_0}}\left[1+\frac{1}{2\mathrm{CN_0}T_\mathrm{I}}\right]} \quad (\mathrm{m}) \tag{4.172}$$

Equation (4.171) represents $\overline{\sigma}_n$ in degrees and Eq. (4.172) represents $\overline{\sigma}_n$ in meters, where λ is the carrier wavelength and the values for different signals λ need to be adjusted.

As a summary of this section, Fig. 4.40 shows the integration time, equivalent noise bandwidth and carrier-to-noise ratio, and the functional curve of $\overline{\sigma}_n$, in which the value of $\overline{\sigma}_n$ is calculated through Eq. (4.171). The unit of the vertical axis is degree, and the horizontal axis represents the input signal carrier-to-noise ratio.

The unit of the horizontal axis is dBHz, and B_n takes the value of 2, 5, 10, and 20 Hz. The left half of the figure is the case when the coherent integration time is 2 ms, and the right half is the coherent integration time of 20 ms. As can be seen from Fig. 4.40, when other conditions are the same, the longer the coherent integration time is, the smaller $\overline{\sigma}_n$ is. The narrower B_n is, the smaller $\overline{\sigma}_n$ is. The stronger the signal strength is, the smaller $\overline{\sigma}_n$ is. To achieve stable tracking of weak signals, we should set the noise bandwidth to be relatively narrower and the coherent integration time longer.

It is worth noting that the root mean square of the random phase difference of the thermal noise calculated through Eq. (4.170) is independent of the order of the tracking loop. In the absence of dynamic stress, the root mean square of the random phase difference of the thermal noise obtained by a Costas loop with the same noise bandwidth is the same, and will not change due to the difference in order. The main purpose of using a higher order loop is to track the dynamic stress more successfully.

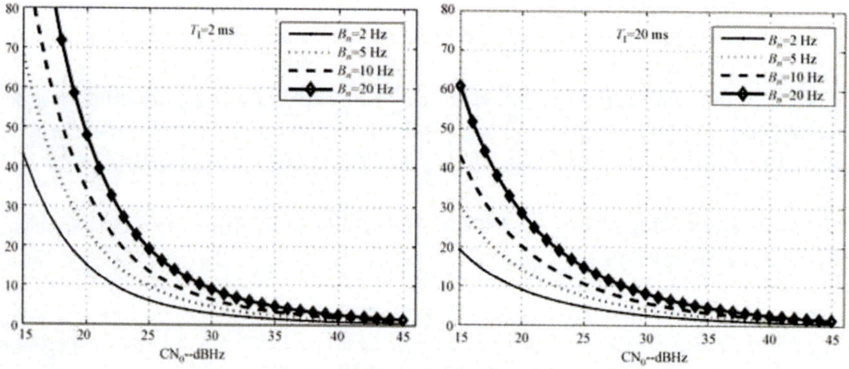

Fig. 4.40 Integration time, equivalent noise bandwidth and carrier-to-noise ratio, and the functional curve of $\tilde{\sigma}_n$

4.2.4 The Pseudo-Code Tracking Loop

In BDS and GPS receivers, the pseudo-code phase must be locked to achieve pseudo-code stripping. The pseudo-code de-spreading of the input signal can only be guaranteed when the phase of the locally generated BDS or GPS pseudo-random code and that of the pseudo-random code in the input signal are consistent. The importance of pseudo-code de-spreading is suggested in the analysis of the carrier loop above. In fact, in Figs. 4.33 and 4.37, the input signal and the local carrier must be multiplied by the local pseudo-code generated by the pseudo-code generator before the I-channel and Q-channel integration, otherwise the subsequent coherent integrator will output a result similar to noise integration, where the local pseudo-code is outputted by the pseudo-code tracking loop. The pseudo-code loop is also responsible for extracting the pseudo-range observation while locking the pseudo-code phase. Therefore, the performance of the pseudo-code tracking loop will directly affect the quality of the pseudo-range observation and the subsequent navigation and positioning results. Therefore, the importance of the pseudo-code tracking loop in the BDS and GPS receivers is obvious.

The pseudo-random code is generally tracked by a delay-locked loop. In this section, the principle of the delay-locked loop will be described in detail, and its performance will be theoretically analyzed.

1. A Coherent Delay-Locked Loop

The word "coherent" here is relative to the input signal carrier phase. Being coherent means that the local carrier phase and the carrier phase of the input signal are fully aligned. In this case, we can only multiply the local carrier signal and the input signal to completely move the signal to the baseband zero frequency, so the loop in this case can also be called a baseband delay-locked loop. Of course, the assumption here is ideal. In fact, it is difficult to achieve full synchronization of the carrier phase, so the non-coherent delay-locked loop is still the most widely used in engineering. This will be explained in the next section.

The input signal is a baseband signal, and it is selected mainly to temporarily avoid the dilemma brought about by the failure of phase alignment of the intermediate frequency carrier during analysis. Meanwhile, the presence or absence of the intermediate frequency carrier does not affect the analysis of the principle of the pseudo-code tracking loop. The basic structure of the baseband delay-locked loop is shown in Fig. 4.41, which includes the multiplier, pseudo-code generator, pseudo-code NCO, loop filter, and integration unit. Careful readers will find that the input signals in the figure do not contain data bits either. This is because the loop integration time is generally within one data bit, so the data bits can be considered as constant during the loop integration time. They are omitted here to simplify the analysis.

The pseudo-code generator in Fig. 4.41 produces three different local phase pseudo-codes: "early", "late", and "prompt" codes, which are denoted by subscripts E, L, and P respectively in the following equations. The coherence interval (referring to the phase difference between the three pseudo-codes) is a very important

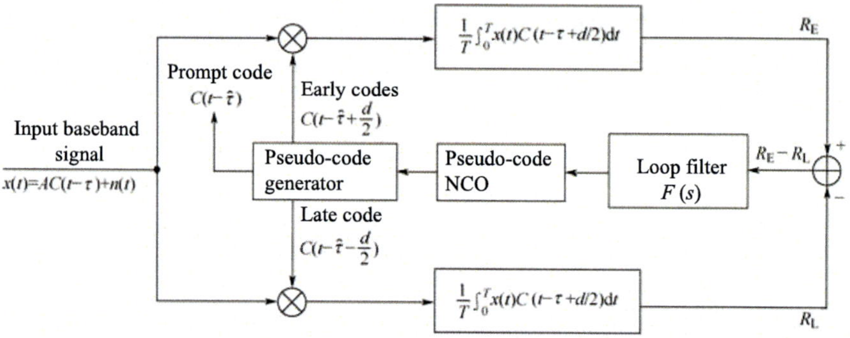

Fig. 4.41 Basic structure of baseband delay-locked loop

parameter. Generally, the phase intervals of the leading, middle, and lag are set to be symmetrical, that is, (the prompt code phase − early code phase) = (late code phase − prompt code phase). The coherence interval in Fig. 4.41 is $d/2$, whose unit is the chip.

We assume that the input signal is

$$x(t) = AC(t - \tau) + n(t) \tag{4.173}$$

where $C(t)$ is a pseudo-random code signal, A is the signal amplitude, and τ is the pseudo-code phase. It is more convenient to adopt the pseudo-code chips as the unit of τ than seconds. For the C/A code in the GPS signal, the starting phase can be between (1 ~ 1023) chips, and for the BDS signal, the range of τ is between (1 ~ 2046). $n(t)$ is Gaussian white noise with a mean of 0 and a unilateral power spectral density of N_0.

The coherence interval is $d/2$, so the three local codes can be expressed as follows:

The early code: $x_E(t) = C(t - \hat{\tau} + d/2)$.

The prompt code: $x_P(t) = C(t - \hat{\tau})$.

The late code: $x_L(t) = C(t - \hat{\tau} - d/2)$.

It can be seen that the phase differences between them are $-d/2$, 0 and $+d/2$, respectively, where d is in units of pseudo-code chips. If $d = 1$, then the pseudo-code phase of the early code will be advanced by 0.5 chips as compared to the pseudo-code phase of the prompt code, and the pseudo-code phase of the late code will be delayed by 0.5 chips. The reason for this setting is that when the early code and the input signal pseudo-code are phase-aligned, its correlation value will reach the rising edge of the pseudo-code autocorrelation function, and that of the late code will drop to the falling edge of the pseudo-code autocorrelation function. The reader only needs to review Sects. 3.1.3 and 3.2.3 to gain a clear understanding of the shape of the autocorrelation function of the GPS and BDS pseudo-codes. According to the above equation, it can be seen that the phase difference between the code phase of the prompt code of the pseudo-code tracking loop and the input signal is $\Delta\tau = (\tau - \hat{\tau})$, and the task of the pseudo-code tracking loop is to make $\Delta\tau \rightarrow = 0$.

The output signals of the early code integrator and the late code integrator in Fig. 4.41 are

$$R_{\mathrm{E}}(\Delta\tau) = \frac{1}{T}\int_{0}^{T} x(t)x_{\mathrm{E}}(t)\mathrm{d}t$$

$$R_{\mathrm{L}}(\Delta\tau) = \frac{1}{T}\int_{0}^{T} x(t)x_{\mathrm{L}}(t)\mathrm{d}t$$

where T is the integration time, which is generally an integer multiple of the pseudo-code period, for example, 1, 5, 10, or 20 ms. It can even exceed 20 ms after bit assist is implemented. It should be noted that when the integration time exceeds 1 ms, it is necessary to strip the secondary code (NH code) to achieve the integration of the BDS D1 code signal, otherwise offset will occur. The early code and the pseudo-code of the input signal differs by $(\Delta\tau - 0.5d)$, and the difference between the late code and the pseudo-code of the input signal is $(\Delta\tau + 0.5d)$. Here, d is pre-determined. $\Delta\tau$ is the only independent variable. Assuming that $d = 1$, i.e. the coherence interval $d/2$ is 0.5 chips, then the correlation functions R_{E} and $-R_{\mathrm{L}}$ of the two integrator outputs are shown in Fig. 4.42 as the two black solid triangles. The figure shows the ideal autocorrelation function of the pseudo-code, meaning that when the phase offset exceeds one chip, the autocorrelation function is 0 and the effect of noise is not considered.

Let us take a look at Fig. 4.42 to analyze $(R_{\mathrm{E}} - R_{\mathrm{L}})$ as a function of $\Delta\tau$.

- When $\Delta\tau \le -1.5T_c$, both R_{E} and R_{L} are 0, so $(R_{\mathrm{E}} - R_{\mathrm{L}}) = 0$;
- When $-1.5T_{\mathrm{c}} < \Delta\tau \le -0.5T_{\mathrm{c}}$, $R_{\mathrm{L}} = 0$, so $(R_{\mathrm{E}} - R_{\mathrm{L}}) = R_{\mathrm{E}}$;

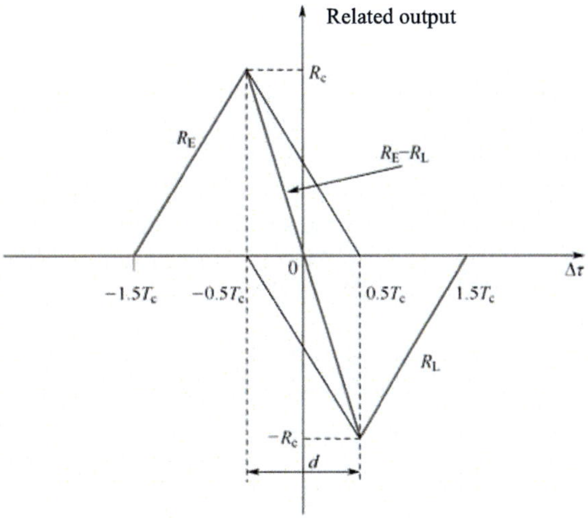

Fig. 4.42 Phase detection curve of the baseband delay hysteresis loop

- When $-0.5T_c < \Delta\tau \leq 0.5T_c$, $(R_E - R_L)$ is a straight line through the origin whose slope is twice the slope of R_E and R_L;
- When $1.0T_c < \Delta\tau \leq 1.5\,T$, $R_E = 0$, so $(R_E - R_L) = -R_L$;
- When $\Delta\tau > 1.5T$, both R_E and R_L are 0, again, so $(R_E - R_L) = 0$.

Here T_c is the pseudo-code chip width. For the GPS C/A code, T_c is 1/1.023 MHz; for the BDS signal, T_c is 1/2.046 MHz. The final $R_E - R_L$ curve is shown by the light-colored thick line in Fig. 4.42. Since its shape is very similar to the italic letter "*S*", it is generally called the *S*-curve.

When the pseudo-code tracking loop is in the tracking state, the capture of the input signal has already been completed, which means that the phase difference between the input and the local pseudo-code is already within a pseudo-code chip. $\Delta\tau$ can be considered as between $[-0.5T_c, +0.5T_c]$, so analysis only needs to focus on the middle part of the thick polyline in Fig. 4.42.

The main purpose of $(R_E - R_L)$ is to obtain the estimator of the pseudo-code phase difference $\Delta\tau$. The above analysis shows that when $\Delta\tau = 0$, $(R_E - R_L)$ is 0, indicating that the pseudo-code phase error is 0 at this time. When $\Delta\tau \neq 0$, there will be an error signal $(R_E - R_L)$ that is not 0 and in a proportional relationship with $\Delta\tau$. The error signal $(R_E - R_L)$ is sent to adjust the frequency of the local NCO after filtering out the high frequency components through the loop filter. When the local pseudo-code phase is early, the pseudo-code NCO will be slowed down. Meanwhile, when the local pseudo-code phase is late, the pseudo-code NCO will be adjusted faster, thus achieving closed loop control of $\Delta\tau$ and gradually stabilizing its value at around 0. When a well-designed pseudo-code tracking loop operates in a steady state, the mean value of $\Delta\tau$ must be near 0. At this time, the pseudo-code phase of the prompt code is aligned with the pseudo-code phase of the input signal, and the output of the prompt code can be used to implement the pseudo-code de-spreading of the input signal, i.e. to be sent to $\hat{C}(t)$ in Figs. 4.33 and 4.37.

If the error signal is analyzed from a more direct perspective, the reason why $(R_E - R_L)$ is 0 when $\Delta\tau = 0$ is that the phase difference between the advance code and the lag code relative to the intermediate code is symmetrical, i.e. ±d/2. Besides, the shape of the BDS and GPS pseudo-code autocorrelation function is also bilaterally symmetric with respect to the autocorrelation maximum. In Fig. 4.42, in the case of $d = 1$ chip, when $\Delta\tau = 0$, R_E takes the intermediate value of the rising edge of the pseudo-code autocorrelation function, and R_L takes the intermediate value of the falling edge of the pseudo-code autocorrelation function, the result of the subtraction is 0. So, as long as the symmetry of the phase of the leading code and the phase of the lag code is guaranteed, there is always a similar result regardless of the value of *d*. In the design of the GPS receiver's pseudo-code tracking loop, *d* is a very important parameter. A relatively small value of *d* will offer some good results, such as the suppression of multipath effects. This technique is called "narrow correlation".

Figure 4.43 shows the effect of different coherence intervals on the *S* curve. The left part is the case when $d = 0.5$ chips, and the right part is the case when $d = 0.25$ chips. The thick line in the figure is still the *S* curve of the phase detector. It can be seen that the shape of the *S* curve at this time is very different from when

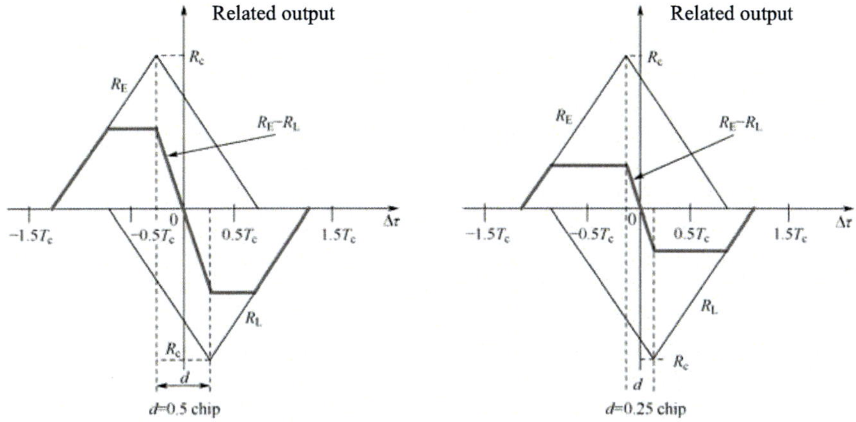

Fig. 4.43 *S*-curve of the phase discriminator with different coherence intervals

$d = 1$ chip, mainly in that the upper and lower tops of the former are flat, and the range of the linear region is reduced correspondingly. When $d = 0.5$, the range of the linear region is [−0.25, +0.25] chips, and when $d = 0.25$, the range of the linear region is [−0.125, +0.125] chips, but the slope of the S curve in the linear region does not change. The upper and lower flat top values of the *S* curve in the figure are ±0.5Rc ($d = 0.5$) and ±0.25Rc ($d = 0.25$) respectively, indicating that when $\Delta\tau$ crosses the linear region, the phase detector output value will remain at the flat top, which is similar to the level clamp effect (or clipping effect). The suppression of the multipath effect is mainly caused by the flat top effect of the *S* curve. For other coherent intervals, the reader can conduct similar analysis according to Fig. 4.43, which will not be difficult.

The highest value of the pseudo-code autocorrelation function in Figs. 4.42 and 4.43 is R_c, which is related to the integration time T and the signal amplitude A. When the value of T is constant, R_c is linearly related to the signal amplitude, so the output value of the correlator changes with the received signal strength. For example, when the receiver moves to open ground, and the received signal is good, then the output of the correlator will be strong. Conversely, when the receiver moves to a city center or forest, and the received signal is occluded, then the correlator output will be weakened. Fortunately, the error signal generation mechanism ensures that the value is always 0 when $\Delta\tau = 0$, regardless of the signal strength. This is because the output of the correlator of the early code and the late code is simultaneously changed with the signal strength, and the change is symmetrical. However, the change in signal strength will affect the tracking of the pseudo-code tracking loop, which is mainly achieved by changing the slope of the *S*-curve around zero.

It has been proven above that the slope of the *S*-curve around the zero point is $k = {}^{2R_c}/_{T_c}$, which does not change with the alteration of the coherence interval. The larger the slope value, the more sensitive the error signal is to $\Delta\tau$, as even a small $\Delta\tau$ can produce a large error signal. On the contrary, if the slope value is relatively

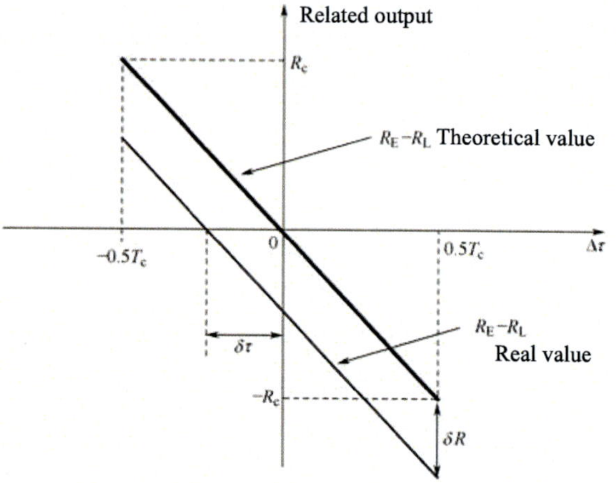

Fig. 4.44 Relationship between slope and noise characteristics in the linear region of the *S*-curve

small, then the generation of the error signal will be slower for $\Delta\tau$. Besides, the value of k also has a significant influence on the noise features of the pseudo-code loop, as can be seen in Fig. 4.44.

The bold solid line in Fig. 4.44 shows the theoretical value of the error signal, while the other solid line shows the actual value of the error signal. It should be noted that the noise here is not the additive Gaussian noise mixed in the IF (Intermediate Frequency) signal, but the additive Gaussian noise equivalent to the phase detector output. The reason for this processing has been explained in Sect. 4.2.2. Due to the presence of noise in the input signal, the output value of the integrator has a deviation of δR from the theoretical value, while the pseudo-code loop always attempts to tune the error signal back to 0 through the feedback loop control, resulting in a deviation between the final phase and the pseudo-code phase of the input signal:

$$\delta\tau = \frac{\delta R}{k} = \frac{\delta R \cdot T_c}{2R_c}$$

where, δR is determined by the additive white Gaussian noise $n(t)$ in the input signal.

$$\delta R = \frac{1}{T}\int_0^T n(t)x_E(t)dt - \frac{1}{T}\int_0^T n(t)x_L(t)dt \tag{4.174}$$

It is easy to prove that $E[\delta R] = 0$. Then, the variance of δR is

$$\mathrm{var}\{\delta R\} = \mathrm{var}\left\{\frac{1}{T}\int_0^T n(t)x_E(t)dt - \frac{1}{T}\int_0^T n(t)x_L(t)dt\right\}$$

$$= E\left\{\left[\frac{1}{T}\int_0^T n(t)x_{\mathrm{E}}(t)\mathrm{d}t - \frac{1}{T}\int_0^T n(t)x_{\mathrm{L}}(t)\mathrm{d}t\right]^2\right\}$$

$$= \frac{1}{T^2}E\left\{\left[\int_0^T n(t)x_{\mathrm{E}}(t)\mathrm{d}t\right]^2\right\} + \frac{1}{T^2}E\left\{\left[\int_0^T n(t)x_{\mathrm{L}}(t)\mathrm{d}t\right]^2\right\}$$

$$- \frac{2}{T^2}E\left\{\int_0^T n(t)x_{\mathrm{E}}(t)\mathrm{d}t \int_0^T n(t)x_{\mathrm{L}}(t)\mathrm{d}t\right\} \quad (4.175)$$

We will now analyze the three items in the last step of Eq. (4.175). First, it is obvious that

$$E\left\{\left[\int_0^T n(t)x_{\mathrm{E}}(t)\mathrm{d}t\right]^2\right\} = E\left\{\left[\int_0^T n(t)x_{\mathrm{L}}(t)\mathrm{d}t\right]^2\right\}$$

And then, we need to calculate $E\{[\int_0^T n(t)x_{\mathrm{E}}(t)\mathrm{d}t]^2\}$ and $E\{\int_0^T n(t)x_{\mathrm{E}}(t)\mathrm{d}t \int_0^T n(t)x_{\mathrm{L}}(t)\mathrm{d}t\}$:

$$\begin{aligned}
E\left\{\left[\int_0^T n(t)x_{\mathrm{E}}(t)\mathrm{d}t\right]^2\right\} &= E\left\{\int_0^T n(t)x_{\mathrm{L}}(t)\mathrm{d}t \int_0^T n(s)x_{\mathrm{L}}(s)\mathrm{d}s\right\} \\
&= E\left\{\int_0^T\int_0^T n(t)n(s)x_{\mathrm{L}}(t)x_{\mathrm{L}}(s)\mathrm{d}t\mathrm{d}s\right\} \\
&= \int_0^T\int_0^T E\{n(t)n(s)\}x_{\mathrm{L}}(t)x_{\mathrm{L}}(s)\mathrm{d}t\mathrm{d}s \\
&= \int_0^T\int_0^T N_0\delta(t,s)x_{\mathrm{L}}(t)x_{\mathrm{L}}(s)\mathrm{d}t\mathrm{d}s \\
&= N_0\int_0^T x_L^2(t)\mathrm{d}t \\
&= N_0 T
\end{aligned} \quad (4.176)$$

while

$$
\begin{aligned}
E\left\{\int_0^T n(t)x_{\mathrm{E}}(t)\mathrm{d}t \int_0^T n(t)x_{\mathrm{L}}(t)\mathrm{d}t\right\} &= \int_0^T\int_0^T E\{n(t)n(s)\}x_{\mathrm{E}}(t)x_{\mathrm{L}}(s)\mathrm{d}t\mathrm{d}s \\
&= \int_0^T\int_0^T N_0\delta(t,s)x_{\mathrm{E}}(t)x_{\mathrm{L}}(s)\mathrm{d}t\mathrm{d}s \\
&= N_0\int_0^T x_{\mathrm{E}}(t)x_{\mathrm{L}}(t)\mathrm{d}t \\
&= N_0(1-d)T
\end{aligned}
\tag{4.177}
$$

The last step of Eq. (4.177) is written because the pseudo-code phase shift of $x_{\mathrm{L}}(t)$ is d relative to $x_{\mathrm{E}}(t)$. Then, the common part in each chip is $(1-d)T_{\mathrm{c}}$, so the integration result of the same part is $(1-d)T$. For the different parts, since the -1, $+1$ of the pseudo-random code adjacent bits are symmetrically distributed, the final integral is zero. Figure 4.45 clearly illustrates the same and different parts of $x_{\mathrm{L}}(t)$ relative to $x_{\mathrm{E}}(t)$, helping readers to understand the integration discussed above.

Finally, we combine Eqs. (4.176) and (4.177), then

$$
\mathrm{var}\{\delta R\} = \frac{2}{T^2}\cdot N_0T - \frac{2}{T^2}\cdot N_0(1-d)T = \frac{2N_0d}{T} \tag{4.178}
$$

According to the relation of δR and $\delta\tau$,

$$
\mathrm{var}\{\delta\tau\} = \frac{\mathrm{var}\{\delta R\}}{k^2}
$$

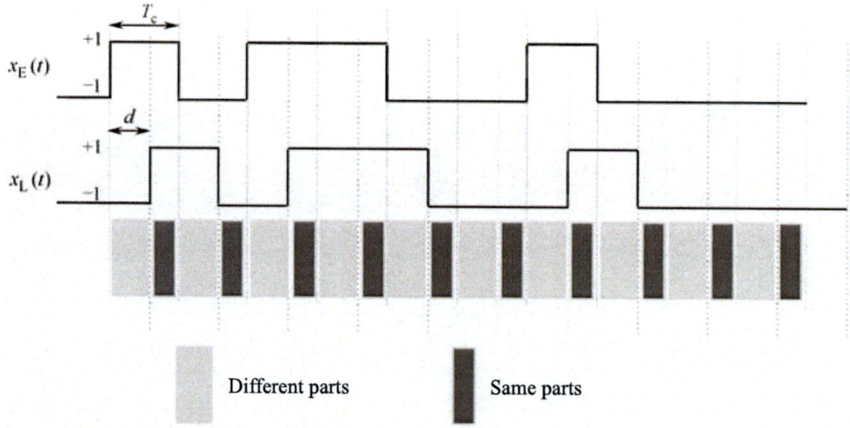

Fig. 4.45 The same and different parts of $x_{\mathrm{L}}(t)$ in relation to $x_{\mathrm{E}}(t)$

$$= \frac{N_0 d}{T} \frac{T_c^2}{2R_c^2} \tag{4.179}$$

Equation (4.179) shows that the variance of the phase noise caused by the input noise is proportional to the input noise power spectral density and the phaser spacing d. R_c^2 represents the power of the input signal, so the variance of the phase noise is inversely proportional to the input signal power, which matches our direct perception that the stronger the input signal is, the less the effect of noise on the final phase noise will be. Equation (4.179) also indicates the relationship between phase noise and integration time. T is the total integration time, which generally appears as the denominator. So, the longer the integration time, the smaller the phase noise will be. Therefore, so as not to cause data bit hopping, the integration time should be used for as long as possible. If the data bits are known, the bit-assisted method can be adopted to eliminate the bit hopping and increase the integration time.

2. Non-coherent Delay-Locked Loop

When the input signal is an intermediate frequency signal, it can be very difficult for the coherent delay-locked loop shown in Fig. 4.41 to be applied in practice, as the coherent integration requires a strict synchronization of both the local carrier phase and the pseudo-code phase with those of the input signal. Strict synchronization of the carrier phase is very difficult to achieve. Even if synchronization can be achieved, the coherent delay-locked loop is very fragile. When dealing with dynamically changing signals, it is often affected by the fact that the carrier phase difference is not zero. By this time, the signal energy will rotate in the I-channel and Q-channel directions, and the pseudo-code phase difference generated by the I-channel integration alone will also be unstable, which will cause the coherent delay-locked loop to lose lock easily. In this case, a non-coherent delay-locked loop as shown in Fig. 4.46 will be adopted.

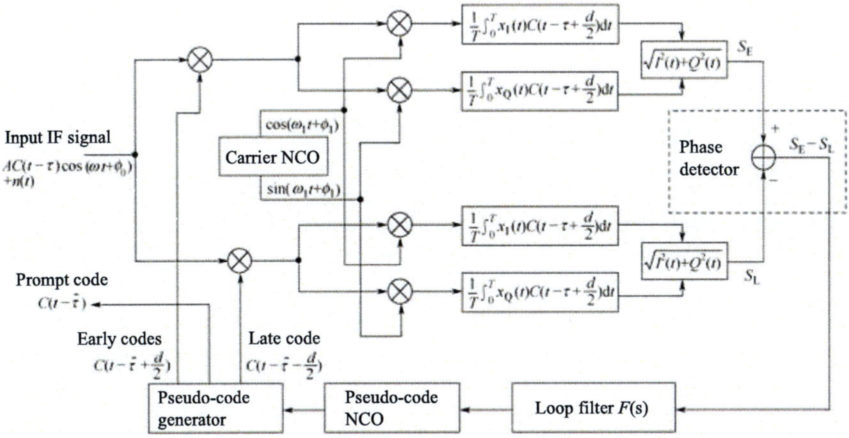

Fig. 4.46 The basic structure of the incoherent delay-locked loop

Figure 4.46 shows that the input signal of the non-coherent delay-locked loop is the intermediate frequency signal, including the carrier and pseudo-code components, and the noise component. A carrier NCO is added to the basic structure of the non-coherent delay-locked loop as compared to the coherent delay-locked loop to generate the I-channel and Q-channel local carrier signals, and the two local carrier signals are respectively multiplied by the signals de-spread by the early and late codes. Therefore, the integrator of these two codes consists of two integrators, which are the in-phase integrator and the quadrature path integrator. The difference between the non-coherent loop and the coherent loop is that the former generates the non-coherent result by taking the square sum of the integral and orthogonal path integration results, while the pseudo-code phase difference of the early and late incoherent results are obtained by the phase detector. The non-coherent operation eliminates the impact of the carrier phase difference on the result.

According to the analysis in previous sections, the mathematical expressions of the four integrator outputs in the early and late codes in Fig. 4.46 can be written easily as

$$\overline{I}_{\mathrm{E}} = A\mathrm{sinc}\left(\frac{\Delta\hat{\omega}T_{\mathrm{I}}}{2}\right)R\left(\Delta\tau - \frac{d}{2}\right)\cos\left[\frac{\Delta\hat{\omega}}{2}T_{\mathrm{I}} + \delta\phi_0\right] + \overline{N}_{\mathrm{E,I}} \tag{4.180}$$

$$\overline{Q}_{\mathrm{E}} = A\mathrm{sinc}\left(\frac{\Delta\hat{\omega}T_{\mathrm{I}}}{2}\right)R\left(\Delta\tau - \frac{d}{2}\right)\sin\left[\frac{\Delta\hat{\omega}}{2}T_{\mathrm{I}} + \delta\phi_0\right] + \overline{N}_{\mathrm{E,Q}} \tag{4.181}$$

$$\overline{I}_{\mathrm{L}} = A\mathrm{sinc}\left(\frac{\Delta\hat{\omega}T_{\mathrm{I}}}{2}\right)R\left(\Delta\tau + \frac{d}{2}\right)\cos\left[\frac{\Delta\hat{\omega}}{2}T_{\mathrm{I}} + \delta\phi_0\right] + \overline{N}_{\mathrm{L,I}} \tag{4.182}$$

$$\overline{Q}_{\mathrm{L}} = A\mathrm{sinc}\left(\frac{\Delta\hat{\omega}T_{\mathrm{I}}}{2}\right)R\left(\Delta\tau + \frac{d}{2}\right)\sin\left[\frac{\Delta\hat{\omega}}{2}T_{\mathrm{I}} + \delta\phi_0\right] + \overline{N}_{\mathrm{L,Q}} \tag{4.183}$$

To simplify the analysis, the noise term is omitted here, so

$$S_{\mathrm{E}} = \sqrt{\overline{I}_{\mathrm{E}}^2 + \overline{Q}_{\mathrm{E}}^2} = A\left|\mathrm{sinc}\left(\frac{\Delta\hat{\omega}T_{\mathrm{I}}}{2}\right)\right|R\left(\Delta\tau - \frac{d}{2}\right) \tag{4.184}$$

$$S_{\mathrm{L}} = \sqrt{\overline{I}_{\mathrm{L}}^2 + \overline{Q}_{\mathrm{L}}^2} = A\left|\mathrm{sinc}\left(\frac{\Delta\hat{\omega}T_{\mathrm{I}}}{2}\right)\right|R\left(\Delta\tau + \frac{d}{2}\right) \tag{4.185}$$

$\Delta\hat{\omega}$ and $\delta\phi_0$ in Eqs. (4.180) to (4.183) are the frequency difference and phase difference between the carrier signal of the local carrier and the input signal. When $\delta\phi_0 \neq 0$, the signal energy rotates between $\overline{I}_{\mathrm{E}}$ and $\overline{Q}_{\mathrm{E}}$. However, after the squaring modulo operation of Eq. (4.184), the influence of $\delta\phi_0$ disappears, and the same result can be obtained through similar analysis of the result of the late code. So even if the carrier phase of the local carrier and the input signal are not strictly synchronized, S_{E} and S_{L} will still generate a steady correlation result, which explains why the non-coherent delay-locked loop is much more stable than the coherent delay-locked loop.

The phase detector in Fig. 4.46 is the model of $(S_E - S_L)$. In fact, the phase detector is not unique. Three other phase detectors for non-coherent delay-locked loops are shown in Table 4.5.

Readers can independently analyze the phase-detection curves of these phase detectors by the method of "coherent delay-locked loop". Here, we focus on the impact of the RF front-end bandwidth on the phase detector. When analyzing the characteristics of the phase detector, it is assumed that the shape of the pseudo-code autocorrelation function is a sharp triangle with bilateral symmetry, which is the ideal shape. A sharp triangle means that all of the higher harmonic components are included. The actual input signal and the local signal are bandwidth-limited, so the shape of the obtained autocorrelation function must have gone through smooth distortion. Figure 4.47 shows the autocorrelation function curve of the front-end of the front-end RF with unlimited bandwidth (see Fig. 4.47a) and the autocorrelation function curve with a limited front-end bandwidth (see Fig. 4.47b). In this case, the angle of the autocorrelation function is no longer sharp, but becomes smooth and soft. The extent of its smoothness is related to the size of the limited bandwidth. The wider the bandwidth, the more obvious the sharp angle will be. Otherwise, the angle will be smoother. When the sharp angle of the autocorrelation function becomes smooth, and the coherence interval becomes smaller, it will be difficult for the phase error outputted by the phase detector to reflect the phase position of the real pseudo-code

Table 4.5 Several phase discriminators commonly used in incoherent delay-locked loops

Phase detector	Property
$\frac{1}{2}\frac{S_E-S_L}{S_E+S_L}$	It adopts the non-coherent lead lag envelope method, and normalizes $(S_E + S_L)$ to make it independent of the signal strength, and the amount of operation is large. When the coherence interval is $d/2$, its phase discrimination is linear at $[-d/2, d/2]$
$\frac{1}{2}\frac{S_E^2-S_L^2}{S_E^2+S_L^2}$	It adopts the non-coherent lead lag power method, which is independent of the signal strength through normalizing $\left(S_E^2 + S_L^2\right)$. The calculation amount is moderate because the square root operation is omitted. When the coherence interval is $d/2$, it is linear at $[-d/2, d/2]$. When the signal is strong, it is close to the phase-detection characteristic of the $\frac{1}{2}\frac{S_E-S_L}{S_E+S_L}$ method, and the square loss is large when the signal is weak
$\frac{1}{2}\left[(\overline{I}_E - \overline{I}_L)\overline{I}_P + (\overline{Q}_E - \overline{Q}_L)\overline{Q}_P\right]$	It adopts the approximate coherent point multiplication power method and utilizes all the integration results of the early, prompt, and late phase pseudo-code signal branches. The calculation amount is small, and it will be greatly affected by the signal strength. It can be normalized by $\overline{I}_P^2$ and $\overline{Q}_P^2$. When the coherence interval is $d/2$, the phase discrimination characteristic is approximately linear at $[-d/2, d/2]$

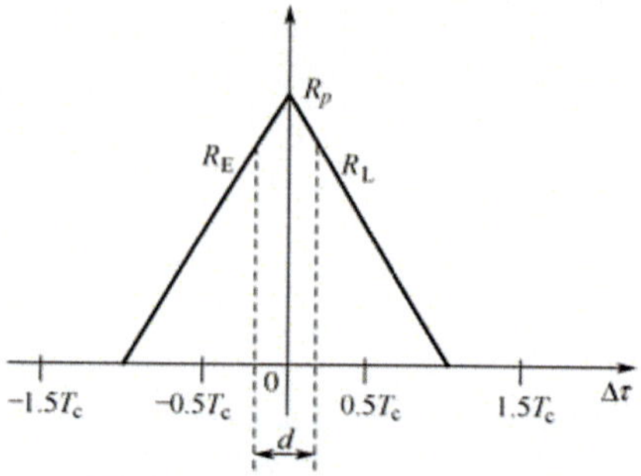

(a) Unlimited RF bandwidth

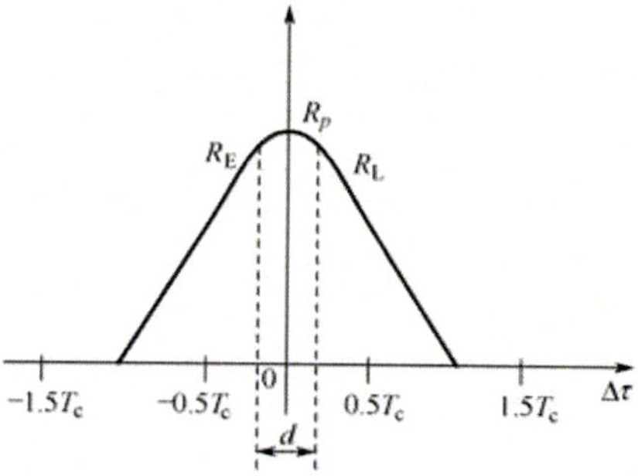

(b) Limited RF bandwidth

Fig. 4.47 Pseudo-code autocorrelation function curves under different bandwidth conditions

main peak, resulting in a larger variance of the pseudo-code phase. Larger errors will also occur in the pseudo-range observation.

Theoretical analysis of the thermal noise characteristics of the non-coherent delay-locked loop is complicated even though the derivation method is the same as in the thermal noise performance analysis in Sect. 4.2.3. Since the statistical features of the phase error outputted by the non-coherent phase detector are difficult to derive, they are skipped in this book, and a relatively reliable conclusion to date will be given instead. It is clarified in Ref. [9] that for the delay-locked loop of the non-coherent phase detector, the root mean square of the pseudo-code random phase difference caused by Gaussian white noise can be expressed as:

$$\sigma_{\text{tDLL}} = \frac{1}{T_c}\sqrt{\frac{B_n\int_{-B_{fe}/2}^{B_{fe}/2} S_s(f)\sin^2(\pi f D T_c)\mathrm{d}f}{(2\pi)^2 \mathrm{CN}_0\left(\int_{-B_{fe}/2}^{B_{fe}/2} f S_s(f)\sin(\pi f D T_c)\mathrm{d}f\right)^2}} \times \sqrt{1+\frac{\int_{-B_{fe}/2}^{B_{fe}/2} S_s(f)\cos^2(\pi f D T_c)\mathrm{d}f}{T\cdot \mathrm{CN}_0\left(\int_{-B_{fe}/2}^{B_{fe}/2} S_s(f)\cos(\pi f D T_c)\mathrm{d}f\right)^2}} \tag{4.186}$$

where B_n is the loop equivalent noise bandwidth in Hertz; $S_s(f)$ is the power spectral density of the desired signal; B_{fe} is the double-band bandwidth of the RF front-end in Hertz, and T_c is the chip width in seconds. For the GPS signal, its frequency is 1/1.023 MHz, while for the BDS signal, it is 1/2.046 MHz. CN_0 is the signal carrier-to-noise ratio, in units of ratio. D is the coherence interval, whose unit is the chip, which is the difference of the late code phase and the early code phase. T is the length of the coherent integration time, whose unit is s.

For BDS and GPS navigation signals, which are BPSK signals, the power spectral density function is $S_s(f) = T_c\text{sinc}^2(\pi f T_c)$, and the value of σ_{tDLL} can be obtained through substituting Eq. (4.186) into it. The following piecewise function approximation is given in Ref. [15] for the non-coherent phase detector using the early minus late power method.

$$\sigma_{\text{tDLL}} = \begin{cases} \sqrt{\frac{B_n}{2\text{CN}_0} D\left[1 + \frac{2}{T \cdot \text{CN}_0(2-D)}\right]}, & D \geq \frac{\pi R_c}{B_{fe}} \\ \sqrt{\frac{B_n}{2\text{CN}_0}\left(\frac{R_c}{B_{fe}} + \frac{B_{fe}}{R_c(\pi-1)}\left(D - \frac{R_c}{B_{fe}}\right)^2\right)\left[1 + \frac{2}{T \cdot \text{CN}_0(2-D)}\right]}, \\ \frac{R_c}{B_{fe}} < D < \frac{\pi R_c}{B_{fe}} \\ \sqrt{\frac{B_n}{2\text{CN}_0}\left(\frac{R_c}{B_{fe}}\right)\left[1 + \frac{1}{T \cdot \text{CN}_0}\right]}, & D \leq \frac{R_c}{B_{fe}} \end{cases} \tag{4.187}$$

It can be seen from Eq. (4.187) that the narrower B_{n}, the smaller D, the higher CN_0, and the longer the coherent integration time T, the smaller σ_{tDLL} will be. The role of the RF front-end bandwidth in the value of σ_{tDLL} is also indicated in Eq. (4.187). It mainly influences the selection of the value of D. It is not true that the value of D shall be as small as possible. When D is less than a certain number, σ_{tDLL} will not continue to grow smaller. Equation (4.187) suggests that the lower limit of D is $\frac{R_c}{B_{fe}}$. When the value of D is less than it, σ_{tDLL} will not continue to improve with the decrease of D. For example, when the RF front-end is 8 MHz, for the GPSC/A code signal, $\frac{R_c}{B_{fe}} \approx \frac{1}{8}$, the coherence interval D can be as small as about 0.125 chips, and any smaller value would not be of much use. For the BDS signal, since its R_c is that of GPS Doubled, the value of D can be as small as 0.25 chips.

The approximate value of D can also be derived from Eq. (4.186). When $\pi f D T_c$ is close to 0, Eq. (4.186) can be approximated as

$$\sigma_{\text{tDLL}} = \frac{1}{T_c}\sqrt{\frac{B_n}{(2\pi)^2 \text{CN}_0 \int_{-B_{fe}/2}^{B_{fe}/2} f^2 S_s(f)\text{d}f}}\sqrt{1 + \frac{1}{T \cdot \text{CN}_0 \int_{-B_{fe}/2}^{B_{fe}/2} S_s(f)\text{d}f}} \tag{4.188}$$

The approximation of $\sin(\pi f D T_c) \approx \pi f D T_c$ and $\cos(\pi f D T_c) \approx 1$ is used in the derivation of Eq. (4.188), from which we can see that when D is small, σ_{tDLL} is close to a limit, which is determined by B_n, CN_0, coherent integration time T and RF front-end bandwidth B_{fe}, where $\sqrt{\int_{-B_{fe}/2}^{B_{fe}/2} f^2 S_s(f)\text{d}f}$ is called the RMS (Root Mean Squared) bandwidth of the signal. The RMS bandwidth measures the amount of high-frequency components contained in the signal, so the degree of sharpness of the signal autocorrelation function is measured quantitatively. The wider the RMS bandwidth of the signal, the sharper the edge of the autocorrelation function can be, thereby generating more accurate pseudo-code phase tracking results. However, it also imposes higher requirements on the RF front-end bandwidth.

The conclusion of the analysis of the thermal noise performance of the pseudo-code tracking loop is similar to the corresponding conclusion of the carrier tracking loop. Although the specific expressions are different (Eqs. 4.170 and 4.186), they all refer to the influence of the additive signal on the random phase difference of the tracking loop with the existence of additive white Gaussian noise. The random phase difference caused by thermal noise is independent of the order of the tracking loop, which mainly affects the dynamic stress performance of the loop. However, the pseudo-code loop is significantly different from the carrier ring in one aspect, which is the pseudo-code loop is often assisted by the dynamic information of the carrier ring. This is because the dynamic stress of the carrier loop is strictly proportional to the dynamic stress of the pseudo-code loop, so the dynamic condition obtained by the carrier loop can be compensated and cancelled in the pseudo-code loop by a certain scale factor, meaning that the pseudo-code loop can offer excellent thermal noise features with low order loop and extremely narrow equivalent noise bandwidth. This technology is called "carrier assistance". The principle and implementation of carrier assistance will be explained in detail in the following sections.

4.2.5 *Problems in Tracking Loop Implementation and Debugging*

The discussion in this section will focus on some practical problems in the specific implementation and debugging of the carrier tracking loop and pseudo-code tracking loop, in which the theoretical analysis in the previous chapters will be integrated. New technologies and new solutions emerging in recent years will be briefly explained, and relevant references will also be given.

1. Carrier NCO and Pseudo-code NCO

According to the previous structural analysis of the carrier tracking loop and pseudo-code tracking loop, the NCO (Numerical Controlled Oscillator) is one of the key units. Its principle and implementation will be explained in detail in this section. The physical source of the NCO should be a VCO (Voltage Controlled Oscillator), but today, with the prevailing of digital technology, the VCO has emerged as an NCO in digital logic circuits and software implementations, the basic structure of which is shown in Fig. 4.48.

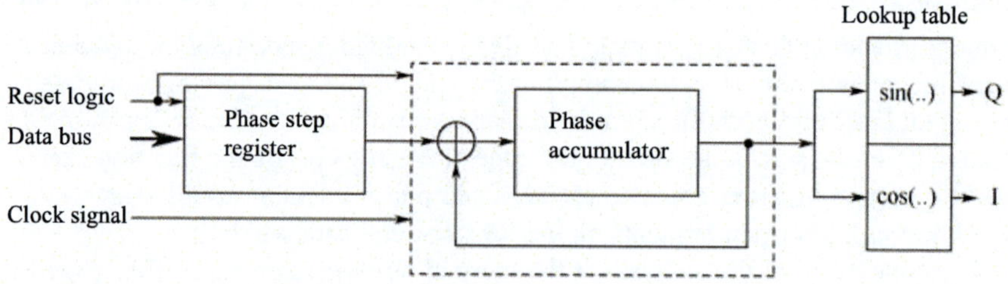

Fig. 4.48 Block diagram of the carrier NCO

The principle of the carrier NCO is shown in Fig. 4.48. Other types of NCOs are very similar, and the only difference is in the lookup table in the end. An NCO basically consists of three parts: a phase step register, a phase accumulator, and a lookup table. There are some other control signals in an NCO, including clock, data lines, and reset signals. The NCO operates at a certain clock frequency, which is generally referred to as the operating frequency of the NCO. At the moment of each working clock, the NCO accumulates the local phase accumulator according to the value of the phase step register. Suppose that at clock k, the value of the phase step register is $\phi_\Delta(k)$, and the value of the phase accumulator is $\phi(k)$, then at clock $k+1$, the value of the phase accumulator is

$$\phi(k+1) = \phi(k) + \phi_\Delta(k) \tag{4.189}$$

$\phi_\Delta(k)$ is the external input to the NCO. So, it can be seen that the phase accumulator is indeed the integral of the external input. If the local phase is used as the output and $\phi_\Delta(k)$ is used as the input, the s domain model of the NCO is $1/s$.

According to Eq. (4.189), it seems that $\phi_\Delta(k)$ will increase without limit. In fact, in the digital system, there is an upper limit for the bit width of the phase accumulator, so $\phi_\Delta(k)$ will not increase without limit. If the number of accumulators is n, then the maximum value of $\phi_\Delta(k)$ is $(2^n - 1)$; when the value is reached, it will return to 0 and start to accumulate anew. By this time, the phase update of one cycle is completed for the NCO. The output of the NCO is determined by the value of the current phase accumulator, and the value of the current phase is used as the index address of the lookup table to generate the output value. For the carrier NCO, since the period of the carrier signal is 2π, the phase range of the corresponding lookup table is $[0, 2\pi]$, and the range of the phase accumulator is $[0, 2^n - 1]$, between which a linear correspondence exists.

Figure 4.49 shows the relationship between the local phase update of the carrier NCO and the output value. The width of the phase accumulator in the figure is 16 bits, and the number in the phase accumulator is added from 0 to 0xFFFF. It then returns to 0, and the cycle is repeated. The lower right part of the figure shows the step shape of the NCO during phase accumulation. The upper right part of the figure is the corresponding sine and cosine lookup table. The value corresponding to the lookup table is output according to the value of the local phase. Thus, the in-phase and quadrature carrier components in the carrier tracking loop explained in the previous section can be obtained.

For pseudo-code NCO, the lookup table is much simpler. Figure 4.50 shows the relationship between the local phase update of the pseudo-code NCO and the output value. As the figure shows, the principles of the pseudo-code NCO and carrier NCO are very similar except for the last lookup table. The pseudo-code NCO outputs a square wave, so it is only necessary to judge whether it is 0 or 1 based on the value of the local phase. According to the figure, when the phase is greater than 0x8000, the output of the pseudo-code NCO is 0. Otherwise, it will be 1, so the lookup table of the pseudo-code NCO is much simpler than the carrier NCO.

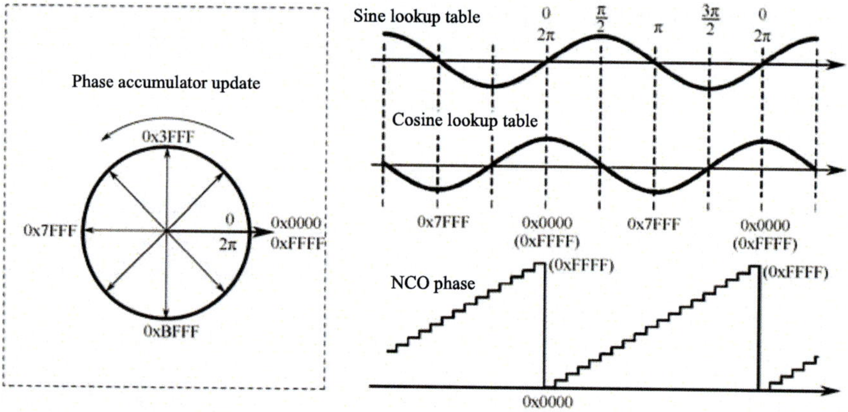

Fig. 4.49 Local phase updating principle and output of the carrier NCO

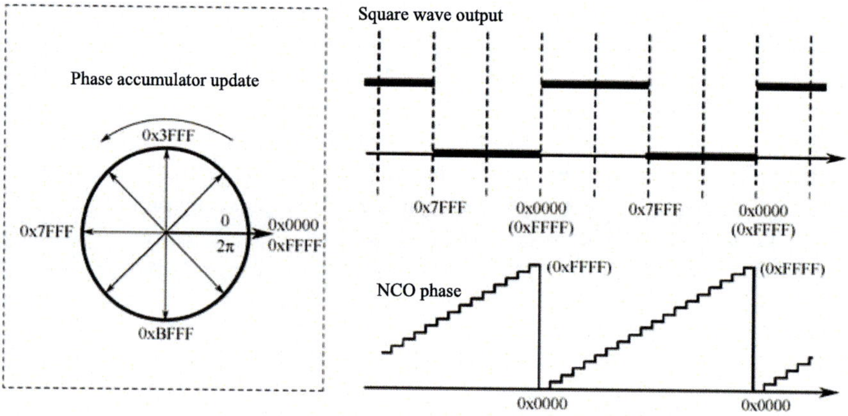

Fig. 4.50 Local phase update principle and output value of the pseudo-code NCO

Clearly, the instantaneous frequency f_{NCO} of the NCO has the following relationship with the phase accumulator bit width n, the input clock frequency f_s, and the phase step value $\phi_\Delta(k)$:

$$f_{\mathrm{NCO}} = \frac{\phi(k+1) - \phi(k)}{2^n T_s} = f_s \frac{\phi_\Delta(k)}{2^n} \tag{4.190}$$

Equation (4.190) shows that f_s and n determine the frequency resolution of the NCO. Generally, in BDS and GPS receivers, f_s is the IF sampling clock, which is determined by the RF front-end; room adjustment is limited, so only the value of n is open to adjustments. The larger n is, the finer the frequency value of the NCO output will be. For example, $f_s = 16.369$ MHz; when $n = 32$ the frequency resolution is 16.369E6/2^{32} = 3.81E−3, which is about 3.8 MHz.

There are two factors to consider when deciding the lookup table for the carrier NCO. One is the quantization bit width of the lookup table, and the other is the number of samples in the $[0, 2\pi]$ cycle. The former can be determined according to the quantization bit width of the radio frequency front-end. For example, the intermediate frequency data quantization bit width is 2 bits, so the lookup table can be designed as 3 bits or 4 bits, because the higher the quantization bit width is, the smaller the quantization loss will be. In cases where the memory capacity overhead is not large, the quantization bit width of the lookup table can be properly increased. If multi-bit sampling is applied to the quantization bit width of the RF front-end, such as the 8–12 bit quantization commonly used in professional mapping receivers, the quantization bit width of the lookup table should be its equivalent, otherwise the advantages brought by high sampling quantization will be lost. The number of samples in the lookup table can be 8 or 16 in one cycle. Experience shows that no higher processing gain will be generated if the number of samples exceeds sixteen. If the number of samples is 16, since $16 = 2^4$, then only the upper 4 bits of the local phase accumulator will be required as the lookup table index of the output value.

2. Conversion from s-Domain to Z-Domain

So far, the domain included in the theoretical derivation of the carrier loop and the pseudo-code loop in the previous chapters is the Laplace domain, i.e. the s-domain. While the loops in the actual system all work on the basis of digital discrete signals, the common tool for digital discrete signal analysis is the Z-transform, with which the analysis of the loop can be more closely integrated with the actual system. The conversion from s-domain to Z-domain will be discussed in this section.

The physical meaning of the s-operator of the s-domain is to generate the derivative of time. Assuming that the s-domain transformation of the time domain signal $x(t)$ is $x(s)$, then $sx(s)$ corresponds to $dx(t)/dt$. If the digital sampling of $x(t)$ is conducted and the sampling interval is T_s, when Ts is small, the derivative of $x(t)$ can be approximated by dividing the difference between adjacent sampling points by the sampling interval, i.e.

$$\frac{dx(t)}{dt} \approx \frac{x(k) - x(k)}{T_s} \xrightarrow{z\,transforms} \frac{x(z) - z^{-1}x(z)}{T_s} = \frac{1 - z^{-1}}{T_s}x(z)$$

where, $x(z)$ is the Z-transform of $x(t)$, so the s operator corresponds to $\frac{1-z^{-1}}{T_s}$ of the Z-domain.

The physical meaning of the $1/s$ operator is to generate the time integration. Taking the above-mentioned time domain signal $x(t)$ into consideration, then $y(s) = \frac{x(s)}{s}$ corresponds to $y(t) = \int_0^{\infty} x(t)dt$. In digital logic circuits and software programming, the integral is often generated by the accumulator, i.e. $y(k) = y(k-1) + x(k)T_s$, then

$$y(k) = y(k-1) + x(k)T_s \xrightarrow{z\ transforms} y(z) = \frac{T_s}{1 - z^{-1}} x(z)$$

$1/s$ operator, the sampling method of the accumulator integral, corresponds to $\frac{T_s}{1-z^{-1}}$ of the Z-domain.

Another method of integrating is to use the average of two adjacent sampling points $x(k)$ and $x(k-1)$ as the integral quantity, i.e. $y(k) = y(k-1) + 0.5[x(k) + x(k-1)]T_s$. The reason for adopting this method is simple: in some cases, more accurate results can be obtained using the average of adjacent samples to replace the single point for the integral operation. So, when Z-transform is applied to both sides of the equation,

$$y(z) - z^{-1}y(z) = 0.5[x(z) + z^{-1}x(z)]T_s \Rightarrow y(z) = 0.5T_s \frac{1 + z^{-1}}{1 - z^{-1}} x(z)$$

The result of this processing is that the 1/s operator corresponds to $\frac{T_s}{2}\frac{1+z^{-1}}{1-z^{-1}}$ of the Z domain. This method is called bilinear transformation.

A comparison of the differential and integral operators in the s-domain and the Z-domain is given in Table 4.6, and a block diagram of the implementation based on the Z-domain expression is also shown in the table.

With the transformation of the s-operator and $1/s$ operator corresponding to the Z-domain, they can be substituted into the s-domain analysis result of the phase-locked loop. The transform of the processing block diagram in Fig. 4.30 from the s-domain to the Z-domain by bilinear transformation is shown in Fig. 4.51. Only the Z-domain transformation of the loop filter is given in the figure, while the integrator in the NCO link is omitted. The transformation from Figs. 4.30, 4.31, 4.32, 4.33, 4.34, 4.35, 4.36, 4.37, 4.38, 4.39, 4.40, 4.41, 4.42, 4.43, 4.44, 4.45, 4.46, 4.47, 4.48, 4.49, 4.50 and 4.51 is straightforward and easy to understand. The only thing to note

Table 4.6 Representation of differential and integral operators in the s- and Z-domains

微分	s	$\frac{1-z^{-1}}{T_s}$	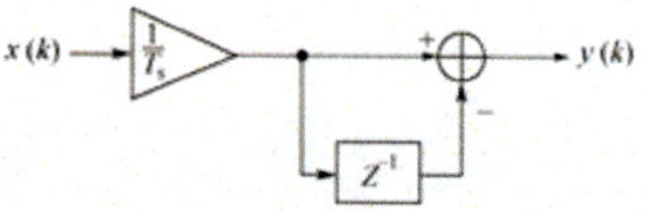
积分 (累加器)	$\frac{1}{s}$	$\frac{T_s}{1-z^{-1}}$	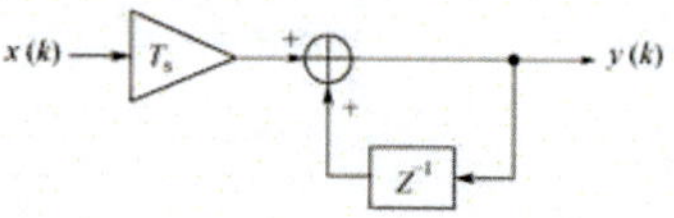
积分 (双线性变换)	$\frac{1}{s}$	$\frac{T_s}{2}\frac{1+z^{-1}}{1-z^{-1}}$	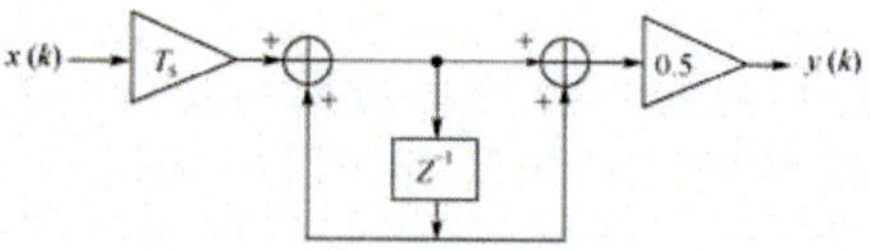

微分 Differential; 积分 (累加器) Integration (accumulator); 积分 (双线性变换) Bilinear transformation

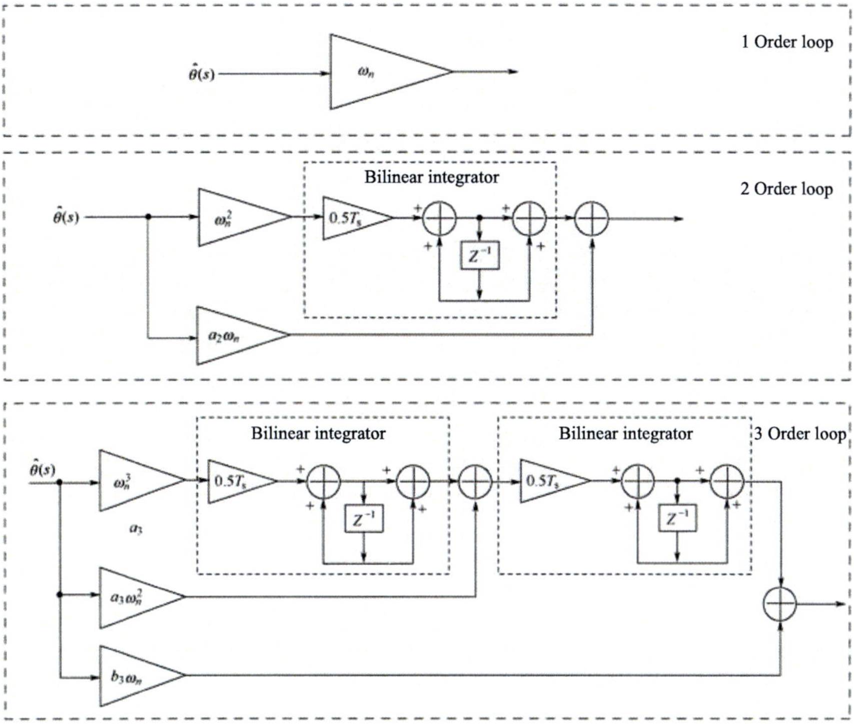

Fig. 4.51 Three kinds of loop filters are implemented in the Z-domain: first-order loop, second-order loop, and third-order loop

is that the input signal gain in the bilinear integrator unit is $0.5T_s$, where T_s is the time interval for loop update, which should not be confused with the IF sampling interval mentioned in the previous section. Generally, the update interval of the loop is the same as the correlator integration time. For example, if the integration time is 1 ms, and the loop is updated every time the integration ends, then $T_s = 0.001$. When the integration time becomes 10 ms, $T_s = 0.01$. In some cases, the loop is not always updated every time the coherent integration ends, that is, when T_s needs to be determined according to the specific time interval of the update loop.

The system transfer function and error transfer function of the phase-locked loop can also be converted by a similar method. Both the steady-state phase difference and the thermal noise analysis can be performed by using the relevant theory in the Z-transform. Readers can consult reference books for signal and system and digital signal processing.

3. Selection of Loop Parameters

Section 4.2.1 has shown that the phase-locked loops of different orders have different tracking competence for dynamic stress. The loop parameters determine its capture capability, response speed, damping coefficient, system stability, equivalent noise bandwidth, and many other factors, where the equivalent noise bandwidth plays a pivotal role in the performance of loop thermal noise, as detailed in Sects. 4.2.2, 4.2.3, and 4.2.4. Even the same loop parameters tend to have opposite effects in terms of dynamic stress and thermal noise performance, thus a comprehensive consideration of the above factors is required in choosing the loop parameters to be used in engineering practice. Sometimes, dynamic adjustment need to be made for different working conditions, focused on system performance.

The second-order loop is the most commonly used in the carrier tracking loop due to its stability and the relatively simple adjustment of parameters. Besides, it can accommodate most dynamic scenes in civilian applications, so it is widely used in the civilian receiver market. In the design of the carrier tracking loop (Costas loop) of BDS and GPS receivers, the parameters of the second-order loop can be selected according to Figs. 4.2 and 4.30, among which are the two most important parameters, namely the damping system ζ and the equivalent noise bandwidth B_n. Of course, ζ and the natural frequency ω_n are also an option. Since B_n and ω_n have a strict linear relationship ($B_n = 0.53\omega n$), the two choices are not contradictory. The damping coefficient determines the speed at which the tracking loop adjusts with the external input change. Most typical is whether the local phase adjustment step can keep up with the input phase change when the input is unit step excitation, namely the performance of the transient response of the loop. Figure 4.52 is the transient response of the second-order Costas loop to the phase step input in different damping systems, where $\omega_n = 10\pi$. In order to highlight the influence of the damping coefficient, the other settings of the loop remain unchanged. It can be seen that when the damping coefficient is small, the system is very sensitive. However, the local phase cannot be adjusted once in place; it should undergo several oscillations before finally

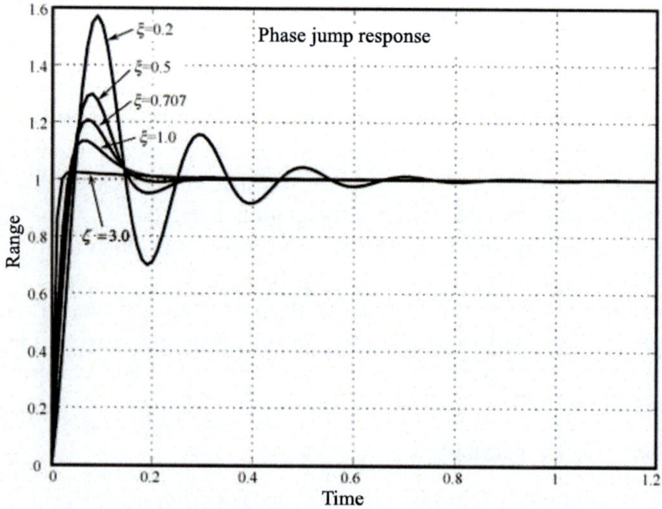

Fig. 4.52 Transient response of the Costas loop to the unit step signal at different ζ values

stabilizing. When the damping system is large, the local phase adjustment is slow, but there is no oscillation and rushing. Generally, a damping coefficient of less than 1.0 is called under-damping. One that is greater than 1.0 is called over-damping. When the damping coefficient is 1.0, it is called critical damping. In engineering, 0.707 is often chosen as the optimized damping coefficient, with which the combination of loop sensitivity and response speed can be achieved.

The selection of the effective noise bandwidth B_n mainly requires consideration of the following factors:

1. The tracking of the dynamic stress of the system. The larger B_n is, the larger the dynamic range the loop can track, as seen in Eq. (4.140).
2. The size of B_n determines the random phase difference caused by the input noise. The smaller B_n is, the smaller the random phase difference will be. However, the random phase difference will be larger. When the random phase difference is large enough, loop instability will also occur.
3. The value of B_n determines the natural frequency ω_n. The larger ω_n is, the sooner the system can be stabilized in the transient response, which means that it can enter the stable state faster.

Therefore, the choice of B_n must be balanced among dynamic performance, response speed, and random phase difference. In this case, different parameters are needed for different application scenarios. For example, in applications with large dynamics, fast response speed, and low positioning accuracy, the value of B_n should be slightly larger; for applications with higher static state and higher positioning accuracy, the value of B_n can be smaller.

The random phase difference and dynamic stress phase difference of the loop analyzed here are applicable to both the carrier tracking loop (Costas loop) and the pseudo-code tracking loop (delay-locked loop). For the carrier tracking loop, an empirical conditional equation considering the phase difference margin for loop stability is given in Ref. [9]:

$$\begin{aligned} &\text{When there is no data modulation } 3\sigma_i + \theta_e \leq 90^\circ, \\ &\text{When there is data modulation } 3\sigma_i + \theta_e \leq 45^\circ \end{aligned} \tag{4.191}$$

In Eq. (4.191), σ_i is the sum of the random phase differences caused by all factors excluding dynamic stress. In addition to the additive white noise in the input signal, it also includes the random phase difference caused by the mechanical vibration of the clock and the Allan variance. The phase differences can be considered as irrelevant to each other, so the total variance can be calculated in the form of squares and sums of the differences, i.e.

$$\sigma_i = \sqrt{\sigma^2_{thermal\ noise} + \sigma^2_{mechanical\ vibration\ of\ clock} + \sigma^2_{Allan\ variance\ of\ clock}}$$

θ_e is the phase difference caused by dynamic stress. The dynamic stress phase difference not the same as the random phase difference. The former is not a kind of random "noise", but more of a "constant bias". Therefore, θ_e in Eq. (4.191) appears as a separate addition. At the outset of the design of the system, we can determine the maximum acceleration and jerk that can be stably tracked first, as well as the limit parameters such as tracking sensitivity. Then we can get the value of σ_i and θ_e in Eq. (4.191) through combining the short-term stability index and phase noise level of the clock to decide if the stability conditions are met, and make adjustments if necessary. When tracking the BDS D1, D2 code signal and GPS C/A code signal, we can refer to the phase difference margin when there is data modulation. For the pilot channel signal in the GPS modernization plan and in the future BDS system, since there is no data modulation, we can refer to the phase difference margin of non-data modulation in Eq. (4.191). The step margin threshold is doubled in the absence of data modulation.

For the pseudo-code tracking loop, the empirical conditional equation for the phase difference for loop stability is given in Ref. [9]:

$$3\sigma_{\text{tDLL}} + R_e \leq D/2 \tag{4.192}$$

In Eq. (4.192), σ_{tDLL} is the random phase difference of the pseudo-code loop due to thermal noise, which can be obtained by Eq. (4.186). R_e is the dynamic stress phase difference, and D is the phase difference between the early code phase and the late code phase in the delay-locked loop. However, it should be noted that the units of σ_{tDLL} and R_e are both chips, so the unit of D is also chips. When the carrier-assisted method is used, the carrier loop can eliminate most of the dynamic stress term in the pseudo-code loop, so only the σ_{tDLL} term needs to be considered in Eq. (4.192).

In modern receivers, multiple tracking channels can often be used to simultaneously process signals of multiple satellites. Each tracking channel has its own tracking loop, and the satellite signal strengths in different channels are also different. In the tracking loop of the strong star signal, a larger B_{n} can be adopted, while a smaller B_{n} is useful for the tracking channel of the weak star signal. The advantage of the processing is that strong dynamic stress processing can be provided by the strong star signal. At the same time, the weak star signal can provide a measurement with less phase noise and higher tracking sensitivity. Furthermore, fuzzy logic can be used to model the error signals outputted by the phase detector and the discriminator, which can better ensure the balance between dynamic characteristics, loop stability, and random phase difference.

4. Application of Lissajous Graphics in Loop Debugging

The Lissajous figure is a synthetic trajectory of the rotation vector in two mutually perpendicular directions, which is often used in the fields of electrical engineering and electronics to determine the frequency multiplier and phase difference between two waveform signals. In the Costas loop and the delay hysteresis loop, the two integral results outputted by the integrator related to the I-channel and Q-channel can be regarded as a rotation vector, so the relevant conclusions of the Lissajous figure can be used to judge the working state of the tracking loop, thereby assisting the debugging of the loop.

Figure 4.53 shows the results of the I-channel and Q-channel integration of the Costas loop in a stable phase-locked state on a two-dimensional plane, which is a typical Lissajous figure. According to the analysis in Sect. 4.2.3, when the carrier phase is stably tracked, the I-channel integration result contains all the energy and noise terms of the signal, while the Q-channel integration result only contains the noise term. Therefore, the Lissajous figure in the two-dimensional plane is two circularly scattered areas separated by a certain interval. The circular scatter area on the left corresponds to the modulated data bit −1, and the circular scatter area on the right corresponds to the modulated data bit +1. In Fig. 4.53, the spacing between the center points of the two scatter areas is represented by *A*, and the scatter diameter of the circular area is represented by *r*. It is clear to see that the ratio of *A* to *r* stands for the signal-to-noise ratio of the input signal.

Figure 4.54 shows the Lissajous figure of the I-channel and Q-channel integration results for signals with four different intensities. The CN_0 of the input signal is 35 dBHz, 40 dBHz, 45 dBHz, and 50 dBHz respectively, and the noise bandwidth of the tracking loop is 20 Hz. The input signal does not contain dynamics higher than the second derivative to ensure the stable locking of the phase. It can be seen from the figure that the signal intensity will cause the interval between the center points of the two circularly dispersed regions to change. The stronger the signal is, the larger the center point interval will be, otherwise it will be smaller. The noise items are normalized during signal tracking, so there is no significant difference in the diameters of the circularly scattered areas in the four images in Fig. 4.54.

There are several factors that can cause the carrier phase of the Costas loop PLL to lose lock during actual operation. The most common ones include the weakening of the signal strength, the large dynamic motion of the carrier, and the sharp change of the RF clock temperature. Figure 4.55 shows the Lissajous figure when the carrier

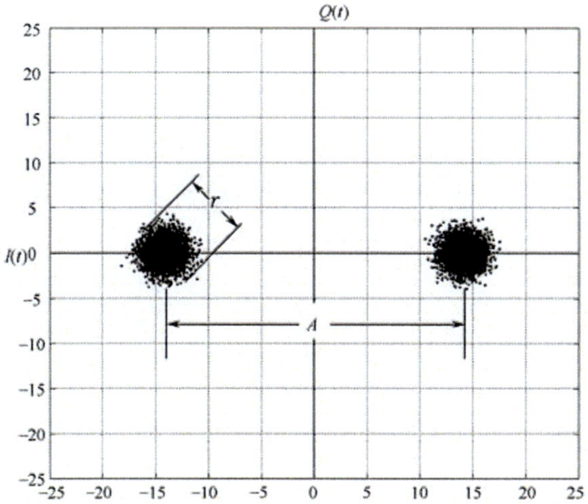

Fig. 4.53 Lissajous figure of typical PLL

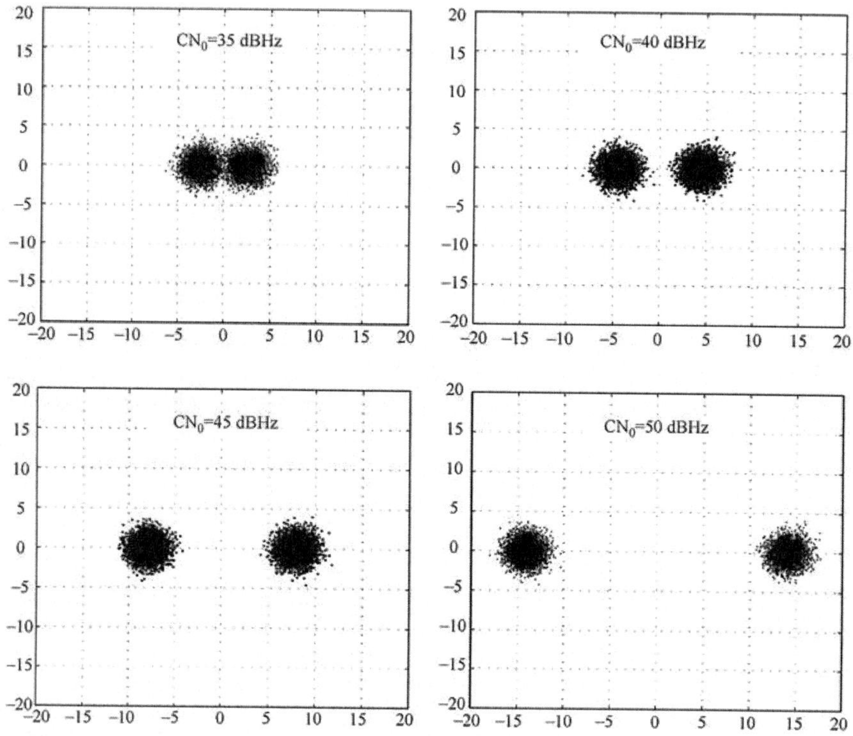

Fig. 4.54 Lissajous figure of PLL at different signal intensities

Fig. 4.55 Lissajous figure of PLL when carrier phase is unlocked

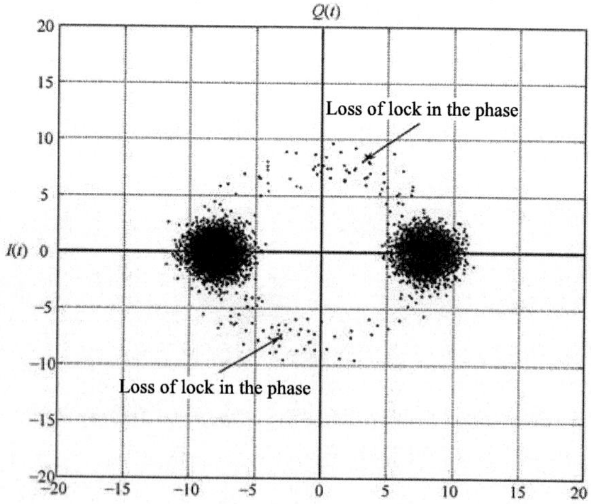

phase is temporarily out of lock. The data in the figure comes from the situation where the carrier phase is temporarily lost due to the artificial addition of the jerk component in the input signal. As can be seen from the figure, when the carrier phase is out of lock, the Lissajous figure will no longer be limited to the left and right circular scattering areas, but will move in the entire quadrant plane. In the actual debugging process, when a similar trajectory as shown in Fig. 4.55 appears in the Lissajous figure conformed by the I-channel and Q-channel integrals of the loop, it is necessary to check the dynamic stress, signal strength, phase detector and discriminator output, as well as the surrounding temperature trends, in order to determine the cause of phase loss in a timely manner, and solve the problem in a targeted way.

It is better to observe the Lissajous figure when the carrier phase is out of lock in the setting of the frequency-locked loop, because the purpose of the Costas loop then is to achieve frequency locking instead of phase locking. When the frequency is locked, the carrier phase can still be in the rotating state, so the Lissajous figure composed of the I-channel and Q-channel integration results at this time will perfectly reveal the shape of the carrier phase rotation. Figure 4.56 is a Lissajous figure for the frequency tracking of the input signal of $CN_0 = 45$ dBHz. The trajectory in the figure is circular, and the shape is determined by two parameters. One is the width of the ring band, which is represented by r in the figure, and the other is the diameter of the central curve of the annular loop, indicated by A in the figure. It is clear that A and r here correspond to those in Fig. 4.53, and the corresponding ratio of A to r also indicates the magnitude of the signal-to-noise ratio of the input signal.

Figure 4.56 shows that after the frequency lock is completed in the frequency-locked loop, the carrier phase is still in the rotating state, so the Lissajous figure

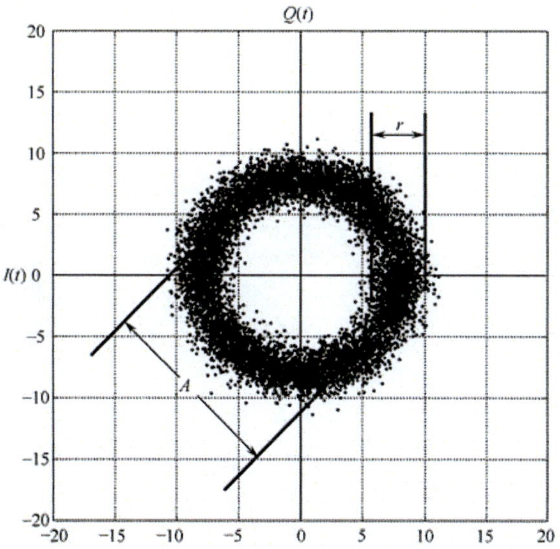

Fig. 4.56 Lissajous figure of a typical frequency-locked loop

of the frequency-locked loop is scattered in four quadrants, and the speed of the phase rotation indicates the magnitude of the frequency error. Of course, the speed of rotation cannot be seen in Fig. 4.56, because only the trajectory map is shown in the Lissajous figure, not the mutual time relationship between the trajectory points. The loop formed by the rotation of the carrier phase has a certain radius, whose size is directly related to the signal strength. Although the carrier phase is rotating, the signal energy is not lost, but is periodically distributed in the in-phase component and the quadrature component. This is also a major feature of the frequency-locked loop, suggesting that it is more difficult to demodulate data bits with a frequency-locked loop.

Figure 4.57 shows the Lissajous figure of the frequency-locked loop at different input signal strengths. The four signal intensities distributed as CN_0 are 35, 40, 45, and 50 dBHz. In the figure, the signal intensity affects the radius of the center curve of the annulus. The stronger the signal is, the larger the radius of the annulus will be, and vice versa. Similar to Fig. 4.54, the noise items are normalized during signal

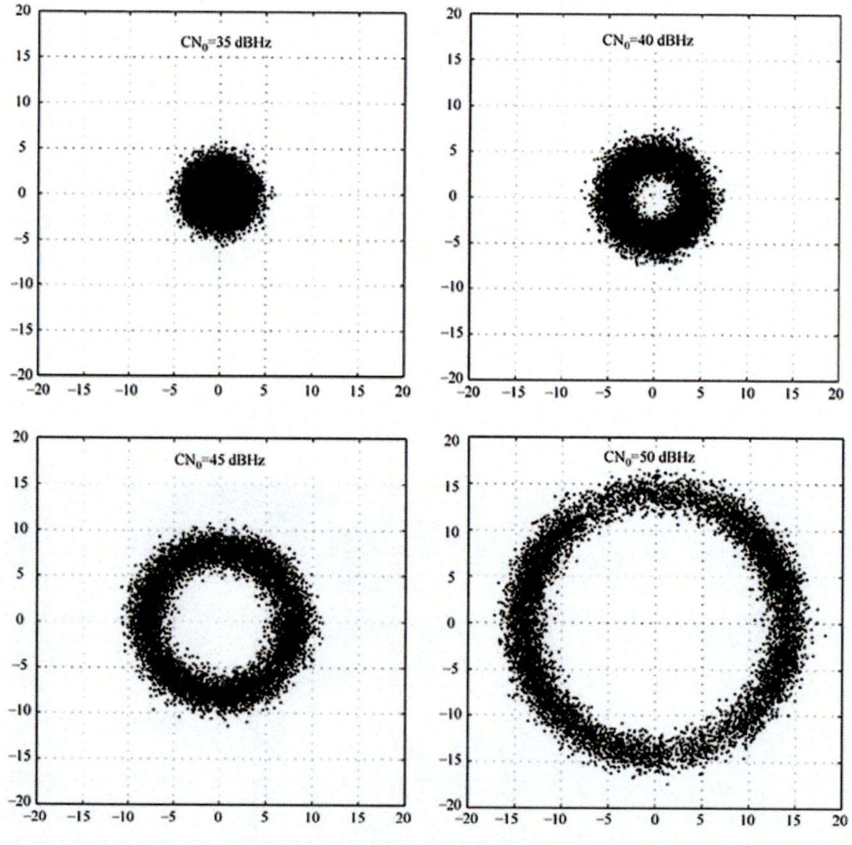

Fig. 4.57 Lissajous figures of frequency-locked loops at different signal intensities

tracking, so there is no significant difference in the width of the loop under the four signal intensities in Fig. 4.57.

For the delay-locked loop, the I-channel and Q-channel integrals corresponding to the early code, the late code, and the prompt code can be drawn together in a two-dimensional plane to form a Lissajous figure. The analytical steps are similar to those described in this section. Readers can independently test and analyze the effects of the code phase error and the carrier frequency phase error on the I-channel and Q-channel integrals outputted by the delay-locked loop.

5. Combination of the Phase-locked Loop, Frequency-locked Loop, and Carrier Assist

According to the theoretical analysis in the previous chapters, the difference between the frequency-locked loop and the phase-locked loop is diverse. In terms of loop implementation, the error signal of the phase-locked loop comes from the phase detector, and that of the frequency-locked loop comes from the discriminator. In terms of the physical meaning of the error signal, the error signal of the phase locked loop is the carrier phase difference between the input signal and the local signal, while that of the frequency-locked loop is the carrier frequency difference between the input signal and the local signal. In terms of the final effect of the loop, the phase-locked loop realizes the locking of the carrier phase of the input signal, and the frequency-locked loop realizes the locking of the carrier frequency of the input signal. In terms of the tracking performance of the loop, since the frequency is the integral of the phase, the frequency-locked loop offers better performance in terms of taking the dynamic stress than the phase-locked loop under the same input signal strength and equivalent noise bandwidth. Based on the characteristics and advantages of the frequency-locked loop over the phase-locked loop, the frequency-locked loop is often used in BDS and GPS receivers to assist the phase-locked loop.

In the process of frequency-locked loop assisted phase-locked loop, the frequency-locked loop often works in the following two situations:

1. When the signal has just been captured, the frequency difference between the local carrier and the input signal is relatively large, and the frequency-locked loop can quickly perform frequency pulling and transfer control to the phase-locked loop.
2. The signal cannot be phase locked because the absolute value of the carrier phase difference exceeds the phase margin given in Eq. (4.191) due to excessive dynamic stress or low signal strength. In cases of phase loss, whether caused by dynamic stress or signal strength, the frequency-locked loop can continue to track the signal as a "backup loop" of the phase-locked loop. Of course, what is tracked by the loop is the signal frequency. Although the frequency-locked loop can keep track of the signal continuously, since the phase lock cannot be realized, the error rate of the demodulated teletext data will be very high, the quality of the observation will be poor, and there will also be serious cycle slips in the carrier phase. Therefore, the frequency-locked loop is not suitable for applications where accuracy is critical.

As shown in Fig. 4.58, since the frequency is the differential of the phase, the frequency-locked loop used to assist the phase-locked loop is often one order lower than the order of the phase-locked loop. The figure shows an example of a first-order frequency-locked loop assisting a second-order phase-locked loop and a second-order frequency-locked loop assisting a third-order phase-locked loop. The key change here is in the synthesizing method of the loop filter, so the NCO unit is omitted. The frequency and phase detector in the figure can be the examples in Figs. 4.3 and 4.4. The frequency-locked loop and the phase-locked loop each have an independent set of loop filters with different noise bandwidths B_n. The frequency error is sent to the control carrier NCO after the high-frequency components in it have been filtered out by the loop filter. However, the input point of the frequency error is later than that of the phase error by one integrator.

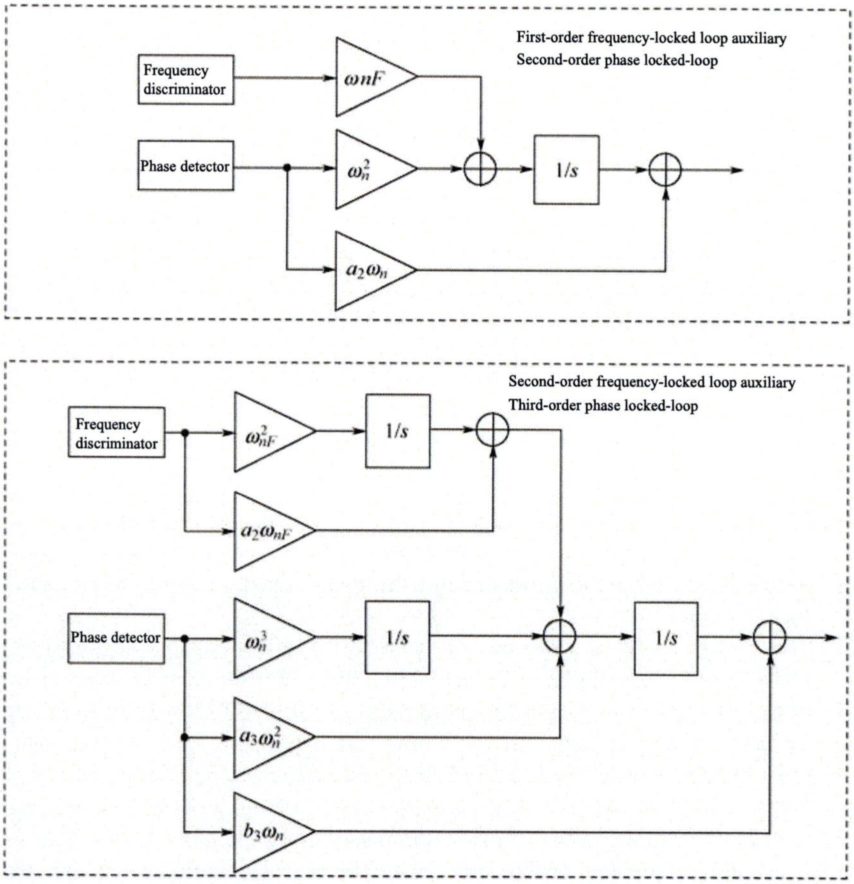

Fig. 4.58 Loop filter block diagram of a frequency-locked loop's auxiliary phase-locked loop

The frequency discriminator and the phase detector in Fig. 4.58 both process the I-channel and Q-channel integration results, so the frequency and phase difference signals outputted by them are correlated. After the phase lock is completed, the output of the discriminator will be 0, and the tracking loop becomes a pure phase-locked loop. Meanwhile, when the phase is out of lock, the discriminator starts to output frequency errors. At this time, the excellent dynamic and thermal noise performance of the frequency-locked loop enables it to lock the frequency, so that the information about the frequency and the code phase of the input signal can be maintained to achieve re-lock of the carrier phase when external conditions permit.

What has been discussed above is the conventional auxiliary mode of the frequency-locked loop and phase-locked loop. There are many other combinations of frequency-locked loops and phase-locked loops, including fuzzy logic control and FFT frequency difference analysis. Readers can refer to the references attached to this chapter for further information.

Because of the motion between the satellite and the receiver, there is a Doppler shift effect in the received signal. The effect has an impact on both the carrier signal and the pseudo-code signal, triggering the same effect in both the received carrier component and the pseudo-code component. Moreover, there is a fixed linear relationship between the two effects. This is not difficult to understand. Simply consider a more general case, namely the relationship between the Doppler shift and relative motion speed and the initial frequency of the signal, which is as follows:

$$f_R = \frac{c}{c - v_{||}} f_S \tag{4.193}$$

where f_S is the transmitted signal frequency; f_R is the received signal frequency; c is the speed of light; and $v_{||}$ is the projection of the relative motion velocity on the direction cosine. Equation (4.193) can be approximated as

$$f_R \approx (1 + \frac{v_{||}}{c}) f_S \tag{4.194}$$

Therefore, the value of the Doppler shift is $\frac{v_{||}}{c} f_S$. For the carrier and the pseudo-code of the satellite signal, $v_{||}$ is the same; the only difference is in the transmission frequency. Therefore, the relationship between the Doppler shift and the carrier is as follows:

$$\frac{f_{\text{Dopp, carrier}}}{f_{\text{Dopp, psedo-code}}} = \frac{f_{\text{frequency of the signal}}}{f_{\text{frequency of the psedo-code}}} = \begin{cases} 1540, & \text{GPS C/A code signal} \\ 763, & \textit{BDS}\ \text{B1 signal} \\ 590, & \textit{BDS}\ \text{B2 signal} \end{cases} \tag{4.195}$$

Equation (4.195) shows that the carrier frequency in the satellite signal is homologous to the dynamic stress of the pseudo-code frequency. According to the proportional relationship of the Doppler shift given in the equation above, the Doppler information obtained by the carrier loop tracking can be input into the pseudo-code loop. Most of the dynamic stress can also be deducted, so a large proportion of the pseudo-code frequency fluctuation caused by the relative motion can be offset, for which the bandwidth of the pseudo-code tracking loop can be set as small, which will significantly improve the noise performance of the loop. This technique is called "carrier assistance". A carrier-assisted tracking loop is shown in Fig. 4.59.

Figure 4.59 shows the most commonly used signal tracking loop model in modern receivers, which includes a carrier tracking loop and a pseudo-code tracking loop. The six coherent integrators correspond to the I-channel and Q-channel integrals of the early, late, and prompt codes respectively. The results generated by the six coherent integrators are sent to the carrier phase detector (or frequency discriminator) and the pseudo-code phase detector respectively to obtain the phase error signal. Then, the loop filter is used to control the NCO through the loop filter, the principle of which has already been described in detail in the previous sections. With the adoption of the carrier assistance technique, the deviation value of the carrier Doppler frequency in the figure is used to deduct the dynamics in the pseudo-code loop after passing through a proportional gain module $1/K$. The value of K is set to be the scale factor in Eq. (4.195) corresponding to different signal types.

The modules on the left side of the dashed line in Fig. 4.59 operate at the IF frequency f_s, and those on the right side operate at the $1/T_I$ frequency, where T_I is the integration time. In general, $f_s >> 1/T_I$, so the modules on the left are generally implemented in hardware, while those on the right are usually implemented in software. In terms of complexity, the functions of NCO, the integrator, the pseudo-code generator, and other modules are relatively simple. The working clock is higher,

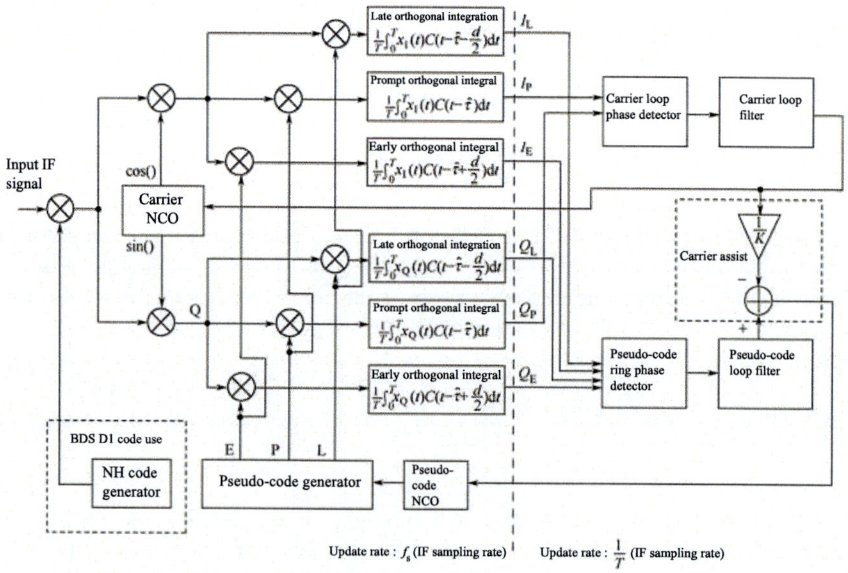

Fig. 4.59 Use of a carrier-assisted carrier tracking loop and pseudo-code tracking loop

making it more suitable for implementation in the hardware, while the functions of the phase detector, loop filter, and tracking loop logic control are more complex. Meanwhile, the requirements for the operating clock are lower, so they are more suitable for software. There is also an NH code generator in Fig. 4.59, which is designed for the BDS D1 code signal. Only upon the achievement of the NH code stripping can the coherent integration be longer than 1 ms.

The basic idea of carrier assistance is to assist the pseudo-code tracking loop with some known dynamic information from the carrier loop. The idea can also be used for carrier loop tracking, for example, to estimate the dynamics of the receiver using inertial sensors. Then, the information is utilized in some way to assist the carrier tracking loop. After deducting the dynamic stress caused by relative motion, the equivalent noise of the carrier loop can be reduced. The tracking loop processed this way is retained with both good dynamics and the thermal noise performance of the loop. This is commonly referred to as 'deep-coupling' baseband processing, and is still a frontier topic in the satellite navigation academic community. Interested readers can consult Refs. [16–21].

6. Carrier-to-Noise Ratio Estimation

After the tracking of the carrier and pseudo-code is achieved, a necessary task of the receiver is to generate a quantitative index of the strength of the currently tracked signal. On the one hand, this will serve as an indication of the working status of the current tracking loop to examine the authenticity of the locked signal; on the other hand, it provides the basis for measuring the reliability of the observation for the subsequent navigation solution task. Information about the signal strength is also needed for some key algorithms such as cross-correlation and pseudo-code lock detection.

Section 4.1.8 shows that the same input signal has different signal strength representations at different signal processing stages. The signal strength will be affected by the length of the coherent integration time and the number of uncorrelated integrals. In order to ensure the consistency of expression, the carrier-to-noise ratio (written as CN_0) is often used in BDS and GPS receivers to measure the relationship between signal strength and noise. CN_0 has appeared many times in this book. Its physical meaning is the ratio of the carrier power to the white noise power spectral density, and its unit is dB/Hz. However, attention should be paid to the difference between CN_0 and CNR. The latter refers to the ratio of carrier power to noise power. The relationship between CNR and CN_0 is as follows:

$$\text{CNR} = \text{CN}_0 - 10\lg\left(B_{\text{w}}\right) \tag{4.196}$$

where the unit of CNR is decibels (dB), and B_w is the signal bandwidth of the antenna and the RF front-end.

CN_0 and CNR are generally used for the expression of signal strength in the radio frequency band. After signal carrier synchronization and baseband demodulation processing are completed, the signal-to-noise ratio SNR is often introduced to represent the relationship between signal strength and noise. The relationship between

SNR and CNR is as follows:

$$\text{SNR} = \text{CNR} + \text{baseband processing gain} \tag{4.197}$$

The baseband processing gain in Eq. (4.197) includes a superposition of coherent integral gain, non-coherent accumulating gain, square loss, and quantization loss, and it can be roughly estimated after the baseband processing algorithm is determined. For example, for GPS signals, the coherent integration time is 10 ms, the ADC quantization loss is 0.5 dB, and there is no uncorrelated accumulation. Then, the baseband processing gain = 10lg (10 × 1023) − 0.5 ≈ 39.6 dB. If the baseband SNR is known, the CNR can be deduced based on the baseband processing gain obtained above, and then the value of CN_0 can be derived from the RF front-end bandwidth according to Eq. (4.196).

The method of estimating the value of CN_0 mentioned above is theoretically feasible, but practical concerns do exist during the implementation of the receiver. First, the calculation of the baseband SNR is not straightforward. Secondly, once the baseband processing algorithm is modified, the entire estimation process needs to be adjusted, which is slightly inconvenient, so this method is not universally adopted in engineering. The method that is commonly adopted for the receiver is to obtain CN_0 with the statistical features of the "signal + noise" observation under different signal processing bandwidths. Specifically, the value of CN_0 is obtained using statistics such as the mean value and variance of "signal + noise" in the broadband and the narrowband.

The principle of coherent integration means that the longer the coherent integration time is, the narrower the equivalent signal bandwidth will be, and the signal bandwidth is inversely proportional to the coherent integration time. Assuming that the coherent integration time is T_{I}, then the I-channel and Q-channel integrator outputs can be expressed as

$$\begin{aligned} s_{\text{I}} &= A\cos(\Phi) + n_{\text{I}} \\ s_{\text{Q}} &= A\sin(\Phi) + n_{\text{Q}} \end{aligned} \tag{4.198}$$

where Φ is the carrier phase difference, and n_{I} and n_{Q} are noise terms, obeying $N(0, \sigma^2)$. It is assumed that the carrier frequency difference has been synchronized and the pseudo-code phase is already aligned, so there is no sine function term and pseudo-code autocorrelation function term. It is easy to prove that the following relationship exists between the signal amplitude A and the noise power σ^2:

$$A = \sqrt{2\text{CN}_0 T_{\text{I}}} \cdot \sigma \tag{4.199}$$

Let us consider two observations: WBP (wideband signal energy), the sum of energy after M times of coherent integration whose integration time is T_{I}; NBP (narrowband signal energy), the energy value with one time of coherent integration, whose integration time is MT_{I}, i.e.

$$\mathrm{WBP} = \sum_{k=1}^{M} (S_{\mathrm{I},k}^2 + S_{\mathrm{Q},k}^2) \tag{4.200}$$

$$\mathrm{NBP} = \left(\sum_{k=1}^{M} S_{\mathrm{I},k}\right)^2 + \left(\sum_{k=1}^{M} S_{\mathrm{Q},k}\right)^2 \tag{4.201}$$

By substituting Eq. (4.198) into Eqs. (4.200) and (4.201), the expressions of WBP and NBP for the signal and noise can be obtained. It can be seen that when there is no signal, WBP obeys a square distribution with a degree of freedom of 2M (χ^2), and NBP obeys a square distribution with a degree of freedom of 2. When there is a signal, WBP and NBP no longer obey the open square distribution, but a superposition of Gaussian distribution and open square distribution due to the existence of signal terms and noise and signal cross terms. Theoretical derivation shows that the mean value, variance, and cross-correlation of WBP and NBP are

$$E(\mathrm{WBP}) = 2M\sigma^2\left(\frac{A^2}{2\sigma^2} + 1\right) \tag{4.202}$$

$$E(\mathrm{NBP}) = 2M\sigma^2\left(\frac{A^2}{2\sigma^2}M + 1\right) \tag{4.203}$$

$$var(\mathrm{WBP}) = 4M\sigma^4\left(\frac{A^2}{\sigma^2} + 1\right) \tag{4.204}$$

$$var(\mathrm{NBP}) = 4M^2\sigma^4\left(\frac{A^2}{\sigma^2}M + 1\right) \tag{4.205}$$

$$COV(\mathrm{NBP}, \mathrm{WBP}) = 4M\sigma^4\left(\frac{A^2}{\sigma^2} + 1\right) \tag{4.206}$$

The derivation process of Eqs. (4.202) to (4.206) is very complicated, so it is presented in detail in Appendix C.

According to a theorem within probability theory for the random variables x and y, the mean value of x/y is

$$E\left(\frac{x}{y}\right) \approx \frac{E(x)}{E(y)} - \frac{\mathrm{COV}(x, y)}{[E(y)]^2} + \frac{E(x)\sigma^2(y)}{[E(y)]^3}$$

Substituting x and y in the above equation with NBP and WBP, and substituting Eqs. (4.202) to (4.206), then

$$Z = E\left(\frac{\mathrm{NBP}}{\mathrm{WBP}}\right) \approx \frac{(A^2/2\sigma^2)M + 1}{(A^2/2\sigma^2) + 1} \tag{4.207}$$

Then, substituting Eq. (4.199), the relational expression of A and σ^2 into Eq. (4.207), we get

$$CN0 = 10\lg\left(\frac{1}{T_{\mathrm{I}}}\frac{Z-1}{M-Z}\right) \tag{4.208}$$

Multiple observations can be made on Z to take the average value for the further reduction of the variance of Z, and thereby ensure the stability of CN_0 estimation. The total time length of the average value needs to be carefully chosen to maintain the balance between accuracy and sensitivity in the estimation. In addition to the estimation of CN_0, the value of Z can also be used as a pseudo-code lock indicator. When the value is lower than a certain line, it can be concluded that the loop has lost lock. The method introduced in Eq. (4.208) is usually called the narrow bandwidth band power ratio method, which is applicable to both GPS and BDS signals. Since there must be no data bit hopping in the NBP calculation process, for the GPS C/A code signal, the coherent integration time can be 1 ms, and the M value can be 20. Meanwhile, for the BDS D1 code signal, the synchronization of the NH code needs to be completed first. For the BDS D2 code signal, since the longest coherent integration time of the D2 code is 2 ms, the M maximum value can be 2, which limits the CN_0 accuracy of the D2 code to some extent.

In addition to the method described above, other common methods for estimating CN_0 are the squared signal-to-noise ratio variance method and the origin moment method [22, 23], in which different combinations of observations are used to calculate the value of CN_0 based on the statistical properties of the coherent integrated signal and signal power. The computational and error characteristics of these methods vary. More details and features can be found in the relevant references attached to this chapter.

4.2.6 Bit Synchronization

When the phase lock is achieved in the carrier tracking loop, the phase difference between the input signal and the local carrier signal is 0. The I-channel integrator starts to output data bits, while the Q-channel integral begins to generate noise components as shown in the I-channel and Q-channel Lissajous figures in Fig. 4.53. At this time, bit synchronization begins to be performed in the receiver, and the left and right scatter areas in the Lissajous figure correspond to the -1 and $+1$ bits respectively. Since the data rate of the GPS signal is 50 bps, the same as the BDS D1 code signal, and the D2 code signal is 500 bps, there can be multiple I-channel and Q-channel integration results in one data bit (if the integration time is 1 ms), so the right data bit start edge must be found to ensure the correctness of data demodulation.

There are mainly three purposes of achieving bit synchronization:

1. Longer coherent integration time can only be achieved after successful bit synchronization. As shown in the above section, with longer coherent integration time comes lower phase noise and higher tracking sensitivity. However, it is impossible to get coherent integration longer than 1 ms before the data bits are implemented, otherwise the positive and negative integrals caused by data bit hopping may be offset.
2. After the bit synchronization is completed, demodulation of the navigation message can be performed, and thereby the satellite ephemeris parameters, ionospheric correction parameters, UTC parameters, and satellite clock correction amount can be obtained to prepare for the positioning solution.
3. After the implementation of bit synchronization, the precise transmission time of the satellite can be obtained to prepare for the extraction of pseudo-range measurements. In fact, if it is merely for the positioning solution, the tracking loop is not indispensable in the second step. The current widely used GNSS assisting (A-GPS and A-BDS) technology can transmit the required information mentioned above to the receiver terminal through the server, thereby greatly shortening the positioning time. In the case of a hot start, the receiver stores the satellite ephemeris, satellite clock correction, and ionospheric information of the latest positioning moment in the non-volatile memory, which can be directly read after the next power up. If the time in between does not exceed 2 h (which is the interval of GPS ephemeris updates), it is no longer necessary to demodulate the message from the real-time tracking signal, thereby the TTFF time can be shortened and the capture sensitivity can be improved. Of course, for an unassisted autonomous receiver, demodulating the navigation message in real time through the tracking loop is still essential.

For GPS signals, the common way of achieving bit synchronization is through the histogram. The basic principle of this method is that when the data bits hop, transition between the positive and negative values may occur in the integration results of the adjacent I-channel and Q-channel integrators. Meanwhile, when the data bits are stable, the integration results of the adjacent two will remain unchanged. The width of data bits of the GPS signal is 20 ms, and bit hopping may occur every 20 ms.

The implementation steps of the histogram method are as follows: after the carrier phase is locked, the tracking loop randomly selects the millisecond time of a certain integration result as time 1. Then, the millisecond time of each integration result is regarded as 2, 3, and so forth, until 20. The subsequent millisecond time corresponding to the integration result is cycled from 1 to 20, and then each one of the milliseconds is attached with a counter. At each subsequent millisecond time, the signs of the current integration and the last integration are compared upon the completion of the I-channel and Q-channel integration to see whether a change has occurred. If there is a change, it indicates that a data bit transition edge might have appeared, so the counter corresponding to the current time should be increased by 1. The value of the 20 counters is compared after a certain time limit (T_{M}) is met, and the millisecond position corresponding to the counter value exceeding the predetermined threshold indicates the time at which the data bit hops.

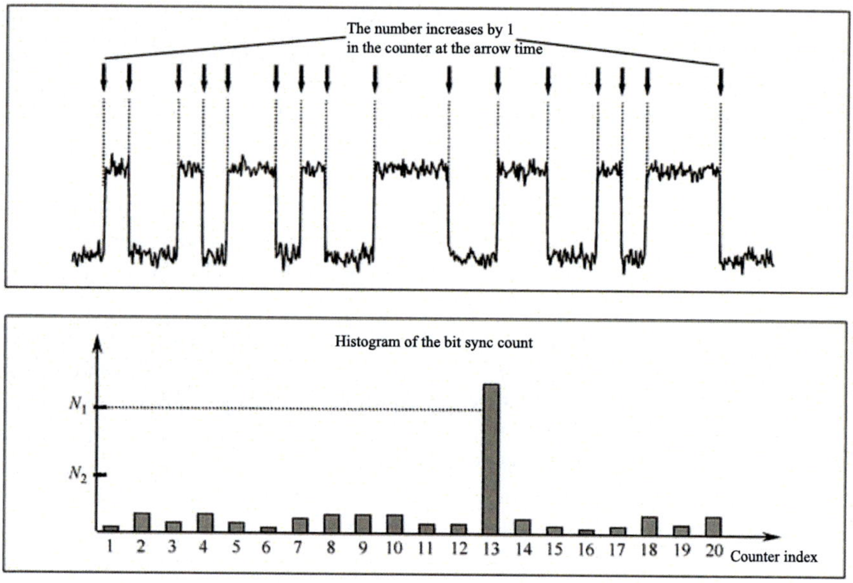

Fig. 4.60 Histogram method for GPS signal bit synchronization

It is easier to understand the above process with the help of Fig. 4.60. The upper part of the figure is the integration result of a certain section of the I-channel. The time points at which the arrows point indicates the hopping of the data bits, when the counter value corresponding to the millisecond time should be increased by 1, and that of the other counters remains unchanged. After a certain period of accumulation, the value distribution in the 20-bit hopping counter is shown in the lower part of the figure. The value of the 13th counter is significantly higher than the other counter values, which means that the millisecond time corresponding to the 13th counter is the bit transition time.

The key of this method is to verify whether there is a bit hopping between the two adjacent integral values. Figure 4.60 shows the processing of the relatively strong signal ($\mathrm{CN}_0 = 45$ dBHz), so the counter value corresponding to the bit transition time is more conspicuous than others. However, if the signal is weak, misjudgment might be made about the hopping of the integral values, and the counter value in the lower part of the figure will be distributed almost evenly. Without a significantly larger counter peak, it will be difficult to work out the transition time of the data bit.

For an input signal with a carrier-to-noise ratio of CN_0, the probability of misjudging the sign of the integral value is

$$P_e = \mathrm{erfc}'(\sqrt{2\mathrm{CN}_0 T_I}) \tag{4.209}$$

where T_{I} is the coherent integration time and $\mathrm{erfc}'(x)$ is the complementary error function whose definition is

$$\mathrm{erfc}'(x) = \frac{1}{\sqrt{2\pi}} \int\limits_{x}^{+\infty} e^{-y^2/2} \mathrm{d}y$$

Therefore, the probability that the hopping of the two adjacent integrals signs is misjudged is

$$P_{\mathrm{esc}} = 2P_e(1 - P_e) \tag{4.210}$$

The coefficient in Eq. (4.210) is 2 because there are two cases of misjudgment. Wither one is correct and the next is wrong, or one is wrong and the next is correct. In both cases, a misjudgment of the bit hopping will occur.

For the GPS signal, the probability that bit hopping will occur is considered to be 50%, so the average number of times the hopping occurs within the observation time T_M is $25T_M$. There will definitely be one bit hopping time that is correct in the 20 counters. Once the bit hopping is confirmed, the value of the counter will be incremented by 1, while the value of the remaining counters still corresponds to the data bits in the stable state. Only when a misjudgment occurs will the value be incremented by 1.

First, we should consider the performance of the counter corresponding to the correct hopping time. In a period of T_M, the data hops $25T_M$ times and remains stable for another $25T_M$ times. So when the data hops, the value of the integrator will increase by 1 if the judgment is correct, and when the data remains stable, the value of the integrator will increase by 1 if the judgment is wrong. The probability of misjudgment (P_{esc}) is given in the above analysis, so it is easy to derive the probability of the correct judgment:

$$P_{\mathrm{right}} = P_e^2 + (1 - P_e)^2 = (1 - 2P_e + 2P_e^2) \tag{4.211}$$

Thus, the mean value of the counter corresponding to the correct data hopping edge is

$$N_{\mathrm{right}} = 25T_M P_{\mathrm{right}} + 25T_M P_{\mathrm{esc}} = 25T_M \tag{4.212}$$

For other counters, since the input data bits are unlikely to hop, their value will only increase if a misjudgment occurs, so the mean value of the other counters is

$$N_{\mathrm{wrong}} = 50T_M P_{\mathrm{esc}} \tag{4.213}$$

Each judgment can be regarded as a Bernoulli experiment, and the probability distribution of failure and success is P_{esc} and $(1 - P_{\mathrm{esc}})$, so the mean square error of each counter value is

$$\sigma_N^2 = 50T_M P_{\mathrm{esc}}(1 - P_{\mathrm{esc}}) \tag{4.214}$$

With the theories above, we can define the two thresholds N_1 and N_2 as

$$N_1 = N_{\text{right}}$$
$$N_{\text{wrong}} \le N_2 \le N_{\text{right}} - 3\sigma_N \tag{4.215}$$

After the observation time of the above histograms exceeds the threshold T_M, the following three situations may occur:

(1) The value of a certain counter exceeds N_1.
(2) The loop is out of lock.
(3) Two or more counter values have exceeded N_2.

- If (1) occurs, it can be concluded that the moment when the data bit hops corresponds to the maximum value of the counter;
- If (2) occurs, the subsequent re-capturing of the signal and re-locking of the carrier phase need to be carried out.
- If (3) occurs, the signal strength will be low, so the confirmed hopping time of the data bit cannot be found. At this time, each counter value needs to be reset in anticipation of the next processing.

For the BDS D2 code signal, since the length of the data bit is 2 ms, the above histogram method is still applicable. However, there are only two counters. Meanwhile, the data bit may hop $250T_M$ times and remain stable $250T_M$ times during a period of T_M, so adjustments need to be made for the above equations.

For the BDS D1 code signal, the situation is slightly more complicated, mainly because it is the 1 kHz NH code that is modulated on the D1 code, which may hop at any millisecond, so the histogram method is not applicable here. Instead, the energy integration method can be more useful. The method does not detect the sign hopping of adjacent millisecond integrals, but performs correlation accumulation of the sliding windows whose coefficients are the NH code on the integration results at two adjacent time points with 20 ms in between. The time corresponding to the maximum energy is then detected. The specific steps are as follows: after the carrier enters the phase lock state, the tracking loop caches the integration result every millisecond. After 20 ms, the previous 20 ms integration result and the sliding window coefficient are multiplied and summed to obtain a result at the end of each millisecond integration. The sliding window coefficient here is set as the BDS NH code, so that after 20 ms, 20 accumulated results can be obtained, one of which must correspond to the correct hopping time of the data bit, so the signal energy can reach the maximum. In other accumulated values, a smaller result will be generated due to the offset of the positive and negative signal. As a matter of fact, the whole process shows that the sliding window multiplication and accumulation is a correlation operation between the integral results of adjacent 20 ms and NH codes, and the integration result of sliding 1 ms is to cover all possible hopping moments of the data bit. The autocorrelation function of the NH code determines the relative magnitude of the largest and the second largest accumulated value throughout the process.

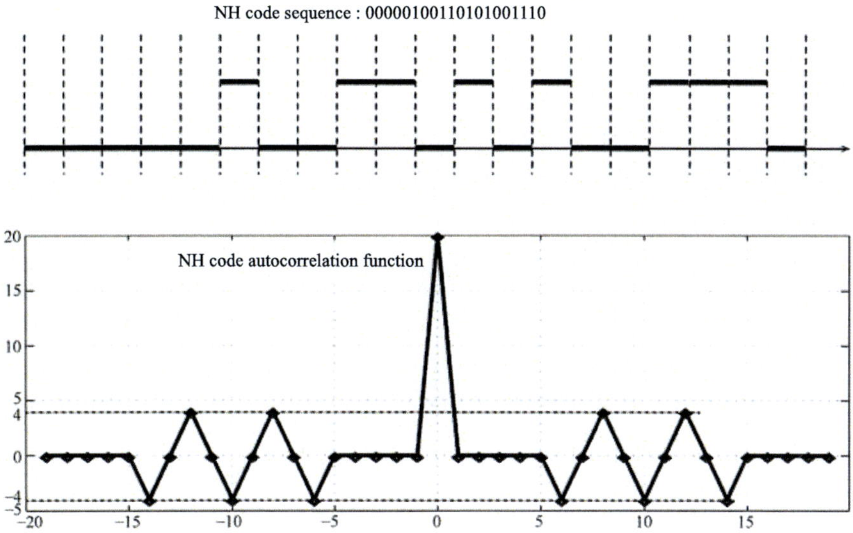

Fig. 4.61 The BeiDou NH code and its autocorrelation function

The sequence value and autocorrelation function of the BDS NH code are shown in Fig. 4.61. There are only four values taken in the autocorrelation function of the NH code: 0, ±4 and 20. It can be seen that the most error-prone one of the 20 accumulated results is the one whose autocorrelation value is ±4, as the time corresponding to it is often regarded as the hopping time of the data bit.

Assuming that the one-way integration result of 1 ms can be expressed as

$$s_{\mathrm{I}} = \sqrt{2\mathrm{CN}_0 T_{\mathrm{I}}}\sigma \cos(\Phi) + n_{\mathrm{I}} \tag{4.216}$$

where 0 or π is taken as the value of Φ, indicating the data bit to be modulated as +1 or −1. n_{I} is the noise term, obeying $N(0, \sigma^2)$. T_{I} is the coherent integration time. Then, after the above sliding window correlation accumulation, the maximum accumulated value can be expressed as

$$s_{\mathrm{I,Max}} = 20\sqrt{2\mathrm{CN}_0 T_{\mathrm{I}}}\sigma \cos(\Phi) + n_{\mathrm{I,Max}} \tag{4.217}$$

where $n_{\mathrm{I,Max}}$ is the noise term, obeying $N(0, 20\sigma^2)$.

According to the autocorrelation function of the BDS NH code shown in Fig. 4.61, the second largest accumulated value can be expressed as

$$s_{\mathrm{I,Sub}} = 4\sqrt{2\mathrm{CN}_0 T_{\mathrm{I}}}\sigma \cos(\Phi) + n_{\mathrm{I,Sub}} \tag{4.218}$$

where, $n_{\mathrm{I,Sub}}$ is the noise term, obeying $N(0, 20\sigma^2)$.

When $s_{\mathrm{I,Sub}} > s_{\mathrm{I,Max}}$, a bit edge misjudgment occurs, so a new variable should be created.

$$s_{\mathrm{I,MS}} = s_{\mathrm{I,Max}} - s_{\mathrm{I,Sub}} = 16\sqrt{2\mathrm{CN}_0 T_{\mathrm{I}}}\sigma + n_{\mathrm{I,MS}} \tag{4.219}$$

Strictly speaking, $n_{\mathrm{I,Max}}$ and $n_{\mathrm{I,Sub}}$ are not completely uncorrelated, as they are from the same sequence of the 1 ms integral noise term. However, due to different combinations of NH code coefficients, it is assumed that the two are approximately irrelevant. Therefore, $n_{\mathrm{I,MS}}$ in Eq. (4.219) obeys $N(0, 40\sigma^2)$, so the probability of misjudgment is

$$P_e = \mathrm{erfc}'\left(\frac{16}{\sqrt{40}}\sqrt{2\mathrm{CN}_0 T_{\mathrm{I}}}\right) \approx \mathrm{erfc}'\left(2.53\sqrt{2\mathrm{CN}_0 T_{\mathrm{I}}}\right) \tag{4.220}$$

It can be seen from Eqs. (4.210) and (4.220) that when the signal CN_0 decreases, the probability of bit synchronization failure increases, and when CN_0 is too low, the correct data bit transition time cannot be found either with the histogram method or the energy integration method. The cost of bit synchronization failure is huge. It can result in a failure to increase the coherent integration time, the incorrect demodulation of data bits, and the wrong satellite transmission time, leading to the false delivery of pseudo-range measurements. Therefore, a protection mechanism must be installed in the receiver to detect and correct the bit synchronization result.

After bit synchronization is achieved, the tracking loop can start demodulating the data bits. For the GPS and the BDS D1 code signal, the I-channel integration results of 20 ms in one data bit are added, and then the threshold decision is performed to obtain +1 and −1 values. For the BDS D2 code signal, the length of one data bit is only 2 ms, so only the two I-channel integration results can be added to perform the threshold decision. Since both are BPSK signals, the bit error rate of the GPS signal and the BDS D1 code signal is

$$P_{e1} = \mathrm{erfc}'\left(\sqrt{40\mathrm{CN}_0 T_{\mathrm{I}}}\right) \tag{4.221}$$

and the error rate of the BDS D2 code signal is

$$P_{e2} = \mathrm{erfc}'\left(\sqrt{4\mathrm{CN}_0 T_{\mathrm{I}}}\right) \tag{4.222}$$

where T_{I} is the length of the coherent integration time, which is 1 ms here. The relationship between P_{e1}, P_{e2}, and CN_0 is shown in Fig. 4.62.

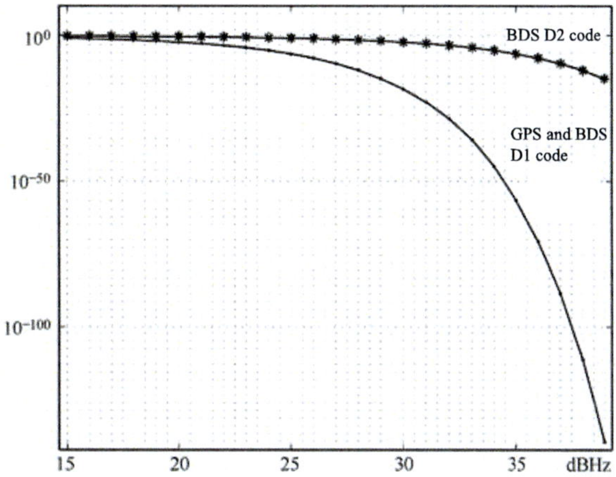

Fig. 4.62 Relationship between the bit error rate and signal strength of the GPS signal, and BeiDou D1 and D2 signals (logarithmic display on the longitudinal axis)

4.2.7 *Sub-frame Synchronization*

After the bit synchronization is completed, the receiver begins demodulating the data bits. From the theoretical analysis of the carrier loop in the previous chapter, it can be derived that when the carrier phase is locked, most of the signal energy will be contained in the I-channel integrator, and the Q-channel integrator will mainly store the noise energy. Therefore, demodulation of data bits is mainly accomplished by performing a 0 and 1 decision on the I-way integration result. For the GPS and BDS D1 code signals, one data bit contains 20 ms, so a 20-ms accumulation of the integration result of the I-channel integrator must be completed before a data bit can be determined. For the BDS D2 code signal, one data bit only contains 2 ms, so a data bit can be obtained through merely 2-ms accumulation of the I-channel result. The software module in the receiver stores the consecutive data bits in a local cache for subsequent processing.

The original data bits are similar to the bit stream in the memory, and are updated continuously over time. The sub-frame synchronization on the bitstream must be performed in the software to organize the stream into electrical fonts. For the BDS and GPS navigation messages, each word contains 30 bits. Only by completing the structural reorganization of the bitstream to the electrical fonts can the subsequent navigation message content be demodulated. Thus, sub-frame synchronization is an indispensable step in the signal processing flow of the receiver. In order to complete the sub-frame synchronization, the GPS and BDS satellite navigation messages are all set with the synchronization word. The first step of the sub-frame synchronization is to find the synchronization word in the continuous bitstream.

The GPS sync word is 10001011, the leading character in the telemetry code mentioned in Sect. 3.3.1. Since the leading character has only 8 bits, assuming that the probability of occurrence of 0, 1 in the bitstream is evenly distributed, then the probability of the emergence of an erroneous leading character is 1/28, which is still relatively high. Therefore, after the synchronization header search is completed, a frame integrity check needs to be performed according to the navigation message frame check logic. Only when this is complete can the navigation message be demodulated.

The synchronization character of BDS is more advanced than GPS in terms of error probability. The synchronization character of BDS is 11100010010, which is 11 bits. Therefore, the probability of an erroneous leading character caused by the random bit stream to emerge is $1/2^{11}$, which is slightly lower than that of GPS. However, it is still necessary to carefully judge the reliability of the receiver, because the influence of erroneous synchronizing characters is huge. Since the subsequent demodulation of the navigation message content needs to be based on the bit position information given by sub-frame synchronization, once an error occurs, it will trigger a series of errors in satellite ephemeris parameters, TOW or SOW, ionospheric correction parameters, and satellite clock correction parameters, which, if used for positioning solutions without examination, will lead to incorrect positioning results.

According to the analysis of the Costas loop, there will be 0, π phase ambiguity in the data bits outputted by it. In reality, if the data bit is 0, the demodulation result of the I-channel integrator output may be 0 or 1 and vice versa, so the real data bits cannot be known from a single demodulated data bit. However, the problem can be solved by the receiver software in a clever way while searching for the sync header. As mentioned earlier, the 8-bit leading character of GPS is 10001011, so there are two possibilities for demodulating the data bits: ① 10001011; ② 01110100. The first result shows that there is no 0, π phase ambiguity, so the subsequent data bits do not need special processing. Phase ambiguity does exist in the second, in which case each subsequent data bit needs to be reversed in phase. The same processing and judgment can be done for the BDS signal, except that the sync character is 11100010010. It is worth mentioning that according to Sect. 3.3.2, the BDS navigation message is interwoven in the organizational form, but observant readers will discover the exception of the first word in each sub-frame of the D1 code and D2 code. This is for convenience in searching for synchronization characters, because de-interleaving must be performed after the sub-frame synchronization is completed.

There is another interesting feature of the navigation message of the GPS signal that helps to solve the problem of phase ambiguity. The definition of the HOW word can be seen in Fig. 3.21. It should be noted that the contents of the 23rd and 24th bits in the HOW word are specially set. The purpose is to keep the lowest two bits of the check digit, namely D29 and D30, at 0. In other words, no matter how the contents of other bits of the HOW word change, the last two check bits of the HOW word are always 0. Therefore, the software design of the GPS receiver can check the two bit values of the leading character backwards (up to the last two check bits of the HOW). If the value is 00, no phase ambiguity occurs, so phase reversion is not needed, otherwise the phase needs to be reversed. This feature can also be used in actual programs to detect if the correct HOW word has been caught. In practice, due to the presence of 0, π phase ambiguity, the received D29 and D30 may be 11 or 00, but not 01 or 10. Hence, if 10 or 01 are detected, then either the current word is not a HOW word, or a bit error has occurred. In both cases, we should re-search for the sync word.

There is no similar arrangement of reserved words in the HOW word as there is for GPS in the BDS message. However, the second word and all subsequent words in the BDS message need to be de-interleaved after the synchronization character is detected, after which the BCH decoding will be performed in every 15 bits. The de-interleaving operation allows the BDS signal to resist continuous burst errors slightly better than the GPS signal. The process of BDS de-interleaving is shown in the decoding section in Fig. 3.26.

The length of one word in both GPS and BDS is 30 bits, of which the GPS check word is 6 bits, the BDS check word is two 4 bits, and the total is 8 bits. Therefore, the probability of the wrong "correct check word" emerging due to the random bitstream in BDS is 1/4 lower than that of GPS, but in terms of the absolute value of probability ($1/2^6$ for GPS and $1/2^8$ for BDS), the error rate for BDS is still relatively high. Therefore, it is necessary to confirm the correctness of the sub-frame synchronization from other aspects.

In addition to the synchronization characters in the GPS TLM word, all other information is marked as "reserved", and not much is available for use. Therefore, after searching for 10001011, only the 6-bit check word in the word is available for further investigation. If the probability of the synchronization word in the random bit stream being correct and the check word passing is $1/2^{14}$ (i.e. $1/2^8 \times 1/2^6$), it is already relatively small. After the TLM is obtained, the HOW word should go through the check. Figure 3.21 shows that the HOW word contains the 17-bit Z-count and the 3-bit sub-frame ID number, which can be demodulated to judge whether the value is in the reasonable range. The GPS time should be between 0 and 604,799 s, and the sub-frame ID number should fall in the range from 1 to 5. In addition, it is necessary to confirm whether the check word of the HOW word is correct. If no problem arises from the above check, the sub-frame synchronization is quite likely to be correct.

Figure 3.32 shows the relationship between the synchronization word of BDS and the sub-frame ID and SOW. Compared with the TLM word of GPS, there is more useful information in the first word of BDS, including synchronization characters, sub-frame ID, and SOW high 8-bit and 4-bit check codes. If another 1 two-bit passes, the SOW low 1 two-bit can be obtained, so the above information can be of great use to judge the correctness of the sub-frame synchronization. However, since the BDS teletext bit needs to be de-interleaved, the actual waiting time is around 24 bits.

In general, if all of the above information is utilized and found to be correct, then the possibility for sub-frame synchronization to be correct is very large. Based on the information of the sub-frame synchronization, the subsequent consecutive 30-bit words can be assembled, and the check word of each word can be checked. If it passes, it can be concluded that the current sub-frame synchronization is correct. If a verification error occurs, it is necessary to check whether it is caused by a bit error or a sub-frame synchronization error.

It is mentioned in Ref. [24] that the correctness of the sub-frame synchronization can be determined by judging whether the TOW or the SOW of two subsequent sub-frames are consecutive. This is a strict criterion, as the two timestamps whose intervals exceed one sub-frame due to the random bits are unlikely to be consecutive.

Table 4.7 Calculating method of verification words for GPS navigation messages

$d1 = D1 \oplus D^{*}30$
$d2 = D2 \oplus D^{*}30$
$d3 = D3 \oplus D^{*}30$
$d4 = D4 \oplus D^{*}30$
…
$d24 = D24 \oplus D^{*}30$
$D25 = D^{*}29 \oplus d1 \oplus d2 \oplus d3 \oplus d5 \oplus d6 \oplus d10 \oplus d11 \oplus d12 \oplus d13 \oplus d14$ $\oplus d17 \oplus d18 \oplus d20 \oplus d23$
$D26 = D^{*}30 \oplus d2 \oplus d3 \oplus d4 \oplus d6 \oplus d7 \oplus d11 \oplus d12 \oplus d13 \oplus d14 \oplus d15$ $\oplus d18 \oplus d19 \oplus d21 \oplus d24$
$D27 = D^{*}29 \oplus d1 \oplus d3 \oplus d4 \oplus d5 \oplus d7 \oplus d8 \oplus d12 \oplus d13 \oplus d14 \oplus d15$ $\oplus d16 \oplus d19 \oplus d20 \oplus d22$
$D28 = D^{*}30 \oplus d2 \oplus d4 \oplus d5 \oplus d6 \oplus d8 \oplus d9 \oplus d13 \oplus d14 \oplus d15 \oplus d16$ $\oplus d17 \oplus d20 \oplus d21 \oplus d23$
$D29 = D^{*}30 \oplus d1 \oplus d3 \oplus d5 \oplus d6 \oplus d7 \oplus d9 \oplus d10 \oplus d14 \oplus d15 \oplus d16$ $\oplus d17 \oplus d18 \oplus d21 \oplus d22 \oplus d24$
$D30 = D^{*}29 \oplus d3 \oplus d5 \oplus d6 \oplus d8 \oplus d9 \oplus d10 \oplus d11 \oplus d13 \oplus d15 \oplus d19$ $\oplus d22 \oplus d23 \oplus d24$

Note $D^{*}29$ and $D^{*}30$ are the last two bits of the former word: $\{D1 \ldots D24\}$ are the first 24 bits of the current word: $\in$ is XOR operation

The cost is that another sub-frame will pass before the completion of the sub-frame synchronization can be announced, which invisibly increases the TTFF.

The check word must go through the check in the process of sub-frame synchronization, and readers refer to Figs. 3.9, 3.26 and 3.28 to have an insight of the checking process. The mechanism of the GPS check word is not explained in the previous chapters, so it is given in Table 4.7, where $\{D25 \ldots D30\}$ is the calculated check words, comparing them with the demodulated 6-bit check word, those that are identical will pass and those that are different will fail.

After sub-frame synchronization is completed, navigation message decoding can be performed for each sub-frame. Specific steps can be performed according to the navigation message structure and content of GPS and BeiDou in Sect. 3.3, or can directly refer to the interface control document of GPS and BeiDou (ICD).

References

1. Tsui JB-Y (2008) Fundamentals of global positioning receivers: a software approach, 2nd edn. Wiley
2. Baum C, Veeravalli V (1994) Hybrid acquisition schemes for direct sequence CDMA system. In: IEEE international conference on communications

3. Gordon J, Povey R (1998) Spread spectrum PN code acquisition using hybrid correlators architectures. Wireless Pers Commun
4. Lawrence RR, Schafer RW, Rader CM (1969) The chirp Z-transform algorithm and its application. Bell Syst Tech J
5. Pei C (2009) Research on key baseband technology of high-performance GPS receiver. Doctoral Dissertation of Graduate School of Chinese Academy of Sciences
6. Spangenberg SM, Scott I, McLaughlin S, Povey GJR, Cruickshank DGM, Grant PM (2000) An FFT-based approach for fast acquisition in spread spectrum communication systems. Wireless Pers Commun
7. Rice SO (1945) Mathematical analysis of random noise. Bell Syst Tech J 24:146–156
8. Misra P, Enge P (2008) Global positioning system—signal, measurement and performance, 2nd edn. Electronic Industry Press, Beijing
9. Kaplan ED (2006) Understanding GPS principles and applications, 2nd edn. Artech House Publishers
10. Van Diggelen F (2009) A-GPS: assisted GPS, GNSS, and SBAS. Artech House
11. Strässle C et al (2007) The squaring-loss paradox. In: Proceedings of ION GNSS 20th international technical meeting of the satellite division, Fort Worth, 25–28 Sept 2007
12. Lindsey WC, Simon MK (1973) Telecommunication systems engineering. Prentice-Hall, Englewood Cliffs
13. Best RE (1984) Phase-locked loops. McGraw-Hill Inc.
14. Taylor D (2002) Introduction to 'synchronous communications,' a classic paper by John P. Costas. In: Proceedings of IEEE, vol 90, pp 1459–1460
15. Betz JW, Kolodziejski KR (2000) Extended theory of early-late code tracking for a bandlimited GPS receiver. Navig J Inst Navig 47(3):211–226
16. Alban S, Akos D et al (2003) Performance analysis and architectures for INS-aided GPS tracking loops. In: Proceedings of ION national technical meeting, ION-NTM 2003, Anahiem
17. Kim H, Bu S, Jee G, Chan-Gook P (2003) An ultra-tightly coupled GPS/INS integration using federated Kalman filter. In: 16th international technical meeting of the satellite division of the U.S. institute of navigation, Portland, 9–12 Sept 2003, pp 2878–2885
18. Gebre-Egziabher D et al (2003) Doppler aided tracking loops for SRGPS integrity monitoring. In: Proceedings of the institute of navigation GNSS conference, Portland, Sept 2003
19. Hamm CR, Flenniken WS et al (2004) Comparative performance analysis of aided carrier tracking loop algorithms in high noise/high dynamic environments. In: 17th international technical meeting of the satellite division of the institute of navigation, Long Beach, 21–24 Sept 2004, pp 523–532
20. Kreye C, Eissfeller B, Winkel JÓ (2000) Improvements of GNSS receiver performance using deeply coupled INS measurements. In: Proceedings of the institute of navigation GNSS conference, Salt Lake City, Sept 2000, pp 844–854
21. Farrell JA (2008) Aided navigation, GPS with high rate sensors. McGraw Hill
22. Falletti E, Pini M, Lo Presti L, Margaria D (2008) Assessment on low complexity C/No estimators based on M-PSK signal model for GNSS receivers. In: Position, location and navigation symposium, 2008 IEEE/ION, 5–8 May 2008, pp 167–172
23. Sharawi M, Akos DM, Aloi DN (2007) GPS C/N_0 estimation in the presence of interference and limited quantization level. IEEE Trans Aerosp Electron Syst 43(1):227–238
24. Gang X (2009) The GPS principle and receiver design. Electronic Industry Press, Beijing
25. Van Dierendonck AJ (1996) GPS receivers. In: Parkinson B, Spiker J, Axelrad P, Enge P (eds) Global positioning system: theory and applications, vol I
26. Spiker J (1996) Fundamental of signal tracking theory. In: Parkinson B, Spiker J, Axelrad P, Enge P (eds) Global positioning system: theory and applications, vol I
27. Spiker J (1996) GPS signal structure and theoretical performance. In: Parkinson B, Spiker J, Axelrad P, Enge P (eds) Global positioning system: theory and applications, vol I
28. China Satellite Navigation System Management Office (2013) Beidou satellite navigation system space signal interface control document (open service signal) version 2.0. Dec 2013
29. Navstar GPS Space Segment /Navigation User Interfaces (2012) IS-GPS-200G, 5 Sept 2012

30. Xurong D, Bin T, De J (2008) Satellite navigation software receiver: the principle and design. National Defense Industry Press, Beijing
31. Rosenfeld D, Duchovny E (2010) Off-board positioning using an efficient GNSS SNAP processing algorithm. In: ION GPS GNSS 2010, Manassas
32. Lowe S (1999) Voltage signal-to-noise ratio SNR nonlinearity resulting from incoherent summations. JPL-NASA, technical report
33. Changxin F, Daoyong Z, Bingxiang X, Chengke W (1995) Communication principles, 4th edn. National Defense Industry Press, Beijing
34. Psiaki ML, Jung H (2002) Extended Kalman filter methods for tracking weak GPS signals. In: ION GPS 2002, Portland, 24–27 Sept 2002
35. Jung H, Psiaki ML (2003) Kalman-filter-based semi-codeless tracking of weak dual-frequency GPS signals. In: Powell SP (ed) ION GPS/GNSS 2003, Portland, 9–12 Sept 2003
36. Kim K-H, Jee G-I, Song J-H (2008) Carrier tracking loop using the adaptive two-stage Kalman filter for high dynamic situations. Int J Control Autom Syst 6(6):948–953
37. Gernot C, O'Keefe K, Lachapelle G (2010) Combined L1/L2 Kalman filter based tracking scheme for weak signal environments. GPS Solut
38. Simon D, El-Sherief H (1994) Fuzzy phase-locked loops. In: Proceedings of the IEEE position location and navigation symposium, Las Vegas, Apr 1994, pp 252–259
39. Kamel AM (2010) Design and testing of an intelligent GPS tracking loop for noise reduction and high dynamic applications. In: Proceedings of the 23rd international technical meeting of the satellite division of the ION, Portland, Sept 2010, pp 3235–3243
40. Mao WL, Tsao HW, Chang FR (2004) Intelligent GPS receiver for robust carrier phase tracking in kinematic environments. In: IEE proceedings radar sonar navigation, vol 151, no 3, June 2004
41. Xiaohui B, Haiyang L, Rui Z, Jie C (2009) A novel efficient tracking algorithm based on FFT for extremely weak GPS signal. In: ION GNSS2009, 22–25 Sept 2009, pp 1700–1706
42. Xiaohui B, Haiyang L, Rui Z, Jie C (2011) An effective method for GNSS receiver airborne noise ratio estimation. J Wuhan Univ Inf Sci Ed 4, 36(4):457–460
43. Mood AM, Graybill EA, Boes DC (1974) Introduction to the theory of statistics, 3rd edn. McGraw-Hill, New York
44. Stephen SA, Thomas JB (1995) Controlled-root formulation for digital phase-locked loops. IEEE Trans Aerosp Electron Syst 78–95
45. Proakis JG (2001) Digital communications. McGraw-Hill Inc.
46. Farrell JA, Barth M (1999) The global positioning system and inertial navigation. McGraw Hill

Chapter 5
Dual Mode Observation Extraction and Error Analysis

The method of bit synchronization and sub-frame synchronization between the BDS and GPS receivers is explained in Chap. 4. After the sub-frame synchronization is completed, the receiver starts to demodulate the content of the navigation message. After the carrier synchronization and pseudo-code synchronization are achieved in the GNSS receiver, the in-phase integrator of the Costas loop begins outputting data bits and extracting the observations. The biggest difference between the BDS and GPS receivers and the general communication system terminals lies in the extraction and processing of the observations. Communication systems commonly seen in life, such as television, radios, mobile phones, and wireless modems, focus on the completion of information transmission and reception. For BDS and GPS receivers, other than the transmission of "information" (for example, the demodulation and storage of navigation messages), it is also their task to extract the position and time information provided by the ranging code and to process it, which is the only way positioning and navigation can be achieved. From this perspective, understanding the various observational principles and generating mechanisms in BDS and GPS receivers is fundamental for understanding the positioning solution algorithm.

The most common observations in BDS and GPS receivers are pseudo-range, Doppler frequency and integral Doppler, and carrier phase. The physical meanings of different observations vary, as does the extraction method, depending on different signal processing and tracking loop states at that time. The positioning accuracy that can be achieved is also different. Therefore, it is also necessary to decide which observations to use specifically according to the practical requirements.

This chapter will elaborate on BDS and GPS dual-mode pseudo-range observations, Doppler frequency and integral Doppler observations, and dual-mode carrier phase observations as well as their physical meanings, in order to lay a theoretical foundation for the subsequent positioning solution algorithm. At the same time, the error properties of the observations will also be discussed to figure out how to reduce or even eliminate the influence of the observation error on positioning accuracy

Y. Lu, *BDS/GPS Dual-Mode Software Receiver*, Navigation: Science and Technology 10, https://doi.org/10.1007/978-981-16-1075-2_5

according to the time correlation and spatial correlation of the error. On this basis, the concept of differential GNSS will be introduced, including the wide-area differential system and local differential system.

5.1 Pseudo-Range Observation

Pseudo-range observation is very simple to conduct. The basic principle is that the propagation speed of electromagnetic waves in a vacuum is the speed of light c (2.99792458 × 108 m/s), so the measurement of distance can be converted into a measurement of time. A basic explanation of the concept of pseudo-range has been explained in Sect. 1.1. In this section, the principles and properties of pseudo-range observation extraction in BDS and GPS receivers will be explained in a more detailed way with solid mathematical and physical theories.

The ranging signal transmitted by the BDS and GPS satellites is an electromagnetic wave signal that travels in the vacuum at the speed of light. The signal travels from satellites to receivers on the Earth's surface in a vacuum most of the time. Only after the signal enters the atmosphere will it no longer travel strictly at the speed of light due to the influence of the ionosphere and the troposphere in the atmosphere, whose delay will be discussed later in the analysis of the error properties of the pseudo-range observation. Here, the traveling speed of the signal can be approximated as fixed, which simplifies the analysis of the principle of pseudo-range observastion.

Both the satellite and the receiver have their own clocks, so each will refer to their own time respectively. Assuming that the signal from the satellite is received at t_{R} of the receiver time, according to the analysis in Sect. 1.1.4, the travelling time of the signal from the satellite antenna is obtained through the timestamp information modulated on the satellite signal, which is recorded as t_{SV}. Thereby, the travelling time ($t_{\mathrm{R}} - \mathrm{t}_{\mathrm{SV}}$) of the signal from the satellite to the receiver can be obtained directly. The distance from the satellite to the receiver is also easy to obtain through multiplying the propagation time by the speed of light, but the situation is not that simple.

The t_{R} and t_{SV} above are the local times obtained according to the clock counts of the receiver and the satellite respectively. However, if the two times are compared with an accurate time reference, for example, the GPS time or the BDS time, then there will be a deviation (written as δt_{R}) between the "true t_{SV}" and the one on the satellite. Similarly, there is also a deviation between the t_{R} on the receiver and the "true t_{R}", as shown in Fig. 5.1.

The time reference in Fig. 5.1 can be either GPS time or BDS time, as long as it is consistent when recording the pseudo-range observations of different satellites. "Consistent" here means that the pseudo-range observations of GPS and BDS satellites should all be based on the same time reference. Both t_{R} and t_{SV} need to be expressed on this reference. The time reference of t_{R} is determined by the receiver designer. Therefore, there is no ambiguity in it. However, t_{SV} is related to different

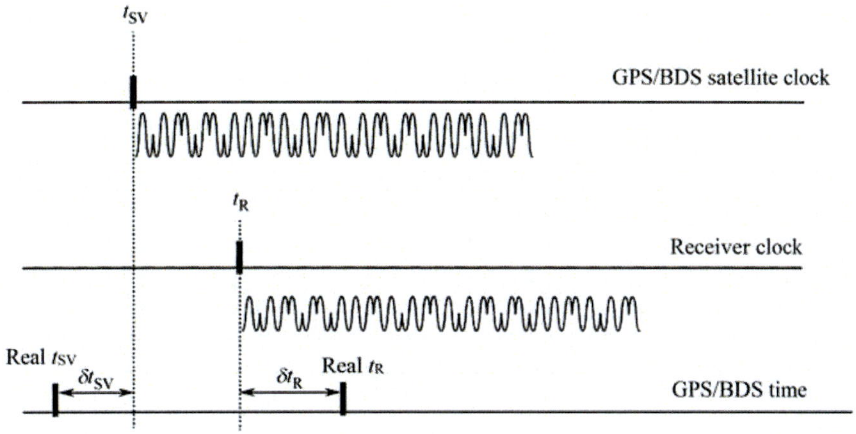

Fig. 5.1 Principle of Pseudo-range observation

satellites. The timestamp of the modulation on the GPS satellite signal is expressed in the GPS time, while that of the modulation on the BDS satellite signal is expressed in the BDS time. Therefore, when extracting the dual-mode pseudo-range observation, it is necessary to convert the transmission time of the GPS satellite to the BDS time, or to convert that of the BDS satellite to the GPS time. The conversion between the GPS and BDS time has already been described in Sect. 1.3.5. In the initial phase of the design, a common time reference can be specified on BDS and GPS receivers, and all time-related observations need to be converted to this time reference. However, since the GPS system has been in operation for more than 30 years, and there are many well-developed GPS receivers the market, in order to maintain compatibility with other products, it is a safe choice to set the time reference in the dual-mode receiver to GPS time.

There is a difference of a full 14 s between BDS and GPS time, which is judged only on the seconds scale. Meanwhile, on the nanosecond level, there is still an unknown time deviation after 14 s is added to the BDS and GPS time. Although small, this cannot be ignored for positioning, because even a difference of microseconds will bring about a positioning error of about 300 m if not processed. T_{GB} is used in this book to indicate the system deviation between BDS and GPS time, which can be subtracted when the former is converted to the latter.

When the time reference is set as GPS time, t_{SV} in the receiver will be obtained according to the timestamp of the satellite's transmitting signal. After the local time t_R is introduced, the GPS satellite pseudo-range observation can be expressed as

$$\rho_G = c(t_R - t_{SV}) \tag{5.1}$$

For the BDS satellite pseudo-range observation, it is necessary to increase the influence of T_{GB}. At this time, the pseudo-range observation can be expressed as

$$\rho_B = c[t_R - (t_{SV} - T_{GB})] \tag{5.2}$$

It should be noted that t_{SV} in Eq. (5.1) represents the GPS satellite transmission time in the GPS time, and t_{SV} in Eq. (5.2) represents the BDS satellite transmission time in BDS time. For the sake of simplicity, it will be used to represent the launch time of both the GPS the BDS satellites. Readers should pay attention to the difference.

According to Eqs. (5.1) and (5.2), in order to ensure the accuracy of the pseudo-range, both t_R and t_{SV} must be precise enough; any inaccuracy in either will lead to a very large distance error. As can be seen in the analysis below, in BDS and GPS receivers, t_{SV} can be measured with sufficient accuracy, while t_R cannot.

A very clever method is adopted in the receiver to obtain the accurate t_{SV}, which is closely related to the satellite navigation signal format. In fact, a deeper insight into the BDS/GPS signal format setting will only be achieved after understanding how to obtain an accurate t_{SV}. In short, the acquisition of t_{SV} is implemented by the software of the receiver and the pseudo-code tracking loop.

The acquisition of t_{SV} is implemented in the following four steps:

First, after the tracking loop enters the stable tracking state, the navigation message data bits can be demodulated, and sub-frame synchronization has also been achieved. For the GPS signal, the Z-count can be known by demodulating the TOW of the current sub-frame. As can be seen from Fig. 3.22 in Sect. 3.3.1, the initial transmission time of the current sub-frame is (Z-count − 1) 6 × s. For the BDS D1 code signal, the initial transmission time of the sub-frame can be directly obtained through the SOW of the current sub-frame, which can make the estimation of the t_{SV} accurate to the order of 6 s, and the accuracy can reach 0.6 s for the BDS D2 code for the length of one sub-frame of the D2 code is only one tenth of the length of the D1 code and the GPS sub-frame. Then, the value of t_{SV} can be approximated as

$$t_{sv} \approx \begin{cases} 6(Z-1), & \text{GPS signal} \\ \text{SOW}, & \text{BDS signal} \end{cases} \tag{5.3}$$

Second, the accuracy of t_{SV} can be further improved to the order of 20 ms (or 2 ms) through the bit count of the navigation message. The sub-frames in the navigation message of the GPS and BDS D1 code signals are composed of 300 bits, and the length of each bit is 20 ms. The navigation message sub-frame of the BDS D2 code signal is also composed of 300 bits, but the length of each bit is 2 ms. After stable tracking is achieved, the receiver's software will maintain a count of N_{bit} for the sub-frame bits whose range is from 1 to 300, and clear at the last bit of the sub-frame. So, the estimation of t_{SV} can then be accurate to the order of 20 ms/2 ms by bit counting, i.e.

$$t_{sv} \approx \begin{cases} 6(Z-1) + N_{\text{bit}} \times 0.02, & \text{GPS signal} \\ \text{SOW} + N_{\text{bit}} \times 0.02, & \text{BDS D1 code signal} \\ \text{SOW} + N_{\text{bit}} \times 0.002, & \text{BDS D2 code signal} \end{cases} \tag{5.4}$$

Third, the accuracy of t_{SV} can be increased to 1 ms by counting the pseudo-code period in the navigation telegram. The telegram bits of the GPS and BDS D1 code signals contain 20 pseudo-code periods, while those of the BDS D2 code signal contain 2 pseudo-code periods, each of which is 1 ms. The pseudo-code tracking loop counts the pseudo-code period (represented here by N_C here), the range of which is 1 to 20 for the BDS D1 code and the GPS signal, and 1 to 2 for the BDS D2 code signal. The count value is cleared at the data bit transition edge. Therefore, the estimation of t_{SV} can be accurate to the order of 1 ms through pseudo-code period counting, i.e.

$$t_{sv} \approx \begin{cases} 6(Z-1) + N_{\mathrm{bit}} \times 0.02 + N_C \times 0.001, & \text{GPS signal} \\ \mathrm{SOW} + N_{\mathrm{bit}} \times 0.02 + N_C \times 0.001, & \text{BDS D1 code signal} \\ \mathrm{SOW} + N_{\mathrm{bit}} \times 0.002 + N_C \times 0.001, & \text{BDS D2 code signal} \end{cases} \tag{5.5}$$

Fourth, the accuracy of t_{SV} can be increased to the order of microseconds by the code phase of the local pseudo-code. When the locking of the pseudo-code phase is achieved in the pseudo-code tracking loop, the tracking loop will generate Φ_C, the phase of the local pseudo-code. In the stable tracking state, it can be considered to accurately match the pseudo-code phase of the input signal. For the GPS signal, the range of Φ_C is 1 ~ 1023, and for the BDS signal, the range is 1 ~ 2046, in the unit of chips. The time width of the GPS chip is about 0.977 5 μs, and that of the BDS chip is about 0.488 7 μs. It can be seen that the estimation of t_{SV} can be accurate to the order of the chip (i.e. the order of microseconds) through the local pseudo-code phase, i.e.

$$t_{sv} \approx \begin{cases} 6(Z-1) + N_{\mathrm{bit}} \times 0.02 + N_C \times 0.001 + 0.9775\Phi_C \times 10^{-6}, & \text{GPS signal} \\ \mathrm{SOW} + N_{\mathrm{bit}} \times 0.02 + N_C \times 0.001 + 0.4887\Phi_C \times 10^{-6}, & \text{BDS D1 code signal} \\ \mathrm{SOW} + N_{\mathrm{bit}} \times 0.002 + N_C \times 0.001 + 0.4887\Phi_C \times 10^{-6}, & \text{BDS D2 code signal} \end{cases} \tag{5.6}$$

The above analysis shows that the acquisition of t_{sv} must be implemented by the software and the pseudo-code tracking loop in the receiver. The importance of the navigation signal structure is also clear. The portion in the t_{sv} whose accuracy is above 20 ms (for the BDS D2 code, the value is 2 ms) is generated by the receiver software's analysis of the satellite navigation message, while the portion whose accuracy is below 20 ms (or 2 ms) is provided by the pseudo-code tracking loop. The signal structure settings of BDS and GPS make this possible: the bit content in the navigation message provides the portion whose accuracy is above the millisecond, while the portion whose accuracy is below the millisecond is provided by the pseudo-code phase modulated by the data bit. It is worth mentioning that t_{sv} here is only the representation of the satellite signal's transmission moment in the satellite time, which is different from the real t_{sv} by a time deviation. As will be seen later, the satellite clock bias can be obtained from the clock correction parameters in the ephemeris data and subtracted from the original pseudo-range observations. The clock correction parameters of the GPS and the BDS satellite are (t_{oc}, a_0, a_1, a_2), which are broadcasted in the satellite ephemeris parameters. See Sect. 3.3 for details.

The acquisition of the reception time of the signal, i.e. the local time t_R of the receiver is simpler, and can be directly achieved by the local clock of the receiver. If the local clock is accurate, then t_R will be accurate. However, a very expensive atomic clock is required to guarantee the accuracy as the local clock, which is unacceptable in terms of cost. In general, the current commercial BDS and GPS receivers generally adopt temperature-controlled crystal oscillators (TCXO) as the core component of local clocks. A fine TCXO has a frequency accuracy of 0.1 to 1 ppm, which means that the local clock may deviate by up to 1 ms after 1000 to 10,000 s. Therefore, there must be an unpredictable time deviation between the local clock and the actual receiving time, which is represented by δt_R in Fig. 5.1.

Since the true signal transmission time is the time difference between "true t_{SV}" and "true t_R" in Fig. 5.1, then according to the above analysis, Eqs. (5.1) and (5.2) can be expressed as

$$\begin{aligned}\rho_G &= c(t_R - t_{SV}) \\ &= \sqrt{(x_u - x_s)^2 + (y_u - y_s)^2 + (z_u - z_s)^2} + c\delta t\end{aligned} \tag{5.7}$$

$$\begin{aligned}\rho_B &= c(t_R - t_{SV}) + cT_{GB} \\ &= \sqrt{(x_u - x_s)^2 + (y_u - y_s)^2 + (z_u - z_s)^2} + c\delta t + cT_{GB}\end{aligned} \tag{5.8}$$

where, (x_s, y_s, z_s) is the position coordinate of the satellite at t_{SV}, and (x_u, y_u, z_u) is the position coordinate of the user, which is unknown. δt is the sum of the signal transmission time deviation and the receiver time deviation. As mentioned previously, the satellite clock deviation can be calculated and deducted, after which only the time deviation of the receiver itself will be included in , i.e. $\delta t \approx \delta t_R$. With the satellite launch time t_{SV}, (x_s, y_s, z_s) can be obtained through ephemeris data, and the specific method will be detailed later. It can be seen that the t_{SV} mentioned above has two functions: one is to calculate the pseudo-range observation, the other is to calculate the satellite position coordinates as an input variable. It should be noted that the BDS satellite coordinates derived from the BDS satellite's ephemeris data are in the CGS2000 coordinate system, while the GPS satellite coordinates derived from the GPS ephemeris data are in the WGS84 coordinate system. According to References [1] and [2], the coordinates of the same point in the two coordinate systems are very similar, so the coordinates of the BDS satellite and the GPS satellite can be directly mixed without conversion. Since this book only covers two GNSS systems (BDS and GPS), the coordinates of their satellites will no longer be marked with the CGS2000 and WGS84 coordinates.

The origin of the term "pseudo-range" can be seen in Eqs. (5.7) and (5.8). It is precisely due to the inaccuracy of the local clock that an unpredictable time deviation exists in the pseudo-range observation. In order to distinguish it from the real distance observation, it is called "pseudo-range" observation.

The existence of the time deviation δt makes the problem a little more complicated, because the unknown quantities are now $(x_u, y_u, z_u, \delta t)$, one more than (x_u, y_u, z_u). Furthermore, for the dual-mode receiver, since the system deviation TGB between the BDS time and the GPS time creates another unknown quantity, the unknown quantities are now $(x_u, y_u, z_u, \delta t, T_{GB})$.

Modern BeiDou and GPS receivers usually have multiple channels, each of which processes the signals of one satellite, so each channel can generate a pseudo-range measurement equation like Eqs. (5.7) or (5.8). We know that solving five unknown quantities requires at least five equations, which is not difficult for modern multi-channel receivers. The existence of TGB seems to increase the minimum number of required equations by one. However, there are currently 14 satellites available in the BDS system, so providing an additional BDS pseudo-range observation is not a tricky task. In fact, there are between eight and 10 BDS satellites available in China and the Asia–Pacific region, so it is not difficult to provide enough BDS + GPS dual-mode pseudo-range observations in the open sky.

There is also an important question that beginners tend to overlook, that is, how to ensure that each δt in the pseudo-range measurement equations is same in value? In fact, inside the receiver, multiple pseudo-range observations are read simultaneously. For example, the timer inside the receiver triggers all channels at a certain time and starts to read the pseudo-range observations, which ensures that the pseudo-range observations collected by all channels (including the GPS tracking channel and the BDS tracking channel) share the same local time at that moment. Therefore, δt can be the same in value in all of the pseudo-distance equations.

In fact, in the pseudo-range equation, in addition to the receiver clock bias, the error E_{eph} of the satellite position caused by the ephemeris data itself is also regarded as an uncertain quantity. After the satellite clock bias E_{eph} is deducted, there is still a small deviation τ_s and a transmission delay of the ionosphere and troposphere, which are represented by T_{iono} and T_{tron} respectively. In addition, the thermal noise inside the receiver will also affect the pseudo-range observations, as well as the multipath effect caused by the environment during signal transmission. MP represents the multipath noise, and n_r is the thermal noise inside the receiver, with which the pseudo-range equation can be further written as

$$\begin{aligned}\rho_G = &\sqrt{(x_u - x_{Gs})^2 + (y_u - y_{Gs})^2 + (z_u - z_{Gs})^2} + c\delta t \\ &+ c\tau_{G,s} + E_{G,eph} + T_{G,iono} + T_{G,tron} + MP_G + n_r\end{aligned} \tag{5.9}$$

$$\begin{aligned}\rho_B = &\sqrt{(x_u - x_{Bs})^2 + (y_u - y_{Bs})^2 + (z_u - z_{Bs})^2} + c\delta t + cT_{GB} \\ &+ c\tau_{B,s} + E_{B,eph} + T_{B,iono} + T_{B,tron} + MP_B + n_r\end{aligned} \tag{5.10}$$

From Eqs. (5.9) and (5.10) the respective GPS and BDS pseudo-range observations can be obtained, where the subscripts "G" and "B" represent GPS and BDS. The ionospheric delay, tropospheric delay, satellite clock error, and multipath effect belong to different satellite signals, so subscripts are added to distinguish them.

$n_\rho = c\tau_s + E_{eph} + T_{iono} + T_{tron} + MP + n_r$ represents the total error term in the pseudo-range observation. Among them, the errors can be divided into two types: the common error and the unique error. The common error is the error term shared by all receivers in a small area (such as 50 km^2), while the unique error is the error term of the receiver's own, which varies from receiver to receiver. Common errors include $c\tau_s$, E_{eph}, T_{iono} and T_{tron}, while unique errors include MP and n_r. Common errors can be eliminated or reduced by the difference.

Equations (5.9) and (5.10) are the dual mode pseudo-range equations of BDS and GPS receivers. They are the mathematical basis of pseudo-range positioning, and must be thoroughly understood by readers.

5.2 Carrier Phase Observation

The carrier phase observation in the BDS and GPS receivers can be understood through Fig. 5.2.

In Fig. 5.2, the satellite transmits navigation signals to the receiver. The pseudo-random code and the navigation message bits are omitted in the figure to highlight the carrier component. Supposing that the recorded carrier phase is ϕ_S the moment the navigation signal leaves the antenna, and ϕ_E after the signal travels all the way to the receiver, then the carrier phase observation in the receiver the difference between the two, written as Φ, i.e.

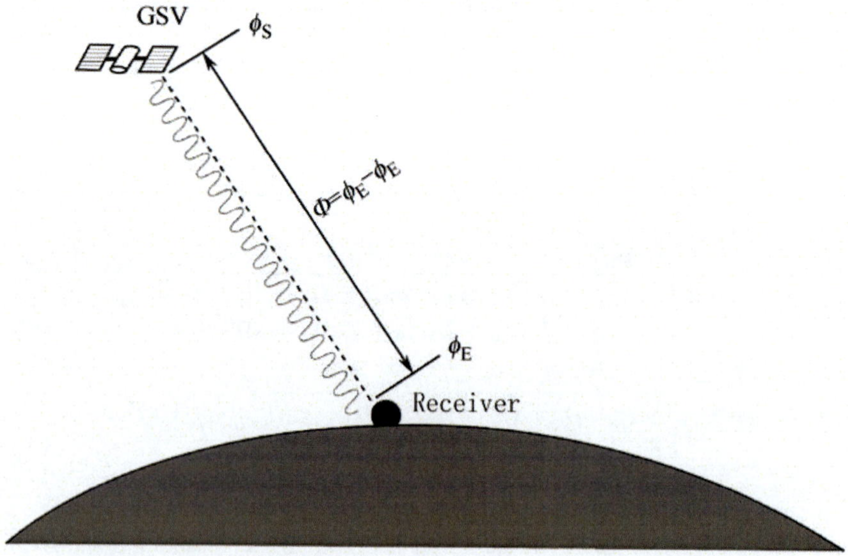

Fig. 5.2 Principle of carrier phase observation

$$\Phi = \phi_E - \phi_S \tag{5.11}$$

Φ is the total number of carrier cycles and the part that is less than one cycle on the propagation path from the satellite antenna to the receiver antenna:

$$\Phi = N_{E,S} + \Delta\phi = \frac{r}{\lambda} \tag{5.12}$$

where $N_{E,S}$ is an integer, and $\Delta\phi$ is the part that is less than one cycle, which is a radian value between 0 and 2π. r is the distance between the satellite antenna and the receiver antenna, and λ is the carrier wavelength.

Equation (5.11) facilitates our understanding of the principle of the carrier phase, but it is difficult to extract the carrier phase observations in practice. First, the receiver cannot measure the carrier phase of the signal at the satellite antenna end. Second, the carrier phase of the received navigation signal can be easily obtained internally in the receiver. However, since the carrier signal is composed of sine waves, and the waveforms in each cycle of the sine wave are identical, how does the local carrier phase measured at a certain moment correspond to the moment when the signal leaves the satellite antenna? In other words, even if the measurement of ϕ_E can be achieved in the receiver, there is no way to conduct the same for ϕ_S.

In cases where ϕ_S cannot be measured, it is impossible to get $N_{E,S}$ and $\Delta\phi$ in Eq. (5.12), so it is only a theoretical equation written to reveal the relationship between the carrier phase and the distance.

When the carrier tracking loop of the receiver enters the phase locked state, in the absence of a cycle-slip, the local carrier phase and the carrier phase of the input signal can be considered as fully synchronized, so the carrier phase measured by the receiver is the increment of the local carrier phase, that is,

$$\Phi_k = \Phi_{k-1} + \int_{T_{k-1}}^{T_k} 2\pi f_{NCO}(t)\mathrm{d}t \tag{5.13}$$

where Φ_k and Φ_{k-1} are carrier phase observations at T_k and T_{k-1}, and the time span between the two moments is T. f_{NCO} is the instantaneous frequency of the carrier NCO, whose relationship with the phase step value of the NCO is shown in Eq. (4.190). f_{NCO} changes with time, so the time integral is introduced to the equation to find the phase increment. When the time interval T is small, it can also be approximated as $2\pi f_{NCO}T$. In general, the observation time is the moment at which the measurement is latched, at which point the pseudo-range, carrier phase, and integral Doppler observations are simultaneously latched to ensure that the observations of all channels share the same local time.

For the commonly used civilian navigation receivers, observation is mostly carried out once per second, that is, one positioning result can be obtained per second. For highly dynamic receivers, the observation frequency is higher: 5 times, 20 times, or 100 times per second, and the corresponding frequency of the positioning result is also higher. In terms of observation extraction, the only change is the shortened observation interval. However, the computational burden in the PVT algorithm of the latter have increased.

Φ_0 is the initial value when observations are extracted from a satellite for the first time. Generally, after the tracking loop corresponding to the satellite enters the carrier phase lock, the carrier phase is extracted, and the initial value Φ_0 can be calculated according to the position of the receiver (such as the positioning result obtained by the least square method) and the distance between the satellites. Obviously, since large deviations exist in the local satellite positions, such initial values are not true carrier phase values in Eq. (5.11), so the initial value can also be simply set to 0. No matter how the initial value is set, it will introduce a random number of whole cycles N' in Eq. (5.12). Because the single-difference, double-difference or even triple-difference processing will generally be adopted for carrier phase positioning, and the initial value is converted into the solution of "ambiguity of whole cycles" as a fixed whole-cycle value, it will not affect the positioning method using the carrier phase.

Φ_k can be obtained through Eq. (5.13), but the values of f_{NCO} and T are deviated because the crystal frequency of the RF front-end of the receiver is deviated, and the calculated NCO instantaneous frequency value and observation are calculated. Thus, it can be calculated that the NCO instantaneous frequency value and the observation interval T are all deviated. The carrier phase in the receiver is generally obtained by accumulating the number of times the carrier NCO overflows. At the end of each coherent integration, the portion between the carrier phase 0 and 2π is obtained through arctan2(Q, I). Then, the local phase of the NCO is counted in whole cycles during the coherent integration to obtain the whole-cycle portion of the carrier phase, and the two components are combined to form the final carrier phase observation. This process is shown in Fig. 5.3.

In Fig. 5.3, after the Φ_{k-1} observation, the full-cycle counter of the carrier phase will be cleared to zero. The phase accumulation register will be updated each time the carrier NCO clock is updated, and the value of the phase accumulation register will also be checked for overflow from 2π to 0. If an overflow occurs, the carrier phase of the signal will advance for a full cycle, so the carrier phase's full-cycle counter will be incremented by 1, and so on, until the observation of Φ_k is latched. In Fig. 5.3, a total of N whole cycles are accumulated. When the Φk observation is latched, in addition to the N full-cycle carrier phases, accumulation of the fractional part of the carrier

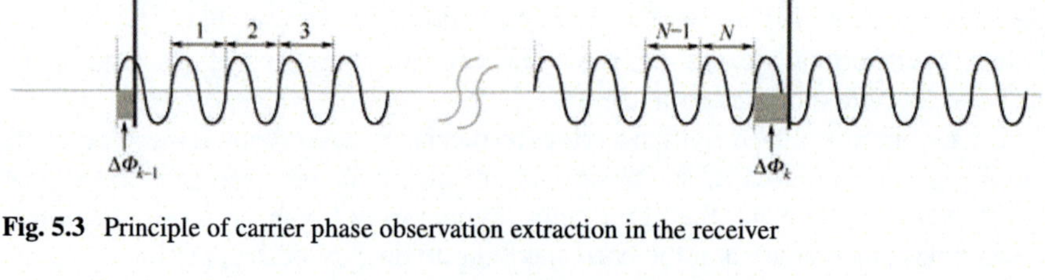

Fig. 5.3 Principle of carrier phase observation extraction in the receiver

phase also needs to be carried out. The fractional part of the current carrier phase is calculated through arctan2(Q, I), and the fractional part of the previous carrier phase needs to be subtracted from it. As shown in Fig. 5.3, the two shaded areas represent the fractional parts of Φ_{k-1} and Φ_k and are denoted as $\Delta\Phi_{k-1}$ and $\Delta\Phi_k$. Then

$$\Phi_k = \Phi_{k-1} + N + (\Delta\Phi_k - \Delta\Phi_{k-1}) \tag{5.14}$$

What is described above is the extraction method of carrier phase observations in the receiver, in which the carrier phase observation obtained will contain a theoretical intermediate frequency accumulation term, i.e. $\omega_{\mathrm{IF}}t$. However, it will be eliminated in the processing of inter-satellite single difference so that the use of the carrier phase will not be impacted. The theoretical intermediate frequency accumulation term will not be included in the discussion of the carrier phase observations later in this section.

The mathematical models of the carrier phase observations of GPS and BDS satellite are as follows:

$$\begin{aligned}\lambda_{\mathrm{G}}\Phi_{\mathrm{G}} = {} & \sqrt{(x_{\mathrm{u}} - x_{\mathrm{Gs}})^2 + (y_{\mathrm{u}} - y_{\mathrm{Gs}})^2 + (z_{\mathrm{u}} - z_{\mathrm{Gs}})^2} + c\delta t + \lambda_{\mathrm{G}} N_{\mathrm{G}} \\ & + c\tau_{\mathrm{G,s}} + E_{\mathrm{G,eph}} - T_{\mathrm{G,iono}} + T_{\mathrm{G,tron}} + \mathrm{MP}_{\mathrm{G}} + \varepsilon_{\mathrm{r}}\end{aligned} \tag{5.15}$$

$$\begin{aligned}\lambda_{\mathrm{B}}\Phi_{\mathrm{B}} = {} & \sqrt{(x_{\mathrm{u}} - x_{\mathrm{Bs}})^2 + (y_{\mathrm{u}} - y_{\mathrm{Bs}})^2 + (z_{\mathrm{u}} - z_{\mathrm{Bs}})^2} + c\delta t + cT_{\mathrm{GB}} \\ & + \lambda_{\mathrm{B}} N_{\mathrm{B}} + c\tau_{\mathrm{B,s}} + E_{\mathrm{B,eph}} - T_{\mathrm{B,iono}} + T_{\mathrm{B,tron}} + \mathrm{MP}_{\mathrm{B}} + \varepsilon_{\mathrm{r}}\end{aligned} \tag{5.16}$$

According to Eqs. (5.15) and (5.16), the GPS and BDS carrier phase observations multiplied by their respective carrier wavelengths are very similar to the mathematical equations for pseudo-range observation, where the common errors E_{eph}, τ_{s}, T_{iono}, and T_{tron} are identical, and the only difference is that the sign of T_{iono} in the carrier phase observation is opposite to the sign in the pseudo-range observation. N_{G} and N_{B} are the unknown carrier whole-cycles in the respective GPS and BDS carrier phase observations, which is unique to carrier phase observation. In cases where the signal-to-noise ratio is sufficiently high, as long as the tracking lock of the carrier phase in the carrier tracking loop remains stable, the values of N_{G} and N_{B} will be fixed. The unique error ($\mathrm{MP} + \varepsilon_{\mathrm{r}}$) in Eqs. (5.15) and (5.16) and ($\mathrm{MP} + n_{\mathrm{r}}$) in the pseudo-range phase observation are identical in form, but the error value of the former is smaller than the latter by two orders of magnitude. In general, the phase noise value of the carrier tracking loop caused by the multipath effect and thermal noise (converted to the distance in meters) is in the order of centimeters—only about 1% of the phase noise value of the pseudo-code tracking loop [3].

If T_{GB} and the common error terms in Eqs. (5.15) and (5.16) can be eliminated by differential processing, then they can be rewritten into

$$\begin{aligned}\lambda_{\mathrm{G}}(\Phi_{\mathrm{G}} - N_{\mathrm{G}}) = {} & \sqrt{(x_{\mathrm{u}} - x_{\mathrm{Gs}})^2 + (y_{\mathrm{u}} - y_{\mathrm{Gs}})^2 + (z_{\mathrm{u}} - z_{\mathrm{Gs}})^2} \\ & + c\delta t + (\mathrm{MP}_{\mathrm{G}} + \varepsilon_{\mathrm{r}})\end{aligned} \tag{5.17}$$

$$\lambda_B(\Phi_B - N_B) = \sqrt{(x_u - x_{Bs})^2 + (y_u - y_{Bs})^2 + (z_u - z_{Bs})^2} + c\delta t + (MP_B + \varepsilon_r) \quad (5.18)$$

It can be seen that if $\{\lambda_G(\Phi_G - N_G), \lambda_B(\Phi_B - N_B)\}$ is used as the observation to conduct positioning solution, since the variance of the noise term is in the centimeter level, the accuracy of the positioning result can reach the same level, which is the advantage of carrier phase positioning. However, at the same time, it is obvious that the carrier phase observation is only applicable after N_G and N_B. The whole-cycle number of the carrier is known, which is a big problem in the carrier phase application. This issue is known as the "ambiguity of whole cycles problem".

5.3 Doppler Frequency and Integral Doppler Observation

The Doppler effect is a very common physical effect that can occur during the transmission and reception of waveform signals. In short, for a waveform signal transmitted at frequency f, if the receiver is in motion with respect to the transmitter, an offset regarding f will appear in the received signal frequency, which is called the phenomenon of Doppler shift. Furthermore, when the relative motion direction between the receiver and the transmitter is consistent with the direction vectors of the two, the received signal frequency will be higher than f. Meanwhile, when the relative motion direction and the direction vector between the two are opposite, the frequency of the received signal will be lower than f.

The Doppler effect is also common in daily life. For example, when a train approaches, its whistle sounds sharp and rapid to the human ear. When the train moves away, its whistle becomes low and elongated. Since the tone of the train whistle does not change at the time of transmission, the different tones heard by the human ear are the frequency changes caused by the Doppler effect. The Doppler effect not only appears in sound waves but also in electromagnetic waves. The well-documented expansion of the universe is evidenced by scientific analysis of the spectrum of distant stars and the discovery of the redshift phenomenon, which is thought to arise from the Doppler effect caused by stars moving away at high speed.

Since GPS satellites, BDS MEOs, and IGSO satellites are all flying at high speeds, the Doppler effect is almost inevitable for receivers on the surface of the Earth. Further analysis shows that the value of the Doppler shift is related to three factors: the angle α between the distance vector between the satellite and the receiver and the relative velocity vector, the absolute value of the relative velocity, and the carrier wavelength λ, which can be written as

$$f_d = \frac{\boldsymbol{v}_u \cdot \boldsymbol{H}}{\lambda} = \frac{|\boldsymbol{v}_u| \cos\alpha}{c} f_{carrier} \quad (5.19)$$

where, "·" is a vector dot product, $\boldsymbol{v}_u$ is the relative velocity vector between the receiver and the satellite, $\boldsymbol{H}$ is the direction cosine vector between the receiver and

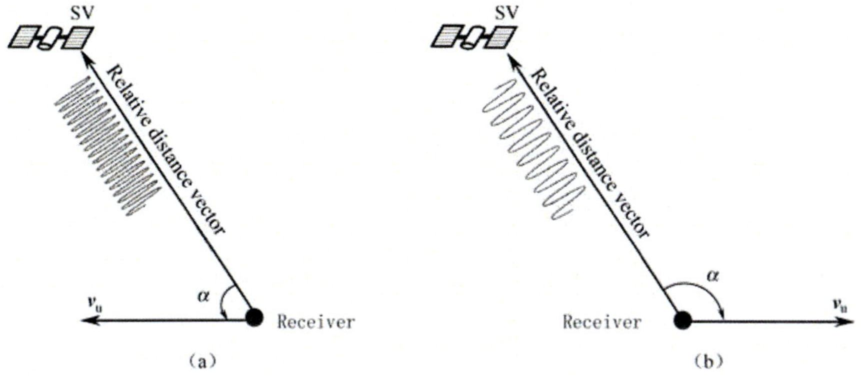

Fig. 5.4 Principle of the doppler effect

the satellite (i.e. the unit vector obtained through normalizing the relative position vector), and λ is the carrier wavelength, $\lambda = c/f_{\text{carrier}}$, where c is the speed of light.

Figure 5.4 shows the distance vector between the satellite and the receiver, the relative velocity vector, the angle α between the two vectors, and the Doppler shift. Figure 5.4a shows a case of $\alpha < 90°$, where the receiver and the satellite are moving in opposite directions, and the Doppler shift is positive, which means the received carrier frequency is higher than the frequency at the time of transmission. Figure 5.4b shows a case of $\alpha > 90°$, where the receiver and the satellite are back-moving, and the Doppler shift is negative, indicating that the received carrier frequency is lower than the frequency at the time of transmission.

Doppler frequency observations in BDS and GPS receivers can be read by the phase step register of the carrier NCO. As mentioned in Chap. 4, the relationship between the phase step register and the NCO instantaneous frequency is shown in Eq. (4.190). After the instantaneous frequency of the NCO is read, there are two ways to obtain the Doppler frequency observation: one is to directly send the instantaneous frequency of the NCO as a Doppler frequency observation, and the other is to deduct the instantaneous frequency of the NCO from the theoretical intermediate frequency f_c and then send it as a Doppler frequency observation. Before a detailed explanation of these two methods, the concept of theoretical intermediate frequency must be described.

The theoretical intermediate frequency is a nominal intermediate frequency value determined by the frequency scheme of the RF front-end of the receiver whose meaning in physics is the carrier frequency of the received GPS or BDS IF signal without the Doppler effect and local clock drift. Its value is determined by the configuration of the RF front-end. The specific principles and case analysis will be described in detail in the subsequent chapters about the RF front-end. Here is a simple example to help readers understand the theoretical IF and why it is necessary for the extraction of the Doppler frequency observation.

Figure 5.5 shows the RF front-end setup for a typical GPS receiver. The GPS signal is received through the antenna. It is amplified by a low noise amplifier (LNA) before

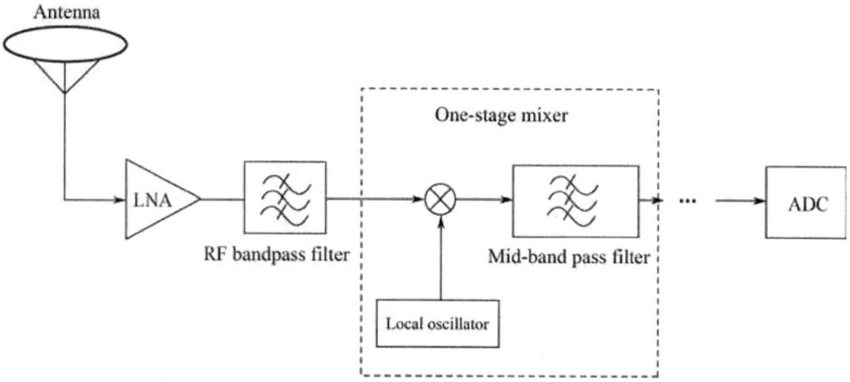

Fig. 5.5 A Simple GPS RF front-end

entering the mixer. The RF front-end of a traditional GPS receiver has one or more stages of mixers. For the sake of convenience, only one will be discussed here. The RF carrier frequency of the GPS is 1575.42 MHz. Assuming that the nominal value of the local oscillator is 1565.42 MHz, then the carrier frequency nominal value f_c of the low frequency component outputted by the mixer is (1575.42–1565.42 MHz), i.e. 10 MHz. This nominal value is the theoretical value without the Doppler shift and the local oscillator frequency deviation. That is to say, only when the satellite and the receiver remain stationary, and the frequency of the TCXO outputted by the local frequency synthesizer is precisely 1565.42 MHz, will the carrier frequency of the IF signal after ADC sampling be 10 MHz. As mentioned above, the relative motion between the satellite and the user during signal transmission leads to a Doppler shift in the received signal. Besides, since the actual oscillation frequency and nominal value of the TCXO will always be deviated, the actual frequency of the local oscillator will not be 1565.42 MHz. Based on the above two factors, the final carrier IF value is determined by the Doppler effect and the local clock bias. Therefore, the actual signal's intermediate frequency will not be 10 MHz. Rather, there will be a deviation whose value is determined by the above two factors. It is worth noting that since the carrier frequencies of the BDS and GPS satellite signals are different and so is their signal bandwidth, the frequency schemes of the two RF front-ends will be different, resulting in different theoretical intermediate frequencies. This requires special attention in algorithm design and software implementation.

The theoretical IF value must be known first before signal acquisition, because although the Doppler shift and clock drift of the TCXO determine the size of the Doppler frequency range that needs to be searched, the position of the Doppler frequency range is determined by the theoretical intermediate frequency. When the signal acquisition is completed and the signal tracking starts, the value of the instantaneous frequency of the carrier NCO is between ± 5 kHz of the theoretical intermediate frequency. If f_c is not subtracted from the instantaneous frequency of the NCO, a theoretical intermediate frequency term will be contained in the Doppler frequency observation, which is universal for Doppler frequency observations for all satellites. It is a common item, and can be eliminated in practice by inter-satellite single-difference operation.

Now we will look at a case where f_c is deducted from the instantaneous frequency of the NCO. In this case, there is no longer a theoretical IF term for Doppler frequency observation. The mathematical expression of the Doppler frequency observation will be very simple, as shown below:

$$f_d = (\nu_s - \nu_u) \cdot H + c\,\dot{\delta t} + n_d \tag{5.20}$$

where $\boldsymbol{H}$ is the direction cosine vector between the receiver and the satellite, $\nu_s = [\nu_{sx}, \nu_{sy}, \nu_{sz}]^T$ is the velocity vector of the satellite, $\nu_u = [\nu_{ux}, \nu_{uy}, \nu_{uz}]^T$ is the velocity vector of the user, $\dot{\delta t}$ is the clock drift, and n_d is the total noise term. Note that the unit of each physical quantity in Eq. (5.20) is m/s, and the Doppler observation in the receiver is measured in Hertz or radians. If it is in Hertz, the value needs to be multiplied by the carrier wavelength, while if it is in radians, the value in Hertz needs to be further multiplied by 2π. There is a slight difference between the L1 frequency of the GPS and the B1 frequency of the BDS, and in the wavelength of the L2 frequency of the GPS and the wavelength of the B2 frequency of the BDS, which can be calculated through $\lambda = c/f_{\text{carrier}}$. $\dot{\delta t}$ is a unitless quantity representing the ratio of the clock deviation drift to the nominal frequency, generally expressed in ppm or 10^{-6}.

It can be seen from Eq. (5.20) that the Doppler frequency observation is a function of the relative motion speed of the satellite and the user, as well as the local clock drift. Therefore, it is necessary for calculating the user's speed. Before using Doppler observation to calculate the user's motion speed, it is necessary to know the direction cosine vector between the satellite and the user. So, generally speaking, the positions of the user and the satellite need to be obtained first before the user's speed and clock drift can be solved using Eq. (5.20).

The physical meaning of the Doppler frequency observation can be regarded as the time derivative of the pseudo-range observation, which can be verified by conducting time derivation of Eqs. (5.9) and (5.10) and making a first-order Taylor approximation for the distance term on the right side of the equations. Note that the TGB in the BDS satellite's pseudo-range observation can be considered as a constant. Therefore, its derivative is zero, which is also easy to understand because the difference between the two atomic times of the BDS and GPS will be almost unchanged. Since $\dot{T}_{GB} = 0$, the Doppler frequency observations of the BDS and GPS satellites have the same mathematical expression.

Doppler frequency observation is also called pseudo-rate change rate observation, which is the difference between pseudo-range observations per unit time. However, if the actual collected pseudo-range data is compared with the Doppler frequency data, it will be noticed that the noise in the pseudo-range observation per unit time is much larger than that in the Doppler frequency observation. This is because the noise terms with high time-correlated in the pseudo-range observation do not appear in the Doppler frequency, including ionospheric delay, tropospheric delay, part of the ephemeris error term, and multipath effects under slight environmental changes, so the Doppler frequency observation is much "cleaner" than the pseudo-range observation.

The integral Doppler observation is the integral of the frequency value of the carrier tracking loop output over a certain period of time, expressed as:

$$\Delta \Phi_{\mathrm{d}}(t_2, t_1) = \Phi_2 - \Phi_1 = \int_{T_1}^{T_2} 2\pi f_{\mathrm{NCO}}(t)\,\mathrm{d}t \tag{5.21}$$

In Eq. (5.21), $\Delta \Phi_{\mathrm{d}}(t_2, t_1)$ is the integral Doppler observation from t_1 to t_2. Comparing Eq. (5.21) with Eq. (5.13), it can be seen that $\Delta \Phi_{\mathrm{d}}(t_2, t_1)$ is actually the difference in carrier phase observations from t_1 to t_2, which is easy to understand because the phase is the integral of the frequency by nature, so the difference of the carrier phases in a certain period of time is the integral of the Doppler frequency in that period, i.e. the integral Doppler observation.

If the time interval from t_l to t_2 is the unit time, then $\Delta \Phi_{\mathrm{d}}(t_2, t_1)/\ (2\pi)$ and the Doppler frequency observation f_d will be very close, but there is a fundamental difference between the two. f_d is the instantaneous value of the Doppler frequency at a certain time, while $\Delta \Phi_{\mathrm{d}}(t_2, t_1)/\ (2\pi)$ is the integral value of the Doppler frequency over a period of time. The difference between the two can be seen in Fig. 5.6.

The curve in Fig. 5.6 is the Doppler frequency value in the satellite signal over a certain period of time. The abscissa is the observation time whose unit is seconds. The unit of the vertical axis is Hertz, and the Doppler frequency observations corresponding to the observation times $T_1, T_2, \ldots, T_N, T_{N+1}$ are $f_1, f_2, \ldots, f_N, f_{N+1}$ respectively. The shaded area in the adjacent observation time in the figure is the integral of the Doppler frequency during this period. In the above analysis, the shaded area is the integral Doppler observation in this period. In terms of values, there is only a slight difference between the Doppler frequency observation and the integral Doppler

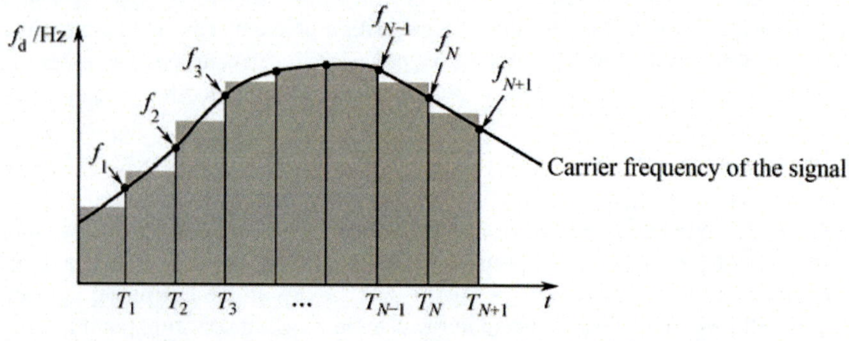

Fig. 5.6 The difference between doppler frequency observation and integral doppler observation

observation. However, when the carrier frequency of the input signal changes drastically, the difference is more obvious, and when it changes only a little, the difference between the two is minor.

In the physical sense, the Doppler frequency observation measures the instantaneous magnitude of the relative motion velocity between the satellite and the receiver, while the integral Doppler observation measures the change of distance between the satellite and the receiver over a period of time. Since the noise variance of the carrier phase observation is small, the integral variation of the Doppler observation is extremely high in precision, which can reach the centimeter or even millimeter level without a carrier phase cycle slip.

5.4 Analysis of Measurement Error Features

As can be seen in Sects. 5.1 and 5.2, both the pseudo-range observations and the carrier phase observations can be considered as the sum of the distance between the satellite and the receiver and several deviations or noise terms. The word "deviation" here refers to a random variable that remains constant or changes slowly, while the word "noise" refers to a random variable that changes rapidly and is difficult to predict. The distinction between noise and deviation is not absolute, and the deviation within different time spans can also be regarded as noise. The cumulative effect of all deviations and noise terms is called the User Equivalent Range Error (UERE). In this section, the UERE of BDS and GPS pseudo-range and carrier phase observations will be analyzed, with an emphasis on the main deviation or error terms, to discover their cause and statistical features in order to reduce or eliminate them.

5.4.1 Satellite Clock Error

A satellite is equipped with an atomic clock to generate the on-board operating frequency. Although the atomic clock has good long-term stability, as a source of time and frequency signals, it will still face problems such as frequency offset and aging rates. Therefore, there are bound to be deviations between the satellite time and the time standard used by the navigation system—a problem that both the BDS satellite and the GPS satellite will face. In order to solve this problem, the ground monitoring station of the BDS and the GPS satellites both monitor their satellite signals in real time and offer parameters for the correction of the satellite clock. In engineering practice, the clock bias of the satellite clock is modeled as the following binomial:

$$\Delta\tau(t) = a_0 + a_1(t - t_{oc}) + a_2(t - t_{oc})^2 \tag{5.22}$$

where, t_{oc} is the correction time of the star clock, a_0 is the zero-bias correction parameter of the star clock, a_1 is the correction parameter for the clock speed (frequency deviation) of the star clock, and a_2 is the correction parameter for the clock speed rate of the star clock. The four parameters are provided by the satellite ephemeris data blocks in the respective navigation messages.

Technically, the independent variable t in Eq. (5.22) should be the transmission time of the BDS signal or the GPS signal in their respective time reference (BDS time or GPS time). To better understand this, we can review Eq. (5.6), in which the satellite launch time t_{SV} of the GPS and BDS signals is given. However, it should be noted here that the deviation of the satellite clock and the respective time base $\Delta\tau(t)$ are included in t_{SV}, for which $[t_{SV}-\Delta\tau(t)]$ should be given as an argument to Eq. (5.22) to calculate the value of $\Delta\tau(t)$, so an iteration might be involved. If the value of $\Delta\tau(t)$ is relatively large, an iteration will be required, but the value of $\Delta\tau(t)$ is usually very small. Since the result of Eq. (5.22) is not closely related to time variation, the value of t_{SV} obtained through Eq. (5.6) can be directly substituted into it, to calculate $\Delta\tau(t)$. It should be noted that the value of $(t_{SV}-t_{oc})$ should be between (–302, 400, 302, 400). If the value of $(t_{SV}-t_{oc})$ is greater than 302, 400, the result needs to be subtracted from 604, 800, and if the value is less than −302, 400, 604, 800 needs to be added.

In addition to the correction of Eq. (5.22), there is also a relativistic correction for the satellite clock error. As seen in Sect. 3.2.1, the BDS and GPS satellite clocks are adjusted due to the combined effects of special relativity and general relativity. However, since the satellite orbit is not a regular circle, the impact of the relativistic effect on the clock varies at different orbital positions. The relativistic corrections given by the BDS and GPS signal interface control documents are

$$\Delta t_r = F e_s \sqrt{A} \sin E_k \tag{5.23}$$

where, e_s is the orbital eccentricity, A is the orbital semi-major axis, and E_k is the satellite's near-point angle, which will be explained in detail in the section about satellite position calculation. F is a constant, which is defined as

$$F = \frac{-2\sqrt{\mu}}{c^2} \tag{5.24}$$

where c is the speed of light and μ is the gravitational constant. The values of the gravitational constants of the BDS and GPS systems have been given in Fig. 1.1. The values of F calculated according to the above information in the BDS and GPS systems are

$$F_{GPS} = -4.442\ 807\ 633\ 4 \times 10^{-10}\ [\mathrm{s/m^{1/2}}]$$
$$F_{BD} = -4.442\ 807\ 309\ 0 \times 10^{-10}\ [\mathrm{s/m^{1/2}}]$$

The F constants of the two systems are very close.

The satellite's clock error correction also includes the group delay correction T_{GD}, which is provided in the satellite navigation message, so the final correction of the satellite clock is

$$\delta t_{\text{SV}} = \Delta\tau(t) + \Delta t_{\text{r}} + T_{\text{GD}} \tag{5.25}$$

By deriving Eq. (5.25) for time, the rate of change of the satellite's clock error correction can be obtained. Since the change of T_{GD} with time is small, the rate of change of the satellite clock can be

$$\delta \dot{t}_{\text{SV}} = a_1 + 2a_2(t - t_{\text{oc}}) + Fe_{\text{s}}\sqrt{A}\dot{E}_k \cos(E_k) \tag{5.26}$$

The residual of the satellite clock difference will affect the measurement accuracy of the pseudo-range and carrier phase observations, and that of the rate of change of the satellite clock will affect the Doppler observation. Before the cancellation of the SA policy in 2000, the US government controlled the error of the pseudo-range observation of civilian receivers by adding a random jitter to the GPS satellite clock, thereby controlling the deviation of the positioning result of the civilian receiver. Since the clock difference of the same satellite is identical for all receivers, regardless of the user's position, it can be eliminated by the differential method.

5.4.2 Ephemeris Error

The satellite ephemeris data includes broadcast ephemeris and precise ephemeris. The broadcast ephemeris is the ephemeris data broadcast by the satellite in real time through the navigation message, including satellite orbit parameters and disturbance items, while the precise ephemeris is the ephemeris data obtained through post-processing of observations from satellite tracking stations all over the world. The most mature versions of precise ephemeris are those of GPS and GLONASS released by IGS (International GNSS Service). The compilation and release of the precise ephemeris of the BDS satellites is also underway.

Satellites in space do not operate in an ideal inertial system. They are affected by various factors such as the force caused by the uneven distribution of the Earth's mass, atmospheric resistance, tides, and solar light pressure. Therefore, the operation trajectory of a satellite is very complicated and irregular. Due to these factors, there are deviations between the true position of the satellite and that calculated by its broadcast ephemeris and precise ephemeris. The position error is three-dimensional. For the convenience of analysis, the error is represented by a radial component, a tangential component, and a normal component, wherein the radial component is directed to the line of the satellite and the receiver, the tangential component is directed to the flight speed of the satellite, and the normal component points to the direction perpendicular to the orbital plane of the satellite.

Table 5.1 GPS position error calculated by GPS broadcast ephemeris data

差分量	Block IIA (σ)/m	Block IIR (σ)/m	Block IIR–M (σ)/m
法向误差	1.03	0.81	0.72
切向误差	2.05	1.34	1.32
径向误差	0.36	0.17	0.16
三维误差	2.32	1.57	1.51

误差分量 Error component; 法向误差 Normal error; 切向误差 Tangential error; 径向误差 Radial error; 三维误差 Three-dimensional error

The position error calculated by the broadcast ephemeris of GPS satellites is given in Reference [4], as shown in Table 5.1.

The position error given by the precision ephemeris is much smaller (generally 5 to 15 cm), but since it can only be obtained afterwards, the scope of use of the precision ephemeris is limited.

Figure 5.7 is a graphical representation of satellite position error, where P_u is the position of the receiver, and P_s and $\hat{P}_s$ are the true position of the satellite and the calculated position respectively. Obviously, the error vector of the satellite is

$$\Delta \boldsymbol{P}_s = \hat{P}_s - P_s$$

The distance $\left\| \hat{R} \right\|$ based on the position of the receiver and the calculated position of the satellite is

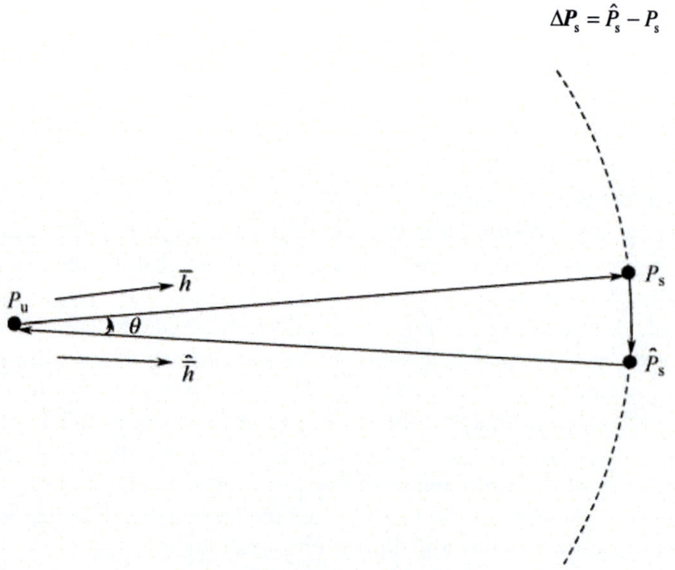

Fig. 5.7 The influence of ephemeris error on pseudo-range observation

$$\left\| \hat{R} \right\| = \hat{h} \cdot (P_{\mathrm{u}} - \hat{P}_{\mathrm{s}}) \tag{5.27}$$

In Eq. (5.27), $\hat{h}$ is the unit distance vector from the receiver to the calculated position of the satellite, as shown in Fig. 5.7. At the same time, the distance vector R between the position of the receiver and the real position of the satellite is

$$R = (P_{\mathrm{u}} - P_{\mathrm{s}}) \tag{5.28}$$

Then, Eq. (5.27) can be rewritten as

$$\begin{aligned} \|R\| &= \hat{h} \cdot (P_{\mathrm{u}} - P_{\mathrm{s}} + P_{\mathrm{s}} - \hat{P}_{\mathrm{s}}) \\ &= \hat{h} \cdot (P_{\mathrm{u}} - P_{\mathrm{s}}) + \hat{h} \cdot \Delta \boldsymbol{P}_{\mathrm{s}} \end{aligned} \tag{5.29}$$

In Eq. (5.29), $\hat{h} \cdot (P_{\mathrm{u}} - P_{\mathrm{s}}) = \|R\| \cos(\theta) \approx \|R\|(1 - \theta^2/2)$, where $\theta \approx \Delta \boldsymbol{P}_{\mathrm{s}}/A$, A is the radius of the satellite orbit. According to the values in Fig. 5.1, $\Delta \boldsymbol{P}_{\mathrm{s}} < 10$ m. Therefore, by substituting the orbital radius of the BDS or GPS satellite into the approximation of θ, we can see that the value of θ is so small that it can be safely considered that $\|R\|(1 - \theta^2/2) \approx \|R\|$. Substituting the result into Eq. (5.29), then

$$\left\| \hat{R} \right\| - \|R\| = \hat{h} \cdot \Delta \boldsymbol{P}_{\mathrm{s}} \tag{5.30}$$

$\|\Delta R\| = (\left\| \hat{R} \right\| - \|R\|)$ in Eq. (5.30) is the error of the pseudo-range and carrier phase observation caused by the satellite position error. It can be seen from the expression on the right side of Eq. (5.30) that the tangential component of the ephemeris error is perpendicular to the normal component and $\hat{h}$, so ΔR is only affected by the radial component of the ephemeris error. At present, the laser reflector on the GPS and the GLONASS satellites is helpful for measuring the radial error of the satellite position. By conducting distance measurement of the GLONASS satellite based on the laser ranging principle, we can see that the normal and tangential errors are both about 1 m, and the radial distance error is within ± 10 cm [5].

Like the star clock error, the ephemeris error is the same for all receivers, so it can also be eliminated through the differential method.

5.4.3 Ionospheric Delay

The ionosphere forms one part of the Earth's atmosphere, and is 50 to 1000 km away from the Earth's surface. Due to cosmic rays, its main purpose is the ionization of solar radiation. It is partially or fully ionized, containing a large amount of free electrons and positively charged ions. The influence of the ionosphere on electromagnetic waves changes its propagation velocity and causes refraction, reflection, and scattering. The ionosphere is a dispersive medium. “Dispersive” means that the dielectric constant of a medium is related to frequency, giving different propagation velocities to electromagnetic waves of different frequencies.

The phase velocity v_p of an electromagnetic wave propagating in a medium can be written as

$$v_p = \lambda f \tag{5.31}$$

where f is the frequency and λ is the wavelength. The speed at which information is modulated on an electromagnetic wave is called the group velocity, and is represented by v_g. More than a century ago, Rayleigh discovered the following relationship between phase velocity and group velocity:

$$v_g = v_p - \lambda \frac{dv_p}{d\lambda} \tag{5.32}$$

In a vacuum, v_g and v_p are both the speed of light, but in the dispersive medium the two are no longer the same.

According to Eq. (5.31), $d\lambda/\lambda = -df/f$. Substitute it into Eq. (5.32) to obtain

$$v_g = v_p + f \frac{dv_p}{df} \tag{5.33}$$

We can then introduce two refractive index factors n_p and n_g, in which n_p is the phase refractive index and n_g is the group refractive index. According to the definition of the refractive index, n_p and n_g have the following relationship with v_p and v_g:

$$n_p v_p = c \tag{5.34}$$

$$n_g v_g = c \tag{5.35}$$

In the above formula, c is the speed of light. The differential of v_p to the frequency f can be derived from Eq. (5.34):

$$\frac{dv_p}{df} = -\frac{c}{n_p^2} \frac{dn_p}{df} \tag{5.36}$$

If we substitute dv_p/df obtained from Eq. (5.33) into Eq. (5.36), then

$$n_g = \frac{n_p}{1 - \frac{f}{n_p} \frac{dn_p}{df}} \tag{5.37}$$

The relationship between n_p and n_g in the dispersive medium is given in Eq. (5.37). After first-order approximation, Eq. (5.37) can be written as

$$n_g = n_p + f\frac{dn_p}{df} \tag{5.38}$$

The relationship between the phase refractive index n_p and the frequency f in the dispersive medium is

$$n_p = 1 + \frac{a_1}{f^2} + \frac{a_2}{f^3} + \cdots \tag{5.39}$$

If we substitute n_p into Eq. (5.38), we can obtain the relationship between group refractive index n_g and frequency f:

$$n_g = 1 - \frac{a_1}{f^2} - \frac{2a_2}{f^3} + \cdots \tag{5.40}$$

For the ionospheric medium, the coefficients $a_1, a_2, \cdots$ in the above two equations are related to the concentration of free electrons whose value can be obtained through actual measurements.

The group delay and phase delay during the propagation of satellite signals in the ionosphere can be obtained through the following equations:

$$\delta r_p = \int_S (n_p - 1)ds = \int_S \left(\frac{a_1}{f^2} + \frac{a_2}{f^3} + \cdots\right)ds \tag{5.41}$$

$$\delta r_g = \int_S (n_g - 1)ds = \int_S \left(\frac{-a_1}{f^2} + \frac{-2a_2}{f^3} + \cdots\right)ds \tag{5.42}$$

S in Eqs. (5.41) and (5.42) is the propagation path of the signal in the ionosphere. If the high-order terms in the two equations are omitted, two conclusions can be drawn:

1. The absolute values of the delay of the group velocity and the phase velocity of the ionosphere to the satellite signal are the same, but the signs are opposite.
2. The influence of the ionosphere on the group velocity and phase velocity is inversely proportional to the square of the frequency, which explains why the equation of the pseudo-range observation in Sect. 5.1 is identical in form but opposite in sign with the ionospheric delay term in the carrier phase observation in Sect. 5.2: the ionosphere influences the pseudo-range on the group velocity, while it influences the carrier phase on the phase velocity.

Further research shows that if the concentration of free electrons on the signal propagation path is represented by N_e, the coefficient a_1 can be written as

$$a_1 = -40.28N_e \tag{5.43}$$

Since the free electron concentration N_e is always positive, the value of δr_g is always positive and that of δr_p is always negative, which indicates that the ionospheric effect delays the group velocity but advances the phase velocity.

Based on the above conclusions, the ionospheric delay term in pseudo-range observation and carrier phase observation can be written as

$$T_{\text{iono}, \, pseudo-range} = \frac{40.28}{cf_c^2} \int_S N_e(s) \mathrm{d}s \text{ (s)} \tag{5.44}$$

$$T_{\text{iono}, \, carrier \, phase} = -\frac{40.28}{cf_c^2} \int_S N_e(s) \mathrm{d}s \text{ (s)} \tag{5.45}$$

The total count of path-free electron TEC is defined as

$$\text{TEC} \triangleq \int_S N_e(s) \mathrm{d}s \tag{5.46}$$

Then, the ionospheric delay term in pseudo-range observation and carrier phase observation can be simplified as

$$T_{\text{iono}, \, pseudo-range} = \frac{40.28 \times \text{TEC}}{cf_c^2}, \quad T_{\text{iono}, \, carrier \, phase} = -\frac{40.28 \times \text{TEC}}{cf_c^2} \tag{5.47}$$

Since the ionospheric delay and the carrier frequency have the above relationship, it is possible to combine them to eliminate the ionospheric delay in the presence of dual frequency observation. After the ionospheric delay in Eq. (5.9) is substituted with $T_{\text{iono,pseudo-range}}$ in Eq. (5.67), the pseudo-range observation equations at two different frequency points f_1 and f_2 can be written. Then, it can be proved that the ionospheric delay will no longer be contained in the following pseudo-range combination.

$$\overline{\rho} = \frac{f_1^2 \rho_1 - f_2^2 \rho_2}{f_1^2 - f_2^2} \tag{5.48}$$

f_1 and f_2 in Eq. (5.48) are two different carrier frequencies, which can be L1 and L2 for GPS signals, and B1 and B2 for BDS signals. L1, L5 or B1, and B3 are also applicable. ρ_1 and ρ_2 are pseudo-range observations corresponding to f_1 and f_2. Equation (5.48) explains why the effects of the ionosphere can be eliminated by the dual-frequency receiver, and why single-frequency receivers cannot perfectly address the problem of ionospheric delays.

If we make a switch and combine the pseudo-range observations at f_1 and f_2 as shown in the following equations, we can obtain the magnitude of the ionospheric delay at f_1 and f_2:

$$I_1 = \frac{f_2^2}{f_1^2 - f_2^2}(\rho_1 - \rho_2) \tag{5.49}$$

$$I_2 = \frac{f_1^2}{f_1^2 - f_2^2}(\rho_1 - \rho_2) \tag{5.50}$$

In the dual-frequency receiver, not only can the pseudo-range observation be combined, but also the carrier phase observation. Readers can replace the ionospheric delay term in Eq. (5.15) with $T_{\text{iono,Carrierphase}}$ in Eq. (5.47), write the carrier phase measurement equations at two frequencies, and practice performing similar combinations of observations independently.

The above analysis is based on the premise that only the a_1 term is included in Eqs. (5.41) and (5.42). The two equations show that the higher the carrier frequency is, the smaller the error caused by omitting the higher-order items will be. For L-band satellite signals, the error caused by omitting the a_2 term and subsequent higher-order terms is acceptable. If there is a three-frequency observation, the delay of the a_2 term can be obtained from the combination of observations, thereby the result of ionospheric delay will be more accurate.

Since the single-frequency receiver can only provide single-frequency observations, it is impossible to eliminate the ionospheric delay with the combination of pseudo-range observations in Eq. (5.48). The alternative solution is to eliminate the ionospheric delay to some extent by establishing a mathematical model. At present, the Klobuchar model is used in both BDS and GPS navigation messages. It includes a total of 8 ionospheric parameters (α_1, α_2, α_3, α_4, β_1, β_2, β_3, and β_4), the specific application of which can be found in the interface control documents for BDS and GPS, or in Appendix E of this book. The Klobuchar model can eliminate ionospheric delay by about 50%.

After the SA policy was abolished, the ionospheric delay became the dominant source of deviation in the GPS pseudo-range observation. The ionospheric delay is generally on the order of a few meters, but during the active period of sunspots, the increase in density of free electrons in the ionosphere can increase the ionospheric delay to the order of tens of meters. Therefore, if the ionospheric delay is ignored, the positioning result will be greatly deviated. According to Eqs. (5.47) and (5.46), the ionospheric delay depends on the specific signal propagation path, so the receivers at different locations on the earth are subject to different ionospheric delays. Receivers that are geographically close can be considered to have similar ionospheric delays. Therefore, the ionospheric delay can be eliminated by the method of "region difference". The effect of differential cancellation is related to the length of the baseline, which will be explained later.

5.4.4 Tropospheric Delay

In atmospheric science, the troposphere is the lowest part of the atmosphere, closest to the Earth's surface. Its height varies depending upon the latitude and season. In terms of latitude, the upper troposphere at low latitudes is 17 to 18 km high, at mid-latitudes 10 to 12 km, and at high latitudes only 8 to 9 km. From a seasonal perspective, it is higher in summer than in winter. In the field of satellite navigation, tropospheric delay is often compared with ionospheric delay. The troposphere here refers to the atmospheric region extending from the surface of the Earth to a height of about 50 km. This is different from the definition in atmospheric science.

In the field of satellite navigation, tropospheric delay is of interest mainly due to its impact on satellite signal propagation. For historical reasons, the parts below the ionosphere are collectively called the troposphere. The main components in the troposphere are neutral atoms and atmospheric molecules, which exhibit non-dispersive characteristics to electromagnetic waves, i.e. the same transmission speed for electromagnetic waves of different frequencies. Therefore, the troposphere can be considered as a non-dispersive medium for satellite signals.

Tropospheric delay is typically 2 to 3 m. A "typical case" is when the satellite is directly above the zenith of the receiver, as shown in Fig. 5.8, where P_u is the location of the receiver. The tropospheric delay is d when the satellite is above the zenith of the receiver; it is $\hat{d}$ when the satellite position deviates from the zenith

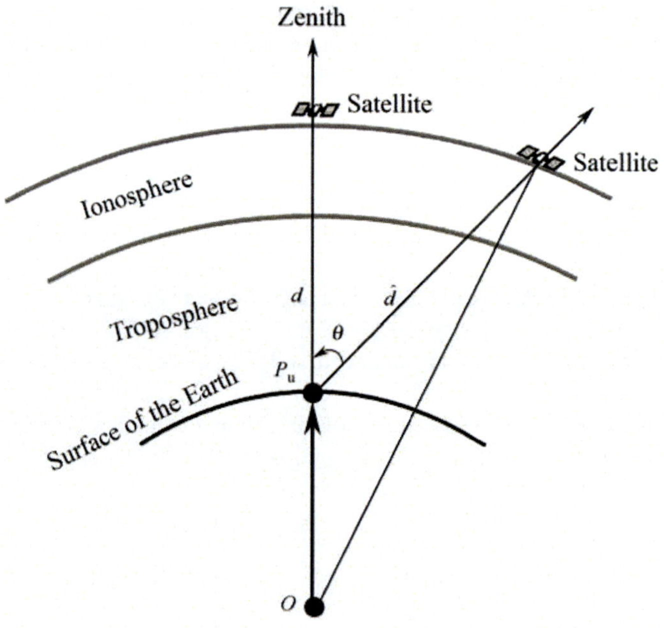

Fig. 5.8 Diagram of tropospheric delay

position. It can be seen that when the satellite elevation angle is low, the tropospheric delay will become significantly larger. In severe cases it can reach 20–30 m, so it must be processed in the receiver to ensure accurate positioning.

Similar to the analysis of ionospheric delay, if the refractive index factor of the troposphere is defined as $n(\mathrm{s})$, the tropospheric delay is

$$\delta r_T = \int_S [n(s) - 1]\mathrm{d}s \tag{5.51}$$

In this equation, S is the signal transmission path.

The above analysis shows that the tropospheric delay has no relationship with the signal frequency, so it cannot be eliminated by the combination of pseudo-range and carrier phase observation in the ionospheric delay elimination method, nor can it be calculated by the observation of the receiver. In practice, it is generally eliminated by means of a mathematical model.

$n(s)$ is mainly affected by temperature, humidity, and height. It can be summarized into two categories: the dry component refractive index (Dry), denoted as $n_\mathrm{d}(s)$; and the wet component refractive index (Wet), denoted as $n_\mathrm{w}(s)$. $n_\mathrm{d}(s)$ is mainly determined by oxygen molecules and nitrogen molecules in the atmosphere, and $n_\mathrm{w}(s)$ is mainly determined by the water vapor content in the atmosphere. Since the atmospheric humidity varies, $n_\mathrm{w}(s)$ is more difficult to predict than $n_\mathrm{d}(s)$. In the total tropospheric delay, $n_\mathrm{d}(s)$ accounts for approximately 90%, and $n_\mathrm{w}(s)$ accounts for approximately 10%.

Currently, there are many mathematical models for tropospheric delays. These models generally include two steps. The first is to estimate the ionospheric delay of the zenith path; the second is to multiply the path delay at different elevation angles by a tilt factor. F. The tropospheric delay model is analyzed by taking the Chao model as an example.

The Chao model needs to know the following physical quantities in advance:

- P: Atmospheric pressure (unit "N/m^2");
- E: Satellite elevation angle (unit "radians")
- T: Temperature (unit "K");
- e_0: Atmospheric pressure caused by water vapor (unit "mb" [1 mb = 100 Pa]);
- α: Temperature change rate with height (unit "k/m").

The Chao model first calculates the dry component delay and tilt factor

$$\delta r_\mathrm{d} = 2.276 \times 10^{-5} P \tag{5.52}$$

$$F_\mathrm{d} = \frac{1}{\sin E + \frac{0.00143}{\tan E + 0.0445}} \tag{5.53}$$

It then calculates the wet component delay and the tilt factor

$$\delta r_{\mathrm{w}} = 4.70 \times 10^2 \frac{e_0^{1.23}}{T^2} + 1.705 \times 10^6 \alpha \frac{e_0^{1.46}}{T^3} \tag{5.54}$$

$$F_{\mathrm{w}} = \frac{1}{\sin E + \frac{0.00035}{\tan E + 0.017}} \tag{5.55}$$

Then, the final total tropospheric delay is

$$\delta r_{\mathrm{Tro}} = F_{\mathrm{d}} \delta r_{\mathrm{d}} + F_{\mathrm{w}} \delta r_{\mathrm{w}} \tag{5.56}$$

The unit of $\delta r_{\mathrm{Tro}} \delta r_{\mathrm{d}}$ δr_{w} in the above formula is m. The Chao model needs to know meteorological data such as atmospheric pressure, temperature, water vapor pressure, and temperature as a function of altitude, and these inputs are difficult to obtain in the independently operated BDS and GPS receivers. Therefore, the Chao model is not widely used in practice. The simplified model only needs to know the satellite height h_s, the receiver height h_r, and the satellite elevation angle E, as the Magnavox and Collins models are respectively

$$\delta r_{\mathrm{M}} = \frac{2.208}{\sin E} \left(\mathrm{e}^{\frac{-h_r}{6\,900}} - \mathrm{e}^{\frac{-h_s}{6\,900}} \right) \tag{5.57}$$

$$\delta r_{\mathrm{C}} = \frac{2.4225}{0.026 + \sin E} \left(\mathrm{e}^{\frac{-h_r}{7\,492.8}} \right). \tag{5.58}$$

The difference in the Magnavox and Chao models, and the Magnavox and Collins models is less than 1 m when the satellite elevation angle is greater than 15°.

Like the dependence of the ionospheric delay on the path, the tropospheric delay is also related to the specific receiver position. Geographical correlation determines that the tropospheric delay can be eliminated by the regional difference method, but the elimination effect is also related to the baseline length. The shorter the baseline length, the more thoroughly it is eliminated.

5.4.5 *The Multipath Effect*

In the process of transmitting a satellite signal from Space to the receiver, it is often affected by the surrounding environment of the receiver in addition to the direct path. There are also one or more reflection paths with scattered wave signals due to atmospheric scattering. The ranging error caused by the superposition of multiple signals is called the "multipath effect". Its principle is shown in Fig. 5.9.

In Figure 5.9, the reflected path signal and the direct path signal caused by the reflective surface of a building near the working environment of the receiver are superimposed to obtain the actual signal entering the receiver antenna. Reflection

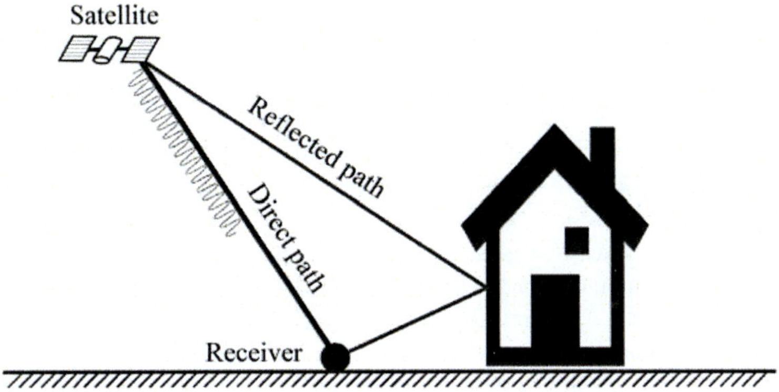

Fig. 5.9 The multipath effect

of the path signal can also be caused by factors such as ground plane, sea level, mountains, or exterior glass walls of tall buildings.

In practice, the formation of multipath signals is very complicated. Figure 5.10 is a schematic diagram of two special cases. Figure 5.10a shows the case where the direct path is occluded, in which case there is only a reflection or scattering path. Figure 5.10b shows a case in which there are multiple reflection paths. In practice, the environment in which the receiver works varies widely. There may be multiple reflection paths, or there may be no direct path. Alternatively, the signal of the direct path may be weakened by the occlusion of a forest. These factors lead to a very complicated theoretical analysis of the multipath effect.

Multipath signals can be modeled as follows:

$$r_{\mathrm{MP}}(t) = \alpha_0 x(t - \tau_0) \mathrm{e}^{-\mathrm{j}\Phi_0} \mathrm{e}^{-\mathrm{j}2\pi f_c t} + \sum_{i=1}^{N} \alpha_i x(t - \tau_i) \mathrm{e}^{-\mathrm{j}\Phi_i} \mathrm{e}^{-\mathrm{j}2\pi f_i t} \tag{5.59}$$

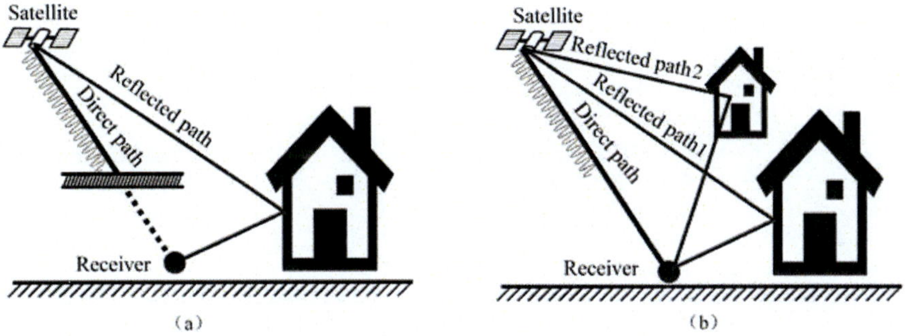

Fig. 5.10 Two complicated instances of the multipath effect

The first term in Equation (5.59) represents a direct signal, and the subsequent term in the accumulating sign represents a multipath reflection signal, which is obviously included in the case of N-way reflected signals. α_0 and α_i represent the respective signal strength of direct and reflected signals. The strength of the direct signal is mainly determined by path attenuation and occlusion condition. The intensity of the reflected signal is related to the reflective surface depending on the reflection coefficient of the reflective surface. Generally, smooth ground, water surfaces, saline-alkali beaches, mining ground, and glass curtain walls have larger emission coefficients, and are able to generate stronger reflection signals. Receivers operating in these areas must have good multipath suppression capability. $x(t)$ is the envelope of the modulated signal that determines the shape of the autocorrelation function. Φ_i is the received carrier phase initial value. f_c is the carrier frequency of the direct signal, and f_i is the carrier frequency of the reflected signal. In the case of small receiver dynamics, the carrier frequency of the direct signal and the reflected signal are close. If the difference between the two is large, the effect of the reflected signal on the final positioning will be small because the autocorrelation function in the pseudo-code loop is affected by the Doppler frequency offset, so the correlation peak caused by the signal of the reflected path can be ignored.

Φ_i in Eq. (5.59) is related to the specific transmission path. Due to the complexity of the surrounding environment and the relative motion of the satellite and the receiver, the carrier phase of the reflected-path signal actually received by the receiver is in the form of a random variable resulting in two multipath effects: when the carrier phase of the Φ_i and direct path signals are in phase, the autocorrelation function of the synthesized signal is enhanced relative to the direct path signal. When the carrier phase of the Φ_i and direct path signals inverts, the autocorrelation function of the composite signal is cancelled relative to the direct path signal. Other multipath effects range between fully enhanced and fully offset.

Figure 5.11 shows the two results of the multipath effect. Figure 5.11a shows the enhancement effect, and Figure 5.11b shows the cancellation effect. The autocorrelation function of the synthetic path signal in the two images is compared with

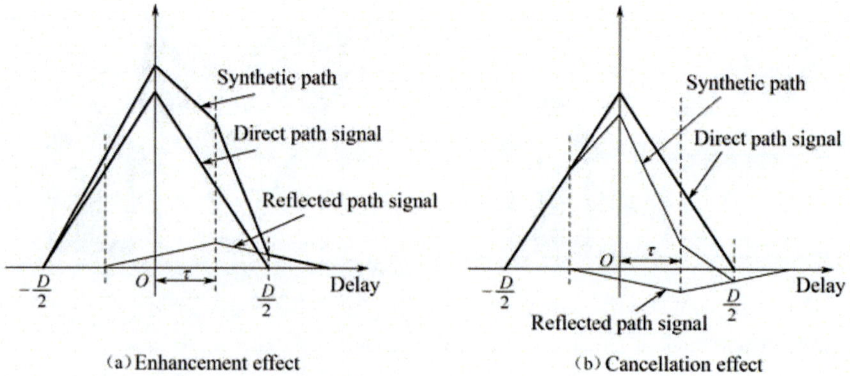

Fig. 5.11 Two results of the multipath effect

the direct path signal. Distortion has occurred. The autocorrelation function in the enhancement effect is larger than the autocorrelation function of the direct signal, and the shape becomes wider. Combined with the knowledge of the DLL loop in Chap. 4, it is not difficult to deduce that the deviation between the intermediate pseudo-code phase and the real pseudo-code phase of the pseudo-code loop at this time is positive. The autocorrelation function in the cancellation effect is smaller than the maximum value of the autocorrelation function of the direct signal. The shape becomes narrower, and it easy to derive that the deviation between the intermediate pseudo-code phase and the true pseudo-code phase of the pseudo-code loop tracking loop is a negative value at this time.

According to the above analysis, combined with the phase detector curve at different correlator intervals in Fig. 4.43 in Chap. 4, the envelope curve of the multipath effect can be derived, under the condition that the coherence interval D of the advance code and the lag code have different values. Figure 5.12 shows the effect of different multipath delays on the phase error of the pseudo-code with D=1.0 chip and D=0.1 chip. The vertical axis is the difference between the pseudo-code phase and the real pseudo-code phase of the pseudo-code loop tracking, with the chip as its unit. The envelope curve of the multipath effect in the figure is obtained based on the multipath signal with the direct path and a reflection path superimposed. When the error value is positive, it corresponds to the enhancement effect; when the error value is negative, it corresponds to the cancellation effect. The figure clearly shows that the narrow correlation technique can indeed suppress the multipath signal.

Since the pseudo-random chip width of the BDS signal is only half of the C/A chip width of the GPS, the curve envelope width and height of the multipath effect of the BDS signal are only half of those of the GPS. This means that when the BDS in

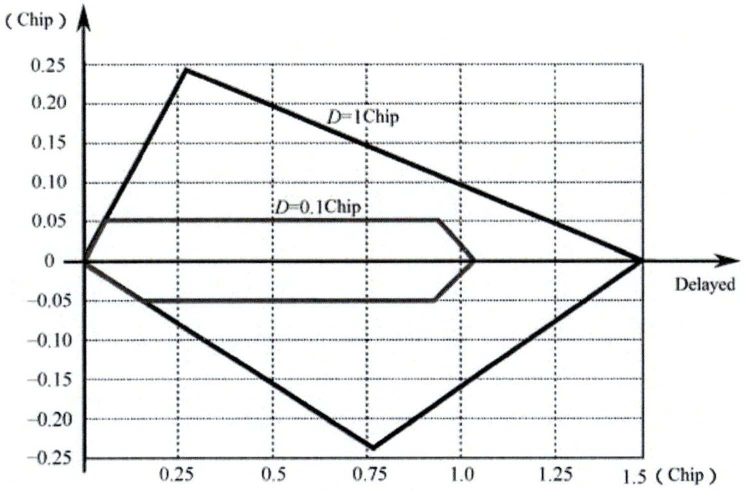

Fig. 5.12 Error envelope curve of the multipath effect at different correlation intervals

the same multipath environment, the multipath effect of the signal is slightly better than GPS.

Figure 5.12 is based on the ideal signal plus several assumptions obtained by computer simulation. It is based on several preconditions, for example, there is only one reflection path signal, there is a main path signal, and the autocorrelation function of the correlator contains infinite-width high-frequency components. It is difficult for the actual system to meet these conditions, but this does not prevent a qualitative analysis of the relationship between correlation intervals and multipath effects.

The influence of the multipath effect on pseudo-code observation is much larger than that of carrier phase observation. Readers only needs to review the difference between the pseudo-code tracking loop and the carrier tracking loop (especially the phase-detector principle of each loop) to understand this conclusion. The experimental data shows that the influence of the multipath effect on pseudo-code observation is 5–10 m, which is greater in extreme cases, and the influence on carrier phase is within ¼ of a carrier wavelength (5 cm).

Multipath effects have a variety of influences on receivers operating in different dynamic scenarios. When a receiver is in a high-speed motion state, the surrounding environment changes abruptly, and the reflection path is in a time-varying state, resulting in a rapid change of the composite signal of the multipath signal. The superposition of different reflection path signals causes the carrier phase of the composite signal to change faster. The amplitude phenomenon of the multipath effect is more random, so the effect is more similar to that of white noise. The Kalman filtering technique in PVT processing can more successfully filter and eliminate multipath errors. When the receiver is at rest, the surrounding environment remains unchanged, and the reflection path is basically invariant. At this time, the amplitude fluctuation and phase change of the multipath signal mainly depend on the geometric characteristics of the satellite motion and the reflection path, and a multipath error is presented. The periodic characteristics create more difficulties for subsequent processing.

After understanding the mechanism of the multipath effect, it can be suppressed from the following aspects:

(1) Choosing an operating environment

Under appropriate conditions, try to avoid operating in strong reflective environments such as smooth ground, static water level, or glass curtain walls. If unfavorable environmental conditions cannot be avoided, consider setting up artificial shielding reflections.

(2) Use hardware to suppress the reflected path signal

For example, a choke coil antenna, as shown in Fig. 5.13. This antenna uses a specially arranged high-frequency isolation loop band. Since the reflected path signal from the surrounding ground tends to have a low or negative elevation angle, the choke coil antenna offers a good shielding effect for the path signal.

Fig. 5.13 A choke coil antenna

(3) Eliminate multipath signals from the perspective of signal processing

Use the narrow correlation technique described in the above (or PAC or MEDLL) to estimate and eliminate the reflection path signal from the perspective of signal processing.

The multipath effect is not a common error. It cannot be eliminated by differential means, as it is related to the specific working environment of the receiver. Therefore, the error caused by the multipath effect in differential GNSS has become a major source of errors.

5.4.6 Receiver Errors

Errors in satellite clocks and ephemeris data come from a satellite's position and time. Ionospheric and tropospheric delays are caused by the signal propagation path. Multipath effects are errors caused by the external environment. Receiver errors are due to internal factors caused by the receiver, including thermal noise at the RF front-end, delays caused by cables and passive components, interference from RF continuous or image frequency, sample quantization errors, tracking loop heat, phase noise, and dynamic stress deviation of the tracking loop. Receiver errors are related to specific receiver settings, and therefore cannot be eliminated by differential means. In the receiver, the signal of a certain satellite is often processed by an independent signal tracking channel. The channels are independent of each other, so receiver errors in different satellite signals can be considered as irrelevant.

5.5 Differential GNSS Technology

According to the analysis of the errors and noise terms of the BDS and GPS observations in Sect. 5.4, it can be seen that the satellite clock bias, satellite ephemeris error, ionospheric delay, and tropospheric delay are common errors for receivers in the same region. In other words, pseudo-range observations and carrier phase observations contain approximately the same highly correlated error terms. Therefore, a common receiver can be used as a reference receiver. A common error term calculated by the reference receiver according to its own position can be broadcasted to other receivers in real time or non-real time through the communication link, and common errors can be eliminated to improve the positioning accuracy of other receivers. This is the basic principle of differential GNSS. The common error term calculated by the base station receiver is customarily referred to as a differential correction amount. In the BDS and GPS systems, the base station and mobile station receivers receive and process the BDS and GPS satellite signals simultaneously. Using "differential GNSS" to describe the differential technology is more precise than a single "differential GPS" or "differential BDS".

Figure 5.14 is a schematic diagram of the differential GNSS technique. The base station receiver in the figure is the reference receiver mentioned above. Its location is generally known and in a fixed state. The "mobile" nature of the mobile station receiver is relative to the base station, and does not mean that it is necessarily in motion. According to the analysis of the multipath effect in Sect. 5.4, differential processing cannot eliminate the multipath effect due to the different surrounding environments of the base station and the mobile station. Since the base station receiver needs to calculate the correction amount according to its own position and the received observation, it is necessary to ensure the quality of the observation received by the base station, otherwise the base station's receiving opportunity will include its own multipath effect in the observation correction amount, and will transmit it to

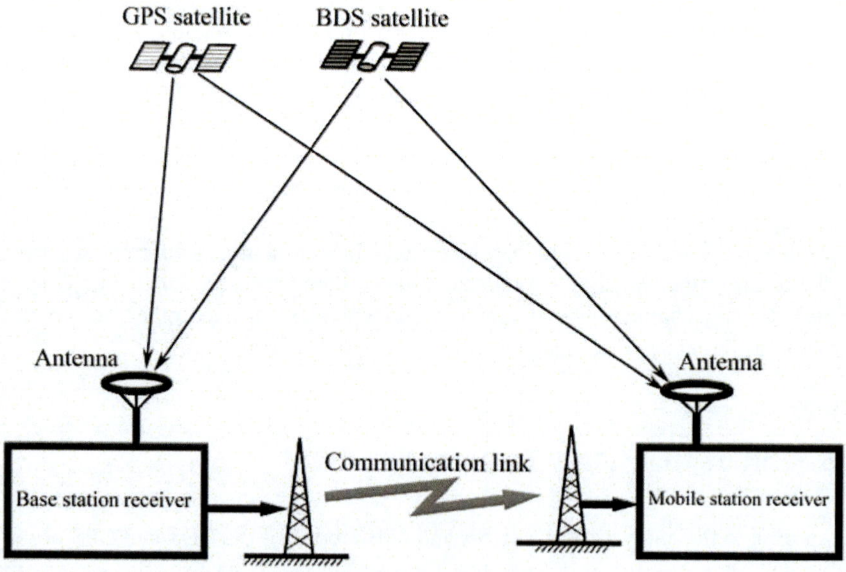

Fig. 5.14 The principle of differential GNSS technology

the mobile station. Therefore, the base station must not be located in an area with serious multipath effects. Generally, the ideal location would have a wide field of view, weak ground reflection, and high terrain. A choke coil antenna and high quality receiver should be used where possible, to further suppress the multipath effect and eliminate any errors in the receiver.

The distance between the base station receiver and the mobile station receiver is referred to as the baseline length. From the previous analysis of the error characteristics, the shorter the baseline length, the stronger the spatial correlation of the error source, and the better the effect of eliminating common errors by differential techniques. When the base station receiver and the mobile station receiver are connected to one antenna, the baseline length is zero. This configuration is called a zero-baseline configuration. When the antenna distance between the base station receiver and the mobile station receiver is very short (several meters or tens of meters), a very accurate physical measurement of the baseline length can be performed, which is called an ultra-short baseline configuration. The zero-baseline configuration and ultra-short baseline configuration ensure that the two receivers have substantially the same ionospheric delay, tropospheric delay, satellite clock error, and ephemeris error, and almost the same multipath effect, so that the receiver can produce an accurate evaluation of the delay length, delay stability, carrier phase, and pseudo-range measurement quality.

There is a communication link between the base station and the mobile station, whose main task is to transmit the differential correction to the mobile station receiver. Generally speaking, the communication connection is unidirectional. The base station receiver is responsible for transmitting, while the mobile station receiver is only responsible for receiving. The specific communication method can be determined according to actual application requirements and hardware and software conditions, such as the digital radio station, Modem, mobile TCP/IP, WIFI, and GPRS/WCDMA/TDS-CDMA. Factors that need to be considered in the communication link include data bandwidth, real-time performance, coverage distance, communication terminal power consumption, and volume. One advantage of a one-way communication link is that differential corrections can be transmitted to multiple mobile stations by broadcast. The number of mobile stations in such a system architecture can be almost unlimited.

Although the differential system using the spatial correlation of the error term must be within the local range (base length of 10–100 km), it can be differentiated over a wide area (up to thousands of kilometers or globally) by other means. The latter is called the wide-area differential system. The main types of wide-area differential systems are briefly explained at the end of this chapter.

Unlike differential correction, the difference technique can be divided into the position differential technique, the pseudo-range differential technique, and the carrier phase differential technique.

1. Position differential technique

The basic principle of this technique is as follows: the base station calculates the position coordinates according to the received pseudo-range observation, and then performs the difference with the known position coordinates to obtain the position correction amount, and transmits it to the mobile station; the mobile station calculates the position correction amount for itself. The position coordinates are corrected to obtain local position coordinates of higher precision.

Assuming that the known location of the base station receiver is (x_b, y_b, z_b), its known location here is generally obtained by precision mapping, so that higher accuracy can be guaranteed. The set of pseudo range observations received by the base station receiver is $\{\rho_i, i{=}1, \ldots, N\}$. The mathematical equation of the pseudo-range observation can be obtained from Eqs. (5.9) and (5.10). In cases of single mode observation (only GPS pseudo-range or only BDS pseudo-range), at least four pseudo-range measurements can solve the position and receiver clock difference; in cases of dual mode observation, at least five pseudo-range observations can be used to calculate the position and receiver clock difference, as well as the system deviation between BDS and GPS $T_{GB.}$ The set of satellites participating in the location solution is determined by the logic of the satellite in the receiver. Here, the set of satellite pseudo-range observations is assumed to be $\{\rho_j, j{=}1, \ldots, M\}$, when $M \leq N$. The position coordinates calculated based on the selected pseudo-range observation set are

$$\left(\widehat{x}_b, \widehat{y}_b, \widehat{z}_b\right) = f(\rho_{j1}, \cdots, \rho_{jM}) \tag{5.60}$$

Then, subtract Eq. (5.60) and (x_b, y_b, z_b) to obtain the position correction amount

$$(\delta x_b, \delta y_b, \delta z_b) = (x_b, y_b, z_b) - \left(\widehat{x}_b, \widehat{y}_b, \widehat{z}_b\right) \tag{5.61}$$

Send $(\delta x_b, \delta y_b, \delta z_b)$ to the mobile station, and calculate the position of the mobile station $\left(\widehat{x}_r, \widehat{y}_r, \widehat{z}_r\right)$ based on the received pseudo-distance measurement set. Modify it by using $(\delta x_b, \delta y_b, \delta z_b)$.

$$(x_r, y_r, z_r) = \left(\widehat{x}_r, \widehat{y}_r, \widehat{z}_r\right) + (\delta x_b, \delta y_b, \delta z_b) \tag{5.62}$$

(x_r, y_r, z_r) obtained by Eq. (5.62) is the position coordinates of the mobile station after the position difference.

Equations (5.60) to (5.62) show the process of position difference, where the subscripts "b" and "r" respectively represent the base station (Base) and the mobile station (Rover). The formulae related to the base station and the mobile station will follow this convention.

The basic premise of the position difference is that the base station and the mobile station each have the same error term in the calculated position coordinates. This conclusion requires preconditions, namely that the satellite pseudo-range observation set used by the base station and the mobile station are consistent. Both are $\{\rho_j, j{=}1,\ldots, M\}$, and the algorithm for calculating the position is also identical; otherwise there is no guarantee that the error terms in the position coordinates calculated by the base station and the mobile station are the same. This precondition is difficult to guarantee because the working environment of the mobile station and the base station are different, and the receivers are not necessarily the same model. Even a receiver of the same model cannot guarantee that the same time can be captured and tracked to the same assembly of satellites. Therefore, the position differential technique is not widely used in practice.

2. Pseudo-range differential technique

The differential correction of the pseudo-range differential transmission is the pseudo-range error term of the visible satellite of the base station. The mobile station uses the pseudo-range error term to correct the locally received pseudo-range observation, and then uses the differentially corrected pseudo-range observation to perform the positioning solution to calculate the local location.

The pseudo-range observation of the i-th GPS and BDS satellite received by the base station can be written as

$$\begin{aligned}\rho_{\mathrm{b}}^{(\mathrm{G}i)} &= r(p_{\mathrm{b}}, p_{\mathrm{s}}^{(\mathrm{G}i)}) + c\delta t_{\mathrm{b}} + c\tau_{\mathrm{b,G,s}} + E_{\mathrm{b,G,eph}} \\ &\quad + T_{\mathrm{b,G,iono}} + T_{\mathrm{b,G,tron}} + MP_{\mathrm{b,G}} + n_{\mathrm{b,r}}\end{aligned} \tag{5.63}$$

$$\begin{aligned}\rho_{\mathrm{b}}^{(\mathrm{B}i)} &= r(p_{\mathrm{b}}, p_{\mathrm{s}}^{(\mathrm{B}i)}) + c\delta t_{\mathrm{b}} + cT_{\mathrm{GB}} + c\tau_{\mathrm{b,B,s}} \\ &\quad + E_{\mathrm{b,B,eph}} + T_{\mathrm{b,B,iono}} + T_{\mathrm{b,B,tron}} + MP_{\mathrm{b,B}} + n_{\mathrm{b,r}}\end{aligned} \tag{5.64}$$

Equations (5.63) and (5.64) represent the pseudo-range observations of GPS and BDS respectively. The superscripts in $\rho_{\mathrm{b}}^{(\mathrm{G}i)}$ and $\rho_{\mathrm{b}}^{(\mathrm{B}i)}$ represent the i-th satellite of GPS and the i-th satellite of BDS; $r(p_{\mathrm{b}}, p_{\mathrm{s}}^{(\mathrm{G}i)})$ and $r(p_{\mathrm{b}}, p_{\mathrm{s}}^{(\mathrm{B}i)})$ are base stations and the distance of i-th GPS satellite, and the distance between the base station and the i-th BDS satellite. p_{b} and p_{s} are the location of the base station and the location of the corresponding satellite respectively. The definitions of other noise terms in Eqs. (5.63) and (5.64) are the same as those in Eqs. (5.9) and (5.10) except that the subscript "b" is added to indicate the base station.

Since the location of the base station is known, the satellite position can be calculated from the ephemeris data, and then $r(p_{\mathrm{b}}, p_{\mathrm{s}}^{(\mathrm{G}i)})$ and $r(p_{\mathrm{b}}, p_{\mathrm{s}}^{(\mathrm{B}i)})$ can also be calculated, thereby obtaining the pseudo-range correction of the GPS and BDS satellites.

$$\begin{aligned}\Delta\rho^{(\mathrm{G}i)} &= \rho_{\mathrm{b}}^{(\mathrm{G}i)} - r(p_{\mathrm{b}}, p_{\mathrm{s}}^{(\mathrm{G}i)}) \\ &= c\delta t_{\mathrm{b}} + c\tau_{\mathrm{b,G,s}} + E_{\mathrm{b,G,eph}} + T_{\mathrm{b,G,iono}} \\ &\quad + T_{\mathrm{b,G,tron}} + \mathrm{MP}_{\mathrm{b,G}} + n_{\mathrm{b,r}}\end{aligned} \tag{5.65}$$

$$\begin{aligned}\Delta\rho^{(\mathrm{B}i)} &= \rho_{\mathrm{b}}^{(\mathrm{B}i)} - r(p_{\mathrm{b}}, p_{\mathrm{s}}^{(\mathrm{B}i)}) \\ &= c\delta t_{\mathrm{b}} + cT_{\mathrm{GB}} + c\tau_{\mathrm{b,B,s}} + E_{\mathrm{b,B,eph}} + T_{\mathrm{b,B,iono}} \\ &\quad + T_{\mathrm{b,B,tron}} + \mathrm{MP}_{\mathrm{b,B}} + n_{\mathrm{b,r}}\end{aligned} \tag{5.66}$$

$\Delta\rho^{(\mathrm{G}i)}$ and $\Delta\rho^{(\mathrm{B}i)}$ are transmitted to the mobile station as pseudo-range correction items, and the mobile station receiver deducts the corresponding pseudo-range correction item from the locally received pseudo-range observation to obtain a differential pseudo-range observation.

$$\tilde{\rho}_{\mathrm{r}}^{(\mathrm{G}i)} = r(p_{\mathrm{r}}, p_{\mathrm{s}}^{(\mathrm{G}i)}) + c\Delta t_{\mathrm{br}} + c\Delta\tau_{\mathrm{G,s}} + \Delta E_{\mathrm{G,eph}} + \Delta T_{\mathrm{G,iono}} + \Delta T_{\mathrm{G,tron}} + \Delta \mathrm{MP_G} + \Delta n_{\mathrm{r}} \tag{5.67}$$

$$\tilde{\rho}_{\mathrm{r}}^{(\mathrm{B}i)} = r(p_{\mathrm{r}}, p_{\mathrm{s}}^{(\mathrm{B}i)}) + c\Delta t_{\mathrm{br}} + c\Delta\tau_{\mathrm{B,s}} + \Delta E_{\mathrm{B,eph}} + \Delta T_{\mathrm{B,iono}} + \Delta T_{\mathrm{B,tron}} + \Delta \mathrm{MP_B} + \Delta n_{\mathrm{r}} \tag{5.68}$$

The definitions and explanations in Eqs. (5.67) and (5.68) are as follows:

- Base station and mobile station clock: $\Delta t_{\mathrm{br}} = \delta t_{\mathrm{r}} - \delta t_{\mathrm{b}}$
- Satellite clock residual: $\Delta\tau_{\mathrm{B/G,s}} = \tau_{\mathrm{r,B/G,s}} - \tau_{\mathrm{b,B/G,s}}$
- Satellite ephemeris error residual: $\Delta E_{\mathrm{B/G,eph}} = E_{\mathrm{r,B/G,eph}} - E_{\mathrm{b,B/G,eph}}$
- Satellite ionospheric error residual: $\Delta T_{\mathrm{B/G,iono}} = T_{\mathrm{r,B/G,iono}} - T_{\mathrm{b,B/G,iono}}$
- Satellite tropospheric error residual: $\Delta T_{\mathrm{B/G,tron}} = T_{\mathrm{r,B/G,tron}} - T_{\mathrm{b,B/G,tron}}$
- Multipath residual: $\Delta\mathrm{MP_B} = \mathrm{MP_{r,B}} - \mathrm{MP_{b,B}}$
- Receiver error residual: $\Delta n_{\mathrm{r}} = n_{\mathrm{r,r}} - n_{\mathrm{b,r}}$

Δt_{br} can be estimated as a system state quantity. The clock residual, ephemeris error residual, ionospheric error residual, and tropospheric error residual variance will be greatly reduced. The multipath residuals and the variance of the receiver error residual will increase over the multipath effect and receiver error of a single base station and mobile station, since there is no temporal or spatial correlation between the two.

After the differential correction, the pseudo-range observation $\tilde{\rho}_{\mathrm{r}}^{(\mathrm{G}i)}$, $\tilde{\rho}_{\mathrm{r}}^{(\mathrm{B}i)}$ can be sent to the least-squares or Kalman filter for PVT calculation. There are two changes compared with the pseudo-range observation without differential correction.

① The variance of most noise terms after differential processing becomes smaller.

② The system deviation T_{GB} of BDS and GPS is deducted, so only four satellite observations are required to complete the positioning solution.

The pseudo-range difference process above is applicable to single-frequency observation, dual-frequency, and even multi-frequency measurement. When generating and using pseudo-distance difference items for pseudo-range observation of different frequencies, attention should be paid to correlate the same frequency and the pseudo-range observation of the same satellite number.

3. Carrier phase differential technique

The processing of carrier phase difference and pseudo-range difference is basically the same, except that the pseudo-range difference correction term is changed to the carrier phase difference correction term.

The carrier phase observation of the i-th GPS and the BDS satellite received by the base station can be written as

$$\lambda_G \Phi_b^{(Gi)} = r(p_b, p_s^{(Gi)}) + c\delta t_b + \lambda_G N_{b,G} + c\tau_{b,G,s} + E_{b,G,eph} - T_{b,G,iono} + T_{b,G,tron} + MP_{b,G} + \varepsilon_{b,r} \tag{5.69}$$

$$\lambda_B \Phi_b^{(Bi)} = r(p_b, p_s^{(Bi)}) + c\delta t_b + cT_{GB} + \lambda_B N_{b,B} + c\tau_{b,B,s} + E_{b,B,eph} - T_{b,B,iono} + T_{b,B,tron} + MP_{b,B} + \varepsilon_{b,r} \tag{5.70}$$

Most of the definitions in Eqs. (5.69) and (5.70) are the same as in Eqs. (5.63) and (5.64). The difference is the presence of the phase integer $N_{b,G}$ and $N_{b,B}$ which are tracked by the GPS satellites and BeiDou satellites base station receivers respectively.

Since the position of the base station is known, the satellite's position can be calculated from the ephemeris data. Then, $r(p_b, p_s^{(Gi)})$ and $r(p_b, p_s^{(Bi)})$ can also be calculated, thereby obtaining the carrier phase correction amount for the GPS and the BDS satellites.

$$\begin{aligned}\Delta\Phi^{(Gi)} &= \lambda_G \Phi_b^{(Gi)} - r(p_b, p_s^{(Gi)}) \\ &= c\delta t_b + \lambda_G N_{b,G} + c\tau_{b,G,s} + E_{b,G,eph} \\ &\quad - T_{b,G,iono} + T_{b,G,tron} + MP_{b,G} + \varepsilon_{b,r}\end{aligned} \tag{5.71}$$

$$\begin{aligned}\Delta\Phi^{(Bi)} &= \lambda_B \Phi_b^{(Bi)} - r(p_b, p_s^{(Bi)}) \\ &= c\delta t_b + \lambda_B N_{b,B} + c\tau_{b,B,s} + E_{b,B,eph} \\ &\quad - T_{b,B,iono} + T_{b,B,tron} + MP_{b,B} + \varepsilon_{b,r}\end{aligned} \tag{5.72}$$

$\Delta\Phi^{(Gi)}$ and $\Delta\Phi^{(Bi)}$ are sent to the mobile station as the differential carrier phase correction terms. The receiver of the mobile station deducts the corresponding carrier phase correction terms from the locally received carrier phase observation quantity to obtain the differential carrier phase observations.

$$\begin{aligned}\lambda_G \tilde{\Phi}_r^{(Gi)} &= r(p_r, p_s^{(Gi)}) + c\Delta t_{br} + \lambda_G N_{br}^{(Gi)} \\ &\quad + c\Delta\tau_{G,s} + \Delta E_{G,eph} - \Delta T_{G,iono} \\ &\quad + \Delta T_{G,tron} + \Delta MP_G + \Delta\varepsilon_{br}\end{aligned} \tag{5.73}$$

$$\begin{aligned}\lambda_G \tilde{\Phi}_r^{(Bi)} &= r(p_r, p_s^{(Bi)}) + c\Delta t_{br} + \lambda_B N_{br}^{(Bi)} + c\Delta\tau_{B,s} \\ &\quad + \Delta E_{B,eph} - \Delta T_{B,iono} + \Delta T_{B,tron} + \Delta MP_B + \Delta\varepsilon_{br}\end{aligned} \tag{5.74}$$

The noise term in Eqs. (5.73) and (5.74) is the same as defined in Eqs. (5.67) and (5.68) except for $\lambda_G N_{br}^{(Gi)}$ and $\lambda_B N_{br}^{(Bi)}$. The definitions and descriptions of $N_{br}^{(Gi)}$ and $N_{br}^{(Bi)}$ are as follows:

- The integral number difference of the i-th GPS satellite carrier phase of the base station and the mobile station: $N_{\mathrm{br}}^{(\mathrm{G}i)} = N_{\mathrm{r,G}i} - N_{\mathrm{b,G}i}$
- The integral number difference of the i-th BDS satellite carrier phase of the base station and the mobile station: $N_{\mathrm{br}}^{(\mathrm{B}i)} = N_{\mathrm{r,B}i} - N_{\mathrm{b,B}i}$.

It can be seen that the carrier phase integer number ($N_{\mathrm{br}}^{(\mathrm{G}i)}$, $N_{\mathrm{br}}^{(\mathrm{B}i)}$) of the differential carrier phase observation is the difference between the original carrier number of the mobile station receiver and the integral number of carriers of the base station. Since there is no cycle slip, the carrier phase integer number ($N_{\mathrm{b,G}i}$, $N_{\mathrm{b,B}i}$) of the base station and the carrier phase of the mobile station ($N_{\mathrm{r,G}i}$, $N_{\mathrm{r,B}i}$) are constant, so $N_{\mathrm{b,G}i}$ and $N_{\mathrm{b,B}i}$ are also constant. This shows that the carrier phase observation after the difference still needs to solve the whole-circumference ambiguity problem, but from solving ($N_{\mathrm{r,G}i}$, $N_{\mathrm{r,B}i}$) to solving ($N_{\mathrm{br}}^{(\mathrm{G}i)}$, $N_{\mathrm{br}}^{(\mathrm{B}i)}$). At the same time, it can be seen that the base station does not need to solve the whole phase of the base station carrier phase before generating the carrier phase differential correction term, and only needs to ensure that ($N_{\mathrm{b,G}i}$, $N_{\mathrm{b,B}i}$) remains unchanged. If it occurs during the differential processing, or if the cycle or phase are out of lock, the receiver of the mobile station needs to know in time, so that the entire number of weeks before the solution is invalid, and the new integral is recalculated from the next time.

After the differential processing of the carrier phase observation, if the whole-circumference ambiguity solution is realized, then observations similar to the Eqs. (5.17) and (5.18) can be obtained $\{\lambda_{\mathrm{G}}(\tilde{\Phi}_{\mathrm{r}}^{(\mathrm{G}i)} - N_{\mathrm{br}}^{(\mathrm{G}i)}), \lambda_{\mathrm{B}}(\tilde{\Phi}_{\mathrm{r}}^{(\mathrm{B}i)} - N_{\mathrm{br}}^{(\mathrm{B}i)})\}$, and the accuracy of the observation can reach centimeters or even millimeters.

Similar to pseudo-range observation, the above differential processing can also be applied to dual-frequency and triple-frequency carrier phase observation. At this time, multi-frequency carrier phase observation can obtain wide lane and narrow lane combinations through different combinations, which can further help to reduce the ambiguity of the overall cycle.

4. Differential Relative Positioning

In addition to transmitting the differential correction term, the base station receiver can also transmit the pseudo-range and carrier phase observations directly to the mobile station receiver. At this time, the mobile station receiver can correlate the local pseudo-range with the carrier phase observation and the base station to observe the difference. This method can calculate the baseline vector, that is, the relative position between the mobile station and the base station, so it is called differential relative positioning. This differential process can also be referred to as inter-station single difference.

The pseudo-range observations received by the base station and the mobile station receiver can be written according to Eqs. (5.63) and (5.64), which are omitted here, and directly calculate the inter-station single difference of the pseudo-range observation.

$$\nabla\rho_{\mathrm{br}}^{(\mathrm{G}i)} = \left[r(p_{\mathrm{b}}, p_{\mathrm{s}}^{(\mathrm{G}i)}) - r(p_{\mathrm{r}}, p_{\mathrm{s}}^{(\mathrm{G}i)})\right] + c\Delta t_{\mathrm{br}} + c\Delta\tau_{\mathrm{G,s}} \\ + \Delta E_{\mathrm{G,eph}} + \Delta T_{\mathrm{G,iono}} + \Delta T_{\mathrm{G,tron}} + \Delta \mathrm{MP}_{\mathrm{G}} + \Delta n_{\mathrm{br}} \tag{5.75}$$

$$\nabla\rho_{\mathrm{br}}^{(\mathrm{B}i)} = \left[r(p_{\mathrm{b}}, p_{\mathrm{s}}^{(\mathrm{B}i)}) - r(p_{\mathrm{r}}, p_{\mathrm{s}}^{(\mathrm{B}i)})\right] + c\Delta t_{\mathrm{br}} + c\Delta\tau_{\mathrm{B,s}} + \Delta E_{\mathrm{B,eph}} + \Delta T_{\mathrm{B,iono}} + \Delta T_{\mathrm{B,tron}} + \Delta \mathrm{MP_B} + \Delta n_{\mathrm{br}} \tag{5.76}$$

Similarly, the single-station difference between carrier phase observations is

$$\nabla\lambda_{\mathrm{G}}\Phi_{\mathrm{br}}^{(\mathrm{G}i)} = \left[r(p_{\mathrm{b}}, p_{\mathrm{s}}^{(\mathrm{G}i)}) - r(p_{\mathrm{r}}, p_{\mathrm{s}}^{(\mathrm{G}i)})\right] + c\Delta t_{\mathrm{br}} + \lambda_{\mathrm{G}}N_{\mathrm{br}}^{(\mathrm{G}i)} + c\Delta\tau_{\mathrm{G,s}} + \Delta E_{\mathrm{G,eph}} - \Delta T_{\mathrm{G,iono}} + \Delta T_{\mathrm{G,tron}} + \Delta \mathrm{MP_G} + \Delta \varepsilon_{\mathrm{br}} \tag{5.77}$$

$$\nabla\lambda_{\mathrm{G}}\Phi_{\mathrm{br}}^{(\mathrm{B}i)} = \left[r(p_{\mathrm{b}}, p_{\mathrm{s}}^{(\mathrm{B}i)}) - r(p_{\mathrm{r}}, p_{\mathrm{s}}^{(\mathrm{B}i)})\right] + c\Delta t_{\mathrm{br}} + \lambda_{\mathrm{G}}N_{\mathrm{br}}^{(\mathrm{B}i)} + c\Delta\tau_{\mathrm{B,s}} + \Delta E_{\mathrm{B,eph}} - \Delta T_{\mathrm{B,iono}} + \Delta T_{\mathrm{B,tron}} + \Delta \mathrm{MP_B} + \Delta \varepsilon_{\mathrm{br}} \tag{5.78}$$

The square brackets in Eqs. (5.75) through (5.78) are the distance differences between satellite-to-base receivers and satellite-to-mobile receivers.

$$\left[r(p_{\mathrm{b}}, p_{\mathrm{s}}) - r(p_{\mathrm{r}}, p_{\mathrm{s}})\right] \approx \boldsymbol{H}^{\mathrm{T}}(p_{\mathrm{b}} - p_{\mathrm{r}}) \tag{5.79}$$

The position of the satellite in Eq. (5.79) is denoted by p_{s}. The superscript is omitted for simplicity, indicating that it is applicable to both BDS and GPS satellites. $(p_{\mathrm{b}} - p_{\mathrm{r}})$ represents the baseline distance vector, $\boldsymbol{H}^{\mathrm{T}}$ is the cosine vector in the unit direction of the satellite to the base station or mobile station, and T represents the transposition. In fact, the unit direction cosine vector from the satellite to the base station is represented by $\boldsymbol{H}_{\mathrm{b}}$, and the unit direction cosine vector from the satellite to the mobile station is represented by $\boldsymbol{H}_{\mathrm{r}}$, as shown in Fig. 5.15. Since the baseline length is small relative to the satellite height, $\boldsymbol{H}_{\mathrm{b}} \approx \boldsymbol{H}_{\mathrm{r}}$, and it can be used uniformly.

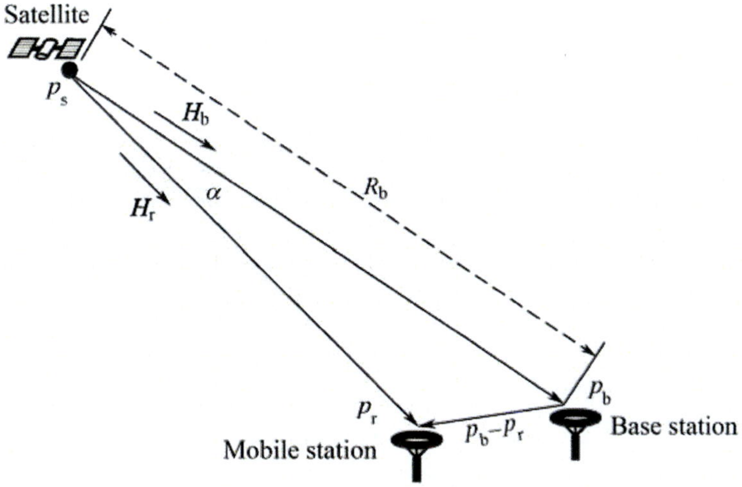

Fig. 5.15 The geometric relationship between the satellite, base station, and mobile station

According to the geometric relationship shown in Fig. 5.15, it can be concluded that

$$\begin{aligned}[r(p_b, p_s) - r(p_r, p_s)] &= \boldsymbol{H}_r^T(p_r - p_s) - \boldsymbol{H}_b^T(p_b - p_s) \\ &= \boldsymbol{H}_r^T(p_r - p_b + p_b - p_s) - \boldsymbol{H}_b^T(p_b - p_s) \\ &= \boldsymbol{H}_r^T(p_r - p_b) + \boldsymbol{H}_r^T(p_b - p_s) - \boldsymbol{H}_b^T(p_b - p_s) \\ &= \boldsymbol{H}_r^T(p_r - p_b) + (\cos\alpha - 1)R_b \\ &\approx \boldsymbol{H}_r^T(p_r - p_b) + \frac{\alpha^2}{2}R_b \end{aligned} \tag{5.80}$$

In the derivation process of Equation $\boldsymbol{H}_r^T(p_b - p_s) = (\cos\alpha)R_b$, Eq. (5.80) is used. α is the angle between the satellite and the base station, and the satellite and the mobile station. R_b is the distance between the satellite and the base station, and is identified in Fig. 5.15.

Comparing Eqs. (5.80) with (5.79), we can see that the error caused by the approximation of Eq. (5.79) is $\frac{\alpha^2}{2}R_b$. α can be approximated by the following equation.

$$\alpha \approx \sin\alpha \approx \frac{\|p_b - p_s\|}{R_b}$$

Then, the approximate distance error is

$$\frac{\alpha^2}{2}R_b \approx \frac{\|p_b - p_r\|^2}{2R_b} \tag{5.81}$$

$\|p_b - p_s\|$ is the baseline length, so Eq. (5.81) indicates that the greater the baseline length, the greater the distance error. For GPS satellites, the R_b of the base station on the Earth's surface is about 20,182 km, and the substitution type (5.81) shows that a baseline length of 1 km will bring about an error of 2.4 cm. For the BDS IGSO/GEO satellite, the Earth base station's R_b is about 35,786 km, at which time the baseline length of 1 km will bring about a 1.4 cm error. For the BDS MEO, the base station on the Earth's surface has an R_b of about 21,528 km, and a baseline of 1 km. The length will cause an error of about 2.3 cm.

Through the above approximation, the pseudo-range observation after the single-station difference between stations can be written as

$$\nabla\rho_{br}^{(Gi)} = \boldsymbol{H}_{(Gi)}^T\Delta p_{br} + c\Delta t_{br} + \Delta MP_G + \Delta n_{br} \tag{5.82}$$

$$\nabla\rho_{br}^{(Bi)} = \boldsymbol{H}_{(Bi)}^T\Delta p_{br} + c\Delta t_{br} + \Delta MP_B + \Delta n_{br} \tag{5.83}$$

where Δp_{br} is the baseline vector, and $\boldsymbol{H}_{(Gi)}^T$ and $\boldsymbol{H}_{(Bi)}^T$ are the unit direction cosine vectors of GPS and BDS satellites respectively. For the sake of brevity, Eqs. (5.82) and (5.83) omit common error terms such as satellite clock error, ephemeris error, and ionosphere and tropospheric delays.

The pseudo-range observation after the single difference between the stations is sent to the PVT solving unit for the $(\Delta p_{br}, \Delta t_{br})$ to be solved. Note that the location information obtained here is the position of the mobile station receiver relative to the base station receiver, not the absolute position of the mobile station receiver. If the location of the base station receiver is known, the absolute position of the mobile station receiver can be obtained by adding Δp_{br} to it.

For carrier phase observation, inter-station single-difference operation can also be performed. The obtained observation expression is similar to Eqs. (5.82) and (5.83), but one more carrier is used. Therefore, the carrier phase observation using the single-station difference between stations must first solve the problem of whole-circumference ambiguity.

5. Wide-area Differential System

The local area difference system uses the base station receiver to calculate the differential correction amount in a certain local area, and then transmits it to the mobile station receiver in the area through the communication link to provide a positioning service with certain accuracy requirements. The coverage area of this architecture is limited by the bias in the observations and the temporal and spatial correlation of the noise terms, typically within 100 km of the square. To cover a wider area, the number of base stations must be increased. Unlike local differential systems, wide-area differential systems cover much larger areas and generally cover international or intercontinental areas. In addition, the generation and format of differential corrections for wide-area differential systems are quite different from local differential systems. At present, wide-area differential systems mostly use synchronous satellite broadcasting, mainly WAAS in the United States, EGNOS in the European Union, MSAS and QZSS in Japan, and GAGAN in India. The commercial systems include StarFire (John Deere) and OmniSTAR (Fugro).

For the sake of brevity, this book will only explain the WAAS wide-area differential system, including its working principle, system architecture, and service content.

WAAS stands for Wide-area Augmentation System. It is a Space-based navigation enhancement and auxiliary system led by the Federal Aviation Administration (FAA), and is used as a supplementary system for GPS. Its main purpose is to improve the accuracy, availability, and reliability of navigation and positioning services. The main contractor for the system is the Raytheon Company.

The construction of WAAS is very similar to GPS. It is also divided into a Space section, a ground control section, and a user terminal. Figure 5.16 shows the Space section and the ground control section. The Space section is composed of three geosynchronous orbit satellites (January 2014 data) at 98°W (PRN133), 133°W (PRN135), and 107.3°W (PRN138). The ground control section consists of three master control stations (WMS) and 38 base stations (WRS). The user terminal is mainly composed of a GPS receiver with a receiving and processing function for

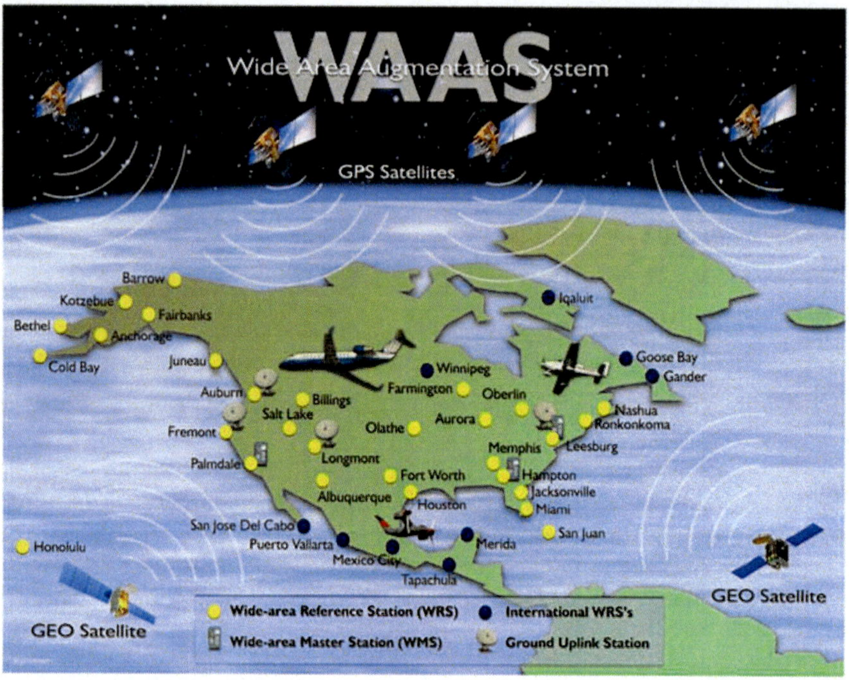

Fig. 5.16 Structure of WAAS

the WAAS signal. Initially, WAAS was used to provide Class I precision landings for aviation flights, so the main users were aircraft and airports. Later, the user base gradually extended to more general industries and users with high requirements for navigation and positioning accuracy.

WAAS has 20 base stations in the United States (seven in Alaska and one in Hawaii), one in Puerto Rico, five in Mexico, and four in Canada. The visible base stations cover the North American continent and a portion of the Pacific. The position of the base station can be accurately measured and mapped, so its exact position coordinates are known. The base station is equipped with a high-quality GPS receiver for real-time acquisition of GPS satellite signal quality and data, as well as for monitoring WAAS space satellite signals. All of this information is transmitted to the WAAS master station, which analyzes the raw data and observations, and is responsible for generating GPS satellite position errors, clock errors, and ionospheric correction parameters.

The WAAS master station processes the data from the base station to obtain a differential correction, and then transmits it through WAAS's geosynchronous orbit satellite. In addition to the above differential correction information, the WAAS navigation message includes GEO satellite position, velocity, acceleration, GEO satellite almanac data, GEO satellite time and UTC time parameters, GPS observation RMS error levels, and GPS satellite integrity information. The organizational structure of the navigation message is carried out in the form of data blocks, which are similar to the concept of data words in GPS navigation messages. One data block contains 250 bits and lasts for 1 second, among which there are 8 bit synchronization words and 6 bit data block types, 212 bits of data content, and 24 verification bits. The verifying mechanism is CRC-24Q. The WAAS document defines a total of 63 data blocks, but currently only 28 are used. Blocks 29-63 are reserved for future expansion of the system.

The carrier frequency of the WAAS signal (1575.42 MHz) is the same as the L1 carrier frequency of the GPS signal. The data rate is 250 bps. The convolutional code is used to obtain a navigation message with a symbol rate of 500 cps. The spreading code rate is also 1.023 MHz. In fact, the structure of the spread code generator is the same as the C/A code of the GPS. The WAAS pseudo-random code can be generated by simply changing the G2 selection logic. For details, refer to Chap. 3 of this book. The WAAS pseudo-random code can be regarded as an extension of the C/A code group of the GPS. Its PRN number is 120 to 138, and 12 PRN codes are currently not allocated. The specific generation method of the WAAS pseudo-random code can use the delay phase method and G2 initial phase method described in Sect. 3.1.2. The specific settings of the phase delay amount and initial phase are not explained further here. They can be found in Reference [6]. This shows that the WAAS signal can be received and processed using the hardware of the existing GPS receiver. The signal acquisition and tracking part is the same as the GPS signal processing. The signal acquisition part is even simpler and easier. This is because in GEO, the doppler shift of satellites is much smaller than that of MEO satellites. Processing WAAS satellite signals with existing GPS receiver hardware only requires appropriate adjustments in the software and baseband processing sections. This greatly simplifies the hardware design of the GPS/WAAS receiver, because the hardware of the existing GPS receiver can process the WAAS satellite signal with no or slight modification; no additional equipment or hardware is required to implement the difference correction of the amount of demodulation.

In addition to receiving and processing GPS signals, the WAAS user terminal receiver can receive and process WAAS satellite signals, demodulate the differential correction information therein, and differentially correct the GPS observations. Therefore, a GPS/WAAS capable receiver can provide more accurate positioning results than a stand-alone GPS receiver. The official WAAS document states that 95% of the coverage area can reach 7.6 m, but the actual operation results show that it can reach 1.5 m vertical and 1 m horizontally in most US states, Canada, and the Alaskan regions of the North American continent. Its positioning accuracy has been able to meet the requirements of Class I precision landing systems.

Figure 5.17 is a global coverage map of the WAAS service, where the light gray to dark areas (center) represent highly reliable service areas (>95%). It should be noted that the coverage of this area does not remain unchanged. When there is an ionospheric anomaly such as a solar storm, or temporary unavailability of some WAAS satellites, the coverage area in the picture will change. According to Reference [6], the definition of the coverage area of the more rigorous WAAS service is determined by the area covered by the coordinate points in Table 5.2. There are 12 points in

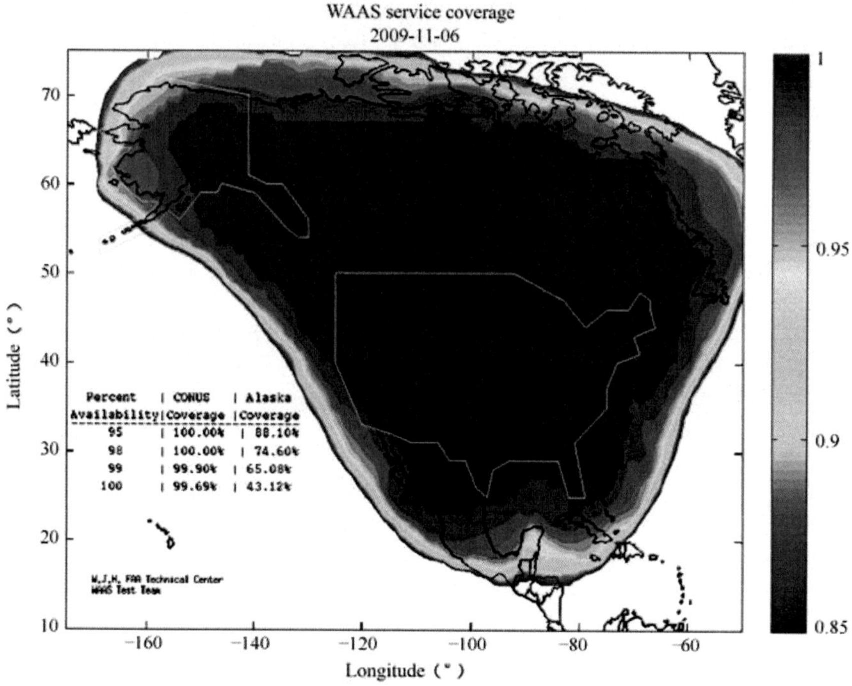

Fig. 5.17 WAAS service coverage

Table 5.2 Boundary point coordinates of the WAAS coverage area

Longitude	Latitude
50°N	61°W
50°N	122°W
70°N	140°W
70°N	165°W
68°N	169°W
20°N	164°W
17°N	160°W
17°N	155°W
30°N	120°W
16°N	75°W
16°N	61°W
50°N	61°W

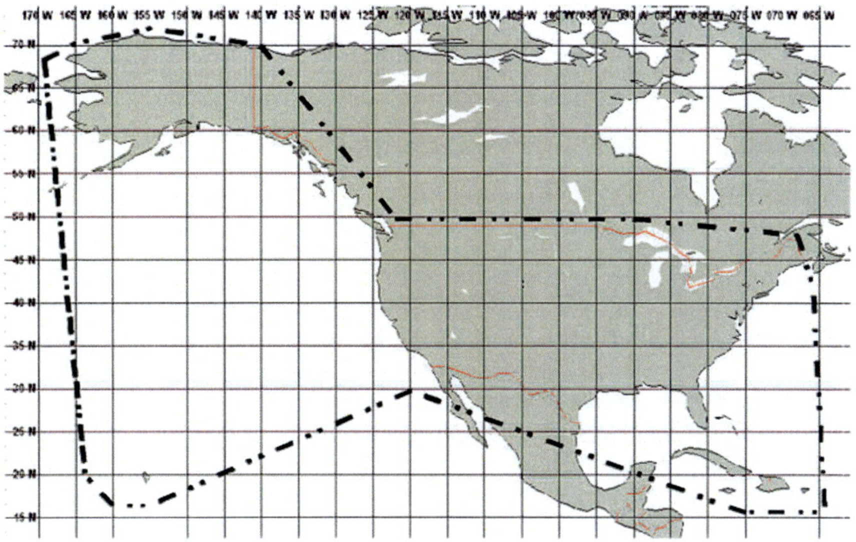

Fig. 5.18 WAAS service coverage

the table, of which the first and last are the same point coordinates. This results in a closed curved surface that extends upwards from the surface to a height of 30,000 m (approximately 100,000 ft), which is the area covered by the WAAS service.

The curve connecting the coordinates in Fig. 5.2 is the dotted box in Fig. 5.18. The WAAS service area is mainly concentrated in the northwest hemisphere, with the easternmost side reaching 61°W, the westernmost side reaching 169°W, and the northernmost side reaching 68°N. The southernmost side reaches 16°N. Although it is possible to receive the WAAS satellite signal outside of the above-mentioned area, there is no monitoring station to provide local GPS satellite error data and ionospheric delay correction, so WAAS cannot be used to improve positioning accuracy.

WAAS divides the error components in GPS observation into two broad categories: fast and slow. The fast variables include satellite position error and clock error, while the slow variables are mainly ionospheric delay. As seen in previous chapters, the fast variables are independent of the receiver position and only have time correlation, so the observations of GPS satellites received within the WAAS coverage can be corrected for the fast variables that are valid during the reference time. For variables, it is necessary first to determine the approximate position of the receiver, and then calculate the ionospheric delay correction amount to the local positioning point through the WAAS ionospheric grid information, and then deduct it. The process is a little complicated. This is also a big difference between WAAS and the local differential system.

Since the ionospheric delay is related to a specific location, and the WAAS service coverage area is vast, treatment of the ionospheric correction cannot be performed by WAAS, as the local differential system produces the ionospheric delay of all GPS

Table 5.3 Division intervals of ionospheric grid points in WAAS

Latitude range	Latitude interval	Longitude interval
85°N	10°	90°
65–75°N	10°	10°
55–55°N	5°	5°
65–75°S	10°	10°
85°S	10°	90°

satellites in each local area. Instead, WAAS divides the Earth into small areas called ionospheric grid points (IGPs). The specific grid density settings are shown in Table 5.3.

Table 5.3 shows that the ionospheric grid is denser at intervals of latitude and longitude in low latitudes. The interval between latitude and longitude is higher in high latitudes, which is consistent with the principle that the distance between the same longitude differences decreases with increasing latitude. The Earth can be divided into 1808 ionospheric grid points, as shown in Fig. 5.19. The data from 1808 grid points cannot be sent in one data packet, so the system divides all grid points into nine bands, numbered 0–8. The interval between adjacent grids is 40°, and each band contains 201 grid points (with the exception of the 8th band, which only has 200 grid points). The number of grid points in a band is 1–201 (or 200). The point in the southwest corner is the first, the north is the second, third, … until the 27th, and then back to the southernmost end of the next column, counting in turn until the

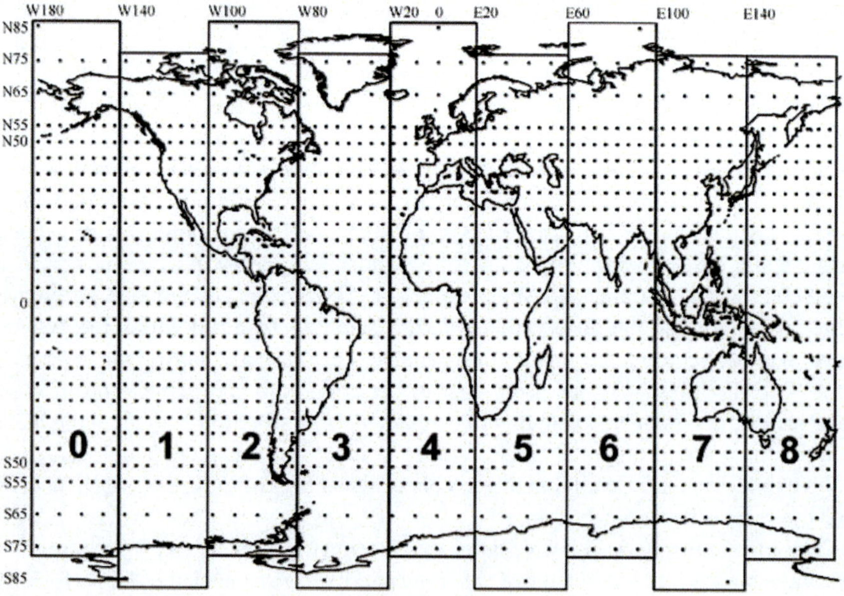

Fig. 5.19 Ionosphere grid point distribution map (WAAS)

north-easternmost corner of the band. The spacing of the grid points within the band is determined according to Fig. 5.3.

After demodulating the ionospheric grid point information, the GPS/WAAS receiver first needs to calculate the local approximate position by using the GPS pseudo-range observation. After obtaining the latitude and longitude of the local position, the GPS satellite corresponding to the pseudo-range observation at the current moment is calculated. The azimuth and elevation angles are then calculated based on their latitude and longitude and the azimuth and elevation of the GPS satellites. They are combined with the Earth's radius and the height of the ionosphere to calculate the intersection of the GPS satellite signal's propagation path and the highest point of the ionosphere. This intersection is called the sub-satellite ionospheric puncture point (IPP), which can be thought of as the intersection of the satellite signal propagating towards the center of the Earth and the highest point of the ionosphere. Figure 5.20 is a schematic diagram of the ionospheric puncture point. The schematic and calculation equation of the tilt factor F_{pp} is also given.

Figure 5.21 shows the ionospheric delay of the WAAS ionospheric grid point at 1:00 UTC time on 9th May 2014. The data comes from the official website of the FAA (USA). The latitude and longitude of the receiver are (φ_u, λ_u). The radius of the earth is *Re,* the typical height of the ionosphere is h_I, and the calculation generally takes *hI* = 350 km. The azimuth and elevation of the GPS satellite are A and E respectively. With the above information, the coordinates of the subsatellite ionospheric puncture point of the GPS satellite (φ_{pp}, λ_{pp}) can be calculated. Then, the receiver calculates three or four ionospheric grid points of (φ_{pp}, λ_{pp}) according to the demodulated ionospheric grid information. Interpolation is used to calculate the ionospheric delay

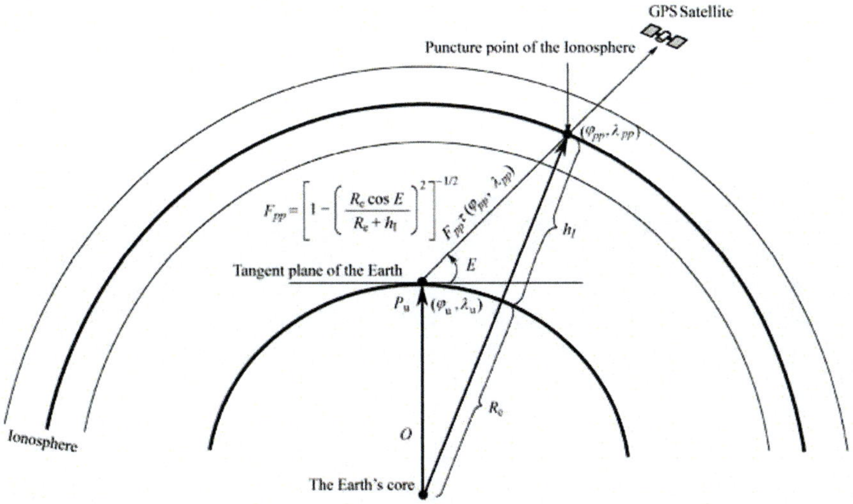

Fig. 5.20 Ionospheric puncture point and tilt factor (WAAS)

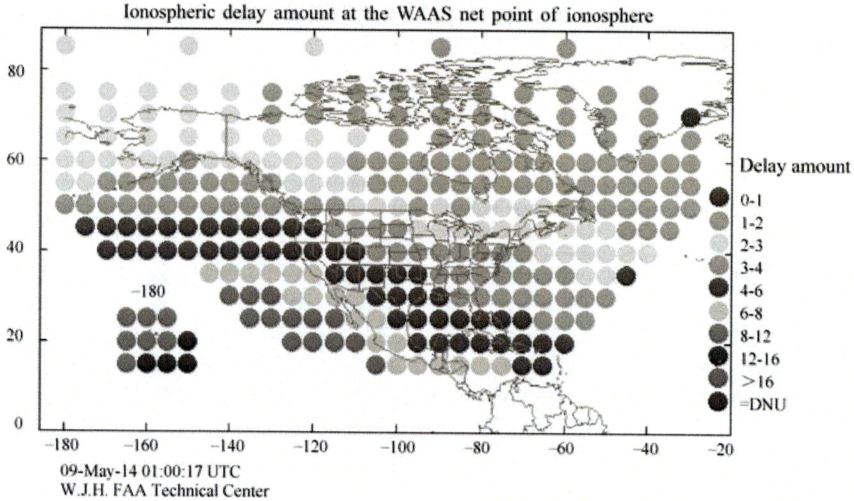

Fig. 5.21 Ionospheric delay in WAAS ionospheric grid points [7]

$\tau_{pp}(\varphi_{pp}, \lambda_{pp})$ of the GPS satellite signal at the ionospheric puncture point, and calculates the tilt factor F_{pp} based on the GPS satellite elevation angle. It finally calculates the ionospheric delay correction amount of the GPS signal $F_{pp}\tau_{pp}(\varphi_{pp}, \lambda_{pp})$ at the receiver position. The above steps are only a brief description of the calculation process. For detailed principles and calculation methods, please refer to Sects. 4.4.9 and 4.4.10 of Reference [6]. After the calculation of $F_{pp}\tau_{pp}(\varphi_{pp}, \lambda_{pp})$, the original pseudo-range observation is deducted from the satellite position correction amount, the satellite clock correction amount, and the ionospheric delay amount, and then the receiver position solution is re-executed to obtain a more accurate positioning result.

Figure 5.21 shows that the ionospheric delay can be greater than 16 m. The ionospheric delay in the high latitudes is smaller than in the vicinity of the Equator, while the maximum ionospheric delay is near the Equator.

The WAAS was put into use in July 2003, and the aviation industry has benefited greatly. Before, precision landing systems required complex ground infrastructure and personnel support. Now, WAAS offers a low-cost, stable, and reliable precision landing system, and the cost of building airports has been greatly reduced. It is estimated that installing a WAAS-based precision landing system costs only about $50,000, while deploying an ILS precision landing system requires $1 million to $1.5 million, plus significant maintenance costs.

However after 10 years of operation, WAAS revealed the following deficiencies:

- WAAS calculates the satellite position error, clock error, and ionospheric delay in an ionospheric grid through the monitoring station. The effect of differential error elimination depends on the geographical density and location of the monitoring station; in areas not covered by the monitoring station, accuracy positioning results cannot be obtained.

- The accuracy of the WAAS receiver is greatly affected by ionospheric delay. Positioning accuracy cannot be guaranteed in the period when sunspot activity is severe.
- WAAS satellites are geostationary orbit satellites that are distributed above the Equator, so at high latitudes, satellite signals or satellite signals are not received;
- The positioning accuracy of the WAAS system is not guaranteed to meet the requirements of Class II and Class III landing systems, so other high-precision landing systems (such as ILS or LAAS) are required for these applications;
- The receivers that support GPS/WAAS signals are costly, especially for receivers with an aviation safety certification, which hinders further market promotion.

References

1. Ziqing W (2008) 2000 Chinese geodetic coordinate system and its comparison with WGS84. Geodesy Geodyn (10)
2. Pengfei C, Hanjiang W, Yingyan C, Hua W (2000) A comparison of ellipsoid parameters of 2000 national geodetic coordinate system with GRS80 and WGS84. J Geomatics (6)
3. Farrell JA (2008) Aided navigation, GPS with high rate sensors. McGraw Hill
4. Cohenour JC (2009) Global positioning system clock and orbit statistics and precise point positioning. Ph. D thesis of Ohio University, Electrical Engineering (Engineering and Technology)
5. Jiyu L (2003) The principle and method of GPS satellite navigation and positioning. Science Press, Beijing
6. Specification for the Wide Area Augmentation System, U.S. Department of Transportation, Federal Aviation Administration, FAA-E-2892b, 13 Aug 2001
7. WAAS Real time data update URL: https://www.nstb.tc.faa.gov/RTData_WaasSatelliteData.htm
8. Huinan W (2003) The principle and application of GPS navigation. Science Press, Beijings
9. Kaplan ED (2006) Understanding GPS principles and applications, 2nd Edn. Artech House Publishers
10. Van Dierendonck AJ (1996) “GPS Receivers.” Global positioning system: theory and applications. vol. 1. In: Parkinson B, Spiker J, Axelrad P, Enge P 1996
11. Baojun F (2007) Applications of relativity theory in modern navigation. National Defense Industry Press, Beijing
12. Navstar GPS Space Segment/Navigation User Interfaces (2012) IS-GPS-200G, 5 Sept 2012
13. China Satellite Navigation System Management Office (2013) Beidou satellite navigation system space signal interface control document (Open Service Signal) Version 2.0. Dec 2013.
14. Yongxing Z, Xiaolin J, Jianfeng J, Qinghua Z (2013) The accuracy analysis of beidou satellite broadcast ephemeris. Geomatics Eng (8):2013
15. Cohenour JC, Van Graas F (2009) Temporal Decor relattion distributions of GPS range measurements due to satellite orbit and clock errors. J Navig 175–182
16. Anonymous. NAVSTAR GPS space segment/navigation user interfaces. Technical Report ICD-GPS-200, ARINC Research Corporation, Apr 1993
17. Anonymous. Phase I NAVSTAR/GPS major field test objective report thermostatic correction. Technical report, Navstar/GPS Joint Program Office, Space & Missile Systems Organization, Los Angeles Air Force Station, Los Angeles, California, 4 May 1979
18. Niell AE (1996) Global mapping functions for the atmosphere delay at radio wavelengths. J Geophys Res 101(B2):3227–3246

19. Shockley J (1984) Consideration of troposheric model corrections for differential GPS. Technical report, SRI International, Feb 1984
20. Spilker JJ (1996) Tropospheric effects on GPS." In: Parkinson B, Spilker J, Axelrad P, Enge P (ed) Global positioning system: theory and applications, vol 1. AIAA, pp 517–546
21. Saastamoinen J (1972) Contribution to the theory of atmospheric refraction. Bull G´eod´esique 105–106
22. Hopfield HS (1970) Torpospheric effect on electromagnetically measured ranges: prediction from surface weather data, Applied Physics Laboratory, Johns Hopkins Univ. Baltimore, MD, July 1970
23. Dierendonck AJV, Fenton P, Ford T (1992) Theory and performance of narrow correlator spacing in a GPS receiver. J Inst Navig 39(3):265–283
24. Dierendonck AJV, Braasch MS (1997) Evaluation of GNSS receiver correlation processing techniques for multipath and noise mitigation. In: Proceedings of the national technical meeting of the institute of navigation (ION NTM '97. Santa Monica, Calif, USA, Jan 1997. pp 207–215
25. Irsigler M, Eissfeller B (2003) Comparison of multipath mitigation techniques with consideration of future signal structures. In: Proceedings of the 16th international technical meeting of the satellite division of the institute of navigation (ION GNSS '03). Portland, Ore, USA, Sept 2003, pp 2584–2592
26. McGraw GA, Braasch MS (1999) GNSS multipath mitigation using gated and high-resolution correlator concepts. In: Proceedings of the national technical meeting of the satellite division of the institute of navigation, San Diego, Calif, USA, Jan 1999
27. Braasch MS (2001) Performance comparison of multipath mitigating receiver architectures. In: Proceedings of the IEEE aerospace conference. Big Sky, Mont, USA, Mar 2001. pp 31309–31315
28. Jones J, Fenton P, Smith B (2004) Theory and performance of the pulse aperture correlator. Tech. Rep. Novatel, Alberta, Sept 2004
29. Sleewaegen JM, Boon F (2001) Mitigating short-delay multipath: a promising new technique. In: Proceedings of the international technical meeting of the satellite division of the institute of navigation (ION GPS '01). Salt Lake City, Utah, Sept 2001, pp 204–213
30. https://www.nstb.tc.faa.gov/Full_VerticalProtectionLevel.htm
31. WAAS website. https://www.faa.gov/about/office_org/headquarters_offices/ato/service_units/techops/navservices/gnss/waas/
32. NSTB/WAAS T&E Team (2006) Wide-area augmentation system performance analysis Report. FAA/William J. Hughes Technical Center, Atlantic City International Airport, NJ 08405, July 2006
33. Services Status of QZSS (2008) The Asia pacific regional space agency forum. Commun Satellite Appl WG 10 Dec 2008
34. EGNOS (2007) A cornerstone of galileo, ESA SP-1303

Chapter 6
Calculating a Satellite's Position and Velocity

The satellite parameters required in the BDS and GPS receivers include the instantaneous position and instantaneous velocity of the satellite. The ephemeris data given by the navigation message is used to calculate the satellite position and velocity, and the ICD document provides these calculation formulae. This chapter not only explains these formulae, but also tries to explain the theory behind them in a concise manner. To make it more understandable, the chapter is divided into two parts. The first part is a detailed theoretical analysis of the satellite orbit; the second part goes on to explain how to use the ephemeris data to calculate the position and speed of the satellite, including GPS satellites, BDS MEO/IGSO satellites, and BDS GEO satellites. BDS GEO satellites are a special case, so there is a section devoted to analyzing its orbital parameters and position calculation principle. In addition to calculating the positional speed of satellites using GPS satellites and broadcast signals from BDS satellites, sophisticated ephemeris and ephemeris extension data are often used in the real-time positioning and post-processing positioning applications of receivers. Precise ephemeris and ephemeris extensions will be explained in the last part of this chapter.

6.1 Satellite Orbit Theory

The trajectory of a satellite in space is called its orbit. The parameter describing the state and position of the satellite's orbit is called the orbital parameter. The satellite orbits the Earth. In addition to being attracted by the Earth's gravitational force pointing to the planet's center, the orbit is also affected by the gravitational forces of the Sun and the Moon, as well as the Sun's light pressure, the Earth's tides, and physical atmospheric phenomena. Therefore, the actual orbit of a satellite is a very complicated curve. If we consider all of the factors that affect satellite motion

Y. Lu, *BDS/GPS Dual-Mode Software Receiver*, Navigation: Science and Technology 10, https://doi.org/10.1007/978-981-16-1075-2_6

from the beginning, we will inevitably fall into very complicated mathematical and physical problems. Therefore, the general analysis method is as follows: firstly, only the satellite orbital under the pull of gravity is considered. At this time, the satellite orbit is called a non-shot orbit. On the basis of no perturbation, the influence of various other disturbance factors is considered. The satellite orbit is called the taken orbit.

When the satellite is operating in a non-shot orbit, it is only affected by centripetal force, so its orbit is an ellipse, and the center of the Earth is a focus of its orbit. It takes only two parameters to determine the shape of the track: one is the long axis radius a and the other is the eccentricity e. The relationship between the minor axis radius b, a and e is

$$b = a\sqrt{1 - e^2} \tag{6.1}$$

Once we know any two of a, b and e, we can calculate the third item. Since satellites only move in a two-dimensional plane, Fig. 6.1 can be used to represent their orbits and motion.

In Fig. 6.1, the position of the satellite is S, the center of mass of the Earth is O, which is a focal point of the elliptical orbit. The distance vector between the satellite and the O point is r. According to the law of universal gravitation, the satellite is affected by the gravity of the Earth, and the gravitational vector can be expressed as

$$f = -\frac{GMm}{r^2}\frac{\boldsymbol{r}}{r} \tag{6.2}$$

At the same time according to Newton's second law, there is

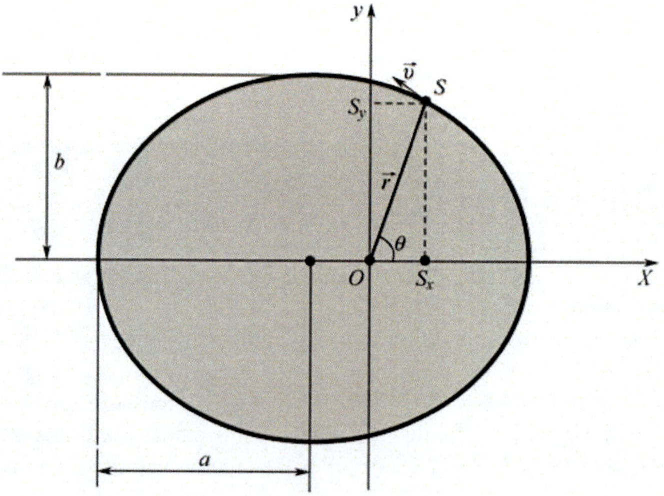

Fig. 6.1 Elliptic curve of a satellite orbit

$$f = ma = m\ddot{\boldsymbol{r}} \tag{6.3}$$

M and m in Eqs. (6.2) and (6.3) are the masses of the Earth and satellites, and G is the universal gravitational constant.

Combining Eqs. (6.2) and (6.3), we can get

$$\ddot{\boldsymbol{r}} = -\frac{GM}{r^2}\frac{\boldsymbol{r}}{r} = -\frac{\mu}{r^2}\frac{\boldsymbol{r}}{r} \tag{6.4}$$

$\mu = \mathrm{GM}$ in Eq. (6.4) is generally referred to as the gravitational constant of the Earth, that is, the GM value in Fig. 1.1.

First consider the satellite's trajectory as three-dimensional, that is, $\boldsymbol{r} = (x, y, z)$ and write Eq. (6.4) in scalar form.

$$\ddot{x} = -\frac{\mu}{r^3}x \tag{6.5}$$

$$\ddot{y} = -\frac{\mu}{r^3}y \tag{6.6}$$

$$\ddot{z} = -\frac{\mu}{r^3}z \tag{6.7}$$

Multiply both sides of Eq. (6.5) by *y*, then multiply both sides of Eq. (6.6) by *x*, and then subtract the two equations.

$$\ddot{x}y - x\ddot{y} = 0 \tag{6.8}$$

Similarly, similar operations are performed for Eqs. (6.5) and (6.7), and Eqs. (6.6) and (6.7). We can get

$$\ddot{x}z - x\ddot{z} = 0 \tag{6.9}$$

$$\ddot{y}z - y\ddot{z} = 0 \tag{6.10}$$

Write Eqs. (6.8), (6.9), and (6.10) in vector forms. We can get

$$\boldsymbol{r} \times \ddot{\boldsymbol{r}} = 0 \tag{6.11}$$

where "*x*" is a vector cross multiplying, and Eq. (6.11) indicates that the acceleration direction of the satellite is parallel to the radial direction. This is also the basic property of centripetal force.

For Eq. (6.11), the integral of time *t* can be obtained.

$$\frac{\mathrm{d}(\boldsymbol{r} \times \dot{\boldsymbol{r}})}{\mathrm{d}t} = \boldsymbol{r} \times \ddot{\boldsymbol{r}} = 0 \Rightarrow \boldsymbol{r} \times \dot{\boldsymbol{r}} = \boldsymbol{h} \tag{6.12}$$

Equation (6.12) shows that $\boldsymbol{h}$ is a constant vector. Since $\boldsymbol{r}$ is necessarily orthogonal to $\boldsymbol{h}$, $\boldsymbol{r}$ is in a two-dimensional plane. At the same time, from the expression of $\mathbf{h}$, the modulus of $\boldsymbol{h}$ is twice the speed of the satellite area. That is, if $\frac{\mathrm{d}s}{\mathrm{d}t}$ is the area velocity of the satellite, then $\frac{\mathrm{d}s}{\mathrm{d}t} = \frac{1}{2}\|\boldsymbol{h}\|$, this is the physical meaning of $\boldsymbol{h}$. Eq. (6.12) shows that the distance vector between the satellite and the center of the Earth sweeps through the equal area per unit time. This property is very useful. Using it, we can infer that the instantaneous angular velocity of the satellite at different locations is necessarily different; the angular velocity is the highest at the perigee, and smallest at the apogee. Further derivation can be proved

$$\|\boldsymbol{h}\| = \sqrt{\mu a(1 - e^2)} \tag{6.13}$$

Equation (6.13) shows that the area velocity of the satellite is a constant determined by the elliptical orbital shape and the gravitational constant. Since the gravitational constant of the Earth is constant, the longer the long semi-axis, the larger the area velocity of the satellite. We can derive the relationship between the speed of the satellite and $\boldsymbol{r}$

$$v(r) = \sqrt{\mu\left(\frac{2}{r} - \frac{1}{a}\right)} \tag{6.14}$$

In the perigee, $r = (1 - e)a$, $v_{perigee} = \sqrt{\frac{\mu(1+e)}{a(1-e)}}$. In apogee, $r = (1 + e)a$, $v_{\text{apogee}} = \sqrt{\frac{\mu(1-e)}{a(1+e)}}$. It can be seen that satellites are fast in perigee and slow in apogee.

Let's review Kepler's laws of planetary motion:

(1) All planets move about the Sun in elliptical orbits, having the Sun as one of their foci.
(2) A radius vector joining any planet to the Sun sweeps out equal areas in equal lengths of time.
(3) The squares of the sidereal periods (of revolution) of the planets are directly proportional to the cubes of their mean distances from the Sun.

Kepler's law is based on the gravitational effect of the planets, and the relationship between the satellite and the Earth is similar. In fact, more broadly, Kepler's law applies to all two-body problems. The above has been well documented by Kepler's first and second laws. Now, let's deduce Kepler's third law.

First, suppose that the satellite takes T to complete an orbit. Then, the area swept by the satellite in diameter $\boldsymbol{r}$ is the area of the entire ellipse $S = \pi ab$, and its area velocity is

$$\frac{S}{T} = \frac{\pi ab}{T} \tag{6.15}$$

From the expression of the area velocity above (6.13), there must be

$$\frac{S}{T}=\frac{1}{2}\|\boldsymbol{h}\| \quad\Rightarrow\quad \frac{\pi ab}{T}=\frac{1}{2}\sqrt{\mu a(1-e^2)} \tag{6.16}$$

Organizing the above equations, we will get

$$\frac{T^2}{a^3}=\frac{4\pi^2}{\mu}. \tag{6.17}$$

Equation (6.17) is actually Kepler's third law, that is, the square of the period of the satellite motion is proportional to the cube of the long axis radius of the elliptical orbit. It can be seen that the satellite motion period is constant when the long axis radius a is constant. The concept of an average angular velocity can be introduced here:

$$n=\frac{2\pi}{T} \tag{6.18}$$

where n is in radians/second. Note that the average angular velocity and the instantaneous angular velocity are separated. The instantaneous angular velocity is a variable related to the position of the satellite. The average angular velocity should be used in the subsequent explanation of the concept of the near-point angle.

In Fig. 6.1, the origin is at the ellipse focus O. θ is the angle between the satellite's geocentric diameter and the long semi-axis. From analytical geometry, the relationship between the norm of $\boldsymbol{r}$, r and the θ is

$$r=\frac{a(1-e^2)}{1+e\cos\theta} \tag{6.19}$$

If the satellite orbit plane's coordinate system takes 0 as the origin, the coordinate of the satellite can be expressed as a coordinate system in which the ellipse's long semi-axis is the X axis and the ellipse's short semi-axis is the Y axis.

$$P_S=\begin{bmatrix}S_x\\S_y\end{bmatrix}=\begin{bmatrix}r\cos\theta\\r\sin\theta\end{bmatrix} \tag{6.20}$$

It can be seen from Eqs. (6.19) and (6.20) that when the shape of the orbit is constant, the coordinates of the satellite are determined by θ. As long as the value of θ is known, r can be obtained, and then $[S_x, S_y]^{\mathrm{T}}$ can be calculated. Since the satellite is continuously moving over time, θ is a function of time. Given a time t, if you know $\theta(t)$, you can calculate the coordinates of the satellite. In reality, the coordinates of the satellite are three-dimensional, but since the satellite only moves

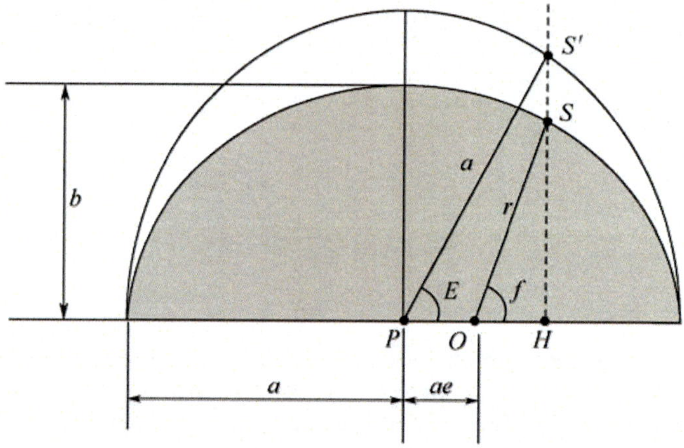

Fig. 6.2 The relationship between the eccentric anomaly and the true anomaly

in a two-dimensional plane, the three-dimensional coordinates[1] of the satellite are $[S_x, S_y, 0]^{\mathrm{T}}$.

In practice, the navigation message sent by the satellite does not directly give the value of θ, but sends other information, and indirectly calculates θ from the receiver software. In order to understand this process, it is necessary to explain the concept of three near-point angles, which are the true anomaly, the eccentric anomaly, and the mean anomaly. Let's first look at Fig. 6.2 to observe the concept of true anomaly and eccentric anomaly.

In Fig. 6.2, O is the focus of the ellipse. P is the geometric center of the ellipse. S is the position of the satellite. f is the angle between the satellite's radial path and the long semi-axis. That is, f is the true anomaly. It is θ in Fig. 6.1, but f is used here to be consistent with the traditional representation. To produce a circle with radius a at the center of P, make a line through S perpendicular to the long axis of the ellipse at the same, intersecting the ellipse at the point S', connecting points S' and P.$S'P$ is the angle E between the long axis of the ellipse, which is the eccentric anomaly.

It is easy to verify, and the following relationship is established (see Appendix D for the detailed derivation process):

$$\|S'H\| = \frac{r \sin f}{\sqrt{1 - e^2}} \tag{6.21}$$

$$r \cos f = a \cos E - ae \tag{6.22s}$$

[1]The three-dimensional coordinate system here is a temporary coordinate system, where the long semi-axis of the orbit is the X-axis, and the short semi-axis is the Y-axis, forming the right hand system. The Z-axis is perpendicular to the satellite track surface. There is a rotation relationship between the temporary coordinate system and the ECEF coordinate system.

$$r \sin f = b \sin E = a\sqrt{1-e^2} \sin E \tag{6.23}$$

From Eqs. (6.22) and (6.23), we can obtain

$$r = a(1 - e\cos E) \tag{6.24}$$

The above equation is a satellite orbit equation with the eccentric anomaly E as a variable.

By combining Eqs. (6.22), (6.23), and (6.24), the following relative equations can be obtained:

$$\cos f = \frac{\cos E - e}{1 - e\cos E} \tag{6.25}$$

$$\sin f = \frac{\sqrt{1-e^2}\sin E}{1 - e\cos E} \tag{6.26}$$

From the above two equations, combined with the half-angle formula, we can obtain

$$\tan\frac{f}{2} = \frac{\sin f}{1 + \cos f} = \frac{\sqrt{1+e}}{\sqrt{1-e}} \tan\frac{E}{2} \tag{6.27}$$

According to the Doppler equation in celestial mechanics, the relationship between the eccentric anomaly E and time t is

$$E - e\sin E = n \cdot (t - t_0) \tag{6.28}$$

where n is the average angular velocity mentioned above; t_0 is the reference time, which can be considered as the moment when the satellite passes the perigee. Equation (6.28) gives the relationship between E and time, and E can be calculated given a time t. This defines the mean anomaly M as follows:

$$M \triangleq n(t - t_0) \tag{6.29}$$

The meaning of M is: given a reference time t_0, the angle at which the satellite turns at the average angular velocity n at the orbital plane at time t, is called the mean anomaly. It should be noted that mean anomaly is not the angle at which the satellite actually turns. Because the angular velocity of an actual satellite is not a constant, but an amount defined by the Doppler equation, the mean anomaly is a linear function of time t. With the definition of the mean anomaly M, Eq. (6.28) can be written as

$$E - e\sin E = M \tag{6.30}$$

Equation (6.30) is a transcendental equation that can be solved by an iterative method.

The GPS satellite transmits the average angular velocity so that the receiver can calculate M, E, and r, and finally obtain the coordinates of the satellite in the orbital plane. So far, we have made clear how to determine the coordinates of the satellite in the orbital plane, and a rotation conversion is required from the satellite's orbital plane to the ECEF coordinate plane.

Figure 6.3 shows the positional relationship between the satellite's orbital plane and the ECEF coordinate system. The figure shows the ECEF coordinate system, that is, the center of the Earth is the origin, the X axis points to the vernal equinox, the Z axis points to the north pole, and the Y axis forms the right hand system together with the X and Z axes. At the same time, the satellite's orbital coordinate system is also given, that is, the center of the Earth is the origin, the long semi-axis of the elliptical orbit is the X_s axis, the normal direction of the elliptical orbit is the Z_s axis, and the Y_s, Xs and Zs axes together form the right-hand coordinate system. To distinguish them, the XYZ axis of the ECEF is represented by a solid line, and the axis of the satellite orbit rectangular coordinate system $XsYsZs$ is indicated by a broken line. From the inertial system, the ECEF coordinate system rotates with the rotation of the Earth, so the ECEF coordinate system in Fig. 6.3 is an instantaneous coordinate system when the X-axis just points to the vernal equinox. The ECEF coordinate system at other times can be rotated through the Earth. The angular velocity is linked to the ECEF coordinate system in the figure.

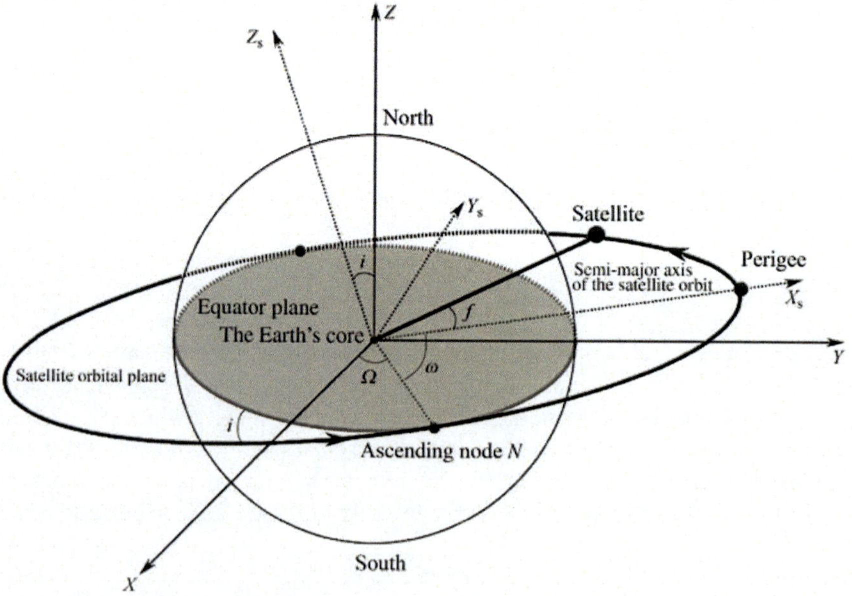

Fig. 6.3 The relationship between a satellite's orbital plane and Earth's ECEF coordinate system

There is an inclination between the satellite's orbital plane and the Earth's equatorial plane, called the orbital plane inclination. i is used here to indicate the angle between the Z-axis and the Zs-axis. The satellite's orbital plane intersects with the equatorial plane in a straight line. This line intersects with the Earth's Equator at two points: one is the intersection when the satellite flies from north to south, and the other is the intersection when the satellite flies from south to north. The former is called the Descending Node, while the latter is called the Ascending Node. Nis used to indicate the ascending node. The angle between the line connecting the center of the Earth and the ascending node N and the X-axis of the ECEF coordinate system is called the ascending node right ascension, which is indicated as Ω in the figure. Obviously, the two parameters of the orbital dip i and the ascending node Ω have been able to determine the position of the satellite's orbital plane in the ECEF coordinate system.

The orbital dip i and the ascending node ascension Ω only determine the relationship between the satellite's orbital plane and the equatorial plane, but the satellite's orbital coordinate system is still undetermined because its long semi-axis remains unchanged while keeping i and Ω constant. It can be rotated at will. Then the angle between the long axis of the satellite orbit and the ascending node N is defined in Fig. 6.3. It is called the perigee angular distance, denoted by ω. If the angular distance of the perigee is certain, the satellite's orbital coordinate system is determined. It can be seen that the five parameters a,e, i, Ω, and ω determine the elliptical orbit of the satellite. Among them, a and e determine the shape of the satellite orbit, while i, Ω and ω determine the position of the satellite orbit.

As we already know, the coordinates of the satellite in the orbital plane coordinate system can be expressed as $[S_x, S_y, 0]^T$. Then the following analysis is how to convert the $[S_x, S_y, 0]^T$ to the position coordinates of the satellite in the ECEF coordinate system.

As can be seen from Fig. 6.3, the satellite's orbital coordinate system is first rotated around the Z_s axis by $-\omega$ angle. At this time, the relative positions of the satellite's orbital plane and the Earth's equatorial plane remain unchanged, except that the geocentric, ascending node N and the satellite orbit are positioned in a line. In the second step, the satellite's orbital coordinate system is rotated around the X_s axis by the $-i$ angle. At this time, the satellite's orbital plane coincides with the equatorial plane, that is, the Z axis of the ECEF coordinate system and the Z_s axis of the orbital rectangular coordinate system coincide. In the third step, the satellite's orbital coordinate system is rotated around the Z_s axis by $-\Omega$ angle, and the satellite's orbital coordinate system coincides with the ECEF coordinate system.

The above three rotations can be represented by the following three rotation matrices $\boldsymbol{R}_1$, $\boldsymbol{R}_2$ and $\boldsymbol{R}_3$:

$$\boldsymbol{R}_1 = \begin{bmatrix} \cos\omega & -\sin\omega & 0 \\ \sin\omega & \cos\omega & 0 \\ 0 & 0 & 1 \end{bmatrix}$$

$$R_2 = \begin{bmatrix} 1 & 0 & 0 \\ 0 & \cos i & -\sin i \\ 0 & \sin i & \cos i \end{bmatrix}$$

$$R_3 = \begin{bmatrix} \cos \Omega & -\sin \Omega & 0 \\ \sin \Omega & \cos \Omega & 0 \\ 0 & 0 & 1 \end{bmatrix}$$

Then, the final rotation matrix is

$$R_{S2E} = R_3 R_2 R_1 \tag{6.31}$$

It should be noted here that the order in which R_3, R_2, and R_1 are multiplied cannot be confused.

The final form of R_{S2E} is

$$R_{S2E} = \begin{bmatrix} C(\Omega)C(\omega) - S(\Omega)C(i)S(\omega) & -C(\Omega)S(\omega) - S(\Omega)C(i)C(\omega) & S(\Omega)S(i) \\ S(\Omega)C(\omega) + C(\Omega)C(i)S(\omega) & -S(\Omega)S(\omega) + C(\Omega)C(i)C(\omega) & -C(\Omega)S(i) \\ S(i)S(\omega) & S(i)C(\omega) & C(i) \end{bmatrix} \tag{6.32}$$

In the equation, for simplification, complete $S(\cdot) = \sin(\cdot)$, $C(\cdot) = \cos(\cdot)$. According to the expression of R_{S2E}, the position coordinates of the satellite in the ECEF coordinate system can be obtained

$$\begin{bmatrix} X \\ Y \\ Z \end{bmatrix} = R_{S2E} \begin{bmatrix} S_x \\ S_y \\ 0 \end{bmatrix} \tag{6.33}$$

It can be seen from Eqs. (6.22) and (6.23) that when the eccentric anomaly E is known,

$$\begin{bmatrix} S_x \\ S_y \end{bmatrix} = \begin{bmatrix} a(\cos E - e) \\ a\sqrt{1 - e^2} \sin E \end{bmatrix}. \tag{6.34}$$

Substituting Eqs. (6.34) into (6.33), the value of [X, Y, Z] can be obtained.

So far, we have analyzed the non-shot orbital motion of the satellite. In actual operation, the satellite is affected by many disturbances. The shape of the Earth is not an exact sphere. Therefore, the Earth's gravity received by the satellite can be regarded as a superposition of central gravity and non-central gravity. The non-central gravity is due to the shape of the Earth, which is about 10^{-3} of the central gravity. In addition, the satellite is subject to the gravitational pull of the Sun and the Moon, the gravitational pull of the Earth's surface, and the effects of solar light pressure and atmospheric drag. Strict analysis of these disturbances is a very complex task that is beyond the scope of this book. Engineers working on receiver design do not need to have a thorough understanding of the rigorous theoretical analysis of these disturbance effects. In short, the impact of the disturbance term on the satellite orbit has the following main aspects:

- The satellite's orbital plane is no longer stationary in the ECEF coordinate system, but rotates slowly. This is mainly because, due to the influence of the Earth's perturbation, the satellite's orbital ascending node N is slowly moving forward on the Equator, so the ascending node Ω is not constant.
- The perigee is no longer stationary in the orbit of the satellite. This indicates that the semi-major axis of the satellite's elliptical orbit is not fixed, which also causes the perigee angular distance ω to change slowly.
- The mean anomaly M is also changing slowly. Due to the influence of multiple perturbations, the average angular velocity of the satellite also changes, resulting in a corresponding change in the mean anomaly.

The BDS and GPS navigation messages contain these disturbances, which influence the corrections of the message. The receiver software only needs to be demodulated and used correctly. In the following section, we will offer a detailed explanation of the receiver using the ephemeris data to calculate the satellite position, and point out how to use the correction of the navigation message to eliminate the disturbance.

6.2 GPS Satellites and BDS MEO/IGSO Satellites

As seen in Sect. 6.1, a satellite's orbit can be completely determined by the five Kepler parameters. The position of the satellite is determined by the eccentric anomaly E. In practice, in order to consider the influence of the disturbance term on the satellite position, the broadcast ephemeris parameters of the satellite broadcast also contain six corrections to rectify the pitch angle of the ascending node, the radius of the orbit, and the inclination of the orbital plane. Since the relationship between the eccentric anomaly and the time is complicated, and the mean anomaly M and the time are simple linear relations, the satellite is actually the correction of the mean anomaly and the angular velocity correction term of the reference time, and then M is calculated. The eccentric anomaly E is solved by Eq. (6.30).

The broadcast ephemeris of the GPS satellites and BDS MEO and IGSO satellites contain the same content, specifically six Kepler parameters (square root $\sqrt{a}$ of the semi-major axis, eccentricity e_s, perigee angular distance ω, near point of reference time angle M_0, the ascending node of the right ascension in the reference time Ω_e, and the satellite orbital inclination in the reference time i_0), and the nine perturbation parameters (the correction of the average angular velocity Δn, the rate of change of the ascending node $\dot{\Omega}$, the rate of change of the orbital inclination i_{dot}, and the six disturbance corrections items C_{uc}, C_{us}, C_{rc}, C_{rs}, C_{ic},C_{is}). The calculation method for a satellite's position is also the same, so they can be combined for explanation and analysis. Table 6.1 shows the ephemeris data parameters included in navigation messages sent by GPS satellites and BDS MEO/IGSO satellites.

Table 6.1 Ephemeris data in the navigation message

Parameter	Description	Type
M_0	Mean anomaly in reference time	Orbital parameter
Δn	A satellite's average angular velocity correction	Perturbation parameter
e_s	Eccentricity of a satellite's elliptical orbit	Orbital parameter
$\sqrt{a}$	Square root of the semi−major axis of the satellite's elliptical orbit	Orbital parameter
Ω_e	Ascending node of the satellite's orbit	Orbital parameter
i_0	Inclination angle of the satellite's orbital plane relative to the Earth's equatorial plane at the reference time	Orbital parameter
ω	Perigee angular distance of the satellite's orbit	Orbital parameter
$\dot{\Omega}$	Rate of change in the ascending node of the satellite's orbit	Perturbation parameter
$i$dot	Satellite's orbit plane inclination change rate	Perturbation parameter
C_{uc}, C_{us}	Harmonic correction of the pitch angle	Perturbation parameter
C_{rc}, C_{rs}	Harmonic correction of a satellite's geocentric diameter	Perturbation parameter
C_{ic}, C_{is}	Harmonic correction of a satellite's orbit inclination	Perturbation parameter
t_{oe}	Reference time of ephemeris data	Orbital parameter

According to the theoretical analysis in Sect. 6.1, combined with the description of the perturbation terms in the navigation message given in the GPS and BDS ICD documents, detailed steps for calculating the satellite position using the ephemeris data in Table 6.1 can be given.

1. Calculate the normalization time

Since the satellite's ephemeris data is relative to the reference time t_{oe}, it is necessary to normalize the observation time t as follows:

$$t_k = t - t_{oe} \tag{6.35}$$

In the equation, the unit of t_k is second, and the absolute value of t_k is controlled within one week, that is: if $t_k > 302{,}400$, $t_k = t_k - 604{,}800$. If $t_k < -302{,}400$, $t_k = t_k + 604{,}800$. 604,800 is the count of seconds in a week.

2. Calculate the average angular velocity of the satellite operation

This step requires the use of disturbance corrections in the navigation message. The theoretical average angular velocity of the satellite operation is

$$n_0 = \sqrt{\frac{\mu}{a^3}}$$

In the equation, μ is the gravity constant of the Earth, which can be found in Fig. 1.1; a is the semi-major axis of the ellipse, which is from the ephemeris data. At the same time, the ephemeris data also transmits the correction term Δn, and the average angular velocity finally used is $n = n_0 + \Delta n$.

2. Calculate the mean anomaly M of the satellite at time t_k

The mean anomaly and time have a linear relationship, which is

$$M = M_0 + nt_k \tag{6.36}$$

where t_k is the normalized time obtained in step 1; n is the corrected average angular velocity obtained in step 2.

4. Calculate the eccentric anomaly E of the satellite at time t_k

Use the previous step to obtain the mean anomaly, and Eq. (6.30) to list the equation

$$E = M + e_s \sin E \tag{6.37}$$

This is a transcendental equation that can be solved using an iterative method. In general, iterations within 10 times are accurate enough.

5. Calculate the geocentric diameter of the satellite

This step needs to be obtained from the previous step using Eq. (6.24)

$$r = a(1 - e_s \cos E) \tag{6.38}$$

6. Calculate the true anomaly of the satellite f at the normalized moment

Combining Equations (6.25) and (6.26), we can get

$$f = \tan^{-1}\left(\frac{\sqrt{1 - e_s^2}\sin E}{\cos E - e_s}\right) \tag{6.39}$$

Or we can use the half-angle formula to get f:

$$\tan\frac{f}{2} = \frac{\sin f}{1 + \cos f} = \frac{\sqrt{1 + e_s}}{\sqrt{1 - e_s}}\tan\frac{E}{2} \tag{6.40}$$

However, the above formula diverges at $E = 180^\circ$ and requires special handling.

7. Calculate the angular distance ϕ of the ascending node

$$\phi = f + \omega \tag{6.41}$$

where f is from the previous step; ω is the perigee angular distance of the satellite orbit from the ephemeris parameters.

8. Calculate the perturbation corrections $\delta\mu$, δr, and δi while correcting ϕ_k, r, and i

Correcting the angular ascending node

$$\delta\mu = C_{uc}\cos(2\phi) + C_{us}\sin(2\phi) \tag{6.42}$$

Correcting the satellite's ground diameter

$$\delta r = C_{rc}\cos(2\phi) + C_{rs}\sin(2\phi) \tag{6.43}$$

Correcting the satellite's orbit tilt

$$\delta i = C_{ic}\cos(2\phi) + C_{is}\sin(2\phi) \tag{6.44}$$

$\{C_{uc}, C_{us}, C_{rc}, C_{rs}, C_{ic}, C_{is}\}$ in the above equation are all from the satellite's ephemeris data, and then use these corrections to update the angular distance of the ascending node ϕ, the satellite's geocentric diameter r, and the satellite's orbit inclination i

$$\phi_k = \phi + \delta\mu \tag{6.45}$$

$$r_k = r + \delta r \tag{6.46}$$

$$i_k = i_0 + i_{\text{dot}} \cdot t_k + \delta i \tag{6.47}$$

9. Calculate the position coordinates of a satellite in an elliptical orbit Cartesian coordinate system

In the elliptical rectangular coordinate system with the center of the Earth as the origin and the long axis of the ellipse as the X axis, the position coordinates of the satellite are

$$\boldsymbol{P}_s = \begin{bmatrix} r_k \cos\phi_k \\ r_k \sin\phi_k \\ 0 \end{bmatrix} \tag{6.48}$$

10. Calculate the ascending node of the satellite's elliptical orbit Ω at the normalized moment

Due to the disturbance term, the ascending node is not a constant, but is determined by

$$\Omega_k = \Omega_e + (\dot{\Omega} - \omega_{ie})t_k - \omega_{ie} t_{oe} \tag{6.49}$$

Among them, Ω_e comes from the ephemeris data. Its meaning is not at the reference point of the ascending point of the right ascension, but the starting point from the Greenwich meridian to the satellite orbit ascending point; $\dot{\Omega}$ is the rate of change of the ascending node; $\omega_{ie} = 7.292\ 115\ 146\ 7 \times 10^{-5}$ is the angular rate of the Earth's rotation.

11. Calculate the coordinates of the satellite in the ECEF coordinate system

This step transforms the coordinates of the satellite in the orbital Cartesian coordinate system to the ECEF coordinate system.

First, we should make $x_k = r_k \cos\phi_k$, $y_k = r_k \sin\phi_k$, then

$$\boldsymbol{P}_e = \begin{bmatrix} E_x \\ E_y \\ E_z \end{bmatrix} = \boldsymbol{R}_z(-\Omega_k)\boldsymbol{R}_x(-i_k)\begin{bmatrix} x_k \\ y_k \\ 0 \end{bmatrix} \tag{6.50}$$

$$= \begin{bmatrix} x_k \cos \Omega_k - y_k \cos i_k \sin \Omega_k \\ x_k \sin \Omega_k + y_k \cos i_k \cos \Omega_k \\ y_k \sin i_k \end{bmatrix} \tag{6.51}$$

Note: The rotation matrix in Eq. (6.51) has only two rotation operations, while the previous Eq. (6.32) has three rotation matrices because the $R(-\omega)$ operation already included in calculating ϕ_k.

The above is the step of calculating the position of the satellite using ephemeris data. With the position of the satellite, its speed can be obtained by making a difference in the satellite position at an adjacent time. This method of calculating satellite speed has the advantages of being simple and perceptible, but it is necessary to calculate the position of the satellite twice to obtain the speed, and the calculation amount is relatively large. In fact, the calculation equation of the satellite speed can be derived directly from Eq. (6.51). The specific derivation steps are as follows:

First, to derive Eq. (6.51), we can get

$$\begin{bmatrix} \dot{E}_x \\ \dot{E}_y \\ \dot{E}_z \end{bmatrix} = \begin{bmatrix} \dot{x}_k \cos \Omega_k - \dot{y}_k \cos i_k \sin \Omega_k + y_k \sin i_k \sin \Omega_k \dot{i}_k - E_y \dot{\Omega}_k \\ \dot{x}_k \sin \Omega_k + \dot{y}_k \cos i_k \cos \Omega_k - y_k \sin i_k \cos \Omega_k \dot{i}_k + E_x \dot{\Omega}_k \\ \dot{y}_k \sin i_k + y_k \cos i_k \dot{i}_k \end{bmatrix} \tag{6.52}$$

For the calculation of Eq. (6.52), we need to know the expressions of $\dot{x}_k$, $\dot{y}_k$, $\dot{i}_k$, and $\dot{\Omega}_k$. To derive t_k Eqs. (6.50), (6.47), and (6.49), we can get

$$\dot{x}_k = \dot{r}_k \cos \phi_k - r_k (\sin \phi_k) \dot{\phi}_k \tag{6.53}$$

$$\dot{y}_k = \dot{r}_k \sin \phi_k + r_k (\cos \phi_k) \dot{\phi}_k \tag{6.54}$$

$$\dot{i}_k = 2[C_{is} \cos (2\phi) - C_{ic} \sin (2\phi)]\dot{\phi} + i_{\text{dot}} \tag{6.55}$$

$$\dot{\Omega}_k = \dot{\Omega} - \omega_{ie} \tag{6.56}$$

Hence, the next step is to get the expressions for $\dot{\phi}$, $\dot{\phi}_k$, and $\dot{r}_k$. The derivations of Eqs. (6.45), (6.46), and (6.41) are

$$\dot{\phi}_k = [1 + 2C_{us} \cos (2\phi) - 2C_{uc} \sin (2\phi)]\dot{\phi} \tag{6.57}$$

$$\dot{r}_k = a e_s \sin E \dot{E} + [2C_{rs} \cos(2\phi) - 2C_{rc} \sin(2\phi)]\dot{\phi} \tag{6.58}$$

$$\dot{\phi} = \dot{f} \tag{6.59}$$

To derive Eq. (6.39), we can obtain

Table 6.2 Steps for calculating a satellite's speed using ephemeris data

Steps	Calculations
1	$\dot{E} = \frac{n_0+\Delta n}{1-e_s \cos E}$
2	$\dot{\phi} = \frac{\sqrt{1-e_s^2}}{1-e_s \cos E}\dot{E}$
3	$\dot{r}_k = ae_s \sin E\dot{E} + (2C_{rs}\cos 2\phi - 2C_{rc}\sin 2\phi)\dot{\phi}$
4	$\dot{\phi}_k = (1 + 2C_{us}\cos 2\phi - 2C_{uc}\sin 2\phi)\dot{\phi}$
5	$\dot{\Omega}_k = \dot{\Omega} - \omega_{ie}$
6	$\dot{i}_k = 2(C_{is}\cos 2\phi - C_{ic}\sin 2\phi)\dot{\phi} + \text{idot}$
7	$\dot{y}_k = \dot{r}_k \sin\phi_k + r_k \cos\phi_k\dot{\phi}_k$
8	$\dot{x}_k = \dot{r}_k \cos\phi_k - r_k \sin\phi_k\dot{\phi}_k$
9	$V_x = \dot{x}_k \cos\Omega_k - \dot{y}_k \cos i_k \sin\Omega_k + y_k \sin i_k \sin\Omega_k\dot{i}_k - E_y\dot{\Omega}_k$
10	$V_y = \dot{x}_k \sin\Omega_k + \dot{y}_k \cos i_k \cos\Omega_k - y_k \sin i_k \cos\Omega_k\dot{i}_k + E_x\dot{\Omega}_k$
11	$V_z = \dot{y}_k \sin i_k + y_k \cos i_k\dot{i}_k$

$$\dot{f} = \frac{\sqrt{1-e_s^2}}{1-e_s\cos E}\dot{E} \tag{6.60}$$

Finally, we calculate the expression of $\dot{E}$. This can be obtained by simultaneously deriving and sorting around Eq. (6.37).

$$\dot{E} = \frac{n_0+\Delta n}{1-e_s\cos E} \tag{6.61}$$

The derivation process of the above formula needs to use $\dot{M} = n_0 + \Delta n$.

Reverse the above process to calculate a satellite's speed using ephemeris data, as shown in Table 6.2.

6.3 BDS GEO Satellite

Although the law of movement of BDS GEO satellites also follows Kepler's law, if we use the equations in Sect. 6.2 to calculate a satellite's position, we will encounter an unexpected problem. To understand the root cause of this problem, observe the two situations in Fig. 6.4.

Figure 6.4 shows the intersection of a satellite's orbital plane and the equatorial plane at two different orbital inclinations, while Fig. 6.4a shows a case where the orbital inclination is large, and Fig. 6.4b shows a case where the orbital inclination is small. The meaning of each indicator in the figure is the same as in Fig. 6.3, and can

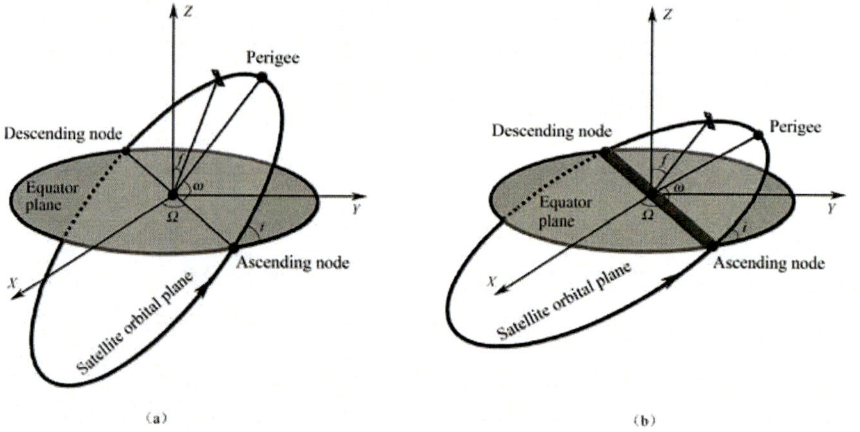

Fig. 6.4 Intersection of a satellite's orbital planes and the Earth's equatorial plane with different orbital dips

be seen as a simplified version of it. The two ends of the intersection of the equatorial plane and the orbital plane on the equator are the descending point and the ascending node. As can be seen from the previous chapters, the ascending point Ω and the perigee angular distance ω need to be determined according to the position of the ascending node. When the orbital inclination i continues to decrease, as shown in Fig. 6.4b, the specific positions of the ascending and descending points will gradually become blurred, and the connection between the two will progressively become a fuzzy band instead of a clear connection. The smaller the track inclination, the larger the blur band. The most extreme case is in the BDS GEO satellite, where the orbital inclination i is 0, and the satellite's orbital plane and the equatorial plane are parallel or nearly parallel. At this time, the position of the ascending and descending points is singular—the satellite's orbital plane and the equatorial plane. There will be no clear line of intersection.

This problem first appeared in the generation of GEO satellite ephemeris data. Because there is no clear ascending and descending point, it will be impossible to determine the ascending node Ω and the perigee angular distance ω, even if it can be difficult to follow the GPS and BDS MEO/IGSO satellites. The ephemeris format generates ephemeris data and also causes large satellite position calculation errors due to the singularity of the geometric relationship between the orbital plane and the equatorial plane.

From the root cause of the GEO satellite, it is not a problem with the satellite orbit, but an issue of how to express it. Looking further, the root of the problem is that the reference plane selection of the orbit is unreasonable. If the satellite's orbital plane is not parallel to the Earth's equatorial plane, this problem can be circumvented. Therefore, when generating the ephemeris data of the BDS GEO satellite, the GEO satellite's orbit is artificially deflected by an angle β, and the satellite can be considered to be running on this virtual orbit. The specific steps are as follows:

① Rotate the ECEF coordinate system $T_{\text{cast}} + \omega_{is}(t - t_{oe})$ around the Z axis so that the rotated ECEF coordinate system's X axis points to the vernal equinox at the reference time.
② Rotate the ECEF coordinate system around the X-axis by the angle β and then calculate the satellite's position in the new coordinate system.
③ Fit the broadcast ephemeris data of the GEO satellite according to the satellite position in the new coordinate system.

The above process is expressed by

$$\underset{k,\text{ rot}}{\vec{r}} = \boldsymbol{R}_x(\beta)\boldsymbol{R}_z[T_{\text{cast}} + \omega_{is}(t - t_{oe})]\underset{k}{\vec{r}} \tag{6.62}$$

where $\boldsymbol{r}_k$ is the position vector of the GEO satellite in the ECEF coordinate system at time t; $\boldsymbol{r}_{k,\text{rot}}$ is the position vector of the satellite in the coordinate system after the above rotation; t_{oe} is the reference time of the ephemeris fitting period; ω_{is} is the angular velocity of the Earth's rotation, T_{cast} is the time of the Greenwich star.

The broadcast ephemeris thus operated also needs to transmit the Greenwich stellar T_{cast} at the reference time when broadcasting, which is more complicated than the original 15 Kepler parameters. The data format is incompatible with the original ephemeris data format. A simplified method proposed in Reference [1] is $T_{\text{cast}} = 0$. It is equivalent to the X-axis of the new coordinate system after the completion of the first-step rotation described above coincides with the X-axis of the ECEF coordinate system at the reference time. The experiment verifies that the processing is performed on the ephemeris. The accuracy of the combination is not affected.

The selection of the rotation angle β of the satellite's orbital plane needs to be carefully considered, because the unreasonable angle will cause the loss of the fitting accuracy loss of some ephemeris parameters, mainly the correction term of the average angular velocity Δn and the rate of change of the inclination of the orbital plane i_{dot}. Reference [2] analyzes the influence of the orbital surface rotation angle on the GEO satellite's broadcast ephemeris fitting. It is proven by numerical simulation that the fitting accuracy loss of the ephemeris parameters does not exceed the limit when $\beta \geq 5°$. At present, the broadcast ephemeris of the BDS GEO satellite adopts the scheme of $\beta = 5°$.

When calculating the position of the GEO satellite at the BDS receiver, it is necessary to deal with the operation of the above-mentioned GEO satellite orbit rotation. The specific performance is to multiply a rotation matrix when calculating the satellite position, so the GEO satellite and the MEO/IGSO satellite processing is slightly different. The difference only occurs after step 9 in Sect. 6.2. The detailed steps for calculating the GEO satellite position vector are listed below.

1–9. The same as BDS MEO/IGSO satellites.
10. Calculate the ascending point of the satellite's elliptical orbit Ω_k at the normalized moment.

$$\Omega_k = \Omega_e + \dot{\Omega} t_k - \omega_{ie} t_{oe} \tag{6.63}$$

Equation (6.63) when compared with Eq. (6.49) does not have a $-\omega_{ie}t_k$ term, because the $-\omega_{ie}t_k$ angle is also rotated around the Z axis, so it can be omitted here.

11. Calculate the coordinates of the satellite in the ECEF coordinate system.

First, calculate the position of the satellite based on the Ω_k and i_k calculated in the previous steps.

$$\boldsymbol{P}_{e,rot} = \begin{bmatrix} E_{x,rot} \\ E_{y,rot} \\ E_{z,rot} \end{bmatrix} = R(-\Omega_k)R(-i)\begin{bmatrix} x_k \\ y_k \\ 0 \end{bmatrix}$$
$$= \begin{bmatrix} x_k \cos\Omega_k - y_k \cos i_k \sin\Omega_k \\ x_k \sin\Omega_k + y_k \cos i_k \cos\Omega_k \\ y_k \sin i_k \end{bmatrix} \tag{6.64}$$

However, the $P_{e,\mathrm{rot}}$ calculated by Eq. (6.64) is the position vector in the virtual ECEF coordinate system after being rotated by 5° around the X axis, so it needs to be rotated back again. That is,

$$\boldsymbol{P}_e = \boldsymbol{R}_z(\omega_{is}t_k)\boldsymbol{R}_x(-5°)\boldsymbol{P}_{e,\mathrm{rot}} \tag{6.65}$$

where $\boldsymbol{R}_z(\omega_{is}t_k) = \begin{bmatrix} \cos(\omega_{is}t_k) & \sin(\omega_{is}t_k) & 0 \\ -\sin(\omega_{is}t_k) & \cos(\omega_{is}t_k) & 0 \\ 0 & 0 & 1 \end{bmatrix}$, $\boldsymbol{R}_x(-5°) = \begin{bmatrix} 1 & 0 & 0 \\ 0 & \cos(5°) & -\sin(5°) \\ 0 & \sin(5°) & \cos(5°) \end{bmatrix}$.

If there is a $-\omega_{ie}t_k$ item in the calculation of Ω_k in the previous step 10, then Eq. (6.65) will have one more item $R_z(-\omega_{is}t_k)$.

If $\boldsymbol{R}_z(\omega_{is}t_k)\boldsymbol{R}_x(-5°)$ is recorded as a uniform rotation matrix $\boldsymbol{\Phi}(t_k)$, that is

$$\boldsymbol{\Phi}(t_k) = \boldsymbol{R}_z(\omega_{is}t_k)\boldsymbol{R}_x(-5°) \tag{6.66}$$

then Eq. (6.65) can be written as

$$\boldsymbol{P}_e = \boldsymbol{\Phi}(t_k)\boldsymbol{P}_{\mathrm{e,rot}} \tag{6.67}$$

For Eq. (6.67), the time t_k is derived. We can get

$$\boldsymbol{V}_e = \frac{\mathrm{d}\boldsymbol{\Phi}(t_k)}{\mathrm{d}t_k}\boldsymbol{P}_{e,\mathrm{rot}} + \boldsymbol{\Phi}(t_k)\boldsymbol{V}_{e,\mathrm{rot}} \tag{6.68}$$

Equation (6.68) is the speed calculation equation of the GEO satellite, where $V_{e,\mathrm{rot}}$ is calculated according to the steps in Fig. 6.2. Note that the $\dot{\Omega}_k = \dot{\Omega} - \omega_{ie}$ of step 5 is changed to $\dot{\Omega}_k = \dot{\Omega}$, and $\frac{\mathrm{d}\boldsymbol{\Phi}(t_k)}{\mathrm{d}t_k}$ is to derive the t_k in the $\Phi(t_k)$ matrix.

Expand Eq. (6.66) to get

$$\boldsymbol{\Phi}(t_k) = \boldsymbol{R}_z(\omega_{is}t_k)\boldsymbol{R}_x(-5^\circ) = \begin{bmatrix} \cos(\omega_{is}t_k) & \sin(\omega_{is}t_k)\cos(5^\circ) & -\sin(\omega_{is}t_k)\sin(5^\circ) \\ -\sin(\omega_{is}t_k) & \cos(\omega_{is}t_k)\cos(5^\circ) & -\cos(\omega_{is}t_k)\sin(5^\circ) \\ 0 & \sin(5^\circ) & \cos(5^\circ) \end{bmatrix}.$$

To derive the above equation, we can get

$$\frac{\mathrm{d}\boldsymbol{\Phi}(t_k)}{\mathrm{d}t_k} = \begin{bmatrix} -\omega_{is}\sin(\omega_{is}t_k) & \omega_{is}\cos(\omega_{is}t_k)\cos(5^\circ) & -\omega_{is}\cos(\omega_{is}t_k)\sin(5^\circ) \\ -\omega_{is}\cos(\omega_{is}t_k) & -\omega_{is}\sin(\omega_{is}t_k)\cos(5^\circ) & \omega_{is}\sin(\omega_{is}t_k)\sin(5^\circ) \\ 0 & 0 & 0 \end{bmatrix} \tag{6.69}$$

The final expression $\boldsymbol{V}_e$ can be derived by substituting Eqs. (6.69) into (6.68).

The above analysis is the special feature of the broadcast ephemeris of the BDS GEO satellite, and is how the BDS receiver uses the GEO broadcast ephemeris to calculate the position and speed of the satellite. The rotating orbit method avoids the singularity problem in the process of GEO satellite broadcasting ephemeris fitting, and also unifies the ephemeris data format and content of GEO, MEO, and IGSO satellites. The generation and playback of the calendar offers great convenience for the generation and playback of BDS satellite broadcasting ephemeris. However, the cost is that the calculation of the GEO satellite position and velocity is slightly larger.

6.4 Interpolation Calculation of Satellite Position and Velocity

The analysis in the first two sections shows that calculating the position and velocity of a satellite with ephemeris data is a rather complicated task. It takes a long time to perform heavy loads of floating-point operations, and takes up more resources in the processor. The problem is especially prominent in low-cost and low-precision portable receivers. There are ways to reduce the amount of computation, one of which is to use polynomial fitting techniques to estimate a satellite's position and velocity to avoid complex, cumbersome operations. The basic idea of this method is to calculate the satellite parameters of merely the starting point, the intermediate point, and the end point within a certain time span, and perform the estimation of the parameters for the remaining points in the time span with polynomial fitting as shown in Fig. 6.5.

In Fig. 6.5, it is assumed that the time span is T, the time of the starting point is 0, the time of the intermediate point is $T/2$, and the time of the end point is also T.

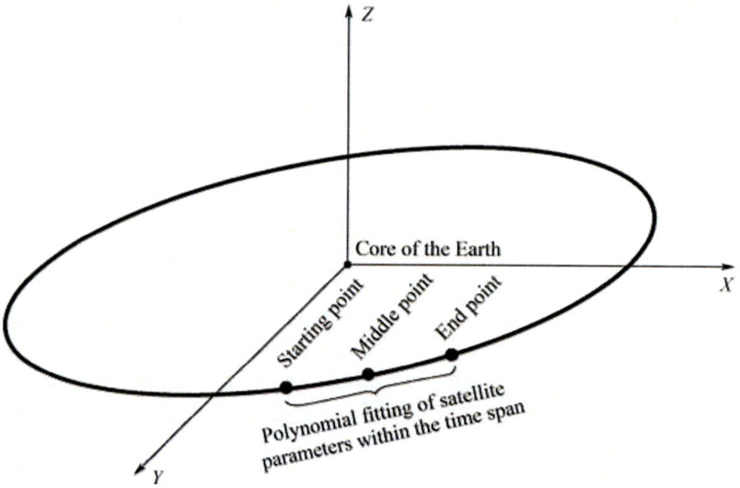

Fig. 6.5 Calculation of a satellite's position and velocity with polynomial fitting

The satellite parameters of these three points are obtained from the ephemeris data and used as three sets of known data to determine the coefficient of the polynomial. Once the polynomial coefficients are determined, the satellite parameters at other times within the time span can be easily calculated with the polynomial fit curve.

A simple rationality analysis of the method of approximate calculation is given below. If we carry out time derivation of Eq. (6.43), we can obtain

$$\begin{bmatrix} V_x \\ V_y \end{bmatrix} = \begin{bmatrix} \dot{S}_x \\ \dot{S}_y \end{bmatrix} = \begin{bmatrix} -a \sin E \cdot \dot{E} \\ a\sqrt{1-e^2} \cos E \cdot \dot{E} \end{bmatrix} \tag{6.70}$$

Equation (6.70) is the two-dimensional velocity component expression of the satellite in the unobserved orbit. Obviously, the satellite's velocity course angle ψ_{sv} can be expressed as

$$\psi_{SV} = \arctan\left(\frac{V_x}{V_y}\right) \approx -E \tag{6.71}$$

According to Eq. (6.28), we get

$$\dot{E} = \frac{n}{1 - e_s \cos E} \approx n \tag{6.72}$$

n in Eq. (6.72) is the orbital angular rate of the satellite, which is approximately 1.45×10^{-4} radians/sec for GPS satellites and BDS MEO satellites. For BDS IGSO/GEO satellites it is approximately 7.27×10^{-5} radians/second. Within 1 min, the heading angle of the satellite is only deflected by less than 0.5°, for which the trajectory of the satellite can be approximated as a straight line. Since the position is the integral of the velocity, the change of the satellite's position can also be considered as approximately linear.

In Fig. 6.5, assuming that the moments of the starting point, the intermediate point, and the end point are t_1, t_2 and t_3 respectively, then it can be defined that $\tau_1 = t_2 - t_1$, $\tau_2 = t_3 - t_1$, and generally, $\tau_2 = 2\tau_1 = T$, in which T is called the time span.

Suppose that the satellite positions at (t_1, t_2, t_3) are

$$\boldsymbol{P}_1 = \begin{bmatrix} x_1 \\ y_1 \\ z_1 \end{bmatrix}, \boldsymbol{P}_2 = \begin{bmatrix} x_2 \\ y_2 \\ z_2 \end{bmatrix}, \boldsymbol{P}_3 = \begin{bmatrix} x_3 \\ y_3 \\ z_3 \end{bmatrix}$$

we can then write

$$\begin{bmatrix} 1 & 0 & 0 \\ 1 & \tau_1 & \tau_1^2 \\ 1 & \tau_2 & \tau_2^2 \end{bmatrix} \begin{bmatrix} a_{1x} & a_{1y} & a_{1z} \\ a_{2x} & a_{2y} & a_{2z} \\ a_{3x} & a_{3y} & a_{3z} \end{bmatrix} = \begin{bmatrix} x_1 & y_1 & z_1 \\ x_2 & y_2 & z_2 \\ x_3 & y_3 & z_3 \end{bmatrix} \tag{6.73}$$

The coefficient matrix can be determined according to Eq. (6.73):

$$\boldsymbol{A} = \begin{bmatrix} a_{1x} & a_{1y} & a_{1z} \\ a_{2x} & a_{2y} & a_{2z} \\ a_{3x} & a_{3y} & a_{3z} \end{bmatrix} = \begin{bmatrix} 1 & 0 & 0 \\ 1 & \tau_1 & \tau_1^2 \\ 1 & \tau_2 & \tau_2^2 \end{bmatrix}^{-1} \begin{bmatrix} x_1 & y_1 & z_1 \\ x_2 & y_2 & z_2 \\ x_3 & y_3 & z_3 \end{bmatrix} \tag{6.74}$$

Once the coefficient matrix is $\boldsymbol{A}$ obtained, the satellite coordinates at any moment t between $[t_1, t_3]$ can be obtained from the following equation:

$$\begin{bmatrix} x & y & z \end{bmatrix} = \begin{bmatrix} 1 & (t - t_1) & (t - t_1)^2 \end{bmatrix} \boldsymbol{A} \tag{6.75}$$

Compared with conventional methods for calculating a satellite's position using ephemeris data, Eq. (6.75) indicates that the method of polynomial fitting can greatly reduce the computation load.

Since a satellite's orbit is a second-order curve, a desirable result can be achieved through second-order polynomial fitting, which is why it is sufficient to select only three satellite parameters at known times. The trade-off between the computational cost and the fitting accuracy must be considered in the selection of the time span T. The larger T is, the smaller the computational cost, but the worse the fitting accuracy will be.

Figure 6.6 depicts the GPS satellite's position error calculated through polynomial fitting and ephemeris calculation for a period of time when $T = 20$ s and $T = 60$ s. The figure shows that when $T = 20$ s, the error between the satellite's position after fitting and that calculated using the ephemeris are in millimeters. When $T = 60$ s, the position error is at the decimeter level. So, for general applications, a time span of 20 s is a good choice.

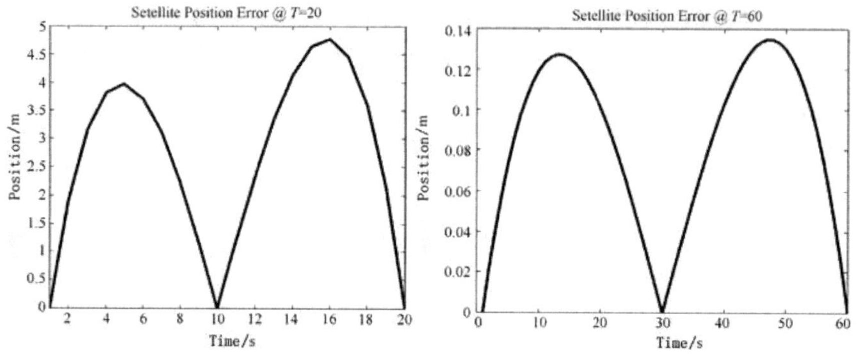

Fig. 6.6 GPS satellite position error in the polynomial fitting method when $T = 20$ and $T = 60$

What is described above is a method of fitting by second-order curve, and other alternative methods include Lagrangian polynomial interpolation, Chebyshev polynomial interpolation and linear successive interpolation, etc. The determination of a specific method requires a measurement between interpolation precision and computational complexity.

6.5 Precise Ephemeris and Ephemeris Expansion

In addition to the broadcast ephemeris data broadcasted by GPS satellites, the International GNSS Service (IGS) also provides precision ephemeris with higher precision. The predecessor of IGS is the International GPS Service (IGS). Established in 1993, it is an international organization of more than 200 research institutions in more than 80 countries and regions around the world, and provides free GPS satellite orbit and clock data with centimeter-level accuracy for engineers and research institutes in satellite navigation and related fields. IGS has more than 350 continuously operating GPS and GLONASS satellite tracking stations around the world. It is responsible for sending raw observations from the pseudo-range and carrier phases of GPS and GLONASS satellites to the IGS Analysis Center Coordinator (ACC) to generate three precise ephemerides: IGS Final, IGS Rapid, and IGS Ultra-Rapid, which are referred to as IGS, IGR, and IGU internationally.

The accuracy of the IGS Final is the highest, with an orbital position error of less than 5 cm, a satellite clock error of less than 0.1 ns, and an update rate of one week. However, the data delay is relatively long at 13 days, meaning that it will take two weeks to obtain the weekly IGS Final data. Longer delays limit the application of IGS Final ephemeris, for example, in situations where real-time processing or high-efficiency applications are required.

The accuracy of the IGR ephemeris data is basically the same as that of the IGS Final, i.e. the orbital position error and the satellite clock error are 5 cm and 0.1 ns respectively. In general applications, the difference in accuracy between the two will not be noticed. The update rate of IGR is one day, and the data delay is 17 h. The IGR ephemeris data of the day can be obtained at 17:00 the next day, which can greatly improve the timeliness of data processing compared with IGS Final.

The IGU ephemeris data is divided into two parts: the real-time satellite ephemeris, and the predicted satellite ephemeris. The length of each is 24 h. The real-time satellite ephemeris needs to be delayed by 3 h. It has an orbital position error of 5 cm and a satellite clock error of 0.2 ns. Based on the satellite ephemeris of the first 24 h, the one for the next 24 h can be predicted. Therefore, the predicted satellite ephemeris is very suitable for applications that require real-time processing, but the trade-off is relatively large errors, in which the satellite position error is 10 cm and the satellite clock error is about 5 ns.

Unlike the broadcast ephemeris, the precise Ephemeris of IGS is not represented by 15 Kepler parameters, but by satellite coordinates at certain time intervals. The satellite coordinate framework is the ITRF framework provided by the International Earth Rotation Organization. The WGS-84 coordinate systems adopted by ITRF and GPS are somewhat different in concept, but in fact, the difference between the coordinates of the same site is very small. The time intervals of IGS Final, IGR, and IGU are all 15 min, and the satellite coordinates within adjacent time intervals can be obtained by interpolation or fitting. The Lagrangian difference method is taken as an example below to explain how to get the satellite coordinates at any moment.

Suppose that the satellite positions $S(t_i)$ of N moments are obtained from the IGS precision ephemeris at moments of t_i, $i = 1, \ldots, N$, then the satellite position $S(t)$ at t can be written as

$$S(t) = \sum_{i=1}^{N} \alpha_i(t) S(t_i) \tag{6.76}$$

where

$$\alpha_i(t) = \prod_{\substack{j=1 \\ i \neq j}}^{N} \frac{t - t_j}{t_i - t_j} \tag{6.77}$$

Generally, $t \in [t_1, t_N]$ and t are set at the middle of $[t_1, t_N]$.

If the main purpose of the precision ephemeris is to improve the positioning accuracy of the broadcast ephemeris, then the ephemeris expansion can be considered as a supplement to the broadcast ephemeris in an effective time span. Ephemeris Extension (EE), also known as long-term ephemeris or long-term satellite orbits, is also essentially a satellite orbital parameter similar to a broadcast ephemeris, yet its effective time is much longer than that of the broadcast ephemeris. It can generally be valid for 1 day, 1 week, or even longer, forming a sharp contrast with that of the broadcast ephemeris, which is merely 2–4 h.

According to the principle of satellite navigation and positioning, the success of a positioning solution has two basic conditions:

1. Sufficient pseudo-range observations;
2. A satellite's position as a spatial reference point.

The broadcast ephemeris broadcasted by a satellite is used to calculate the position of the satellite, but it is often impossible to demodulate the broadcast ephemeris when the satellite signal is blocked or disturbed by the environment. It can be seen from the text format of GPS and BDS satellite that the entire set of broadcast ephemeris data is sent once every 30 s. The broadcast ephemeris cannot be demodulated if a serious error appears in any bit in the ephemeris message, in which case there will be a need to wait for another 30 s. The consequence is a prolonged positioning time or even a failure of positioning. Besides, since the broadcast ephemeris is only valid for 2 to 4 h, new ephemeris data must be demodulated once the receiver's working time exceeds the validity period of the broadcast ephemeris.

The main purpose of the ephemeris expansion is to compensate for the short-lived defects of the broadcast ephemeris. Because the ephemeris extension is usually valid for a few days or weeks, if an ephemeris expansion is achieved in the receiver at the current time, there is no need to demodulate the ephemeris data in the next few days or weeks. What is required are just enough pseudo-range observations for the positioning solution to begin immediately, which can greatly shorten the positioning time and enhance user experience. Moreover, in general, the minimum signal strength required for reliable demodulation of teletext data is much higher than the minimum signal strength required to obtain pseudo-range observations, so the use of ephemeris extension can also improve the positioning sensitivity of the receiver. From the above two aspects, the adoption of ephemeris expansion has contributed tremendously to improving the performance of the receiver.

Although the ephemeris extension is longer than the broadcast ephemeris in terms of validity, its positioning accuracy is not as good, for which it initially appeared in the navigational receiver rather than the professional ones. On the one hand, the professional surveying and mapping receivers usually work in a good environment such as the open sky and unobstructed areas, thus there is no problem with the ephemeris data being demodulated. On the other hand, the requirements of the professional surveying and mapping receivers for positioning accuracy is much higher than that for the navigation receiver. Therefore, ephemeris extensions are more widely accepted and applied in navigation receivers.

Ephemeris expansion appeared with the A-GPS architecture. Since ephemeris expansion is often generated by a third-party agency or organization, it can only be obtained through an appropriate communication link in the receiver. For the A-GPS system architecture, a communication link is required to get the auxiliary information, so receivers with the A-GPS function are equipped with the physical media to achieve ephemeris extension, especially receivers that are powered up occasionally. Because the working time is short, it is impossible for them to continuously demodulate the navigation message. Meanwhile, after the ephemeris extension is obtained through the communication link, it is possible to achieve positioning in a short time. A typical example is the E911-enabled mobile phone positioning required by the US FCC.

The broadcast ephemeris is the real-time ephemeris data, with which the satellites' position currently or within 2 h in the future can be calculated. Meanwhile, the ephemeris expansion can be regarded as "future" ephemeris data, which can be used to calculate the satellite position in the future for a longer time span. According to Reference [3], there are three ways that ephemeris extension can be generated:

1. Obtaining it solely from the reference network;
2. Obtaining it from the broadcast ephemeris generated by the reference network and self-demodulated by the receiver;
3. Obtaining it solely from the broadcast ephemeris self-demodulated by the receiver.

The first method requires reference network nodes all over the world to record the broadcast ephemeris of the GPS satellite over 24 h before the ephemeris extension can be calculated through a certain algorithm and finally sent to the receiver through the network. The third method does not require the participation of any reference network. The receiver itself can demodulate the broadcast ephemeris of the GPS satellite by itself, and then calculate the ephemeris extension through a certain algorithm. The second method is a combination of the first and the third. When the broadcast ephemeris can be demodulated, it can be used to test or correct the ephemeris expansion. When the signal strength is not strong enough to demodulate the broadcast ephemeris, the ephemeris expansion is used to achieve satellite positioning. In general, the extended ephemeris obtained through the first method is highest in accuracy, but it needs the support of reference networks all over the world. The accuracy of the extended ephemeris obtained in the third method is limited, and the validity is also much shorter than the first one. Another setback of the third method is that it is impossible to acquire the extended ephemeris of all orbiting satellites, because the receiver at a certain location on Earth can only receive signals from orbiting satellites that transit through the zenith. Therefore, it cannot demodulate the navigation messages from all satellites in the space constellation. The second method combines the advantages of the first and third, but the requirements for the receiver are higher, not only in terms of the hardware and software resources and the reference network communication, but also the function of correcting the ephemeris expansion through the broadcast ephemeris.

It is worth mentioning that there is no clear boundary for the expiration date of the ephemeris expansion, because the accuracy of the position of the satellite calculated with the ephemeris expansion will gradually deteriorate over time, and it is difficult to say whether the ephemeris extension will lose its accuracy when the position error of the satellite exceeds 10 m or 100 m. It depends on the acceptable range of error for positional accuracy for different applications. Figure 6.7 shows the satellite position error and clock error in a function of time based on one type of ephemeris extension. The ordinate is in meters and the abscissa is in days. The ephemeris expansion is calculated by a certain algorithm based on the two-month

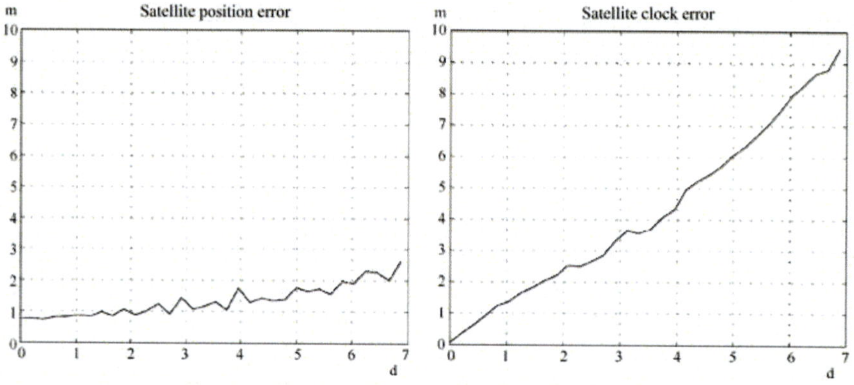

Fig. 6.7 Satellite positioning and clock error calculated by an ephemeris extension

ephemeris data in Reference [3]. It should be noted that the curve in Fig. 6.7 is only to illustrate the trend of ephemeris expansion error with time. The specific value is not very significant, because the ephemeris expansion obtained by different algorithms will give different error curves. Even the same ephemeris expansion algorithm based on satellite data of different time spans will also obtain different error curves.

There are many ways to calculate ephemeris extensions, which can be condensed into two broad categories. The first is to establish a satellite dynamics equation, where the gravitational term without the orbit and the perturbation terms are considered as the force of the satellite. This means that the motion trajectory of the satellite is affected by the following equation.

$$\ddot{\boldsymbol{r}} = -\frac{\mu}{r^2}\frac{\boldsymbol{r}}{r} + \boldsymbol{a}_{\text{earth}} + \boldsymbol{a}_{\text{sun}} + \boldsymbol{a}_{\text{moon}} + \boldsymbol{a}_{\text{solar}} + \boldsymbol{a}_{\text{tide}} + \boldsymbol{a}_{\text{atm}} + \boldsymbol{a}_{\text{other}} \tag{6.78}$$

where $\boldsymbol{a}_{\text{earth}}$ is the perturbation caused by the irregular shape of the Earth, $\boldsymbol{a}_{\text{sun}}$ is the solar gravitation, $\boldsymbol{a}_{\text{moon}}$ is the moon gravitation, $\boldsymbol{a}_{\text{solar}}$ is the solar pressure, $\boldsymbol{a}_{\text{tide}}$ is the tidal force, $\boldsymbol{a}_{\text{atm}}$ is the atmospheric drag effect, and $\boldsymbol{a}_{\text{other}}$ represents other forces.

In this method, the satellite's orbit model is established first, which mainly refers to the perturbation models mentioned above. Then, the models are solved according to the satellite's position and velocity information over a period of time to generate the functional relationship of each perturbation with changes in time and $\boldsymbol{r}$. According to Eq. (6.78), aside from the gravity of the Earth, the satellite is also affected by many perturbations, among which the one caused by the Earth's irregular shape (a non-standard sphere) $\boldsymbol{a}_{\text{earth}}$ is the greatest. The other perturbations are not as large, yet the long-term effect cannot be ignored. Further theoretical analysis shows that due to the influence of perturbations, the satellite's ascending node gradually moves westward on the Equator, the orbital perigee is also moving, and the angular velocity of the satellite is also slowly increasing. Some of these perturbations can be represented by accurate mathematical models, and some only by relatively accurate empirical equations.

Once the perturbation models are established, the respective perturbation items at a certain moment in the future can be integrated according to the reference time to obtain the satellite's position and velocity at a certain moment in the future:

$$\boldsymbol{r}(t_r) = \boldsymbol{r}(0) + \int_{t_0}^{t_r} \left[\dot{\boldsymbol{r}}(t) + \int_{t_0}^{t} a[t, \boldsymbol{r}(\tau)]\mathrm{d}\tau \right] \mathrm{d}t \tag{6.79}$$

where a $[t, \boldsymbol{r}(\tau)]$, a function of time and satellite position, represents the sum of forces acting on the satellite at the right side of Eq. (6.78). The equation shows that the calculation process is complicated and cumbersome, and needs a CPU with a strong processing capability. Representative examples of this method include GIPSY software from JPL Laboratories, GEODYN software from NASA, and Micro-Cosm software from Van Martin Systems.

In the second method, it is not necessary to establish the satellite dynamics equation as in Eq. (6.79). Instead, the data format of the broadcast ephemeris is still used, i.e. the Kepler parameter $\boldsymbol{X} = (\sqrt{a}, e_s, \omega, M_0, \Omega_e, i_0, \dot{\Omega}, \Delta n, i_{\mathrm{dot}}, C_{uc}, C_{us}, C_{rc}, C_{rs})$, is used to represent the satellite orbit. The satellite position can then be written as a function of the Kepler parameter and time:

$$\boldsymbol{r}(t) = \boldsymbol{f}(t, \sqrt{a}, e_s, \omega, M_0, \Omega_0, i_0, \dot{\Omega}, \Delta n, i_{dot}, C_{us}, C_{uc}, C_{rs}, C_{rc}, C_{is}, C_{ic}) \tag{6.80}$$

Based on the known position of a satellite during a certain time period, the residual vector of the satellite's position can be obtained by calculating the positions of different epochs $\hat{\vec{r}}(t)$ according to the initial Kepler parameter X_0:

$$\Delta\boldsymbol{r}(t) = \tilde{\boldsymbol{r}}(t) - \hat{\boldsymbol{r}}(t) \approx \sum_{i=1}^{M} \frac{\partial \boldsymbol{f}}{\partial x_i} \Delta x_i + \text{h.o.t. (high phase)} \tag{6.81}$$

The Taylor series expansion on the position residual is performed in Eq. (6.81), where only the first-order approximation is given. x_i is the Kepler parameter term, and M is the number of the Kepler parameter argument. Based on the satellite position residual $\Delta\boldsymbol{r}(t)$ during a certain epoch of Eq. (6.81), the correction amount of the initial Kepler parameter can be obtained through the least-squares method. Thereby, the ephemeris extension data can be obtained, and the process is as shown in Eqs. (6.82) and (6.83):

$$\Delta\boldsymbol{x} = (\boldsymbol{B}^{\mathrm{T}}\boldsymbol{B})^{-1}\boldsymbol{B}^{\mathrm{T}}\Delta\boldsymbol{r}(t) \tag{6.82}$$

$$\boldsymbol{X} = \boldsymbol{X}_0 + \Delta\boldsymbol{x} \tag{6.83}$$

where $\boldsymbol{B}$ is the matrix of partial derivatives of $\boldsymbol{f}(t, X)$ to the Kepler parameter term in different epochs. It should be noted that the result of $\boldsymbol{f}(t, X)$ is a three-dimensional vector representing the coordinates of a satellite's position on the X axis, Y axis, and Z axis respectively. Hence, $f(t, X)$ in one epoch can provide three coefficient vectors, and the expression of $\boldsymbol{B}$ is as follows:

$$\boldsymbol{B} = \begin{bmatrix} \left.\frac{\partial f(t_0, X)}{\partial x_1}\right|_{X=X_0} & \cdots & \left.\frac{\partial f(t_0, X)}{\partial x_M}\right|_{X=X_0} \\ \vdots & & \vdots \\ \left.\frac{\partial f(t_N, X)}{\partial x_1}\right|_{X=X_0} & \cdots & \left.\frac{\partial f(t_N, X)}{\partial x_M}\right|_{X=X_0} \end{bmatrix} \tag{6.84}$$

M in Eq. (6.84) indicates the number of independent variables of the Kepler parameter, which may be 15, or lower if some less influential disturbance parameters are omitted. N is the number of epochs. According to the above analysis, $\boldsymbol{B}$ in Eq. (6.84) will be a $3N \times M$ matrix. Of course, what is described here is merely the basic principles of ephemeris expansion generation. In real situations, there are many more factors to consider. Interested readers can refer to related literature or patent documents.

The residual vector of a satellite's position can be calculated through a set series of initial parameters of the satellite orbit X_0. Based on X_0, the position of the satellites in various epochs can be obtained. The position of the reference satellite can be selected from the positions of the satellite obtained from the broadcast ephemeris, or from the satellite ephemeris provided by the IGS precision ephemeris. The advantage of using the IGS precision ephemeris to obtain the reference satellite position is that the accuracy of the satellite position is high. However, the method is only applicable for GPS satellites at present, as there is no formal and reliable supply channels of precision ephemeris for the BDS satellites yet. Besides, uninterrupted internet connection is also a necessity. The effectiveness of the broadcast ephemeris should be guaranteed by reference stations all over the world so as to ensure the demodulation of the broadcast ephemeris on all GPS and BDS satellites in the full cycle. Reference [3] shows that with this scheme, the satellite position error within 7 days will be on the order of a few meters, and the pseudo-range observation error calculated thereby will be within 10 m.

References

1. Changxu G (2010) GPS: theory, algorithms and applications (2nd Edn). Springer, Berlin
2. Xianqiang C, Yuanxi Y, Xianbing Wu (2012) The impact of orbit rotation on the fitting of GEO satellite broadcast ephemeris. J Astronaut 33(5):590–596
3. Spiker J (1996) "GPS signal structure and theoretical performance." In: Parkinson IB, Spiker J, Axelrad P, Enge P (1996) Global positioning system: theory and applications
4. Navstar GPS Space Segment/Navigation User Interfaces (2012) IS-GPS-200G, 5 Sept 2012
5. China Satellite Navigation System Management Office (2013) Beidou satellite navigation system space signal interface control document (Open Service Signal) Version 2.0. Dec 2013
6. Tsui JBY (2008) Fundamentals of global positioning receivers: a software approach, 2nd Edn. Wiley

7. Misra P, Enge P (2008) GPS signal, measurement and performance (Second Edition). Electronic Industry Press Apr 2008
8. Yudong G, Xiaoning Xi, Wei W (2007) The Algorithm design of improved fitting for broadcast ephemeris of GEO navigation satellite. J National Univ Defense Sci Technol 29(5):18–22
9. Rengui R, Xiaolin J, Xianbing W et al. (2011) A discussion on the fitting of the broadcast ephemeris of the regional stationary orbit navigation satellite with the coordinate rotation method. J Geomatics, 40:145–150
10. FengH, Gang W, Li L, Cheng CL et al. (2011) The fitting and experimental analysis of broadcast ephemeris of the earth stationary orbit satellites. J Geomatics 40(5)
11. Guangming L, Ying L, Yuanlan W, Liwei Z (2008) Research on the parameter fitting algorithm of the broadcast ephemeris of the navigation satellite. J Nat Univ Defense Sci Technol 30(3)
12. Cheng CL, Bo T (2006) The impact of reference system selection on the fitting accuracy of kepler broadcast ephemeris. J Airc Meas Control 25(4):19–25
13. Van Diggelen F (2009) A-GPS: assisted GPS, GNSS, and SBAS. Artech House
14. Shouxin Z (1996) The theory and application of gps satellite surveying and positioning. National University of Defense Science and Technology Press, Beijing
15. Qifeng X (2001) Space geodesy: satellite navigation and precise positioning. PLA Press, Beijing
16. Wang H-S (2012) GPS ephemeris extension using method of averaging. Recent Pat Space Technol 2(2)
17. DLM Warren, Raquet JF (2002) Broadcast versus precise GPS ephemerides: a historical perspective. In: Proceedings of the 2002 national technical meeting of the institute of navigation, San Diego, CA, 28–30 Jan 2002
18. Kouba J (2009) Geodetic survey division (2009). In: A guide to using international GNSS service (IGS) products. https://acc.igs.org/UsingIGSProductsVer21.pdf
19. IGS Central Organization website. https://www.igs.org/components/prods.html
20. Zhenghang Li, Jinsong H (2005) GPS Survey and data processing. Wuhan University Press, Wuhan
21. Zhang J, Zhang K, Grenfell R, Deakin R (2006) GPS satellite velocity and acceleration determination using the broadcast ephemeris. J Navig 59(02):293–305
22. Shiqiang H, Bin W, Junping C (2011) Using IGS ephemeris to forecast GPS satellite orbit. Annu J Shanghai Observatory, Chin Acad Sci (32)
23. Seppänen M, Perälä T, Piché R (2011) Autonomous satellite orbit prediction
24. Garin L (2012) Ephemeris extension method for GNSS applications, U.S. Patent 8, 274,430 B2, 25 Sep 2012
25. Mcburney P, I'Qezaei S (2011) Long term compact satellite models. U.S. Patent 2011/ 0234456 A1, 29 Sep 2011
26. Yijun C (2005) A GPS satellite orbit analysis based on IGS precise ephemeris. Master Thesis of Wuhan University
27. van Diggelen F, Abraham C, LaMance J (2001) Method and apparatus for generating and distributing satellite tracking information in a compact format. U.S. Patent 6,651,000, 25 July 2001
28. Weng CT, Chien YC, Fu CL, Yau WG, Tsai YJ (2009) A broadcast ephemeris extension method for standalone mobile apparatus. In: Proceedings of the 22nd international technical meeting of the satellite division of the institute of navigation (ION GNSS 2009), 22–25 Sept 2009

Chapter 7
Position, Velocity, and Time Solutions

In this chapter, a detailed explain will be given of how to use the pseudo-range, carrier phase, and Doppler observations discussed in Chap. 5 to calculate the position, velocity, and time deviation of the receiver. This chapter will be divided into two parts. The first deals with calculating the user position, velocity, and time difference using the least-squares solution (PVT solution). The second part explains how to use the Kalman filter to perform the PVT solution. A conventional solution, the least-squares method demonstrates its significance both in reality and in theory, and is also fundamental for understanding the Kalman filter. After the PVT solution is completed through the least-squares method, it is possible to obtain some important information in the receiver, such as the geometric precision factor and the elevation angle of the satellite. This is explained alongside the least-squares method, so that readers can develop a holistic concept of the navigation algorithm inside the receiver. The Kalman filtering solution is relatively difficult to understand, as it necessitates abundant background knowledge of the signal estimation theory and adaptive filtering theory. However, it is an indispensable part of modern receiver design, so a large portion of the second part is given over to explaining it based on the least-squares method. However, since this book is not centered on the theories of signal estimation and adaptive filtering, the explanation of relevant theories will not be comprehensive. Interested readers can refer to other accredited literature in this area.

In this chapter, the mathematical models for BDS and GPS dual-mode observations will be integrated in order to analyze the algorithm of dual-mode joint positioning and compare it to the single-mode positioning algorithm, in order to analyze the performance advantages as well as the problems of the former. The least-squares and Kalman filter algorithms need to be adjusted according to the characteristics of dual-mode observations. Since single-mode and dual-mode observations coexist, a positioning algorithm that is compatible with both must be applied in the actual implementation of the BDS/GPS dual-mode receiver.

Y. Lu, *BDS/GPS Dual-Mode Software Receiver*, Navigation: Science and Technology 10, https://doi.org/10.1007/978-981-16-1075-2_7

This chapter offers some information about the stochastic process and state parameter estimation. Since readers' knowledge of these topics will vary, the basics of parameter estimation and the stochastic process will be explained before an in-depth discussion of the positioning algorithm. Interested readers can refer to literature on state parameter estimation and signal estimation theory for a better understanding.

7.1 The Least-Squares Solution

7.1.1 Basic Principles

Let's consider that the state observation equation containing the noise term is as follows:

$$\boldsymbol{Y} = \boldsymbol{A}\boldsymbol{x} + \boldsymbol{n} \tag{7.1}$$

where $\boldsymbol{x}$ is the quantity of the system state to be estimated. $\boldsymbol{x} \in \mathbb{R}^k$, in which $\mathbb{R}^k$ represents the k-dimensional vector space and k is the number of system state variables. Although $\boldsymbol{x}$ cannot be directly observed, we can observe Y and use $\tilde{Y}$ to represent the actual observation. A is a systematic observation matrix that reflects the conversion from state quantity to observation measurement. Specifically, each row vector of A converts the state variable to an observation. It is inevitable that noise will be included in the observation. In Eq. (7.1), $\boldsymbol{n}$ is used to represent the noise vector in the observation. Suppose that m observations can be obtained, then Y and n are a matrix (vector) of $m \times 1$, and A is a matrix of $m \times k$.

The question now is how to estimate $\boldsymbol{x}$ in an optimal way according to observation $\tilde{Y}$, which raises another issue: how do we define "optimal"?

Assuming that the estimated value of state x is $\hat{\boldsymbol{x}}$, the cost function $J(\hat{\boldsymbol{x}})$ can be defined as follows:

$$J(\hat{\boldsymbol{x}}) = (\tilde{\boldsymbol{Y}} - \boldsymbol{A}\hat{\boldsymbol{x}})^{\mathrm{T}}(\tilde{\boldsymbol{Y}} - \boldsymbol{A}\hat{\boldsymbol{x}}) \tag{7.2}$$

The cost function $J(\hat{\boldsymbol{x}})$ in Eq. (7.2) can be understood like this: as mentioned earlier, each line of $\boldsymbol{A}$ converts a state variable into an observation. Because $\hat{\boldsymbol{x}}$ is an estimate of the state quantity of the system, each line of $\boldsymbol{A}$ will convert $\hat{\boldsymbol{x}}$ to the estimate of an observation. If the actual observation $\tilde{\boldsymbol{Y}} = [\tilde{y}_1, \tilde{y}_2, \cdots, \tilde{y}_m]^{\mathrm{T}}$, then $\boldsymbol{A}\hat{\boldsymbol{x}} = [\hat{y}_1, \hat{y}_2, \cdots, \hat{y}_m]^{\mathrm{T}}$. Obviously, $\boldsymbol{A}\hat{\boldsymbol{x}}$ is the estimated value of Y in Eq. (7.1), $(\tilde{\boldsymbol{Y}} - \boldsymbol{A}\hat{\boldsymbol{x}})$ is the difference between the actual observation and the estimated observation, so $J(\hat{\boldsymbol{x}})$ is the square of the 2-norm of $(\tilde{\boldsymbol{Y}} - \boldsymbol{A}\hat{\boldsymbol{x}})$. Obviously $J(\hat{\boldsymbol{x}})$ is a quadratic form of $\hat{\boldsymbol{x}}$, which measures the difference between the measured observations and the estimated observations. When $(\tilde{\boldsymbol{Y}} - \boldsymbol{A}\hat{\boldsymbol{x}})$ is an 0 vector, $J(\hat{\boldsymbol{x}}) = 0$, indicating that a perfect estimate of $\boldsymbol{x}$ is achieved. With the definition of the cost function, we can call $\hat{\boldsymbol{x}}$ (which minimizes the cost function) "optimal". In this sense, the technique is defined as "least-squares".

If we expand Eq. (7.2), then:

$$\begin{aligned} J(\hat{\boldsymbol{x}}) &= \tilde{\boldsymbol{Y}}^{\mathrm{T}}\tilde{\boldsymbol{Y}} - \hat{\boldsymbol{x}}^{\mathrm{T}}\boldsymbol{A}^{\mathrm{T}}\tilde{\boldsymbol{Y}} - \tilde{\boldsymbol{Y}}^{\mathrm{T}}\boldsymbol{A}\hat{\boldsymbol{x}} + \hat{\boldsymbol{x}}^{\mathrm{T}}\boldsymbol{A}^{\mathrm{T}}\boldsymbol{A}\hat{\boldsymbol{x}} \\ &= \tilde{\boldsymbol{Y}}^{\mathrm{T}}\tilde{\boldsymbol{Y}} - 2\tilde{\boldsymbol{Y}}^{\mathrm{T}}\boldsymbol{A}\hat{\boldsymbol{x}} + \hat{\boldsymbol{x}}^{\mathrm{T}}\boldsymbol{A}^{\mathrm{T}}\boldsymbol{A}\hat{\boldsymbol{x}} \end{aligned} \tag{7.3}$$

The second step of Eq. (7.3) occurs because $\hat{\boldsymbol{x}}^{\mathrm{T}}\boldsymbol{A}^{\mathrm{T}}\tilde{\boldsymbol{Y}}$ and $\tilde{\boldsymbol{Y}}^{\mathrm{T}}\boldsymbol{A}\hat{\boldsymbol{x}}$ are both scalars, and are equal.

Then, we can derive $\hat{\boldsymbol{x}}$ from Eq. (7.3). According to the definition of vector derivation in Appendix A,

$$\frac{\partial J(\hat{\boldsymbol{x}})}{\partial \hat{\boldsymbol{x}}} = -2\boldsymbol{A}^{\mathrm{T}}\tilde{\boldsymbol{Y}} + 2\boldsymbol{A}^{\mathrm{T}}\boldsymbol{A}\hat{\boldsymbol{x}} \tag{7.4}$$

Readers can refer to the definition of vector derivation in the scalar function in Section A.5 of Appendix A to derive the above equation.

The extreme point of $J(\hat{\boldsymbol{x}})$ can be obtained when $\frac{\partial J(\hat{\boldsymbol{x}})}{\partial \hat{\boldsymbol{x}}} = 0$, i.e.

$$-\boldsymbol{A}^{\mathrm{T}}\tilde{\boldsymbol{Y}} + \boldsymbol{A}^{\mathrm{T}}\boldsymbol{A}\hat{\boldsymbol{x}} = 0 \tag{7.5}$$

If we solve Eq. (7.5), then

$$\hat{\boldsymbol{x}} = (\boldsymbol{A}^{\mathrm{T}}\boldsymbol{A})^{-1}\boldsymbol{A}^{\mathrm{T}}\tilde{\boldsymbol{Y}} \tag{7.6}$$

Since $\frac{\partial^2 J(\hat{\boldsymbol{x}})}{\partial \hat{\boldsymbol{x}}^2} = 2\boldsymbol{A}^{\mathrm{T}}\boldsymbol{A}$ is a positive definite matrix, the value of $J(\hat{\boldsymbol{x}})$ must reach the minimum at this time. Equation (7.6) is the least-squares estimate of the state parameter, referred to as the LS estimation or the LS solution.

It should be noted that the least-squares solution in Eq. (7.6) is only applicable when matrix $(\boldsymbol{A}^{\mathrm{T}}\boldsymbol{A})$ is reversible, which means that at least k of the m row vectors in A are linearly independent, i.e. $\mathrm{rank}(\boldsymbol{A}) = k$, where $\mathrm{rank}(A)$ is the rank of the A matrix, or in other words, the number of vectors in the maximum linearly independent groups of row vectors or column vectors in the A matrix. In general applications, usually $m > k$, so this condition can be met. If $m < k$, the number of observations in Eq. (7.1) will be less than the number of system state variables, and Eq. (7.1) will be an indefinite (or underdetermined) equation, so the least-squares method will no longer be applicable.

When the estimate of $\boldsymbol{x}$ is $\hat{\boldsymbol{x}}$, the estimation error $\delta\boldsymbol{x}$ of the state quantity can be defined as:

$$\delta\boldsymbol{x} = \boldsymbol{x} - \hat{\boldsymbol{x}} \tag{7.7}$$

If we substitute Eq. (7.6) into Eq. (7.7), then:

$$\begin{aligned}\delta \boldsymbol{x} &= \boldsymbol{x} - (\boldsymbol{A}^{\mathrm{T}}\boldsymbol{A})^{-1}\boldsymbol{A}^{\mathrm{T}}\tilde{\boldsymbol{Y}} \\ &= \boldsymbol{x} - (\boldsymbol{A}^{\mathrm{T}}\boldsymbol{A})^{-1}\boldsymbol{A}^{\mathrm{T}}(\boldsymbol{A}\boldsymbol{x} + \boldsymbol{n}) \\ &= \boldsymbol{x} - \boldsymbol{x} - (\boldsymbol{A}^{\mathrm{T}}\boldsymbol{A})^{-1}\boldsymbol{A}^{\mathrm{T}}\boldsymbol{n} \\ &= -(\boldsymbol{A}^{\mathrm{T}}\boldsymbol{A})^{-1}\boldsymbol{A}^{\mathrm{T}}\boldsymbol{n}\end{aligned} \tag{7.8}$$

Equation (7.8) shows that the estimation error $\delta\boldsymbol{x}$ of the system state vector at this time is independent of the system state quantity, and is only affected by the observed noise vector n.

If the mean value of the noise vector n is 0, i.e. $E\{\boldsymbol{n}\} = \boldsymbol{0}$, then:

$$E\{\delta\boldsymbol{x}\} = -(\boldsymbol{A}^{\mathrm{T}}\boldsymbol{A})^{-1}\boldsymbol{A}^{\mathrm{T}}E\{\boldsymbol{n}\} = \boldsymbol{0} \tag{7.9}$$

Equation (7.9) shows that the LS estimation is unbiased.

The co-variance matrix of $\delta\boldsymbol{x}$ can be calculated according to Eq. (7.8):

$$\begin{aligned}\mathrm{var}(\delta\boldsymbol{x}) &= E\{(\delta\boldsymbol{x})(\delta\boldsymbol{x})^{\mathrm{T}}\} \\ &= (\boldsymbol{A}^{\mathrm{T}}\boldsymbol{A})^{-1}\boldsymbol{A}^{\mathrm{T}}E\{\boldsymbol{n}\boldsymbol{n}^{\mathrm{T}}\}\boldsymbol{A}(\boldsymbol{A}^{\mathrm{T}}\boldsymbol{A})^{-1} \\ &= (\boldsymbol{A}^{\mathrm{T}}\boldsymbol{A})^{-1}\boldsymbol{A}^{\mathrm{T}}\boldsymbol{R}\boldsymbol{A}(\boldsymbol{A}^{\mathrm{T}}\boldsymbol{A})^{-1}\end{aligned} \tag{7.10}$$

where $\boldsymbol{R} = E\{\boldsymbol{n}\boldsymbol{n}^{\mathrm{T}}\}$ is the co-variance matrix of $\boldsymbol{n}$. Obviously, $\boldsymbol{R}$ is positive. Furthermore, when the respective noise vectors are irrelevant to each other, $\boldsymbol{R}$ is a diagonal matrix, i.e. $\boldsymbol{R} = \mathrm{diag}\{\sigma_1, \sigma_2, \cdots, \sigma_m\}$.

The above analysis is based on the definition of the cost function, and the leas-squares algorithm is derived from the extreme point of the cost function. The definition of least-squares will now be discussed from the perspective of linear space.

We write the system observation matrix $\boldsymbol{A}$ in the form of a column vector, i.e.

$$\boldsymbol{A} = [\boldsymbol{a}_1, \boldsymbol{a}_2, \cdots, \boldsymbol{a}_k]$$

We then mark the linear space composed of $\boldsymbol{A}$'s column vectors as $S(\boldsymbol{A})$.

For estimate $\hat{\boldsymbol{x}}$ of any state quantity, $\boldsymbol{A}\hat{\boldsymbol{x}}$ can be regarded as a vector in $S(\boldsymbol{A})$, because

$$\boldsymbol{A}\hat{\boldsymbol{x}} = \sum_{i=1}^{k} \boldsymbol{a}_i x_i \tag{7.11}$$

That is, $\boldsymbol{A}\hat{\boldsymbol{x}}$ can be represented by a linear combination of $\boldsymbol{A}$'s column vectors. Thus, the measured linear distance between $\tilde{\boldsymbol{Y}}$ and $\boldsymbol{A}\hat{\boldsymbol{x}}$ is

$$D(\tilde{\boldsymbol{Y}}, \boldsymbol{A}\hat{\boldsymbol{x}}) = \|\tilde{\boldsymbol{Y}} - \boldsymbol{A}\hat{\boldsymbol{x}}\| \tag{7.12}$$

For any two vectors $\boldsymbol{s}_1$ and $\boldsymbol{s}_2$, if the spatial distance is defined as $\|\boldsymbol{s}_1 - \boldsymbol{s}_2\| = (\boldsymbol{s}_1 - \boldsymbol{s}_2)^{\mathrm{T}}(\boldsymbol{s}_1 - \boldsymbol{s}_2)$, then according to Eq. (7.12), $D(\tilde{\boldsymbol{Y}}, \boldsymbol{A}\hat{\boldsymbol{x}})$ is actually the cost function $J(\hat{\boldsymbol{x}})$ defined earlier in this section.

Visually, when $\boldsymbol{A}\hat{\boldsymbol{x}}$ is a projection of $\tilde{\boldsymbol{Y}}$ on $S(\boldsymbol{A})$, the value of $D(\tilde{\boldsymbol{Y}}, \boldsymbol{A}\hat{\boldsymbol{x}})$ will reach the minimum, because $(\tilde{\boldsymbol{Y}} - \boldsymbol{A}\hat{\boldsymbol{x}})$ and $S(\boldsymbol{A})$ are perpendicular, and $\hat{\boldsymbol{x}}$ is the least-squares estimate of $\boldsymbol{x}$.

This process can be illustrated with a simple example. If $\boldsymbol{x} = [x_1, x_2]^{\mathrm{T}} \in \mathbb{R}^2$, $\tilde{\boldsymbol{Y}} = [\tilde{y}_1, \tilde{y}_2, \tilde{y}_3]^{\mathrm{T}} \in \mathbb{R}^3$, $\boldsymbol{A} = [\boldsymbol{a}_1\, \boldsymbol{a}_2]$, where $\boldsymbol{a}_1, \boldsymbol{a}_2 \in \mathbb{R}^3$, then the state observation equation can be written as:

$$\begin{bmatrix} \tilde{y}_1 \\ \tilde{y}_2 \\ \tilde{y}_3 \end{bmatrix} = \boldsymbol{A}\boldsymbol{x} = \begin{bmatrix} \boldsymbol{a}_1\, \boldsymbol{a}_2 \end{bmatrix} \begin{bmatrix} x_1 \\ x_2 \end{bmatrix} \tag{7.13}$$

We can now assume that (a_1, a_2) forms a two-dimensional plane, and when $\boldsymbol{A}\hat{\boldsymbol{x}}$ is a projection of $\tilde{\boldsymbol{Y}}$ on this plane, $\hat{\boldsymbol{x}}$ is the least-squares estimate of $\boldsymbol{x}$. For any other $\boldsymbol{x}^*$, $\boldsymbol{A}\boldsymbol{x}^*$ will have the following property:

$$\|\tilde{\boldsymbol{Y}} - \boldsymbol{A}\boldsymbol{x}^*\| > \|\tilde{\boldsymbol{Y}} - \boldsymbol{A}\hat{\boldsymbol{x}}\| \tag{7.14}$$

The property shown in Eq. (7.14) can be visually represented in Fig. 7.1.

7.1.2 *Weighted Least-Squares*

During the extraction of observation $\tilde{\boldsymbol{Y}}$, noise is inevitable, meaning that some of the observations are more reliable than others. Therefore, in the estimation of the state parameters using the least-squares, it is logical to maximize the adoption of

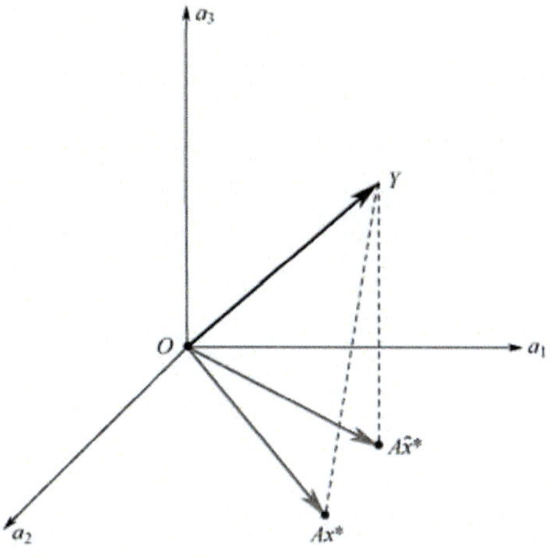

Fig. 7.1 Understanding the least-squares solution from the perspective of linear space

the highly reliable observations and minimize that of the lower ones. This is why the concept of weight was introduced. The improved least-squares solution is called weighted least-squares, or WLS.

Compared with conventional least-squares, the uniqueness of WLS is first reflected in the cost function, i.e.

$$J(\hat{\boldsymbol{x}}) = (\tilde{\boldsymbol{Y}} - \boldsymbol{A}\hat{\boldsymbol{x}})^{\mathrm{T}}\boldsymbol{W}(\tilde{\boldsymbol{Y}} - \boldsymbol{A}\hat{\boldsymbol{x}}) \tag{7.15}$$

where

$$\boldsymbol{W} = \begin{bmatrix} w_1 & 0 & \cdots & 0 \\ 0 & w_2 & \cdots & 0 \\ \vdots & \vdots & \ddots & 0 \\ 0 & 0 & 0 & w_m \end{bmatrix}$$

is called the weighting matrix. This is a diagonal matrix of $m \times m$, in which each element on the diagonal corresponds to the weight of an observation. When $w_i = w_j, \forall i, j \in [1, m]$, the WLS will degenerate into the regular LS.

Following the derivation method of the extreme point $J(\hat{\boldsymbol{x}})$ introduced in Sect. 7.1.1, we can first extend Eq. (7.15), then:

$$J(\hat{\boldsymbol{x}}) = \tilde{\boldsymbol{Y}}^{\mathrm{T}}\boldsymbol{W}\tilde{\boldsymbol{Y}} - 2\tilde{\boldsymbol{Y}}^{\mathrm{T}}\boldsymbol{W}\boldsymbol{A}\hat{\boldsymbol{x}} + \hat{\boldsymbol{x}}^{\mathrm{T}}\boldsymbol{A}^{\mathrm{T}}\boldsymbol{W}\boldsymbol{A}\hat{\boldsymbol{x}} \tag{7.16}$$

If we derive $\hat{\boldsymbol{x}}$ in Eq. (7.16), according to the definition of vector derivation in Appendix A, we get

$$\frac{\partial J(\hat{\boldsymbol{x}})}{\partial \hat{\boldsymbol{x}}} = -2\boldsymbol{A}^{\mathrm{T}}\boldsymbol{W}\tilde{\boldsymbol{Y}} + 2\boldsymbol{A}^{\mathrm{T}}\boldsymbol{W}\boldsymbol{A}\hat{\boldsymbol{x}} \tag{7.17}$$

The value of $J(\hat{\boldsymbol{x}})$ will reach the extreme when $\frac{\partial J(\hat{\boldsymbol{x}})}{\partial \hat{\boldsymbol{x}}} = 0$. So if Eq. (7.17) is equal to 0, we can obtain the WLS estimation for x as:

$$\hat{\boldsymbol{x}}_{\mathrm{WLS}} = (\boldsymbol{A}^{\mathrm{T}}\boldsymbol{W}\boldsymbol{A})^{-1}\boldsymbol{A}^{\mathrm{T}}\boldsymbol{W}\tilde{\boldsymbol{Y}} \tag{7.18}$$

Equation (7.18) shows that the WLS estimation of the system state parameters can be performed only if matrix $(\boldsymbol{A}^{\mathrm{T}}\boldsymbol{W}\boldsymbol{A})$ is reversible.

If we continue to follow the definition of the system state estimation error in Eq. (7.7), then

$$\begin{aligned} \delta\boldsymbol{x}_{\mathrm{WLS}} &= \boldsymbol{x} - (\boldsymbol{A}^{\mathrm{T}}\boldsymbol{W}\boldsymbol{A})^{-1}\boldsymbol{A}^{\mathrm{T}}\boldsymbol{W}\tilde{\boldsymbol{Y}} \\ &= \boldsymbol{x} - (\boldsymbol{A}^{\mathrm{T}}\boldsymbol{W}\boldsymbol{A})^{-1}\boldsymbol{A}^{\mathrm{T}}\boldsymbol{W}(\boldsymbol{A}\boldsymbol{x} + \boldsymbol{n}) \\ &= (\boldsymbol{x} - \boldsymbol{x} - (\boldsymbol{A}^{\mathrm{T}}\boldsymbol{W}\boldsymbol{A})^{-1}\boldsymbol{A}^{\mathrm{T}}\boldsymbol{W}\boldsymbol{n} \\ &= -(\boldsymbol{A}^{\mathrm{T}}\boldsymbol{W}\boldsymbol{A})^{-1}\boldsymbol{A}^{\mathrm{T}}\boldsymbol{W}\boldsymbol{n} \end{aligned} \tag{7.19}$$

Equation (7.19) shows that the WLS estimation error of $\delta\boldsymbol{x}_{\text{WLS}}$ is still independent of $\boldsymbol{x}$ and is only related to A, W and the observed noise vector n. The WLS estimation is also unbiased when the mean value of the observed noise is zero.

The co-variance matrix of $\delta\boldsymbol{x}_{\text{WLS}}$ estimated through WLS then becomes:

$$\text{var}\{\delta\boldsymbol{x}_{\text{WLS}}\} = (\boldsymbol{A}^{\text{T}}\boldsymbol{W}\boldsymbol{A})^{-1}\boldsymbol{A}^{\text{T}}\boldsymbol{W}\boldsymbol{R}\boldsymbol{W}\boldsymbol{A}(\boldsymbol{A}^{\text{T}}\boldsymbol{W}\boldsymbol{A})^{-1} \tag{7.20}$$

where $\boldsymbol{R}$ is still the co-variance matrix of the noise vector $\boldsymbol{n}$. In cases where the respective noise quantities of $\boldsymbol{n}$ are not related to each other, $\boldsymbol{R}$ can be written as $\text{diag}\{\sigma_1, \sigma_2, \cdots, \sigma_m\}$.

Since the observed noise measures the credibility of the observation, it is logical to consider the connection between the weighted matrix and the noise co-variance matrix. Generally,

$$\boldsymbol{W} = \boldsymbol{R}^{-1} = \begin{bmatrix} \frac{1}{\sigma_1} & 0 & \cdots & 0 \\ 0 & \frac{1}{\sigma_2} & \cdots & 0 \\ \vdots & \vdots & \ddots & 0 \\ 0 & 0 & 0 & \frac{1}{\sigma_m} \end{bmatrix} \tag{7.21}$$

A direct interpretation of Eq. (7.21) is that the larger σ is, the less reliable the observation will be, so a relatively small weight should be given to it. Conversely, the smaller σ is, the more reliable the observation will be, and it thus deserves a large weight.

If we substitute Eq. (7.21) into Eq. (7.18), (7.19) and (7.20), then:

$$\hat{\boldsymbol{x}}_{\text{WLS}} = (\boldsymbol{A}^{\text{T}}\boldsymbol{R}^{-1}\boldsymbol{A})^{-1}\boldsymbol{A}^{\text{T}}\boldsymbol{R}^{-1}\tilde{\boldsymbol{Y}} \tag{7.22}$$

$$\delta\boldsymbol{x}_{\text{WLS}} = -(\boldsymbol{A}^{\text{T}}\boldsymbol{R}^{-1}\boldsymbol{A})^{-1}\boldsymbol{A}^{\text{T}}\boldsymbol{R}^{-1}\boldsymbol{n} \tag{7.23}$$

$$\begin{aligned} \text{var}\{\delta\boldsymbol{x}_{\text{WLS}}\} &= (\boldsymbol{A}^{\text{T}}\boldsymbol{R}^{-1}\boldsymbol{A})^{-1}\boldsymbol{A}^{\text{T}}\boldsymbol{R}^{-1}\boldsymbol{R}\boldsymbol{R}^{-1}\boldsymbol{A}(\boldsymbol{A}^{\text{T}}\boldsymbol{R}^{-1}\boldsymbol{A})^{-1} \\ &= (\boldsymbol{A}\boldsymbol{R}^{-1}\boldsymbol{A}^{\text{T}})^{-1} \end{aligned} \tag{7.24}$$

Now, if $\boldsymbol{P} = \text{var}\{\delta\boldsymbol{x}_{\text{WLS}}\}$, then Eq. (7.22) becomes:

$$\hat{\boldsymbol{x}}_{\text{WLS}} = \boldsymbol{P}\boldsymbol{A}^{\text{T}}\boldsymbol{R}^{-1}\tilde{\boldsymbol{Y}} \tag{7.25}$$

Because $\boldsymbol{P}^{-1} = \boldsymbol{A}\boldsymbol{R}^{-1}\boldsymbol{A}^{\text{T}}$, $\boldsymbol{P}$ is positively related to $\boldsymbol{R}$. The more accurate the observation is, the smaller $\boldsymbol{R}$ and $\boldsymbol{P}$ will be, and the higher the accuracy of the estimation will be. In this sense, $\boldsymbol{P}^{-1}$ can be called an "information matrix."

Equation (7.24) indicates that choosing $\boldsymbol{R}^{-1}$ as the weighting matrix greatly simplifies the calculation of the co-variance matrix of the estimation error. Besides, it is also clear that the co-variance matrix of $\delta \boldsymbol{x}_{\mathrm{WLS}}$ is independent of the state parameter x and is only related to the co-variance matrix of the observed noise and the systematic observation matrix.

At the end of this section, the LS and WLS techniques can be summarized with a simple example. In a single-state parametric system with $[x]^{\mathrm{T}}$ for the system state vector, each time a state parameter is observed, an observation $\tilde{y}$ will be obtained. A total of N observations can be expressed in the following equations:

$$\begin{aligned} \tilde{y}_1 &= x + v_1 \\ \tilde{y}_2 &= x + v_2 \\ &\vdots \\ \tilde{y}_N &= x + v_N \end{aligned}$$

If we compare the single state parameter system with the state observation equation in Eq. (7.1), then:

$$\boldsymbol{Y} = \begin{bmatrix} \tilde{y}_1 \\ \tilde{y}_2 \\ \vdots \\ \tilde{y}_N \end{bmatrix}, \boldsymbol{A} = \begin{bmatrix} 1 \\ 1 \\ \vdots \\ 1 \end{bmatrix}, \boldsymbol{n} = \begin{bmatrix} \tilde{v}_1 \\ \tilde{v}_2 \\ \vdots \\ \tilde{v}_N \end{bmatrix} \tag{7.26}$$

Firstly, if we assume that the noise power contained in all the observations is equal and the noises are all white noises, then the covariance matrix is:

$$\boldsymbol{R} = \begin{bmatrix} \sigma^2 & 0 & \cdots & 0 \\ 0 & \sigma^2 & \cdots & 0 \\ \vdots & \vdots & & \vdots \\ 0 & 0 & \cdots & \sigma^2 \end{bmatrix}$$

Then, according to Eq. (7.6), the least-squares estimate for x is:

$$\hat{x}_{\mathrm{LS}} = (\boldsymbol{A}^{\mathrm{T}}\boldsymbol{A})^{-1}\boldsymbol{A}^{\mathrm{T}}\boldsymbol{Y} = \frac{1}{N}\sum_{i=1}^{N}\tilde{y}_i \tag{7.27}$$

Obviously, the least-squares estimate now is actually the result of the arithmetic average, which is common in practice. If the observed noise average is 0, the mean of $\hat{x}_{\mathrm{LS}}$ will approach $[x]^{\mathrm{T}}$, which means $\hat{x}_{\mathrm{LS}}$ is unbiased.

At the same time, according to Eq. (7.10), the variance of the estimated result is:

$$\begin{aligned}\mathrm{var}\{\delta x_{\mathrm{LS}}\} &= (\boldsymbol{A}^{\mathrm{T}}\boldsymbol{A})^{-1}\boldsymbol{A}^{\mathrm{T}}\boldsymbol{R}\boldsymbol{A}(\boldsymbol{A}^{\mathrm{T}}\boldsymbol{A})^{-1} \\ &= \frac{1}{N}\left(\sum_{i=1}^{N}\sigma^2\right)\frac{1}{N} \\ &= \frac{\sigma^2}{N}\end{aligned} \tag{7.28}$$

According to Eq. (7.28), the larger N is, the smaller $\mathrm{var}\{\delta x_{\mathrm{LS}}\}$ and the more accurate $\hat{x}_{\mathrm{LS}}$ will be, which tallies with the actual situation.

Now, let's consider a case where the noise power in each observation is not equal. The co-variance matrix of $\boldsymbol{n}$ will be:

$$\boldsymbol{R} = \begin{bmatrix} \sigma_1^2 & 0 & \cdots & 0 \\ 0 & \sigma_2^2 & \cdots & 0 \\ \vdots & \vdots & & \vdots \\ 0 & 0 & \cdots & \sigma_N^2 \end{bmatrix} \tag{7.29}$$

If the weighting matrix is set as $\boldsymbol{W} = \boldsymbol{R}^{-1}$, the WLS estimation for x obtained from Eq. (7.22) will be:

$$\hat{x}_{\mathrm{WLS}} = (\boldsymbol{A}^{\mathrm{T}}\boldsymbol{R}^{-1}\boldsymbol{A})^{-1}\boldsymbol{A}^{\mathrm{T}}\boldsymbol{R}^{-1}\boldsymbol{Y}$$

If we substitute A and R into the above equation, then:

$$\hat{x}_{\mathrm{WLS}} = \frac{1}{\sum_{i=1}^{N}(1/\sigma_i^2)}\sum_{i=1}^{N}\left[(1/\sigma_i^2)\tilde{y}_i\right] \tag{7.30}$$

The result of Eq. (7.30) is the weighted average that we know well. According to Eq. (7.24), the variance of the WLS state estimation error is:

$$\begin{aligned}\mathrm{var}\{\delta x_{\mathrm{WLS}}\} &= (\boldsymbol{A}^{\mathrm{T}}\boldsymbol{R}^{-1}\boldsymbol{A})^{-1} \\ &= \frac{1}{\sum_{i=1}^{N}(1/\sigma_i^2)}\end{aligned} \tag{7.31}$$

Equation (7.31) shows that if the number of tests is increased (meaning that the value of N in the equation becomes larger), then the value of $\mathrm{var}\{\delta x_{\mathrm{WLS}}\}$ will be monotonically decreasing with N, indicating that the accuracy of the estimation can be improved by increasing the number of measurements. When the values of σ_i^2 are identical to each other, the WLS estimation will become an LS estimation.

7.1.2.1 Solution of Position and Time Difference Using Pseudo-Range Observation

The pseudo-range observations of the BDS and GPS dual-mode receivers have already been described in detail in Sect. 5.1. Equations (5.9) and (5.10) are mathematical models of GPS and BDS pseudo-range observations respectively. Each GPS or BDS satellite signal in the receiver tracking channel can provide a pseudo-range observation as shown in Eqs. (5.9) or (5.10). In this section, the calculation of the receiver position and local clock bias based on pseudo-range observations will be explained in detail.

Dual-mode receivers can receive and process two types of pseudo-range observations simultaneously, but sometimes only the single-mode ones are applicable. For example, in certain cases, only one kind of satellite observation, either in GPS or BDS mode, is available. Typical examples are positioning carried out in geographical locations beyond the coverage of the BDS signal area, or situations where either the BDS mode or GPS mode is acceptable due to special considerations (such as strategic security or political factors), and the positioning solution can only be carried out with single-mode pseudo-range observations. Therefore, in this section, the single-mode and dual-mode solution of the position and clock offset will be discussed separately.

Single-mode Observations

In this mode, the pseudo-range observations involved in the calculation come from the same GNSS system, so Eq. (5.9) with removed subscripts can represent the mathematical model of the single-mode pseudo-range observation.

$$\tilde{\rho}_1(\boldsymbol{x}_{\mathrm{u}}) = \sqrt{(x_{\mathrm{u}} - x_{\mathrm{s}_1})^2 + (y_{\mathrm{u}} - y_{\mathrm{s}_1})^2 + (z_{\mathrm{u}} - z_{\mathrm{s}_1})^2} + cb + n_{\rho_1} \tag{7.32}$$

$$\tilde{\rho}_2(\boldsymbol{x}_{\mathrm{u}}) = \sqrt{(x_{\mathrm{u}} - x_{\mathrm{s}_2})^2 + (y_{\mathrm{u}} - y_{\mathrm{s}_2})^2 + (z_{\mathrm{u}} - z_{\mathrm{s}_2})^2} + cb + n_{\rho_2} \tag{7.33}$$

$$\tilde{\rho}_m(\boldsymbol{x}_{\mathrm{u}}) = \sqrt{(x_{\mathrm{u}} - x_{\mathrm{s}_m})^2 + (y_{\mathrm{u}} - y_{\mathrm{s}_m})^2 + (z_{\mathrm{u}} - z_{\mathrm{s}_m})^2} + cb + n_{\rho_m} \tag{7.34}$$

Equations (7.32) to (7.34) contain pseudo-range observations with the amount of m. $[x_{\mathrm{u}}, y_{\mathrm{u}}, z_{\mathrm{u}}]$ is the position of the user and b is the deviation between the user's local clock and GPST/BDT, whose actual value depends on the GNSS system to which the pseudo-range observations belong. $n_\rho = c\tau_{\mathrm{s}} + E_{\mathrm{eph}} + T_{\mathrm{iono}} + T_{\mathrm{tron}} + \mathrm{MP} + n_{\mathrm{r}}$ represents the total error term in the pseudo-range observations. What is known is the pseudo-range observation $\tilde{\rho}_i$ and the satellite coordinates $[x_{\mathrm{s}_i}, y_{\mathrm{s}_i}, z_{\mathrm{s}_i}]$. The tilde "~" above the observation term indicates that it is an actual observation, different from the subsequent prediction. The satellite and user coordinates are both represented in the ECEF coordinate system here, where the acquisition of pseudo-range observations is detailed in Sect. 5.1. The coordinates of the satellite are calculated with the ephemeris data at the transmission time of the signal, which has already been discussed in detail in Chap. 6.

Comparing the above pseudo-range observation equation with the state observation equation with least-squares, we can see that the pseudo-range equation is a nonlinear equation and cannot be directly solved through the LS estimation described in Sect. 7.1.1. Therefore, we must first linearize the pseudo-range equation to perform the LS estimation. In practice, the most common linearization is based on first-order Taylor series expansion.

If we assume that there is a starting value $\boldsymbol{x}_0 = [x_0, y_0, z_0, b_0]$ in the user's coordinate and the local clock, then we can obtain the following equation by making a first-order Taylor series expansion of the pseudo-range equation based on the starting value:

$$\begin{aligned}\tilde{\rho}_i(x_u) = \rho_i(x_0) + \frac{\partial \rho_i}{\partial x_u}|_{x_0}(x_u - x_0) + \frac{\partial \rho_i}{\partial y_u}|_{y_0}(y_u - y_0) \\ + \frac{\partial \rho_i}{\partial z_u}|_{z_0}(z_u - z_0) + \frac{\partial \rho_i}{\partial b}|_{b_0}(b - b_0) + h.o.t. + n_{\rho_i}\end{aligned} \tag{7.35}$$

where $i = 1, 2, \cdots, m$; and h.o.t. are high-order Taylor series terms.

$$\rho_i(\boldsymbol{x}_0) = \sqrt{(x_0 - x_{s_i})^2 + (y_0 - y_{s_i})^2 + (z_0 - z_{s_i})^2} + cb_0$$

$$\frac{\partial \rho_i}{\partial x_\mathrm{u}}|_{x_0} = -\frac{x_0 - x_{s_i}}{\sqrt{(x_0 - x_{s_i})^2 + (y_0 - y_{s_i})^2 + (z_0 - z_{s_i})^2}}$$

$$\frac{\partial \rho_i}{\partial y_\mathrm{u}}|_{y_0} = -\frac{y_0 - y_{s_i}}{\sqrt{(x_0 - x_{s_i})^2 + (y_0 - y_{s_i})^2 + (z_0 - z_{s_i})^2}}$$

$$\frac{\partial \rho_i}{\partial z_\mathrm{u}}|_{z_0} = -\frac{z_0 - z_{s_i}}{\sqrt{(x_0 - x_{s_i})^2 + (y_0 - y_{s_i})^2 + (z_0 - z_{s_i})^2}}$$

$$\frac{\partial \rho_i}{\partial b}|_{b_0} = 1$$

$\rho_i(\boldsymbol{x}_0)$ can be derived from the current position, clock difference, and satellite position. It is often referred to as the predicted pseudo-range, and should be distinguished from the true pseudo-range observations. The subscript "i" here refers to different satellites that the observation is from.

We can define the following vector:

$$\boldsymbol{u}_i \triangleq [\frac{\partial \rho_i}{\partial x_\mathrm{u}}|_{x_0}, \frac{\partial \rho_i}{\partial y_\mathrm{u}}|_{y_0}, \frac{\partial \rho_i}{\partial z_\mathrm{u}}|_{z_0}, 1], \mathbf{d}\boldsymbol{x}_0 \triangleq [(x_\mathrm{u} - x_0), (y_\mathrm{u} - y_0), (z_\mathrm{u} - z_0), (b - b_0)]^\mathrm{T}$$

The vector formed by the first three elements of $\boldsymbol{u}_i$ is generally called the "direction cosine vector". Denoted as $\boldsymbol{DC}_i = [\frac{\partial \rho_i}{\partial x_\mathrm{u}}|_{x_0}, \frac{\partial \rho_i}{\partial y_\mathrm{u}}|_{y_0}, \frac{\partial \rho_i}{\partial z_\mathrm{u}}|_{z_0}]$, it is a representation of the unit direction vector from the user position to the satellite in the ECEF coordinate system. Each satellite has its own direction cosine vector. The vector can also be represented in other coordinate systems with its physical meaning unchanged.

If we sort out Eq. (7.35) and omit the high-order terms, then:

$$\rho_i(\boldsymbol{x}_\mathrm{u}) - \rho_i(\boldsymbol{x}_0) = \boldsymbol{u}_i \cdot \mathrm{d}\boldsymbol{x}_0 + n_{\rho_i} \tag{7.36}$$

On the left side of Eq. (7.36) is the subtraction of the predicted pseudo-range quantity on the initial point from the observed pseudo-range quantity. The difference is generally referred to as the pseudo-range residual, which is represented by δ_{ρ_i}. The mathematical expression of the pseudo-range residual is shown on the right side of the equation. The equation is already linear, but it should be noted that the linearization here is only an approximation in the sense of the first-order Taylor series. Strictly speaking, the sign of equality should be replaced with that of approximate equality in Eq. (7.36). It can be seen in later analysis that as the number of iterations increases, the result of linearization becomes more and more accurate.

Equation (7.36) is the linearization of the pseudo-range observations of a satellite. If the linearization of m observations is carried out simultaneously, the following linear equations can be obtained:

$$\begin{aligned}
\delta\rho_1 &= \boldsymbol{u}_1 \cdot \mathrm{d}\boldsymbol{x}_0 + n_{\rho_1}\\
\delta\rho_2 &= \boldsymbol{u}_2 \cdot \mathrm{d}\boldsymbol{x}_0 + n_{\rho_2}\\
&\vdots\\
\delta\rho_m &= \boldsymbol{u}_m \cdot \mathrm{d}\boldsymbol{x}_0 + n_{\rho_m}
\end{aligned}$$

If we transform the above equations into a matrix, then we can obtain:

$$\delta\boldsymbol{\rho} = \boldsymbol{H}\mathrm{d}\boldsymbol{x}_0 + \boldsymbol{n}_\rho \tag{7.37}$$

where, $\delta\boldsymbol{\rho} = [\delta\rho_1, \delta\rho_2, \cdots, \delta\rho_m]^\mathrm{T}$, $\boldsymbol{H} = [\boldsymbol{u}_1^\mathrm{T}, \boldsymbol{u}_2^\mathrm{T}, \cdots, \boldsymbol{u}_m^\mathrm{T}]^\mathrm{T}$, $\boldsymbol{n}_\rho = [n_{\rho_1}, n_{\rho_2}, \cdots, n_{\rho_m}]^\mathrm{T}$, the number of dimensions of which are $m \times 1$, $m \times 4$ and $m \times 1$ respectively.

According to the analysis in Sect. 7.1.1, the LS estimation of Eq. (7.37) is:

$$\mathrm{d}\boldsymbol{x}_0 = (\boldsymbol{H}^\mathrm{T}\boldsymbol{H})^{-1}\boldsymbol{H}^\mathrm{T}\delta\boldsymbol{\rho} \tag{7.38}$$

If matrix $(\boldsymbol{H}^\mathrm{T}\boldsymbol{H})^{-1}\boldsymbol{H}^\mathrm{T}$ is written as follows:

$$(\boldsymbol{H}^\mathrm{T}\boldsymbol{H})^{-1}\boldsymbol{H}^\mathrm{T} = \begin{bmatrix} h_{11} & h_{12} & \cdots & h_{1m}\\ h_{21} & h_{22} & \cdots & h_{2m}\\ h_{31} & h_{32} & \cdots & h_{3m}\\ h_{41} & h_{42} & \cdots & h_{4m} \end{bmatrix}$$

then it can be proved that:

$$\sum_{i=1}^{m} h_{1i} = 0, \sum_{i=1}^{m} h_{2i} = 0, \sum_{i=1}^{m} h_{3i} = 0, \sum_{i=1}^{m} h_{4i} = 1 \tag{7.39}$$

This means that in matrix $(\boldsymbol{H}^{\mathrm{T}}\boldsymbol{H})^{-1}\boldsymbol{H}^{\mathrm{T}}$, the sum of the elements in the first three row vectors is 0, and the sum of the elements in the fourth row vector is 1. Since the first three row vectors in $(\boldsymbol{H}^{\mathrm{T}}\boldsymbol{H})^{-1}\boldsymbol{H}^{\mathrm{T}}$ are x, y, z, which are used to calculate the user position, according to Eq. (7.39), the common error term in the pseudo-range observation does not affect the accuracy of the positioning solution. Meanwhile, the fourth row vector of $(\boldsymbol{H}^{\mathrm{T}}\boldsymbol{H})^{-1}\boldsymbol{H}^{\mathrm{T}}$ is to calculate the user's clock difference, so the common error term in the pseudo-range observation will affect the accuracy of the solution of the clock difference. This conclusion is theoretically derived, and is consistent with our visual observation. For example, if all pseudo-range measurements are added with the same value Δt, there will be no change in the positioning result, and the clock difference b will deviate by Δt.

Equation (7.38) generates the correction amount between the initial point and the real point after one linearization, which, if used to update the initial point, can help to generate the corrected solution, i.e.,

$$\boldsymbol{x}_1 = \boldsymbol{x}_0 + \mathrm{d}\boldsymbol{x}_0 \tag{7.40}$$

Then, the derivation from Eq. (7.36) to Eq. (7.40) can be repeated with $\boldsymbol{x}_1$ as the starting point, and the new correction amount $\mathrm{d}\boldsymbol{x}_1$ can be used to update the previous solution.

The above process can be described in a general way, and for the k-th update, the process can be written as:

$$\mathrm{d}\boldsymbol{x}_{k-1} = (\boldsymbol{H}_{k-1}^{\mathrm{T}}\boldsymbol{H}_{k-1})^{-1}\boldsymbol{H}_{k-1}^{\mathrm{T}}\delta\boldsymbol{\rho}_{k-1} \tag{7.41}$$

$$\boldsymbol{x}_k = \boldsymbol{x}_{k-1} + \mathrm{d}\boldsymbol{x}_{k-1} \tag{7.42}$$

Both $\boldsymbol{H}$ and $\delta\boldsymbol{\rho}$ are subscripted in Eq. (7.41) and (7.42) because the direction cosine vector of each satellite and its corresponding pseudo-range residuals are recalculated after each update of $\boldsymbol{x}_k$.The update will end when $\|\mathrm{d}\boldsymbol{x}_k\|$ is determined, i.e.,

$$\|\mathrm{d}\boldsymbol{x}_k\| < \text{predetermined threshold} \tag{7.43}$$

where the "predetermined threshold" is referred to as a preset threshold. When $\|\mathrm{d}\boldsymbol{x}_k\|$ is less than the threshold, it is considered that the update can be stopped. In general, if the starting point is set to be the center of the earth, then only about five iterations are needed for the accuracy of the convergence to reach a satisfactory level. As the number of iterations increases, the value of $\|\mathrm{d}\boldsymbol{x}_k\|$ becomes smaller and smaller, and the linearization of Eq. (7.36) becomes more and more accurate.

At the end of the iteration, the value of $\|\mathrm{d}\boldsymbol{x}_k\|$ can be very small, for example, less than 1 cm. However, this does not mean that the error of the user position and clock bias is less than 1 cm. The convergence of $\|\mathrm{d}\boldsymbol{x}_k\|$ only indicates that the solution that minimizes the cost function with least-squares has been found, not that the cost function has approached zero, so the error in the user position and the clock bias may still be large. Therefore, it is impractical to improve the accuracy of the solution by increasing the number of iterations. Errors in the position and clock bias will be further explained in the subsequent introduction to the geometric accuracy factor.

During an iteration, the effects of the Earth's rotation must be taken into consideration. It is already known that it takes 60–80 ms for signals from GPS satellites and BDS MEO satellites to travel from Space to the Earth's surface, and 110–130 ms for signals from BDS IGSO and GEO satellites. During transmission, the Earth has rotated by a certain angle, which, though small, cannot be ignored considering its vast radius.

The angular velocity of the Earth's rotation is ω_{ie}. The time of signal transmission Δt_{tr} can be obtained through dividing the predicted pseudo-range obtained from each iteration by the speed of light. Therefore, the angle of the Earth's rotation during signal transmission is

$$\alpha_k = \omega_{\mathrm{ie}} \Delta t_{\mathrm{tr}} \tag{7.44}$$

where the subscript of α_k indicates the k-th time iteration, which means α_k needs to be recalculated after each iteration.

Since the earth rotates around the z-axis of the ECEF coordinate system, the rotation matrix can be written as follows based on α_k:

$$\boldsymbol{R}_{\alpha_k} = \begin{bmatrix} \cos(\alpha_k) & -\sin(\alpha_k) & 0 \\ \sin(\alpha_k) & \cos(\alpha_k) & 0 \\ 0 & 0 & 1 \end{bmatrix} \tag{7.45}$$

By multiplying the rotation matrix and the satellite coordinates obtained from the ephemeris data, the satellite coordinates with the presence of the rotation effect can be obtained:

$$\begin{bmatrix} x'_{\mathrm{sv}} \\ y'_{\mathrm{sv}} \\ z'_{\mathrm{sv}} \end{bmatrix} = \begin{bmatrix} \cos(\alpha_k) & -\sin(\alpha_k) & 0 \\ \sin(\alpha_k) & \cos(\alpha_k) & 0 \\ 0 & 0 & 1 \end{bmatrix} \begin{bmatrix} x_{\mathrm{sv}} \\ y_{\mathrm{sv}} \\ z_{\mathrm{sv}} \end{bmatrix} \tag{7.46}$$

Then, the direction cosine vector of the satellite and the predicted pseudo-range amount can be calculated with $[x'_{\mathrm{sv}}, y'_{\mathrm{sv}}, z'_{\mathrm{sv}}]^{\mathrm{T}}$.

Dual-Mode Observation

When a dual-mode observation is involved in the calculation, the GPS and the BDS pseudo-range observations will be mixed. According to the analysis in Sect. 5.1, if the GPS time (GPST) is selected as the time reference of the receiver, there will be a system time deviation T_{GB} between BDT and GPST when the pseudo-range observations of the BDS satellite are compared with Eqs. (7.32) to (7.34). The analysis of the dual-mode is centered on the handling of T_{GB}.

One way is to read T_{GB} from the system settings, thereby solving the system deviation of the BDS pseudo-range observations to the GPST. Specifically, the time synchronization parameter between BDT and GPST is given in the BeiDou Navigation message, which is on the fifth page of the fifth sub-frame in the D1 code message, and the 101st page of the fifth sub-frame in the D2 code message. There are two time synchronization parameters between BDT and GPST:

$A_{0\text{GPS}}$: the clock difference of BDT relative to GPST;

$A_{1\text{GPS}}$: the clock speed of BDT relative to GPST.

The equation for the system deviation between BDT and GPST is:

$$T_{GB} = A_{0\text{GPS}} + A_{1\text{GPS}} \times t_E \tag{7.47}$$

where, t_E is the BDT when the receiver extracts the pseudo-range observation.

After the T_{GB} is obtained, the term cT_{GB} corresponding to T_{GB} can be deducted from Eq. (5.10), the expression for the BDS pseudo-range observation. Now, the state of the system to be solved is $\boldsymbol{x} = [x, y, z, b]$, so the BDS and GPS pseudo-range observations can be converted into single-mode ones and thereby be calculated according to the solution for the single-mode observation described above.

The advantage of this method is that the number of state variables will not increase compared to those in the single-frequency mode. Therefore, only the data from at least four BDS or GPS satellites is needed to calculate the receiver's position and local clock difference, for which it is more applicable when there are only a few observations to process. However, the main disadvantage of this method is the need for time synchronization parameters, which means that the data demodulation of the BDS navigation message is a necessity. Generally, there will be a certain time cost for the receiver to be powered on and complete the acquisition and tracking of the BDS satellite signal, the demodulation of the message, and the reading of time synchronization parameters, which depend on the signal acquisition speed, signal strength, and the difference between the startup time and the time parameter. The time cost will increase the TTFF time. What's more, in cases of poor signal quality, the data demodulation cannot even be successfully completed, and this method will no longer be applicable.

Another method is to solve the problem on the user side. The main idea is to regard T_{GB} as the system state quantity to be estimated. Then, the state quantity of the system to be solved is $\boldsymbol{x} = [x, y, z, b, T_{GB}]$, and the four single-mode unknown amounts change to 5. Thus the dual-mode pseudo-range observation can be written as

$$\tilde{\rho}_{G1}(\boldsymbol{x}_u) = \sqrt{(x_u - x_{Gs_1})^2 + (y_u - y_{Gs_1})^2 + (z_u - z_{Gs_1})^2} + cb + n_{\rho_{G1}} \tag{7.48}$$

$$\tilde{\rho}_{G2}(\boldsymbol{x}_u) = \sqrt{(x_u - x_{Gs_2})^2 + (y_u - y_{Gs_2})^2 + (z_u - z_{Gs_2})^2} + cb + n_{\rho_{G2}} \tag{7.49}$$

$$\tilde{\rho}_{\mathrm{G}m}(\boldsymbol{x}_{\mathrm{u}}) = \sqrt{(x_{\mathrm{u}} - x_{\mathrm{Gs}_m})^2 + (y_{\mathrm{u}} - y_{\mathrm{Gs}_m})^2 + (z_{\mathrm{u}} - z_{\mathrm{Gs}_m})^2} + cb + n_{\rho_{\mathrm{G}m}} \quad (7.50)$$

$$\tilde{\rho}_{\mathrm{B}1}(\boldsymbol{x}_{\mathrm{u}}) = \sqrt{(x_{\mathrm{u}} - x_{\mathrm{Bs}_1})^2 + (y_{\mathrm{u}} - y_{\mathrm{Bs}_1})^2 + (z_{\mathrm{u}} - z_{\mathrm{Bs}_1})^2} + cb + cT_{\mathrm{GB}} + n_{\rho_{\mathrm{B}1}} \quad (7.51)$$

$$\tilde{\rho}_{\mathrm{B}2}(\boldsymbol{x}_{\mathrm{u}}) = \sqrt{(x_{\mathrm{u}} - x_{\mathrm{Bs}_2})^2 + (y_{\mathrm{u}} - y_{\mathrm{Bs}_2})^2 + (z_{\mathrm{u}} - z_{\mathrm{Bs}_2})^2} + cb + cT_{\mathrm{GB}} + n_{\rho_{\mathrm{B}2}} \quad (7.52)$$

$$\tilde{\rho}_{\mathrm{B}n}(\boldsymbol{x}_{\mathrm{u}}) = \sqrt{(x_{\mathrm{u}} - x_{\mathrm{Bs}_n})^2 + (y_{\mathrm{u}} - y_{\mathrm{Bs}_n})^2 + (z_{\mathrm{u}} - z_{\mathrm{Bs}_n})^2} + cb + cT_{\mathrm{GB}} + n_{\rho_{\mathrm{B}n}} \quad (7.53)$$

Equations (7.48) to (7.53) show m GPS pseudo-range observations and n BDS pseudo-range observations, where Eqs. (7.48) to (7.50) are GPS pseudo-range observations, Eqs. (7.51) to (7.53) are BDS pseudo-range observations, and the subscripts G and B in $\tilde{\rho}_{\mathrm{G}i}$ and $\tilde{\rho}_{\mathrm{B}i}$ represent GPS and BDS.

The principle of linearizing the dual-mode pseudo-range observation is similar to that of the single-mode one, except that the GPS and BDS satellite signals should be distinguished:

$$\boldsymbol{u}_{\mathrm{G}i} \triangleq [\frac{\partial \rho_i}{\partial x_{\mathrm{u}}}|_{x_0}, \frac{\partial \rho_i}{\partial y_{\mathrm{u}}}|_{y_0}, \frac{\partial \rho_i}{\partial z_{\mathrm{u}}}|_{z_0}, 1, 0]$$

$$\boldsymbol{u}_{\mathrm{B}i} \triangleq [\frac{\partial \rho_i}{\partial x_{\mathrm{u}}}|_{x_0}, \frac{\partial \rho_i}{\partial y_{\mathrm{u}}}|_{y_0}, \frac{\partial \rho_i}{\partial z_{\mathrm{u}}}|_{z_0}, 1, 1]$$

$$\mathrm{d}\boldsymbol{x}_0 \triangleq [(x_{\mathrm{u}} - x_0), (y_u - y_0), (z_{\mathrm{u}} - z_0), (b - b_0), (T_{\mathrm{GB}} - T_{\mathrm{GB0}})]^{\mathrm{T}}$$

where the subscripts "B" and "G" are used to distinguish BDS and GPS pseudo-range observations. The linearized pseudo-range residual can be expressed as:

$$\delta\rho_{\mathrm{G}1} = \boldsymbol{u}_{\mathrm{G}1} \cdot \mathrm{d}\boldsymbol{x}_0 + n_{\rho_{\mathrm{G}1}} \quad (7.54)$$

$$\delta\rho_{\mathrm{G}2} = \boldsymbol{u}_{\mathrm{G}2} \cdot \mathrm{d}\boldsymbol{x}_0 + n_{\rho_{\mathrm{G}2}} \quad (7.55)$$

$$\vdots$$

$$\delta\rho_{\mathrm{G}m} = \boldsymbol{u}_{\mathrm{G}m} \cdot \mathrm{d}\boldsymbol{x}_0 + n_{\rho_{\mathrm{G}m}} \quad (7.56)$$

$$\delta\rho_{\mathrm{B}1} = \boldsymbol{u}_{\mathrm{B}1} \cdot \mathrm{d}\boldsymbol{x}_0 + n_{\rho_{\mathrm{B}1}} \quad (7.57)$$

$$\delta\rho_{\mathrm{B}2} = \boldsymbol{u}_{\mathrm{B}2} \cdot \mathrm{d}\boldsymbol{x}_0 + n_{\rho_{\mathrm{B}2}} \quad (7.58)$$

$$\vdots$$

$$\delta\rho_{\mathrm{B}n} = \boldsymbol{u}_{\mathrm{B}n} \cdot \mathrm{d}\boldsymbol{x}_0 + n_{\rho_{\mathrm{B}n}} \quad (7.59)$$

Equations (7.54) to (7.59) correspond to Eqs. (7.48) to (7.53). It can be seen that the number of $\boldsymbol{u}_i$ vector elements becomes five, and needs to be adjusted according to the difference between BDS and GPS satellites. Besides, the dimension of $d\boldsymbol{x}_i$—the corresponding correction of the system state—is also five.

The calculation of the state correction $d\boldsymbol{x}_i$ and the realization of multiple iterations using the least-squares method are the same as those in the single-frequency mode. Generally, for the k-th update, the process can be written as:

$$d\boldsymbol{x}_{k-1} = (\boldsymbol{H}_{k-1}^{\mathrm{T}}\boldsymbol{H}_{k-1})^{-1}\boldsymbol{H}_{k-1}^{\mathrm{T}}\delta\boldsymbol{\rho}_{k-1}$$
$$\boldsymbol{x}_k = \boldsymbol{x}_{k-1} + d\boldsymbol{x}_{k-1}$$

The expression of the above steps is exactly the same as in the single mode. It should be noted that $\boldsymbol{H}_{k-1}$ is composed of $\boldsymbol{u}_{\mathrm{G}i}$ and $\boldsymbol{u}_{\mathrm{B}i}$, i.e. $\boldsymbol{H} = [\boldsymbol{u}_{\mathrm{G}1}^{\mathrm{T}}, \boldsymbol{u}_{\mathrm{G}2}^{\mathrm{T}}, \cdots, \boldsymbol{u}_{\mathrm{G}m}^{\mathrm{T}}, \boldsymbol{u}_{\mathrm{B}1}^{\mathrm{T}}, \boldsymbol{u}_{\mathrm{B}2}^{\mathrm{T}}, \cdots, \boldsymbol{u}_{\mathrm{B}n}^{\mathrm{T}}]^{\mathrm{T}}$.

The convergence of the dual-mode iteration is judged identically to that of the single mode iteration, and the influence of the Earth's rotation also needs to be considered in the calculation of the satellite position. The specific process is referred to in Eqs. (7.44) to (7.46).

The problem of system clock bias can be solved through increasing the amount of the system state on the user's end, the advantage of which is that it does not require time synchronization parameters provided in the system setting, and the position and clock bias can be determined as long as there are enough dual-mode pseudo-range observations. Therefore, when the signal reception of the satellite is good, the TTFF will be shorter than in the system settings. However, the disadvantage is that the minimum number of satellites required is one more than that in the single mode, thus the applicability of the method will be affected when the signal is occluded.

In real GPS/BDS dual-mode receivers, the two methods can be adopted simultaneously, which means that when the quality of the BDS satellite signal is good, the time synchronization information can be demodulated through the system settings. Otherwise, the method of increasing the state of the system can be adopted after the dual-mode observations of more than five satellites are captured. The specific process is shown in Table 7.1.

Table 7.1 only shows the case when the number of pseudo-range observations is sufficient. When this is less than four, the method described in this section will be impractical. Therefore, elevation assistance or other redundant information assistance can be adopted to achieve positioning.

The coordinate of the receiver's position $P_{\mathrm{u}} = [x, y, z]$ obtained above is in the ECEF coordinate system, yet generally speaking, the actual positioning result needs

Table 7.1 Positioning solutions determined by satellite categories and numbers

Satellite type	Satellite number	Solution
GPS	≥4	Single−mode solution
BDS	≥4	Single−mode solution
GPS+BDS	≥5	Dual−mode user side solution or system level solution
GPS+BDS	=4s	Dual−mode system level solution

to be converted into latitude and longitude coordinates (ϕ, λ, h), in which ϕ is the latitude, λ is the longitude, and h is the altitude. For the specific conversion, please refer to Sect. 1.2.3 in Chap. 1.

7.1.3 Calculation of Speed and Clock Drift Using Doppler Observations

Section 5.3 explains the acquisition of Doppler observations and its mathematical model expressions. In this section, the solution of the receiver's speed and the clock drift through the least-squares method will be elaborated.

The difference between Doppler frequency and integral Doppler observations has been explained in detail in Sect. 5.3. The Doppler frequency reflects the projection of the instantaneous relative velocity vector of the satellite and the receiver on the direction cosine at a certain moment, while the integral Doppler reflects the change of distance between the satellite and the receiver over a period. Since integral Doppler observations cannot usually be generated by low-cost navigation receivers, the Doppler frequency will be adopted instead in this section to calculate the receiver's speed and the clock drift. If not specifically clarified, the Doppler frequency in this section will be regarded as the Doppler observation. For receivers that can generate integral Doppler observation, the observations of adjacent time elements can be differentiated, and the result can be used in the calculation described in this section.

Since GPST and BDT are maintained by atomic clocks in their own systems, the frequency stability of the atomic clock can reach $10^{-12} \sim 10^{-13}$, which, if converted to speed, is $3 \times 10^{-4} \sim 3 \times 10^{-5}$ m/s. The time difference between GPST and BDT can be considered as constant compared to the frequency error variance of the carrier tracking loop, i.e. $\dot{T}_{\mathrm{GB}} = 0$. Therefore, although the dual-mode receiver can generate different Doppler observations of both GPS and BDS satellites, the mathematical expression is the same, namely Eq. (5.20) in Sect. 5.3. When the receiver can acquire Doppler observations of m GPS and BDS satellites, the following equations can be written:

$$f_{\mathrm{d}_1} = (\boldsymbol{v}_{\mathrm{s}_1} - \boldsymbol{v}_{\mathrm{u}}) \cdot \boldsymbol{DC}_1 + c\dot{b} + n_{\mathrm{d}_1} \tag{7.60}$$

$$f_{\mathrm{d}_2} = (\boldsymbol{v}_{\mathrm{s}_2} - \boldsymbol{v}_{\mathrm{u}}) \cdot \boldsymbol{DC}_2 + c\dot{b} + n_{\mathrm{d}_2} \tag{7.61}$$

$$\vdots$$

$$f_{\mathrm{d}_m} = (\boldsymbol{v}_{\mathrm{s}_m} - \boldsymbol{v}_{\mathrm{u}}) \cdot \boldsymbol{DC}_m + c\dot{b} + n_{\mathrm{d}_m} \tag{7.62}$$

where f_{d_i} is the Doppler observation of the i-th satellite, whose unit is m/s. $\boldsymbol{DC}_i$ is the direction cosine vector of the same satellite, which is derived from the position solution process in the previous section. It is usually obtained when the iteration of the positioning solution converges. $\boldsymbol{v}_{\mathrm{s}_i}$ is the speed of the i-th satellite, which is obtained according to the method explained in in Sects. 6.2 and 6.3.

As mentioned above, GPS and BDS satellites are no longer distinguished in Eqs. (7.60) to (7.62). It should be noted that the unit of f_{d_i} is m/s, but the Doppler frequency outputted by the baseband tracking loop is in Hertz or radians. Hence, if f_{d_i} is expressed in Hertz, it needs to be multiplied by the carrier wavelength, and if it is in radians, it needs to be multiplied by 2π. The carrier wavelength can be calculated with $\lambda = c/f_{\text{carrier}}$, and the carrier frequency of GPS or BDS should be chosen according to the signal type of the satellite.

There are four unknown terms to be solved in Eqs. (7.60) to (7.62), which are the user's clock drift $c\dot{b}$ and velocity vector $\boldsymbol{v}_\mathrm{u} = [v_x, v_y, v_z]^\mathrm{T}$ respectively. For convenience of description, they are represented by a uniform vector $\boldsymbol{x}_v = [v_x, v_y, v_z, c\dot{b}]^\mathrm{T}$. If we arrange the satellite speed items in Eqs. (7.60) to (7.62) and move them to the left of the equal sign, we can obtain:

$$f_{d_1} - \boldsymbol{DC}_1 \cdot \boldsymbol{v}_{s_1} = -\boldsymbol{DC}_1 \cdot \boldsymbol{v}_u + c\dot{b} + n_{d_1} \tag{7.63}$$

$$f_{d_2} - \boldsymbol{DC}_2 \cdot \boldsymbol{v}_{s_2} = -\boldsymbol{DC}_2 \cdot \boldsymbol{v}_u + c\dot{b} + n_{d_2} \tag{7.64}$$

$$\vdots$$

$$f_{d_m} - \boldsymbol{DC}_m \cdot \boldsymbol{v}_{s_m} = -\boldsymbol{DC}_m \cdot \boldsymbol{v}_u + c\dot{b} + n_{d_m} \tag{7.65}$$

Equations (7.63) to (7.65) are easy to understand. GPS and BDS satellites move at high speed, which leads to most of the Doppler shift, so this part must be removed. The rest of the shift results completely from the user.'s movement and the local clock drift. Of course, the noise term remains. $(f_{d_i} - \boldsymbol{DC}_i \cdot \boldsymbol{v}_{s_i})$ is generally referred to as a linearized Doppler observation.

When the user stays still, because $\boldsymbol{v}_u = [0, 0, 0]^\mathrm{T}$,

$$(-\boldsymbol{DC}_i \cdot \boldsymbol{v}_u) = 0, \quad i = 1, \cdots, m \tag{7.66}$$

At this time, the right side of Eqs. (7.63) to (7.65) is completely determined by the local clock drift and the noise term. The local clock drift is a common term for all satellite observations, while the noise term is relatively small. If the linearized Doppler observations of multiple satellites are observed at the same time, it is easy to see that their values are very close, which is very useful for actual system debugging.

If we define:

$$\boldsymbol{f}'_d \triangleq \begin{bmatrix} f_{d_1} - \boldsymbol{DC}_1 \cdot \boldsymbol{v}_{s_1} \\ f_{d_2} - \boldsymbol{DC}_2 \cdot \boldsymbol{v}_{s_2} \\ \vdots \\ f_{d_m} - \boldsymbol{DC}_m \cdot \boldsymbol{v}_{s_m} \end{bmatrix}, \boldsymbol{H} \triangleq \begin{bmatrix} -\boldsymbol{DC}_1^\mathrm{T} & 1 \\ -\boldsymbol{DC}_2^\mathrm{T} & 1 \\ \vdots & \\ -\boldsymbol{DC}_m^\mathrm{T} & 1 \end{bmatrix}, \boldsymbol{n}_d \triangleq \begin{bmatrix} n_{d_1} \\ n_{d_2} \\ \vdots \\ n_{d_m} \end{bmatrix} \tag{7.67}$$

then Eqs. (7.63) to (7.65) can be converted into a matrix, which is:

$$\boldsymbol{f}'_d = \boldsymbol{H}\boldsymbol{x}_\mathrm{v} + \boldsymbol{n}_d \tag{7.68}$$

Comparing Eq. (7.68) with Eq. (7.1), we see that Eq. (7.68) is already a standard linear state observation equation, the solution of which is different to that of the position. The solution of the velocity does not need to be iterated. It can be conducted directly using the least-squares method:

$$\boldsymbol{x}_v = (\boldsymbol{H}^{\mathrm{T}}\boldsymbol{H})^{-1}\boldsymbol{H}^{\mathrm{T}}\boldsymbol{f}'_d \tag{7.69}$$

The property of $(\boldsymbol{H}^{\mathrm{T}}\boldsymbol{H})^{-1}\boldsymbol{H}^{\mathrm{T}}$ embodied in Eq. (7.39) is still true here. Therefore, the common error term in all Doppler observations will not affect the solution of a user's speed, but only that of the clock drift.

It can be deduced that during the PVT solution through the least-squares method inside the receiver, the general order is to obtain the position and clock difference of the user by using the Newton iteration method first, and then construct an H matrix with the satellite direction cosine vector obtained in the process. At the same time, the Doppler observation is subtracted from the satellite motion velocity to obtain a linearized Doppler observation. Finally, the user's speed and clock drift can be obtained from Eq. (7.69).

Obtaining the user's PVT solution through the least-squares method has been explained thoroughly in this section and the previous section. As the most common solution, this method has a wide range of applications in real-world receivers, and is the theoretical basis for understanding other solutions. The solution of each least square is based on the observation of a certain time element, which is irrelevant for the observation of the previous time element. If the observed noise of different time elements is regarded as white noise, the error contained in the PVT result obtained through the least-squares method can also be considered as uncorrelated. In actual engineering, this the existence of jumps in the position error and the velocity error obtained from the least-squares solution, which is similar to white noise. Of course, the noise contained in the pseudo-range observation and the Doppler frequency observations is not all white noise. For example, the ionospheric and tropospheric delays show the characteristics of low-frequency slowly-varying signals in time. The position error calculated at this time is also time-correlated. A variety of methods have been developed to filter the results of the least-squares or to make the positioning result smoother. The Kalman filter, which will be described later, is one of them.

7.1.4 Satellite Elevation and Azimuth

After the user's location is obtained, it is important to calculate the elevation and azimuth of the satellite (also known as the argument), which are related to the user's current position and satellite position.

As shown in Fig. 7.2, the user location is P_{u} and the satellite location is P_{sv}. The vector $\boldsymbol{P}_{\mathrm{u-sv}}$, which marks the distance between the user and the satellite, is projected to a point $P_{\perp}$ on the plane between the N axis and the E axis in the user

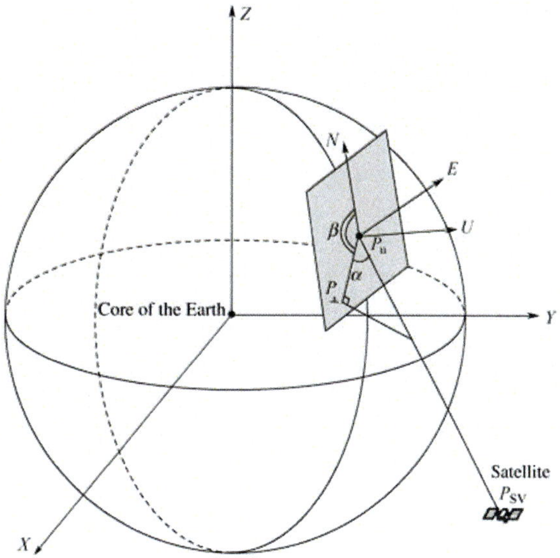

Fig. 7.2 Satellite elevation and azimuth

center's coordinate system, which is actually the tangent plane of the Earth along its ellipsoid from the user position. The N axis points due north and the N axis points due east. The angle between the vector $P_{\perp}P_u$ and the N axis is the azimuth angle, which is generally counterclockwise. Meanwhile, the angle between the vector $\boldsymbol{P}_{u-SV}$ and the tangent plane is the elevation angle. In Fig. 7.2, the elevation angle is represented by α and the azimuth angle by β.

The satellite's elevation information lays the foundation for the receiver's internal logic to predict the visible satellite set. The receiver needs to predict the visible satellite above the zenith and calculate the elevation angle of each visible satellite during operation, so that the downlink satellites that has already reached the horizon can be replaced in time, and the uplink satellites in the horizon direction can be prepared for signal acquisition and tracking. Therefore, the elevation angle of the satellite is an important basis for the receiver's internal star control logic. An important use of the predicted visible satellite set is to work out the approximate orientation of the satellite in order to calculate the elevation and depression angles of the satellite according to the local approximate azimuth when the receiver is started, and begin to save valid ephemeris or almanac data through non-volatile memory such as EEPROM or Flash. Only when the elevation angle of the satellite is greater than 0 can the satellite's signal be received, so the current set of satellites that may be tracked in the sky can be predicted. Thereby, the search space can be reduced, and the power-on positioning of the satellite can be accelerated.

The lower the elevation angle of the satellite, the worse the quality of the received signal will be, because the satellite signal needs to travel extra distance (mostly through the atmosphere) to reach the ground. According to Sect. 5.4.3 and 5.4.4, satellite signals with low elevation angles are more susceptible to the influence of the ionosphere and troposphere, the delays of which are also greater. At the same time, a satellite signal with a lower elevation angle is more likely to have a multipath effect than one with a high elevation angle. In order to ensure accurate positioning, minimum elevation angle requirements (Elevation Mask) for positioning satellites have been set in modern GPS receivers. When the elevation angle of the satellite is lower than the preset threshold, the observations of the satellite will be excluded from the positioning solution. In practice, the elevation threshold is generally set to 5~10 degrees, and the indicator can be adjusted according to actual requirements.

The satellite's elevation is also used to quantify the quality of the pseudo-range and carrier phase observations in the PVT solution. The higher the elevation angle, the better the observations provided by the satellite will be. In weighted least-squares, the quality of the observation is often used to set the weight matrix of the observation or the R matrix of the observation in the Kalman filter algorithm. A suitable R matrix will improve the accuracy and smoothness of the positioning result.

Figure 7.2 shows that the unit vector of $\boldsymbol{P}_{\mathrm{u-SV}}$ is the cosine vector from the user to the satellite, i.e. $\boldsymbol{DC}_i$ defined in Sects. 7.1.3 and 7.1.4, and i stands for different satellites, where $\boldsymbol{DC}_i$ is represented in the ECEF coordinate system. According to the physical meaning of the above elevation and azimuth, it is necessary to convert $\boldsymbol{DC}_i$ into the ENU coordinate system when the elevation and azimuth are under analysis.

According to the definition of the elevation and azimuth, the direction cosine is expressed in the ENU coordinate system as:

$$\boldsymbol{DC}_{\mathrm{ENU}} = [-\sin\beta\cos\alpha, \cos\beta\cos\alpha, \sin\alpha] \tag{7.70}$$

and if the value DC in the ECEF coordinate system is obtained, it can be converted into the ENU coordinate system, i.e.,

$$\boldsymbol{DC}_{\mathrm{ENU}} = \boldsymbol{R}_{e2t}\boldsymbol{DC} \tag{7.71}$$

Here, $\boldsymbol{R}_{e2t}$ is the rotation matrix from ECEF to ENU. Based on Sect. 1.2.4,

$$\boldsymbol{R}_{e2t} = \begin{bmatrix} -\sin(\lambda) & \cos(\lambda) & 0 \\ -\cos(\lambda)\sin(\phi) & -\sin(\lambda)\sin(\phi) & \cos(\phi) \\ \cos(\lambda)\cos(\phi) & \sin(\lambda)\cos(\phi) & \sin(\phi) \end{bmatrix}$$

where, ϕ is the latitude of the receiver's position and λ is the longitude of the receiver's position.

Suppose that the result obtained from Eq. (7.71) can be written as $[\kappa_e, \kappa_n, \kappa_u]$, and compare it with Eq. (7.70), then:

$$\alpha = \tan^{-1}\left(\frac{\kappa_u}{\sqrt{\kappa_e^2 + \kappa_n^2}}\right) \tag{7.72}$$

$$\beta = \tan^{-1}\left(-\frac{\kappa_e}{\kappa_n}\right) \tag{7.73}$$

Equations (7.72) and (7.73) are for the calculation of the elevation and azimuth respectively, and are applicable to both GPS and BDS satellites. The range of elevation angles is [−90°, +90°], and the range of azimuth is [−180°, +180°] or [0°, +360°]. In practice, since the receiver can only receive signals from satellites above the zenith, the satellite elevation calculated from the actual received satellite signals can only be between 0° and 90°. Figure 7.3 shows the 24-h GPS and BDS satellite elevation and azimuth trajectories. The receiver position is (39.9005 °N, 116.4135 °E). For easier identification, the trajectories of GPS satellites and BDS satellites are divided into two diagrams. The left shows the trajectories of GPS satellites, and the right shows those of BDS satellites. The receiver's own positions are on the center of the diagrams, and the elevation and azimuth angles of different satellites are represented by the distances and angles from the center point. As shown in Fig. 7.3, the trajectories of different GPS satellites are similar in shape, while those of the BDS satellite can be easily divided into three categories based on their shapes. B1 to B5 are GEO satellites, so the elevation and azimuth do not change much. B6 to B10 are IGSO satellites whose trajectories are similar to two-thirds of a figure "8", as IGSO satellites are untraceable when they fly to the southern hemisphere. B11 to B14 are MEO satellites, so only signals from a portion of their operating cycle can be received at the observation site.

7.1.5 Geometric Dilution of Precision

The co-variance matrix of the position error can be obtained through analyzing the algorithm for the pseudo-range observation and positioning in Sect. 7.1.3 and Eq. (7.10) of the co-variance matrix of the state error in the least-squares method:

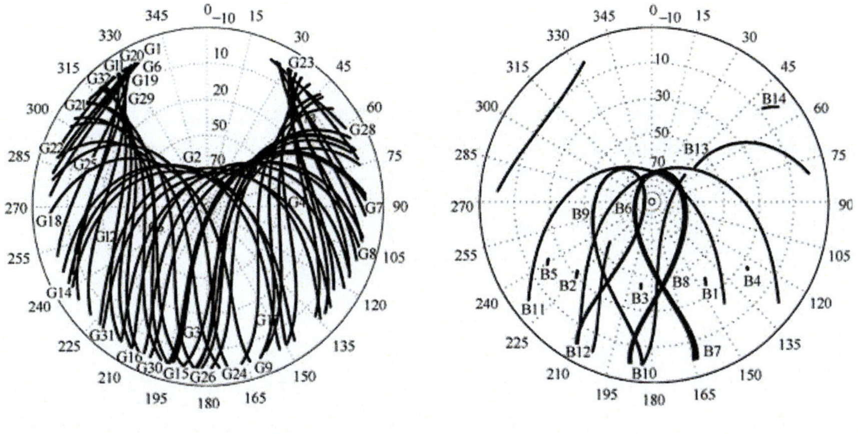

Fig. 7.3 Elevation and azimuth distribution of GPS and BDS satellites (Receiver position: 39.9005 °N, 116.4135 °E)

$$\mathrm{var}\{\delta \boldsymbol{x}_u\} = (\boldsymbol{H}^{\mathrm{T}}\boldsymbol{H})^{-1}\boldsymbol{H}^{\mathrm{T}}\boldsymbol{R}\boldsymbol{H}(\boldsymbol{H}\boldsymbol{H}^{\mathrm{T}})^{-1} \tag{7.74}$$

$\delta \boldsymbol{x}_{\mathrm{u}}$ is used here to distinguish it from the update amount $\mathrm{d}\boldsymbol{x}_i$ for each iteration in Sect. 7.1.3. R is the covariance matrix of the noise vector in the pseudo-range observations. If $\delta \boldsymbol{x}_{\mathrm{u}}$ is to be calculated with Eq. (7.74), the value of R must be known. Now, let's consider the specific form of R.

It is reasonable to assume that the observed noises of different satellites are independent of each other, so R is a diagonal matrix represented by $\mathrm{diag}\{\sigma_1^2, \sigma_2^2, \cdots, \sigma_m^2\}$. Obviously, σ_i^2 is the noise power, which evaluates the pseudo-range observations of the i-th satellite. In general, $\sigma_i^2 \neq \sigma_j^2, \forall i, j \in [1, m]$. There are many factors that determine σ_i^2. According to the analysis in Sect. 5.4, the noise term in pseudo-range observations is affected by a wide variety of factors. For example, as the elevations of different satellites vary, a higher elevation angle will result in a higher carrier-to-noise ratio of the signal. The ionospheric and tropospheric delay during signal transmission will also be smaller, so the observations extracted from it will be cleaner, and smaller noise components can be identified. Correspondingly, the observations provided by the satellite with a lower elevation tend to have a larger noise component, for which σ_i^2 will be larger. In addition, the factors affecting the pseudo-range observation noise are ephemeris errors, satellite clock errors, multipath delays, and receiver errors. However, before the termination of the SA policy for GPS, the most significant error was caused by the random jitter of the satellite clock that was artificially added in the SA policy. All GPS satellites were affected by the same amplitude of the clock error. Thus, in the early-stage GPS error analysis of the receiver's pseudo-range observations, it can be assumed that all satellites provide the same (or similar) pseudo-range measurement noise power, so $\boldsymbol{R} = \sigma^2\boldsymbol{I}$, where $\boldsymbol{I}$ is a matrix of $m \times m$. This assumption can be proved in the equation below. Firstly, it is assumed that the pseudo-range observation error is represented by ρ_{URE}, where the subscript $_{\mathrm{URE}}$ represents the User Ranging Error. If each error term is independent the others, ρ_{URE} can be written as:

$$\rho_{\mathrm{URE}} = \sqrt{\sigma_{\mathrm{Eph}}^2 + \sigma_{\mathrm{SA}}^2 + \sigma_{\mathrm{iono}}^2 + \sigma_{\mathrm{tron}}^2 + \sigma_{\mathrm{MP}}^2 + \sigma_{\mathrm{r}}^2} \tag{7.75}$$

Before the SA policy was abolished for GPS, $\sigma_{\mathrm{SA}} \approx 33$. When σ_{SA} is much larger than the other terms, $\rho_{\mathrm{URE}} \approx \sigma_{\mathrm{SA}}$. Therefore, it is reasonable to assume that the pseudo-range observation noise provided by all satellites has the same (or similar) power. If we substitute $\boldsymbol{R} = \sigma^2\boldsymbol{I}$ into Eq. (7.74), then:

$$\begin{aligned}\mathrm{var}\{\delta \boldsymbol{x}_{\mathrm{u}}\} &= (\boldsymbol{H}^{\mathrm{T}}\boldsymbol{H})^{-1}\boldsymbol{H}^{\mathrm{T}}\sigma^2\boldsymbol{I}\boldsymbol{H}(\boldsymbol{H}^{\mathrm{T}}\boldsymbol{H})^{-1} \\ &= \sigma^2(\boldsymbol{H}^{\mathrm{T}}\boldsymbol{H})^{-1}\end{aligned} \tag{7.76}$$

For the convenience of later theoretical analysis, $(\boldsymbol{H}^{\mathrm{T}}\boldsymbol{H})^{-1}$ will be written as $[\hbar_{i,j}]$, where $\hbar_{i,j}$ represents the element of the i-th row and j-th column of the matrix. Then,

$$\mathrm{var}\{\delta x_\mathrm{u}\} = \sigma^2 \hbar_{1,1} \tag{7.77}$$

$$\mathrm{var}\{\delta y_\mathrm{u}\} = \sigma^2 \hbar_{2,2} \tag{7.78}$$

$$\mathrm{var}\{\delta z_\mathrm{u}\} = \sigma^2 \hbar_{3,3} \tag{7.79}$$

$$\mathrm{var}\{\delta b\} = \sigma^2 \hbar_{4,4} \tag{7.80}$$

Equations (7.77) to (7.80) show that the elements on the diagonal of the $(\boldsymbol{H}^\mathrm{T}\boldsymbol{H})^{-1}$ matrix reflect the accuracy of the positioning results only to a certain degree, because the derivation of the above equations is based on the assumption that the noise power of each satellite is the same, which is difficult to achieve in engineering practice. When the SA policy was operational in the USA, since the pseudo-range observation always had a fixed-range satellite clock component, the magnitude of the positioning error was always proportional to the value of $\hbar_{i,i}$. However, since the abolition of the SA policy, there is no significant error source in the error term of the pseudo-range observations. In this case, if the pseudo-range error is small, it is still possible to obtain a more accurate positioning result even though the value of $\hbar_{i,i}$ is relatively large. Even so, since the early stages of the design and performance analysis of the GPS receiver, the following precision dilutions has been defined:

$$\text{Positional dilution of precision (PDOP)} = \sqrt{\hbar_{1,1} + \hbar_{2,2} + \hbar_{3,3}}$$
$$\text{Time dilution of precision (TDOP)} = \sqrt{\hbar_{4,4}}$$
$$\text{Geometric dilution of precision (GDOP)} = \sqrt{\hbar_{1,1} + \hbar_{2,2} + \hbar_{3,3} + \hbar_{4,4}}$$

The precision dilutions can be regarded as a linear mapping from the measurement errors in the observations to the errors in the state estimation. In cases where the observation measurement errors are the same, a relatively large precision dilution will evoke a larger state estimation error, while a relatively small precision dilution makes the state estimation error smaller. As can be seen from the definition of the precision factor, the precision dilution is independent of the actual observation noise. It is only related to $(\boldsymbol{H}\boldsymbol{H}^\mathrm{T})^{-1}$, which can be directly calculated from the matrix. By observing the form of the matrix, we know that each row vector of the H matrix is composed of the cosine vector of a satellite and the user's direction and 1, so $(\boldsymbol{H}\boldsymbol{H}^\mathrm{T})^{-1}$ must be related to the geometric distribution of multiple satellites.

According to the definition of the precision dilutions above,

$$\begin{aligned}\sigma_\mathrm{Pos} &= \sqrt{\sigma_x^2 + \sigma_y^2 + \sigma_z^2} \\ &= \sqrt{\hbar_{1,1} + \hbar_{2,2} + \hbar_{3,3}} \cdot \sigma_\mathrm{URE} \\ &= \mathrm{PDOP} \cdot \sigma_\mathrm{URE}\end{aligned} \tag{7.81}$$

where σ_{Pos} is the standard deviation of the receiver's position calculated with the least-squares. Equation (7.81) shows the relationship between the ranging error and the final positioning error in the pseudo-range observation. The example in Fig. 7.4 explains why the geometric accuracy factor is related to the geometric distribution of the satellite. Although displayed on a two-dimensional plane, the example can be easily converted to the three-dimensional space of reality.

There are two constellations in Fig. 7.4, each with two stars, which is enough for positioning on a two-dimensional plane. Of course, if the instability of the local clock is taken into consideration, one more satellite will be needed. However, to highlight the key point, it is assumed that the local clock is precise enough here. A circle can be drawn for the pseudo-range equation of each satellite on the plane. Considering the measurement error in the pseudo-range equation, this circle is not a curve, but a concentric zone whose thickness reflects the magnitude of the pseudo-range measurement error. In theory, there will be two intersections of two circles, and the position of the user is where one of the intersections meets the Earth's surface. However, due to the influence of measurement errors, the juncture of the two concentric zones is not a point but an area, which is shown by the shaded part in the figure. Inside the area are all the possible positioning results obtained through the least-squares method. The constellation arrangement on the right side of Fig. 7.4 is a case where the two stars and the user's connection are almost in line; the one on the left is a case where the two stars and the user's connection are almost perpendicular to each other. The thickness of the concentric zone of the same satellite in the two constellation arrangements is constant, and the only change is the positions of the two satellites relative to the user. The intersection area on the right is relatively large, while the one on the left is relatively small. This means that the error of the positioning result caused by the right constellation is relatively large, while the that caused by the left is relatively small.

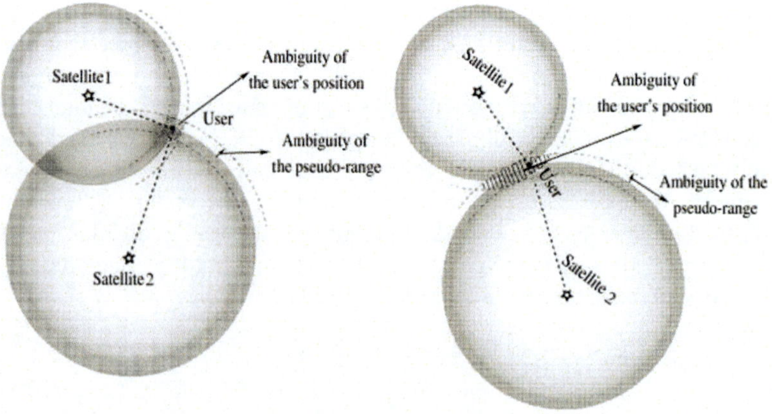

Fig. 7.4 The influence of different constellation arrangements on positioning results

When introduced to a more general case in the three-dimensional space, the example shown in Fig. 7.4 matches the concept of geometric dilution of precision. Since GPS positioning requires at least four satellites, it can be theoretically proven that in the case of four satellites involved in the process, the best constellation arrangement is to have one satellite located at the zenith and the other three scattered at a lower elevation.

The precision dilutions obtained from Eqs. (7.77) to (7.80) are from the ECEF coordinate system. In practice, people tend to be less familiar with the ECEF coordinate system, preferring the Earth-fixed coordinate system, i.e. ENU (or NED). Thus, to get the expression for the GDOP in the ENU coordinate system, a rotation is needed.

First, we write the linearized pseudo-range equation:

$$\delta \boldsymbol{\rho} = \boldsymbol{H}_e \delta \boldsymbol{x}_e + \boldsymbol{n}_\rho \tag{7.82}$$

where $\boldsymbol{H}$ and $\delta \boldsymbol{x}$ are subscripted with "e", indicating that they are in the ECEF coordinate system. In the ENU coordinate system, the state error is written as $\delta \boldsymbol{x}_t$. According to the rotation relationship of vectors in different coordinate systems, $\delta \boldsymbol{x}_t$ in the ENU coordinate system and $\delta \boldsymbol{x}_e$ in the ECEF coordinate system have the following relationship:

$$\delta \boldsymbol{x}_t = \boldsymbol{R}_L \delta \boldsymbol{x}_e \tag{7.83}$$

where

$$\boldsymbol{R}_L = \begin{bmatrix} \boldsymbol{R}_{e2t} & \boldsymbol{0}_{3\times 1} \\ \boldsymbol{0}_{1\times 3} & 1 \end{bmatrix}$$

In the equation, $\boldsymbol{R}_{e2t}$ is the rotation matrix in which the three-dimensional vector $[\delta x_{\mathrm{u}}, \delta y_{\mathrm{u}}, \delta z_{\mathrm{u}}]^{\mathrm{T}}$ in the ECEF coordinate system is rotated $[\delta E_{\mathrm{u}}, \delta N_{\mathrm{u}}, \delta U_{\mathrm{u}}]^{\mathrm{T}}$ in the ENU coordinate system. The expression of $\boldsymbol{R}_{e2t}$ has already been given in Sect. 7.1.5, where, since the clock difference and the specific coordinate system are irrelevant, the term related to time in $\boldsymbol{R}_L$ is 1. As can be seen from the expression of $\boldsymbol{R}_L$, as an orthogonal matrix, $\boldsymbol{R}_L$ has the following properties:

$$\boldsymbol{R}_L^{\mathrm{T}} = \boldsymbol{R}_L^{-1} \quad \Rightarrow \quad \boldsymbol{R}_L^{\mathrm{T}} \boldsymbol{R}_L = \boldsymbol{I} \tag{7.84}$$

So, Eq. (7.82) can be rewritten as

$$\begin{aligned} \delta \boldsymbol{\rho} &= \boldsymbol{H}_e \boldsymbol{R}_L^{\mathrm{T}} \boldsymbol{R}_L \delta \boldsymbol{x}_e + \boldsymbol{n}_\rho \\ &= \boldsymbol{H}_t \delta \boldsymbol{x}_t + \boldsymbol{n}_\rho \end{aligned} \tag{7.85}$$

where $\boldsymbol{H}_t = \boldsymbol{H}_e \boldsymbol{R}_L^{\mathrm{T}}$, $\delta \boldsymbol{x}_t = [\delta E_{\mathrm{u}}, \delta N_{\mathrm{u}}, \delta U_{\mathrm{u}}, \delta b]^{\mathrm{T}}$.

According to an analysis similar to that in the ECEF coordinate system, in the theoretical derivation for the ENU coordinate system, $(\boldsymbol{H}^{\mathrm{T}}\boldsymbol{H})^{-1}$ in Eq. (7.76) turns into:

$$\begin{aligned}(\boldsymbol{H}_t^{\mathrm{T}}\boldsymbol{H}_t)^{-1} &= (\boldsymbol{R}_L\boldsymbol{H}^{\mathrm{T}}\boldsymbol{H}\boldsymbol{R}_L^{\mathrm{T}})^{-1} \\ &= \boldsymbol{R}_L(\boldsymbol{H}^{\mathrm{T}}\boldsymbol{H})^{-1}\boldsymbol{R}_L^{\mathrm{T}}\end{aligned} \tag{7.86}$$

After the value of $(\boldsymbol{H}_t^{\mathrm{T}}\boldsymbol{H}_t)^{-1}$ is obtained, $[\hbar'_{i,j}]$ can be used to mark element at the i-th row of the j-th column. Since $\delta\boldsymbol{x}_t = [\delta E_u, \delta N_u, \delta U_u, \delta b]^{\mathrm{T}}$ now represents the position errors in the east, north, and up directions, the elements on the four diagonal lines of $(\boldsymbol{H}_t^{\mathrm{T}}\boldsymbol{H}_t)^{-1}$ correspond to the position errors in the east, north, and up directions as well as a clock bias, hence we can define:

$$\text{Horizontal dilution of precision (HDOP)} = \sqrt{\hbar'_{1,1} + \hbar'_{2,2}}$$

$$\text{Vertical dilution of precision(VDOP)} = \sqrt{\hbar'_{3,3}}$$

$$\text{Time dilution of precision(TDOP)} = \sqrt{\hbar'_{4,4}}$$

$$\text{Geometric dilution of precision (GDOP)} = \sqrt{\hbar'_{1,1} + \hbar'_{2,2} + \hbar'_{3,3} + \hbar'_{4,4}}$$

Since the GPS satellite is always in motion, the geometric dilution of precision will change dynamically in accordance. The GPS receiver must constantly monitor the condition of the GDOP and employ appropriate strategies when it is degraded, or alert the user about its state. In the early GPS receivers, due to the limitations of integrated circuit technology and the capabilities of the signal processor, only a few satellites could be tracked at the same time, which resulted in less precise GDOPs. Modern GPS receivers can generally track 12 or more satellites at the same time, and can include their pseudo-range observations in the positioning solution, so the precision of GDOP has become less of a problem. Occasionally, when there are obstacles blocking satellites or indoor positioning, GDOPs with less accuracy will appear, which must be considered during use.

In cases where positioning is carried out jointly in the BDS and GPS mode, if the system setting is adopted, the H matrix of the least-squares is still in $m \times 4$ dimensions, for which the definition and calculation of the GDOP mentioned above still apply. If T_{GB} is considered as the system state quantity to be estimated in the user's equipment, the H matrix becomes $m \times 5$-dimensional, and a system time deviation term is added in addition to three position terms and one clock difference term. Then, apart from the above concentration, there will be an additional systematic time dilution of precision for the GDOP:

$$\text{System time deviation dilution of precision(STDOP)} = \sqrt{\hbar'_{4,4}} \tag{7.87}$$

Equation (7.87) is valid when:

$$H = \begin{bmatrix} DC_{G1} & 0 & 1 \\ \vdots & \vdots & \vdots \\ DC_{Gm} & 0 & 1 \\ DC_{B1} & 1 & 1 \\ \vdots & \vdots & \vdots \\ DC_{Bn} & 1 & 1 \end{bmatrix} \tag{7.88}$$

which means that the fourth column of the H matrix corresponds to the systematic time deviation T_{GB}, where $DC_{G}/_{Bi}$ represents the direction cosine vector of the BDS/GPS satellite. Equation (7.88) shows that the $(H_t^T H_t)^{-1}$ matrix at this time is 5×5 dimensional, so the definitions of physical quantities such as PDOP, TDOP, HDOP, VDOP, and GDOP are similar to those in the single GPS mode, and it is only necessary to adjust the corresponding physical meaning according to the diagonal elements of $(H_t^T H_t)^{-1}$.

Figures 7.5, 7.6, 7.7, and 7.8 show the PDOP, HDOP, VDOP, and TDOP values in 24 h in the single GPS solution, the single BDS solution, and the GPS + BDS joint solution respectively. Among them, the receiver's position is (39.9005 °N, 116.4135 °E), the sampling interval is 30 s, and the total number of GPS and BDS satellites tracked most of the time is about 20. All of the satellites tracked during calculation are not involved in the calculation of the GDOP. Judging from the calculation results, the value of VDOP is slightly larger than that of HDOP. This is because all satellites

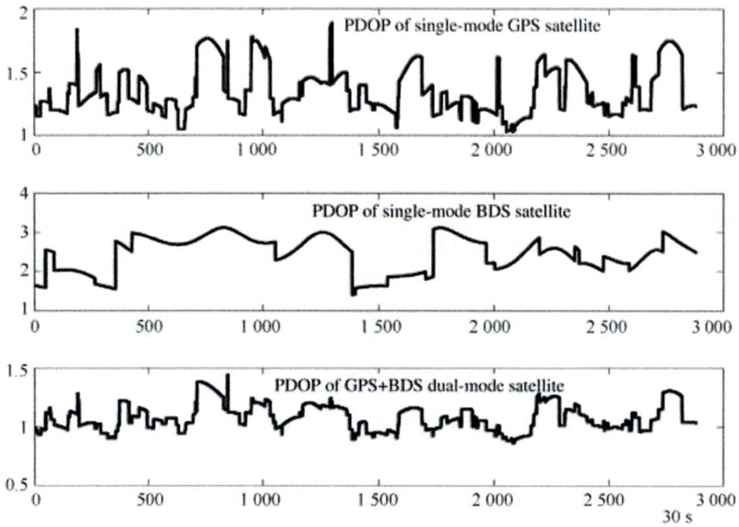

Fig. 7.5 Value of PDOP in 24 h

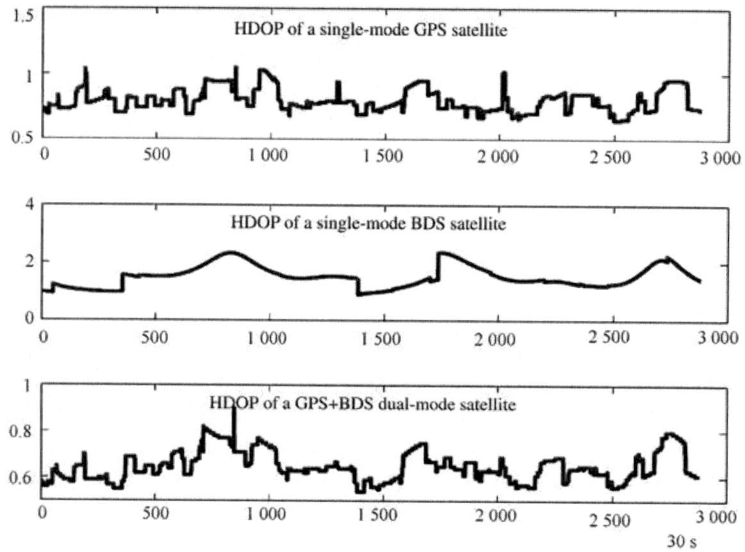

Fig. 7.6 Value of HDOP in 24 h

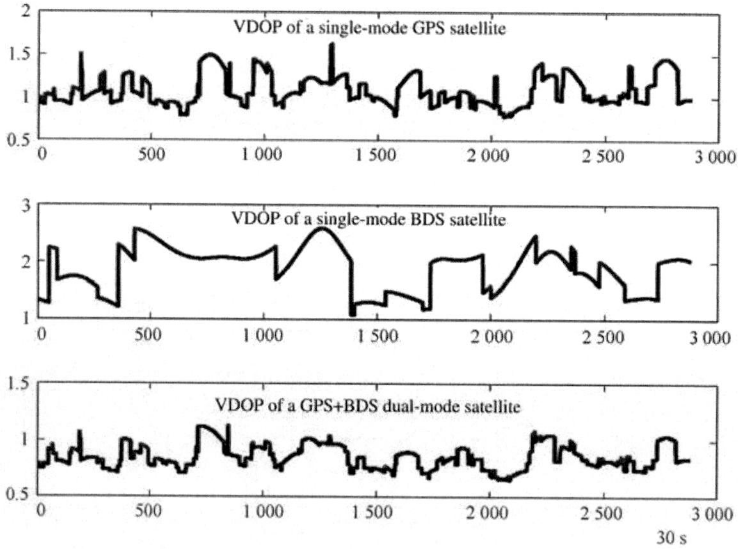

Fig. 7.7 Value of VDOP in 24 h

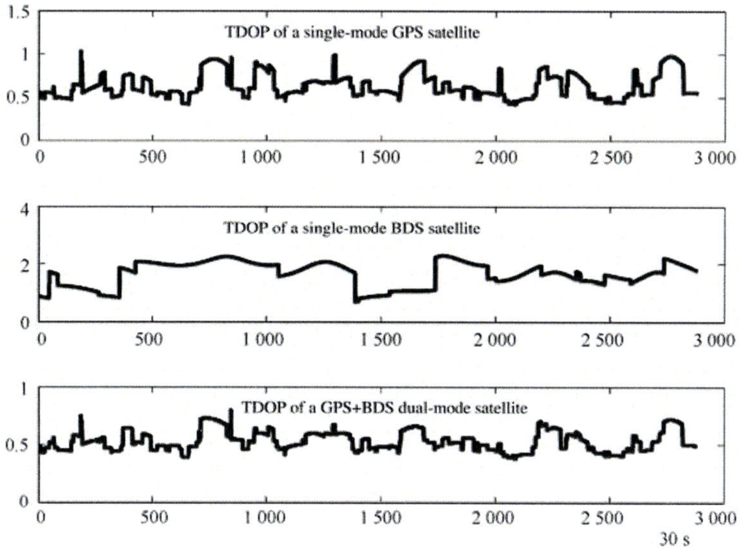

Fig. 7.8 Value of TDOP in 24 h

are zenithal, for which the restraint of the positioning result in the vertical direction is not as strong as it in the horizontal direction. As can be seen from Fig. 7.5, there is a large probability that the PDOP in the dual-mode solution is less than 1, which, according to Eq. (7.81), will result in a smaller σ_{Pos}, a benefit brought by dual-mode positioning.

In Figs. 7.5, 7.6, 7.7, and 7.8, the change of DOP is slower in the single mode of BDS, while in the single mode of GPS it changes faster. This is because the current BDS satellites are mainly composed of GEO and IGSO satellites, and the MEO satellites are few, which makes the geometric distribution of satellites change more slowly than that of GPS satellites. With the increase in BDS MEO satellites, the change of DOP in the BDS single mode tends to be similar to that in the GPS mode.

7.1.6 Receiver Autonomous Integrity Monitoring (RAIM)

The purpose of Receiver Autonomous Integrity Monitoring (RAIM) is to judge the validity of the positioning result after least-squares method by the redundant constraint between different observations. According to the conclusions in Sect. 7.1.3, the number of unknown quantities in the least-squares method is four (or five in cases of dual-mode solution, but the number will be defaulted to four in the following analysis). The receiver is usually able to track more satellites than the number of unknown quantities, so the excess pseudo-range observation will provide redundant information. Based on whether the redundant information matches the least-squares solution, it can be judged whether an obvious error or deviation exists in the current least-squares solution.

Pseudo-range residual decision

One or more false pseudo-range measurements can lead to erroneous positioning results. However, these results cannot be detected during the iteration process in Sect. 7.1.3. To illustrate, for a set of pseudo-range measurements in a certain time element—for example, the measurements of the following seven available GPS satellites—if a length of 300 km is artificially added to the pseudo-range observation of one of the satellites, there will be a significant deviation in the simulated pseudo-range observation, which is very common when erroneous bit synchronization occurs in the receiver. Now we can conduct an iteration on the erroneous pseudo-range observations. The iterative update and pseudo-range residual results are shown in Table 7.2 and Table 7.3 respectively.

Although the pseudo-range observation is artificially destroyed, the update amount can still approach 0 after six iterations. Since the update amount is less than the predetermined threshold, it can still be considered that the iteration reaches its end in the least-squares method. However, it is obvious that the positioning result obtained this way is wrong. The termination of the iterative update only indicates that the minimum value of the cost function defined by Eq. (7.2) with the given pseudo-range observation has been reached. However, the pseudo-range residuals given in Table 7.3 are very large (all ranging from 20 to 100 km). Therefore, the result of the pseudo-range residual can be used to estimate whether the positioning result is correct or not. It should be pointed out that only in redundant equations, where the number of observations is greater than four, the above method can be used to judge the correctness of the positioning result. If there are only four pseudo-range

Table 7.2 Update amount after six iterations of the erroneous pseudo-range observation (unit: meters)

Number of iterations	Δx_u	Δy_u	Δz_u	Δb
1	−3,482,231.70	5,717,628.68	3,985,718.62	1,404,937.22
2	551,235.04	−855,330.66	−6,31,169.52	−1,255,831.43
3	15,528.43	−19759.21	−17,773.41	−36,677.45
4	−12.36	49.56	−0.48	5.82
5	0.07 564	−0.139,474	−0.134,227	−0.102,945
6	−0.239,181E−3	0.527,204E−3	0.103,028E−3	0.266,970E−3

Table 7.3 The pseudo-range residual of each satellite during the iterative convergence of the erroneous pseudo-range observation (unit: meters)

Satellite index	1	2	3	4	6	7
Pseudo−range residual	−1.1036E5	0.4787E5	−0.3670E5	0.2917E5	1.1923E5	0.4653E5

observations, even if an observation is wrong, the pseudo-range residual (close to 0) will still be small after the iteration. This can be verified independently by readers.

The validity of the current least-squares solution can be judged through the calculation of the pseudo-range residual discussed above. This RAIM method is generally referred to as pseudo-range residual decision. First, the current pseudo-range observation should be linearized based on the previous time-based positioning result. The result of the linearization is shown in the following equation:

$$\boldsymbol{\rho} = \boldsymbol{H}\boldsymbol{x} + \boldsymbol{\varepsilon} \tag{7.89}$$

where

ρ is the linearization result of pseudo-range observations of n satellites, $\boldsymbol{\rho} \in \mathbb{R}^n$;

$\boldsymbol{H} \in \mathbb{R}^{n\times 4}$, the row vector in each row of H is $[e_x, e_y, e_z, 1]$, where $[e_x, e_y, e_z]^{\mathrm{T}}$ is the cosine vector from the satellite to the receiver;

$\boldsymbol{x} \in \mathbb{R}^4$ is the system state vector to be estimated, usually including 3 quantities of position and 1 quantity of clock difference;

$\varepsilon \in \mathbb{R}^n$ is the observed noise vector, which usually includes signal transmission noise, satellite ephemeris error, satellite clock bias and receiver error. Generally, the noise term is considered to be Gaussian white noise with a mean of zero, and the noise between different satellites is independent of each other, i.e. $E[\boldsymbol{\varepsilon}] = \boldsymbol{0},\ \mathrm{cov}[\boldsymbol{\varepsilon}] = \sigma_\varepsilon^2 \boldsymbol{I}_n$.

As can be seen from the previous chapter, the least-squares solution of Eq. (7.89) is:

$$\boldsymbol{x}_{\mathrm{ls}} = (\boldsymbol{H}^{\mathrm{T}}\boldsymbol{H})^{-1}\boldsymbol{H}^{\mathrm{T}}\boldsymbol{\rho} \tag{7.90}$$

According to Eq. (7.90), the pseudo-range residual vector is:

$$\boldsymbol{z} = [\boldsymbol{I} - \boldsymbol{H}(\boldsymbol{H}^{\mathrm{T}}\boldsymbol{H})^{-1}\boldsymbol{H}^{\mathrm{T}}]\boldsymbol{\rho} \tag{7.91}$$

If matrix S is defined as $\boldsymbol{S} \triangleq \boldsymbol{I} - \boldsymbol{H}(\boldsymbol{H}^{\mathrm{T}}\boldsymbol{H})^{-1}\boldsymbol{H}^{\mathrm{T}}$, then Eq. (7.91) can be rewritten as:

$$\boldsymbol{z} = \boldsymbol{S}\boldsymbol{\rho} \tag{7.92}$$

It can be verified that the S matrix is a symmetric matrix and idempotent matrix $\left(\boldsymbol{S}^2 = \boldsymbol{S}\right)$. Therefore, as long as the H matrix is known, the S matrix can be obtained, and the z vector can be directly calculated without the least-squares solution $\boldsymbol{x}_{1\mathrm{s}}$. After the z vector is obtained, the pseudo-range residual sum of squares SSE can be obtained:

$$\mathrm{SSE} = \boldsymbol{z}^{\mathrm{T}}\boldsymbol{z} \tag{7.93}$$

Obviously, rather than a vector, SSE is a plain number obtained through summing the squares of the elements of the z vector. If the observed noise in the pseudo-range observations of n satellites are zero-mean Gaussian noises and independent of each other, it can be proved that SSE is an open-square distribution with a degree of freedom of $n-4$, thereby the detection threshold under a given false alarm probability can be calculated according to the probability distribution of the SSE. When the SSE is greater than the threshold, it can be determined that there is an error in the current pseudo-range observation, and correspondingly, the least-squares solution of the current time element has a problem and needs to be discarded. Some scholars suggest that the normalized pseudo-range residual square sum and $\sqrt{\mathrm{SSE}/(n-4)}$ can be applied to make the decision. Obviously, at least 4 pseudo-range observations are required for this method. And if the number of pseudo-range observations is less than or equal to 4, a valid sum of squares of pseudo-range residuals cannot be obtained.

The advantage of pseudo-range residual decision is that the expression form of SSE is simple and the calculation is not complex. Since it is a scalar value, the decision threshold is single. Besides, SSE is independent of the geometric form of the satellite, and is related to the noise distribution of the pseudo-range observation. During the validity period of the SA policy, the noise distribution of the pseudo-range observation was relatively easy to determine, and since noise power of the pseudo-range observation from each GPS satellite is basically the same, it fits the previous theoretical assumption of ε. However, after the SA policy was abolished, the pseudo-range observation from each GPS satellite is no longer the same, so the probability distribution of SSE is relatively more complicated. In practice, the empirical value is often used to determine the decision threshold. One of the shortcomings of the pseudo-range residual decision method is that it can only indicate the existence of problems or errors in the current pseudo-range observation set but cannot determine which satellite it is on. This is because all pseudo-range observations are used to calculate the least-squares solution, and even error occurs in only one pseudo-range observation, it will lead to an entirely erroneous result.

Pseudo-range comparison

In the case where the number of pseudo-range observations is greater than 4, we can assume that the number is n, and the pseudo-range observation set can be divided into two parts, wherein the first part includes 4 pseudo-range observations, and the second part includes the remaining n-4 pseudo-range observations. In the first step, the least-squares solution is obtained based on the four pseudo-range observations of the first part, and the second step is to use the pseudo-range observations in the second part as the source of decision. Specifically, the position of each satellite in the second part needs to be calculated first, then, the pseudo-range can be predicted according to the least-squares solution obtained in Eq. (7.32) and the first step. Subtract the predicted pseudo-range from the pseudo-range obtained in the second part, and if the absolute values of all $\mathrm{n}-4$ pseudo-ranges are very small, then there should be no error in this set of pseudo-range observations, and the least-squares solution obtained in the first step is also reliable. Otherwise, errors do exist in this set of pseudo-range observations, which may be in the first part of the pseudo-range observation, or in the second part as the decision of pseudo-range observations. This process can be regarded as a binary hypothesis test, that is, H_0 and H_1 decisions. When the error occurs, it is the H_1 decision, and when there is no error, it is the H_0 decision. And when H1 decision occurs, if the absolute values of all n − 4 pseudo-ranges are relatively large, it means that the probability of error in the least-squares solution calculated with the four pseudo-range observations of the first part is large. While if only one of the n − 4 pseudo-ranges has a larger absolute value and the others are small, the least-squares solution is more probable to be correct, and the erroneous pseudo-range observation is precisely the one with the bigger pseudo-distance difference. This RAIM method is called pseudo-range comparison, whose difference from the pseudo-range residual decision is that it does not use all pseudo-range observations to calculate the least-squares solution. Thus, if only one pseudo-range observation is erroneous, as long as it is not among the first part of the four pseudo-range observations, then the least-squares solution is still correct. At this time, the erroneous pseudo-range observation can be found through the above logic decision. Besides, the two parts of the pseudo-range observation set need to be regrouped if necessary, and then the least-squares solution and the pseudo-range values need to be recalculated and re-compared. But of course, this will make the computation and logic decisions more complex.

Vector checking

Vector checking is another RAIM method widely applied in practical engineering. It calculates the check vector p with a check matrix $\boldsymbol{P}$:

$$\boldsymbol{p} = \boldsymbol{P}\boldsymbol{\rho} \tag{7.94}$$

where, the check matrix $\boldsymbol{P}$ has the following properties:

① $\boldsymbol{P} \in \mathbb{R}^{(n-4)\times n}$, i.e. the dimension of $\boldsymbol{P}$ is $(n-4) \times n$.
② $\text{rank}(\boldsymbol{P}) = n - 4$, i.e. the rank of $\boldsymbol{P}$ is $n - 4$.
③ $\boldsymbol{P}\boldsymbol{P}^{\text{T}} = \boldsymbol{I}_{n-4}$, the row vectors of $\boldsymbol{P}$ are orthogonal to each other.
④ $\boldsymbol{P}\boldsymbol{H} = \boldsymbol{0}$.

According to the properties of matrix $\boldsymbol{p}$, the check vector is (n − 4)-dimensional and

$$\boldsymbol{p}^{\text{T}}\boldsymbol{p} = z^{\text{T}}z \tag{7.95}$$

where z is the pseudo-range residual vector, i.e. the result of Eq. (7.92), which confirms the consistency of the check vector and pseudo-range residual, thus either of them can be used in the calculation of SSE.

According to the fourth property of matrix $\boldsymbol{p}$, the column vector of $\boldsymbol{P}$ is in the zero space of $\boldsymbol{H}^{\mathrm{T}}$. Substitute Eq. (7.89) into Eq. (7.94) and base on the fourth property of $\boldsymbol{P}$,

$$\boldsymbol{p} = \boldsymbol{P}\boldsymbol{\varepsilon} \tag{7.96}$$

From Eq. (7.96) we can know that the system state vector $\boldsymbol{x}$ becomes 0 vector after leftward multiplication with $\boldsymbol{P}$, and only the erroneous observation vector will be "converted" into the check vector $\boldsymbol{p}$, which is a very useful property. We can understand this property more thoroughly by writing Eq. (7.90) and (7.94) together, as shown in Eq. (7.97).

$$\begin{bmatrix} \boldsymbol{x}_{\mathrm{ls}} \\ \cdots\cdots\cdots \\ \boldsymbol{p} \end{bmatrix} = \begin{bmatrix} (\boldsymbol{H}^{\mathrm{T}}\boldsymbol{H})^{-1}\boldsymbol{H}^{\mathrm{T}} \\ \cdots\cdots\cdots\cdots\cdots \\ \boldsymbol{P} \end{bmatrix} [\boldsymbol{\rho}] \tag{7.97}$$

where, $[\boldsymbol{\rho}]$ corresponds to the observation vector space, $[\boldsymbol{x}_{\mathrm{ls}}]$ corresponds to the system state vector space and $[\boldsymbol{\rho}]$ corresponds to the check vector space. The conversion from the observation space to the system state space is completed through matrix $(\boldsymbol{H}^{\mathrm{T}}\boldsymbol{H})^{-1}\boldsymbol{H}^{\mathrm{T}}$, and the conversion from the observation space to the verification space can be performed with matrix $\boldsymbol{P}$. According to Eq. (7.96), since the check vector $\boldsymbol{P}$ and the observed noise vector $\boldsymbol{\varepsilon}$ are homogeneously linear, the detection and screening of the erroneous observation can be completed with the help of the check vector $\boldsymbol{P}$. Taking n = 6 pseudo-range observations as an example, matrix $\boldsymbol{H}$ in this case is 4 × 6-dimensional and matrix $\boldsymbol{P}$ is 2 × 6-dimensional. Assume that matrix $\boldsymbol{P}$ takes the following form:

$$\boldsymbol{P} = \begin{bmatrix} p_{11} & p_{12} & p_{13} & p_{14} & p_{15} & p_{16} \\ p_{21} & p_{22} & p_{23} & p_{24} & p_{25} & p_{26} \end{bmatrix}$$

When the j-th of the 6 pseudo-range observations is erroneous, i.e. $\rho_j = \boldsymbol{H}_j\boldsymbol{x} + b$, where b is the significant deviation in the pseudo-range observations, if the noise of other observations is negligible, then the check vector is:

$$\boldsymbol{p} \approx \begin{bmatrix} p_{1j} \\ p_{2j} \end{bmatrix} b \tag{7.98}$$

Two conclusions can be drawn from Eq. (7.98). First, the module of $\boldsymbol{p}$ can be used to determine whether there is an observation error. Second, vector $\boldsymbol{p}$ can be used to decide which observation has an error, and the principle is to compare the slope of the line between the origin (0,0) and the element of $\boldsymbol{p}$ with p_{1i}/p_{2i} $(i = 1, \cdots, 6)$. And when i takes the values that makes the slopes closest to each other, the pseudo-range observation of the satellite corresponding to that position is most possible to be erroneous. In practice, the elements of each column vector in matrix $\boldsymbol{P}$ can be regarded as a deviation slope, and each slope corresponds to a pseudo-range observation. In the above example, six deviation slopes can be obtained from six pseudo-range observations. As shown in the left part of Fig. 7.9, there are 6 straight lines, and the slope of each line is determined by p_{1i}/p_{2i}. The radius of the circle in the figure represents the decision threshold. Only when the module of p is larger than the threshold can the observation be judged as erroneous. At this time, vector p will be determined to be parallel with that line so that the wrong pseudo-range observation can be found.

When the number of pseudo-range observations is larger than six, the slope of the deviation needs to be represented in a space with higher dimensions. The right part of Fig. 7.9 shows the slope of the deviation when the number of pseudo-range observations is 7. The check vector space at this point is three-dimensional. The shape determined by the corresponding decision threshold is a sphere instead of a circle. Vector $\boldsymbol{P}$ is a straight line in a three-dimensional space, yet the criteria for decision is still the parallelism of vector $\boldsymbol{p}$ with the deviation slopes.

The properties of matrix $\boldsymbol{P}$ indicates that it is not unique, it can be obtained through the QR decomposition of matrix H, i.e. if

$$\boldsymbol{H} = \boldsymbol{QR} \tag{7.99}$$

then the n − 4 row vectors at the bottom of the transposed matrix of Q is matrix P.

Through vector checking, not only the determination of erroneous pseudo-range observations can be achieved, but the screening of erroneous observations through comparing the slopes of deviation can be completed, that is why it is widely used in practice. However, the check vector also has certain deficiencies. For example, if the deviation slopes of the two satellites are very close, misjudgment may occur. And when there are more than one erroneous observations, the check vector p will be composed of a plurality of straight lines with different slopes, then it will be difficult to determine which observation is wrong. Besides, the decision threshold in Fig. 7.9, i.e. the radius of the circle and the sphere in the figure, is difficult to be determined

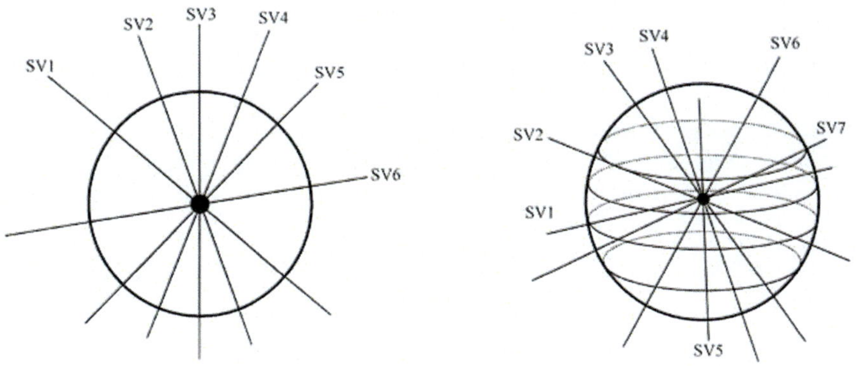

Fig. 7.9 Diagram of the deviation slopes of 6 and 7 pseudo-range observations

accurately. When the module of p is large, it is easy to determine the erroneous observation, and when the mode of p is small, there is no need to determine the erroneous observation. However, when c, the module of p, is in between, the situation will be the trickiest, because although b is not large at this time, such observation can lead to a non-negligible deviation of the positioning result, and it is also difficult to screen out the erroneous observation.

7.2 Maximum Solution Separation

The basic idea of maximum solution separation is to divide N pseudo-range observations into N sets, and the i-th observation is not included in each set, $i = 1, \ldots, N$, which means the number of observations contained in each set is $N - 1$. And if $N - 1$ is large enough ($N - 1 \geq 4$), a valid least-squares solution can still be obtained. In this way, the positioning result of N and the distance between two positioning results can be obtained. If the maximum distance is less than the preset threshold, none of the N pseudo-range observations have errors. On the contrary, errors are present in the other one or more pseudo-range observations. This method is based on the principle that when there is only one erroneous pseudo-range observation, there must be a set of observations that do not contain the erroneous one after the above-mentioned grouping. The positioning result calculated with this observation set must be near to the actual location. The other sets must contain false observations, due to which the positioning result will not be correct, so there must be a large distance between the correct and false positioning results. There are several variations of this method, one of which is to calculate the observations in different groups. In this method, N observations are divided into C_N^M groups, and each group contains M observations. Obviously, if $M \geq 4$, a least-squares solution can be obtained in each group. Comparing the distances among the C_N^M solutions, we can determine whether there is an error in all observations. The erroneous observation can be screened out through further detailed analysis of the C_N^M solutions. The cost of this method is that the computation burden will increase due to the grouping. Complex logic processing is required to screen out the erroneous observation, and decision error may also occur.

The RAIM methods analyzed above all use redundant information from pseudo-range observations. If there is no redundant pseudo-range observation, the accuracy of the least-squares solution cannot be guaranteed. From this perspective, more redundant information can be utilized. This is not necessarily limited to satellite signals, but also comes from inertial sensors, barometers, magnetometers, Roland-C navigation results, and vehicle odometers.

7.3 Kalman Filtering

The emergence of Kalman filtering in the 1960s sparked a wave of research and discussion among scholars all over the world. Kalman filtering is widely used in many fields, especially in target tracking, guidance, control, and navigation. Its salient features include efficiency in recursive operations and operations, and it is optimal in the sense of minimum mean square error. In general, each update of the Kalman filter requires two steps: the first is to update the status, and the second is to observe the measurement update. The state update is based on the system state transition equation, and predicts the time of the system state quantity. The observation update occurs after the system observations are received. The predicted state quantity and the observation are used as input to estimate the system state quantity through the least-squares method. The estimate obtained is used as the starting point for the system update at the next moment. Therefore, Kalman filtering has similarities with the Recursive Least Square (RLS) method in a certain sense. This section will start with RLS and offer a detailed explanation of the principle of Kalman filtering and its application in the BDS/GPS receiver. It also includes engineering problems to discuss the specific implementation of Kalman filtering.

7.3.1 Recursive Least-Squares

As seen in Sect. 7.1.2, when it comes to the system that is described in state transition equation $\boldsymbol{Y} = \boldsymbol{A}\boldsymbol{x}+\boldsymbol{n}$, the weighted least-squares estimate of its system state quantity is:

$$\hat{\boldsymbol{x}} = (\boldsymbol{A}^{\mathrm{T}}\boldsymbol{W}\boldsymbol{A})^{-1}\boldsymbol{A}^{\mathrm{T}}\boldsymbol{W}\tilde{\boldsymbol{Y}} \tag{7.100}$$

In this equation, the definitions of $\boldsymbol{n}$, $\boldsymbol{W}$, $\tilde{\boldsymbol{Y}}$ are the same as they are in Sect. 7.1.2. It is assumed that all observations from the 0 moment are included in $\tilde{\boldsymbol{Y}}$, and such processing is performed simultaneously for all observations. It is generally referred to as batch processing.

Since modern digital systems operate in discrete time domains, the following analysis will be based on digital sampling instants t_{k}, where k = 0, 1, …, ∞. If we assume that there are m observations at time t_m, $\tilde{\boldsymbol{Y}}_m = [\tilde{y}_1, \tilde{y}_2, \cdots, \tilde{y}_m]^{\mathrm{T}}$, then the state transition matrix A is a $m \times k$ matrix, and k is the dimension of the system state vector $\boldsymbol{x}$. Then, the WLS estimate obtained from the equation is based on all the elements in the $\tilde{\boldsymbol{Y}}_m$. Obviously, the entire batch process needs one multiplication of matrices $k \times m$ and $m \times k$, one inversion of matrix $k \times k$, one multiplication of matrices $k \times m$ and $m \times 1$, and one multiplication of matrices $k \times k$ and $k \times 1$.

At the moment t_{m+1}, $\tilde{y}_{m+1}$ is the new observation value. The vector of this is:

$$\begin{aligned}\tilde{\boldsymbol{Y}}_{m+1} &= [\tilde{y}_1, \tilde{y}_2, \cdots, \tilde{y}_m, \tilde{y}_{m+1}]^{\mathrm{T}} \\ &= [\tilde{\boldsymbol{Y}}_m^{\mathrm{T}}, \tilde{y}_{m+1}]^{\mathrm{T}}\end{aligned}$$

If the WLS estimation of the system state is continued in an ongoing batch mode of Eq. (7.100), the new estimate is based on all elements in $\tilde{\boldsymbol{Y}}_{m+1}$. Similar to the previous analysis, the calculation is one multiplication of matrices $k \times (m+1)$ and $(m+1) \times k$, one inversion of the matrix $k \times k$, one multiplication of matrices $k \times (m+1)$ and $(m+1) \times 1$, and one multiplication of matrices $k \times k$ and $k \times 1$.

It seems that there is nothing wrong with doing this. However, after careful analysis, it emerges that all the information contained in $\tilde{\boldsymbol{Y}}_m$ has been used at time t_m. At t_{m+1}, there is no need to recalculate from the $\tilde{y}_1$. This is because, comparing $\tilde{\boldsymbol{Y}}_{m+1}$ with $\tilde{\boldsymbol{Y}}_m$, the new information is only included in $\tilde{y}_{m+1}$. It is only necessary to consider this new observation based on the previous estimate.

The approach that is simply based on Eq. (7.100) has a fatal problem, which is system resource overhead, especially memory overhead. As time passes, new observations will be obtained, and the system needs to open up new storage space to store new observations. That will lead to a moment when the memory in the actual system is exhausted and stops working. Therefore, we have to find a way to avoid these problems. If $\tilde{\boldsymbol{Y}}_m = [\tilde{y}_1, \tilde{y}_2, \cdots, \tilde{y}_m]^{\mathrm{T}}$ based on the least-squares estimate is expressed as $E^*[x|\tilde{y}_1, \tilde{y}_2, \cdots, \tilde{y}_m]$, the relationship between $E^*[x|\tilde{y}_1, \tilde{y}_2, \cdots, \tilde{y}_m, \tilde{y}_{m+1}]$ and $E^*[x|\tilde{y}_1, \tilde{y}_2, \cdots, \tilde{y}_m]$ will be analyzed from the perspective of recursion.

Suppose that at time t_m, based on $\tilde{\boldsymbol{Y}}_m$, the WLS estimate for the system state $\boldsymbol{x}$ is $\hat{\boldsymbol{x}}_m$, the deviation from $\hat{\boldsymbol{x}}_m$ and the true value $\boldsymbol{x}$ is expressed as $\delta\hat{\boldsymbol{x}}_m$, that is

$$\hat{\boldsymbol{x}}_m = \boldsymbol{x} + \delta\hat{\boldsymbol{x}}_m \tag{7.101}$$

At the same time, the co-variance matrix of the error $\mathrm{var}\{\delta\hat{\boldsymbol{x}}_m\} = \boldsymbol{P}_m$ is known, which is the equation in Sect. 7.1.2.

At time t_{m+1}, a new observation $\tilde{y}_{m+1}$ occurs, assuming that the observed noise variance is R_{m+1}. We can combine the WLS estimators at t_m time, $\hat{\boldsymbol{x}}_m$ and $\tilde{y}_{m+1}$ into a new observation vector $[\hat{\boldsymbol{x}}_m^{\mathrm{T}}, \tilde{y}_{m+1}]^{\mathrm{T}}$, and the state transition equation between the observation and the system state parameters becomes:

$$\begin{bmatrix} \hat{\boldsymbol{x}}_m \\ \tilde{y}_{m+1} \end{bmatrix} = \begin{bmatrix} \boldsymbol{I} \\ \boldsymbol{A}_{m+1} \end{bmatrix} \boldsymbol{x} + \begin{bmatrix} \delta\hat{\boldsymbol{x}}_m \\ n_{m+1} \end{bmatrix} \tag{7.102}$$

In this equation, A_{m+1} is the state transition matrix at time t_{m+1}. The subscript "$_{m+1}$" indicates that the state transition matrix can be time-varying. n_{m+1} is the observation noise at time t_{m+1}, constituting a new measured noise vector with $\delta\hat{\boldsymbol{x}}_m$. The n_{m+1} and $\delta\hat{\boldsymbol{x}}_m$ are relatively independent, and the co-variance matrix of the new observation error can be expressed by R_{m+1}. According to the above analysis, R_{m+1} can be expressed as:

$$\boldsymbol{R}_{m+1} = \text{cov}\{\delta\hat{\boldsymbol{x}}_m^{\text{T}}, n_{m+1}\} = \begin{bmatrix} \boldsymbol{P}_m & \boldsymbol{0} \\ \boldsymbol{0} & R_{m+1} \end{bmatrix} \tag{7.103}$$

With the expression of R_{m+1}, the new observation vector $[\hat{\boldsymbol{x}}_m^{\text{T}}, \tilde{y}_{m+1}]^{\text{T}}$ adopts the WLS method. The new state transition matrix and the observation noise are given by the equation, and the weighting matrix is taken as $\boldsymbol{W} = \boldsymbol{R}_{m+1}^{-1}$, then the new state parameter is estimated as:

$$\hat{\boldsymbol{x}}_{m+1} = \boldsymbol{P}_{m+1}\left[\boldsymbol{I}\ \boldsymbol{A}_{m+1}^{\text{T}}\right]\begin{bmatrix} \boldsymbol{P}_m & \boldsymbol{0} \\ \boldsymbol{0} & R_{m+1} \end{bmatrix}^{-1}\begin{bmatrix} \hat{\boldsymbol{x}}_m \\ \tilde{y}_{m+1} \end{bmatrix} \tag{7.104}$$

in which:

$$\boldsymbol{P}_{m+1} = \text{var}\{\delta\hat{\boldsymbol{x}}_{m+1}\}\left(\left[\boldsymbol{I}\ \boldsymbol{A}_{m+1}^{\text{T}}\right]\begin{bmatrix} \boldsymbol{P}_m & \boldsymbol{0} \\ \boldsymbol{0} & R_{m+1} \end{bmatrix}^{-1}\begin{bmatrix} \boldsymbol{I} \\ \boldsymbol{A}_{m+1} \end{bmatrix}\right)^{-1} \tag{7.105}$$

To expand the equation,

$$\boldsymbol{P}_{m+1}^{-1} = \boldsymbol{P}_m^{-1} + \boldsymbol{A}_{m+1}^{\text{T}}R_{m+1}^{-1}\boldsymbol{A}_{m+1} \tag{7.106}$$

In the above equation, $\boldsymbol{A}_{m+1}^{\text{T}}R_{m+1}^{-1}\boldsymbol{A}_{m+1}$ is always non-negative, so it can be seen that the information matrix is always incremented as new observations occur. According to the definition and nature of the information matrix in Sect. 7.1.2, when new observations occur, the information in the information matrix increases. The next step is to see how to use the new information to make the local estimate be more accurate.

We simultaneously multiply $\boldsymbol{P}_{m+1}$ and collate both sides of Eq. (7.106),

$$\boldsymbol{P}_{m+1}\boldsymbol{P}_m^{-1} = \boldsymbol{I} - \boldsymbol{P}_{m+1}\boldsymbol{A}_{m+1}^{\text{T}}R_{m+1}^{-1}\boldsymbol{A}_{m+1} \tag{7.107}$$

The equation will be used in the following derivation.

If we expand Eq. (7.104) and apply the result of Eq. (7.107) to the second step of the following equation, then:

$$\begin{aligned}\hat{\boldsymbol{x}}_{m+1} &= \boldsymbol{P}_{m+1}\boldsymbol{P}_m^{-1}\hat{\boldsymbol{x}}_m + \boldsymbol{P}_{m+1}\boldsymbol{A}_{m+1}^{\text{T}}R_{m+1}^{-1}\tilde{y}_{m+1} \\ &= \hat{\boldsymbol{x}}_m + \boldsymbol{P}_{m+1}\boldsymbol{A}_{m+1}^{\text{T}}R_{m+1}^{-1}(\tilde{y}_{m+1} - \boldsymbol{A}_{m+1}\hat{\boldsymbol{x}}_m)\end{aligned} \tag{7.108}$$

If the RLS gains matrix, then:

$$\boldsymbol{k}_{m+1} = \boldsymbol{P}_{m+1}\boldsymbol{A}_{m+1}^{\text{T}}R_{m+1}^{-1} \tag{7.109}$$

The equation can be simplified as:

$$\hat{\boldsymbol{x}}_{m+1} = \hat{\boldsymbol{x}}_m + \boldsymbol{k}_{m+1}(\tilde{y}_{m+1} - \boldsymbol{A}_{m+1}\hat{\boldsymbol{x}}_m) \tag{7.110}$$

In this equation, $\hat{y}_{m+1} = \boldsymbol{A}_{m+1}\hat{\boldsymbol{x}}_m$ is predicted based on the observation of $\hat{\boldsymbol{x}}_m$ at t_{m+1}. $(\tilde{y}_{m+1} - \boldsymbol{A}_{m+1}\hat{\boldsymbol{x}}_m)$ is the residual of the actual observation and prediction observation.

The equation can be understood as follows: We assume that at time t_m, we have obtained the WLS estimation of the system state $\hat{\boldsymbol{x}}_m$, and have also obtained the estimated error variance matrix $\boldsymbol{P}_m$. At t_{m+1}, new observation $\tilde{y}_{m+1}$ occurs, along with the system transfer matrix $\boldsymbol{A}_{m+1}$ and observed noise variance of the new observations R_{m+1}. First, $\boldsymbol{P}_{m+1}$ is calculated according to Eq. (7.105), and then the gain matrix $\boldsymbol{k}_{m+1}$ is calculated by using $\boldsymbol{P}_{m+1}$, $\boldsymbol{A}_{m+1}$,and R_{m+1}^{-1}. Then, we multiply the residual $(\tilde{y}_{m+1} - \hat{y}_{m+1})$ and $\boldsymbol{k}_{m+1}$, and the result is the updated $\hat{\boldsymbol{x}}_m$. The updated result is the WLS estimation $\hat{\boldsymbol{x}}_{m+1}$ of the system state at time t_{m+1}. The state estimation $\hat{\boldsymbol{x}}_{m+1}$ and estimation error variance matrix of the current time $\boldsymbol{P}_{m+1}$ is used as the starting point of the next iteration.

The process does not start with the initial observations, but cleverly utilizes the results of the last estimate and the current observations. Theoretically, it can be proven that the result of the iterative process is the same as the batch mode from the initial observation. However, the amount of computation is much smaller, as is the overhead of storage space.

The whole process is explained in the schematic diagram in Fig. 7.10. $\hat{\boldsymbol{x}}_m$ and $\boldsymbol{P}_m$ in the picture is saved in local memory. Pieces of information such as $\tilde{y}_{m+1}$,$\boldsymbol{A}_{m+1}$, and

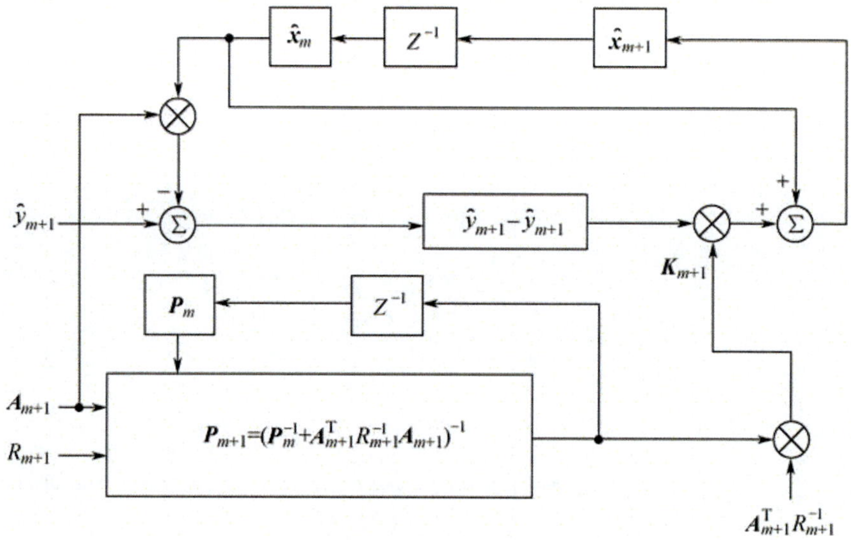

Fig. 7.10 Iterative principle of recursive least-squares (RLS) [1]

R_{m+1} are derived from observations. The results of each processing are updated with $\hat{x}_m$ and P_m. The computational analysis and storage analysis of the RLS algorithm are shown in Table 7.4, [1] in which the variable n in the calculation of the computational quantity and the memory overhead is the number of system state parameters, that is $x \in \mathbb{R}^n$. The observation is a scalar, so $H_{m+1} \in \mathbb{R}^{1\times n}$, in which $\tilde{y}_{m+1}$ and R_{m+1} are scalars. F_{m+1} is actually an intermediate variable matrix whose physical meaning is the information matrix.

Table 7.4 shows that the amount of computation and memory overhead required by the RLS algorithm is only related to the dimension n of the state vector, so its overhead is a fixed number. The amount of computation and memory overhead required for the corresponding batch mode of Eq. (7.100) increases as the processing time increases, so when m is very large, the system will be overwhelmed and crash.

Finally, a simple example is used to summarize the RLS method, making it easier for readers to understand the basic idea of the RLS algorithm.

Considering the last example in Sect. 7.1.2, when the measured noise power is equal, that is, $\text{var}\{\tilde{y}\} = \sigma^2$, its WLS estimation is to average all the observations $[\tilde{y}_1, \tilde{y}_2, \cdots, \tilde{y}_m]$:

$$\hat{x}_m = \frac{1}{m}\sum_{i=1}^{m}\tilde{y}_i,\ \text{var}\{\delta x\} = \frac{\sigma^2}{m}$$

At the moment $m+1$, a new observation $\tilde{y}_{m+1}$ occurs, and the new state is estimated to be:

$$\begin{aligned}\hat{x}_{m+1} &= \frac{1}{m+1}\sum_{i=1}^{m+1}\tilde{y}_i \\ &= \hat{x}_m + \frac{1}{m+1}(\tilde{y}_{m+1} - \hat{x}_m)\end{aligned} \tag{7.111}$$

Table 7.4 Comparison of RLS computation and memory overhead

Calculation item	FLOPS	Scratchpad storage	Global storage
$\boldsymbol{A}_{m+1}$	0	n	n
$r = \tilde{y}_{m+1} - \boldsymbol{A}_{m+1}\hat{\boldsymbol{x}}_m$	n	1	0
$\boldsymbol{d} = \boldsymbol{A}^T_{m+1}\boldsymbol{R}^{-1}_{m+1}$	N	n	0
$\boldsymbol{F}_{m+1} = \boldsymbol{F}_m + \boldsymbol{d}\boldsymbol{A}_{m+1}$	$\frac{1}{2}(n+1)n$	0	$\frac{1}{2}(n+1)n$
$\boldsymbol{P}_{m+1} = \boldsymbol{F}^{-1}_{m+1}$	$n^3 + \frac{1}{2}(n+1)n$	$\frac{1}{2}(n+1)n$	0
$\boldsymbol{K} = \boldsymbol{P}_{m+1}\boldsymbol{d}$	n^2	n	0
$\hat{\boldsymbol{x}}_{m+1} = \hat{\boldsymbol{x}}_m + \boldsymbol{K}r$	n	0	n
The amount of computation	$n^3 + 2n^2 + 4n$	$\frac{1}{2}n^2 + \frac{5}{2}n + 1$	$\frac{1}{2}n^2 + \frac{5}{2}n$

In the above derivation process, the new state estimate $\hat{x}_{m+1}$ is written in a recursive manner. Note that the recursion method here is just a mathematical variant of the batch method, not the RLS method described in this section. The following will derive the corresponding state estimation expression from the principle of the RLS algorithm and compare it with Eq. (7.111).

Now, let's consider the approach from RLS. From the moment 0, the observation is $\tilde{y}_0$, and the observed noise variance is $\text{var}\{\tilde{y}_0\} = \sigma^2$. Then, the WLS of the x is estimated to be $\hat{x}_0 = \tilde{y}_0$ and $P_0 = \text{var}\{\delta\hat{x}_0\} = \sigma^2$.

At time 1, the observation $\tilde{y}_1$ occurs, and P_1, K_1 and $\hat{x}_1$ are obtained according to Eqs. (7.105), (7.109), and (7.110). They are as follows:

$$P_1 = (P_0^{-1} + A_1 R_1^{-1} A_1)^{-1} = \frac{\sigma^2}{2}$$

$$K_1 = P_1 A_1 R_1^{-1} = \frac{1}{2}$$

$$\hat{x}_1 = \hat{x}_0 + K_1(\tilde{y}_1 - A_1\hat{x}_0) = \hat{x}_0 + \frac{1}{2}(\tilde{y}_1 - \hat{x}_0)$$

$A_i = 1, R_i = \sigma^2, \forall i$ is used in the above various derivation processes.

If we continue to use the estimation results of the equation at the previous moment, P_m, K_m and $\hat{x}_m$, then we can obtain the following $P_\text{m+1}$, $K_\text{m+1}$ and $\hat{x}_{m+1}$ at the following moment $m + 1$:

$$\begin{aligned}
P_{m+1} &= (P_m^{-1} + A_{m+1} R_{m+1}^{-1} A_{m+1})^{-1} = \frac{\sigma^2}{m+1} \\
K_{m+1} &= P_{m+1} A_{m+1} R_{m+1}^{-1} = \frac{1}{m+1} \\
\hat{x}_{m+1} &= \hat{x}_m + K_{m+1}(\tilde{y}_{m+1} - A_{m+1}\hat{x}_m) = \hat{x}_m + \frac{1}{m+1}(\tilde{y}_{m+1} - \hat{x}_m)
\end{aligned} \tag{7.112}$$

The equations show that the result of RLS can be regarded as a time-varying low-pass filter for observation $\tilde{y}_m$. The filter coefficient $1/(m + 1)$ gradually decreases with time, indicating that the system's update to the local state will change over time, and is more inclined to retain the estimated results of the previous moment. Comparing Eq. (7.112) with (7.111), the results of RLS are in complete agreement with the results obtained by the batch method. This confirms that the RLS method and the batch method have the same result if they are properly initialized, but it is clear that the RLS method has less total computational complexity and less memory overhead than the batch method.

7.3.2 The Basic Kalman Filter

The RLS algorithm subtly utilizes iterations so that the current system state can be weighted by the least-squares based on the estimation of the time *tm* and the observation of the time $t_{\mathrm{m+1}}$. Therefore, the RLS algorithm requires that the system state remains unchanged at time t_{m} and time $t_{\mathrm{m+1}}$, thereby recursing to all future moments; the RLS algorithm requires that the system state must be constant, otherwise it cannot be used. In practical applications, the system state often needs to change with time, and the RLS algorithm is inappropriate.

Considering a state time linear system, it is described by discrete time difference equations as follows:

$$\boldsymbol{x}_m = \Phi_{m-1}\boldsymbol{x}_{m-1} + \boldsymbol{G}_{m-1}\boldsymbol{u}_{m-1} + \boldsymbol{w}_{m-1} \tag{7.113}$$

$$\boldsymbol{y}_m = \boldsymbol{H}_m\boldsymbol{x}_m + \boldsymbol{v}_m \tag{7.114}$$

where x_m is the system state vector; Φ_m is the state transition matrix; u_m is the input signal vector; H_m is the system observation equation; and w_m and v_m are the system processing noise and the observed noise respectively. Equation (7.113) is called the state transition equation, and Eq. (7.114) is called the systematic observation equation.

Observers cannot directly measure x_m, but can only observe Y_m. The observed Y_m is called an observation, and is written as $\tilde{\boldsymbol{Y}}_m$ to distinguish it. Since w_m and v_m are random variables, the specific value cannot be obtained, but the statistical properties can be analyzed. It is assumed that both w_m and v_m are white noise with an average value of 0 and independent of each other. That is

$$E\{\boldsymbol{w}\} = 0, E\{\boldsymbol{w}_j\boldsymbol{w}_l^{\mathrm{T}}\} = \boldsymbol{QD}_j\boldsymbol{\delta}_{j,l} \tag{7.115}$$

$$E\{\boldsymbol{v}\} = 0, E\{\boldsymbol{v}_j\boldsymbol{v}_l^{\mathrm{T}}\} = \boldsymbol{R}_j\boldsymbol{\delta}_{j,l} \tag{7.116}$$

$$E\{\boldsymbol{w}_j\boldsymbol{v}_l^{\mathrm{T}}\} = 0 \tag{7.117}$$

In which,

$$\boldsymbol{\delta}_{j,l} = \begin{cases} \boldsymbol{I}, \text{ when } l = j \\ \boldsymbol{0}, \text{ when } l \neq j \end{cases} \tag{7.118}$$

Equation (7.118) states that the typical features of white noise are that it is irrelevant in time. Since the power spectrum of a random signal is the FFT of its autocorrelation function, the characteristic of white noise in the frequency domain is that the power spectrum is constant over the entire frequency range. This is similar to the uniform distribution of white light over the entire spectral frequency range.

Equations (7.115), (7.116), and (7.117) are assumptions about system noise. This assumption is true under ideal conditions. There are two main cases in the actual system that do not conform to this assumption. The first is when w_m and v_{m} are colored noise; the second is when w_m and v_{m} are correlated. The countermeasures for these two cases are described in the following sections. Here, in order to derive the basic equation of the Kalman filter, we assume that Eqs. (7.115), (7.116), and (7.119) are established.

Suppose we have an estimate $\hat{\boldsymbol{x}}_{m-1}$ of the state of the system $\boldsymbol{x}_{m-1}$ at the moment $m-1$, and we also know the estimated error variance matrix $\boldsymbol{P}_{m-1} = \mathrm{var}\{(\delta\hat{\boldsymbol{x}}_{m-1})(\delta\hat{\boldsymbol{x}}_{m-1})^{\mathrm{T}}\}$. At the moment m, on the basis of knowing $\boldsymbol{y}_m$, $\boldsymbol{QD}_m$ and $\boldsymbol{R}_m$, how can we use the iterative method to obtain the estimation $\hat{\boldsymbol{x}}_m$ of the state of the system $\boldsymbol{x}_m$? At the moment m, the state of the system x_m changes, but this change is determined by the state transition equation, so we can first predict the time $\boldsymbol{x}_m$ according to Eq. (7.113),

$$\hat{\boldsymbol{x}}_m^- = \Phi_{m-1}\hat{\boldsymbol{x}}_{m-1}^+ + \boldsymbol{G}_{m-1}\boldsymbol{u}_{m-1} \tag{7.119}$$

The superscript "−" is used to indicate that it is before the observation update. The superscript " + " is used to indicate that it is after the observation update. The subsequent content will follow the same representation. It is common to use $\hat{\boldsymbol{x}}_m^-$ as the time update of $\hat{\boldsymbol{x}}_{m\ 1}^+$.

Since $\hat{\boldsymbol{x}}_{m-1}^+$ is unbiased and the mean of $\boldsymbol{w}_{m-1}$ is 0, it can be proven that $\hat{\boldsymbol{x}}_m^-$ is unbiased,

$$\delta\hat{\boldsymbol{x}}_m^- = \hat{\boldsymbol{x}}_m^- - \boldsymbol{x}_m,\ \text{且}\ E\{\delta\hat{\boldsymbol{x}}_m^-\} = \boldsymbol{0} \tag{7.120}$$

We subtract Eq. (7.113) from Eq. (7.119),

$$\delta\hat{\boldsymbol{x}}_m^- = \Phi_{m-1}\delta\hat{\boldsymbol{x}}_{m-1}^+ + \boldsymbol{w}_{m-1} \tag{7.121}$$

We use $\boldsymbol{P}_m^-$ to describe the co-variance matrix of the estimated error of $\hat{\boldsymbol{x}}_m^-$.

$$\begin{aligned}
\boldsymbol{P}_m^- &= E\{[\delta\hat{\boldsymbol{x}}_m^-][\delta\hat{\boldsymbol{x}}_m^-]^{\mathrm{T}}\} \\
&= \Phi_{m-1}E\{[\delta\hat{\boldsymbol{x}}_{m-1}^+][\delta\hat{\boldsymbol{x}}_{m-1}^+]^{\mathrm{T}}\}\Phi_{m-1}^{\mathrm{T}} + E\{\boldsymbol{w}_{m-1}\boldsymbol{w}_{m-1}^{\mathrm{T}}\} \\
&= \Phi_{m-1}\boldsymbol{P}_{m-1}^+\Phi_{m-1}^{\mathrm{T}} + \boldsymbol{QD}_{m-1}
\end{aligned} \tag{7.122}$$

The derivation of the above equation uses Eq. (7.121). There is no cross term in the second step in the derivation process because $E\{(\delta\hat{\boldsymbol{x}}_{m-1}^+)(\boldsymbol{w}_{m-1}^{\mathrm{T}})\} = \boldsymbol{0}$. In $\boldsymbol{w}_{m-1}^{\mathrm{T}}$ only the influence $\hat{\boldsymbol{x}}_m^-$, $\boldsymbol{w}_{m-1}^{\mathrm{T}}$ and $\delta\hat{\boldsymbol{x}}_{m-1}^+$ are irrelevant. Therefore $E\{(\delta\hat{\boldsymbol{x}}_{m-1}^+)(\boldsymbol{w}_{m-1}^{\mathrm{T}})\} = E\{\delta\hat{\boldsymbol{x}}_{m-1}^+\}E\{\boldsymbol{w}_{m-1}^{\mathrm{T}}\}$. When $E\{\boldsymbol{w}_{m-1}^{\mathrm{T}}\} = 0$ so $E\{(\delta\hat{\boldsymbol{x}}_{m-1}^+)(\boldsymbol{w}_{m-1}^{\mathrm{T}})\} = 0$.

Now that we have a time prediction estimate $\hat{\boldsymbol{x}}_m^-$ for $\boldsymbol{x}_m$, the estimated error covariance matrix $\boldsymbol{P}_m^-$ is also known, so we can write the following equation, which is similar to Eq. (7.102):

$$\begin{bmatrix} \hat{\boldsymbol{x}}_m^- \\ \tilde{\boldsymbol{y}}_m \end{bmatrix} = \begin{bmatrix} \boldsymbol{I} \\ \boldsymbol{H}_m \end{bmatrix} \boldsymbol{x}_m + \begin{bmatrix} \delta\hat{\boldsymbol{x}}_m^- \\ \boldsymbol{v}_{m+1} \end{bmatrix} \tag{7.123}$$

So far, we can directly use the method of RLS analysis in Sect. 7.3.1, which is similar to the derivation of Eq. (7.110), to get the $\hat{\boldsymbol{x}}_m^+$ estimate of the x_{m}, as shown below.

$$\hat{\boldsymbol{x}}_m^+ = \hat{\boldsymbol{x}}_m^- + \boldsymbol{k}_m(\tilde{\boldsymbol{y}}_m - \boldsymbol{H}_m\hat{\boldsymbol{x}}_m^-) \tag{7.124}$$

The $\boldsymbol{k}_m = \boldsymbol{P}_m^+\boldsymbol{H}_m^{\mathrm{T}}\boldsymbol{R}_m^{-1}$ in Eq. (7.124) is called the Kalman gain matrix. It is used to weight the observed residuals and then update $\hat{\boldsymbol{x}}_m^+$. It is the value of the system state after the observation update. The error covariance matrix is $\boldsymbol{P}_m^+$. From Eq. (7.106), we know

$$(\boldsymbol{P}_m^+)^{-1} = (\boldsymbol{P}_m^-)^{-1} + \boldsymbol{H}_m^{\mathrm{T}}\boldsymbol{R}_m^{-1}\boldsymbol{H}_m \tag{7.125}$$

An intuitive analysis of the k_{m} expression shows that when the observation contains more noise, that is, when $R^{-1}{}_{\mathrm{m}}$ becomes larger, the corresponding k_{m} becomes smaller, indicating that the current system is more inclined to maintain the original estimation result. When the observation credibility is greater, $R^{-1}{}_{\mathrm{m}}$ becomes smaller, then k_{m} becomes larger, and the system allows more updates from the observations. The Kalman filtering process is actually an intelligent adaptive adjustment process. The estimation of the state parameters is based on the compromise between its current state parameter confidence and the reliability of external measurement.

Based on the above analysis, the Kalman filter can be divided into two steps. The first is to update the system state according to the system state transition equation, and to update the state co-variance matrix. The second step is to observe the measurement update. After the observation quantity occurs, the state co-variance matrix is first updated to calculate the Kalman gain and finally update the system state. The system state and co-variance matrix at the current moment become the initial conditions for the next iteration. In summary, the whole process is represented by the following equations:

$$\hat{\boldsymbol{x}}_m^- = \Phi_{m-1}\hat{\boldsymbol{x}}_{m-1}^+ + \boldsymbol{G}_{m-1}\boldsymbol{u}_{m-1} \tag{7.126}$$

$$\hat{\boldsymbol{y}}_m = \boldsymbol{H}_m\hat{\boldsymbol{x}}_m^- \tag{7.127}$$

$$\boldsymbol{P}_m^- = \Phi_{m-1}\boldsymbol{P}_{m-1}^+\Phi_{m-1}^{\mathrm{T}} + \boldsymbol{QD}_{m-1} \tag{7.128}$$

$$(\boldsymbol{P}_m^+)^{-1} = (\boldsymbol{P}_m^-)^{-1} + \boldsymbol{H}_m^{\mathrm{T}}\boldsymbol{R}_m^{-1}\boldsymbol{H}_m \tag{7.129}$$

$$\boldsymbol{k}_m = \boldsymbol{P}_m^+\boldsymbol{H}_m^{\mathrm{T}}\boldsymbol{R}_m^{-1} \tag{7.130}$$

$$\hat{\boldsymbol{x}}_m^+ = \hat{\boldsymbol{x}}_m^- + \boldsymbol{k}_m(\tilde{\boldsymbol{y}}_m - \hat{\boldsymbol{y}}_m) \tag{7.131}$$

Equations (7.126) to (7.128) are the operations for time update.Equations (7.129) to (7.131) are the operations for observation update. These are the basic equations of Kalman filtering.

In all of the above operations, Eq. (7.129) requires the highest amount of computation in all operations. It is necessary first to find the inverse of matrix $\boldsymbol{P}_m^-$ and then the inverse of $(\boldsymbol{P}_m^-)^{-1} + \boldsymbol{H}_m^{\mathrm{T}}\boldsymbol{R}_m^{-1}\boldsymbol{H}_m$. Matrix inversion has high requirements on the computing power of the processor, especially when the latitude of the matrix is large, so other calculation methods have been developed. Appendix A proves that $\boldsymbol{k}_m$ and $\boldsymbol{P}_m^+$ can be calculated by the following method:

$$\boldsymbol{k}_m = \boldsymbol{P}_m^-\boldsymbol{H}_m^{\mathrm{T}}(\boldsymbol{H}_m\boldsymbol{P}_m^-\boldsymbol{H}_m^{\mathrm{T}} + \boldsymbol{R}_m)^{-1} \tag{7.132}$$

$$\boldsymbol{P}_m^+ = (\boldsymbol{I} - \boldsymbol{k}_m\boldsymbol{H}_m)\boldsymbol{P}_m^- \tag{7.133}$$

There may be doubts here. From Eq. (7.132), this alternative method also needs to reverse the matrix $(\boldsymbol{H}_m\boldsymbol{P}_m^-\boldsymbol{H}_m^{\mathrm{T}}+\boldsymbol{R}_m)$, so what are the advantages of the new method? If we carefully compare Eqs. (7.132) and (7.129), we can see that the original method requires the inverse of $(\boldsymbol{P}_m^-)^{-1} + \boldsymbol{H}_m^{\mathrm{T}}\boldsymbol{R}_m^{-1}\boldsymbol{H}_m$ and $\boldsymbol{P}_m^-$. The dimension of the two matrices is determined by the dimension of the system state vector. When the system state parameters are constant, the dimensions of the two matrices are invariant, and Eq. (7.132) requires the inverse of $(\boldsymbol{H}_m\boldsymbol{P}_m^-\boldsymbol{H}_m^{\mathrm{T}} + \boldsymbol{R}_m)$. The dimension of the matrix is determined by the number of system observations, which may be different at different times. Section 6.2.6 shows that each observation order can be processed by means of sequential processing. At this time, each observation update only processes one observation, and the inverse of the matrix changes into calculating the reciprocal of the logarithm, thereby reducing the amount of calculation.

From Eqs. (7.126) ~ (7.132), the one-step prediction equation of Kalman filtering can also be derived, that is to use $\hat{\boldsymbol{x}}_m^-, \boldsymbol{P}_m^-$ to calculate the equations of $\hat{\boldsymbol{x}}_{m+1}^-, \boldsymbol{P}_{m+1}^-$.

$$\hat{\boldsymbol{x}}_{m+1}^- = \Phi_m\hat{\boldsymbol{x}}_m^- + \Phi_m\boldsymbol{k}_m(\tilde{\boldsymbol{y}}_m - \boldsymbol{H}_m\hat{\boldsymbol{x}}_m^-) + \boldsymbol{G}_m\boldsymbol{u}_m \tag{7.134}$$

$$\boldsymbol{P}_{m+1}^- = \Phi_m(\boldsymbol{I} - \boldsymbol{k}_m\boldsymbol{H}_m)\boldsymbol{P}_m^-\Phi_m^{\mathrm{T}} + \boldsymbol{Q}\boldsymbol{D}_m \tag{7.135}$$

The above derivation process shows that the Kalman filtering process does not have to store all the observations due to the recursive algorithm, which saves memory overhead and computational complexity. Although each observation update only utilizes the observation information at the current time, the system state quantity estimation $\hat{\boldsymbol{x}}_m^+$ includes all the observation information from the initialization time, and the information concentration in $\hat{\boldsymbol{x}}_m^+$ increases as the update time increases.

Another positive feature of Kalman filtering is that each filtering process is based on the system state vector of the previous moment Φ_m to "predict" the current time according to the system state transition matrix. This allows the system state vector to be non-stationary during the observation process. Only the statistical features of the system noise and the observed noise need to be known in the overall derivation process. The first-order and second-order moments of the estimated quantity do not need to be known. Therefore, a significant advantage of Kalman filtering over RLS is that it can estimate non-stationary quantities. Of course, the premise is that the state equation of the system is accurately known, while the system noise and observed noise are smooth white noise processes, and the statistical features do not change with time.

7.3.3 From Continuous Time Systems to Discrete Time Systems

In Sect. 7.3.2, all analyses are based on Eqs. (7.113) and (7.114), both of which are clearly discrete-time systems. All observations and state estimates are based on a certain time interval. In an actual discrete system, the time interval is often a fixed clock cycle T_s. However, in reality, a linear system presented to the system designer's original mathematical model is often continuous in time but not discrete. It is convenient for modern computer processing to convert a continuous time system into a discrete time system. Therefore, converting a continuous time system to a discrete time system is an important and fundamental task for system designers. This section describes how to complete this conversion.

Consider a continuous-time linear system, which is described by the following state observation difference equations:

$$\dot{\boldsymbol{x}}(t) = \boldsymbol{F}(t)\boldsymbol{x}(t) + \boldsymbol{G}(t)\boldsymbol{w}(t) \tag{7.136}$$

$$\boldsymbol{y}(t) = \boldsymbol{H}(t)\boldsymbol{x}(t) + \boldsymbol{v}(t) \tag{7.137}$$

where $x(t)$ is the dimensional system state vector of $(n \times 1)$; $\boldsymbol{F}(t)$ is the state transition matrix of $(n \times n)$; $\boldsymbol{G}(t)$ is the input transformation matrix of $(n \times l)$ (l is the dimension of the input variable); $\boldsymbol{y}(t)$ is the observation vector of $(m \times 1)$ (m is the number of observations); and $\boldsymbol{H}(t)$ is the observation matrix of $(m \times n)$. The system's processing noise and observed noise distribution are stochastic variables $\boldsymbol{w}(t)$ and $\boldsymbol{v}(t)$. All of the above variables are in the continuous time category.

In order to correspond Eqs. (7.136) and (7.137) with (7.113) and (7.114) relatively simply, the input variable $\boldsymbol{u}(t)$ is omitted here.

A reasonable assumption about stochastic variables $\boldsymbol{w}(t)$ and $\boldsymbol{v}(t)$ is to consider them as white noise with a mean of 0. The variance is known, and they are independent of each other:

$$E\{\boldsymbol{w}(t)\} = 0, \quad \text{var}\{\boldsymbol{w}(t)\boldsymbol{w}(t+\tau)^{\mathrm{T}}\} = \boldsymbol{Q}(t)\delta(\tau) \tag{7.138}$$

$$E\{\boldsymbol{w}(t)\} = 0, \quad \mathrm{var}\{\boldsymbol{w}(t)\boldsymbol{w}(t+\tau)^{\mathrm{T}}\} = \boldsymbol{Q}(t)\delta(\tau) \tag{7.139}$$

$$E\{\boldsymbol{w}(t)\boldsymbol{v}(t+\tau)^{\mathrm{T}}\} = \boldsymbol{0} \tag{7.140}$$

In the above equation, $\delta(\tau)$ is the Dirac function, that is, the representation in the continuous time domain $\delta_{j,l}$ in the Eqs. (7.115) and (7.116).

From the knowledge of linear systems, if there is a matrix $\Phi(t)$ of $(n \times n)$, the conditions are met:

$$\Phi(0) = I, \text{ and } \Phi(t) = F(t)\Phi(t), \forall t > 0$$

Then,

$$\boldsymbol{x}(t) = \Phi(t, t_0)\boldsymbol{x}(t_0) + \int_{t_0}^{t} \Phi(t, \lambda)\boldsymbol{G}(\lambda)\boldsymbol{w}(\lambda)\mathrm{d}\lambda \tag{7.141}$$

Among them, $\Phi(t, t_0) = \Phi(t)\Phi^{-1}(t_0)$ is called a state transition matrix on continuous time, and the matrix reveals the conversion of the system state from time t_0 to time (t). It is very easy to verify that $\Phi(t, t_0)$ has the following properties:

$$\Phi(t, t) = \boldsymbol{0} \tag{7.142}$$

$$\Phi(t, t_0) = \Phi^{-1}(t_0, t) \tag{7.143}$$

$$\Phi(t, \alpha)\Phi(\alpha, t_0) = \Phi(t, t_0) \tag{7.144}$$

$$\frac{\mathrm{d}\Phi(t, t_0)}{\mathrm{d}t} = \boldsymbol{F}(t)\Phi(t, t_0) \tag{7.145}$$

$$\frac{\mathrm{d}\Phi(t, t_0)}{\mathrm{d}t_0} = -\Phi(t, t_0)\boldsymbol{F}(t) \tag{7.146}$$

In order to examine whether Eq. (7.141) satisfies Eq. (7.136), we will derive Eq. (7.141) from the time t, and obtain:

$$\begin{aligned}
\dot{\boldsymbol{x}}(t) &= \frac{\mathrm{d}}{\mathrm{d}t}(\Phi(t, t_0)\boldsymbol{x}(t_0)) + \frac{\mathrm{d}}{\mathrm{d}t}\left(\int_{t_0}^{t} \Phi(t, \lambda)\boldsymbol{G}(\lambda)\boldsymbol{w}(\lambda)\mathrm{d}\lambda\right)\pi \\
&= \boldsymbol{F}(t)\Phi(t, t_0)\boldsymbol{x}(t_0) + \int_{t_0}^{t} \boldsymbol{F}(t)\Phi(t, \lambda)\boldsymbol{G}(\lambda)\boldsymbol{w}(\lambda)\mathrm{d}\lambda + \Phi(t, t)\boldsymbol{G}(t)\boldsymbol{w}(t) \\
&= \boldsymbol{F}(t)(\Phi(t, t_0)\boldsymbol{x}(t_0) + \int_{t_0}^{t} \Phi(t, \lambda)\boldsymbol{G}(\lambda)\boldsymbol{w}(\lambda)\mathrm{d}\lambda) + \boldsymbol{G}(t)\boldsymbol{w}(t) \\
&= \boldsymbol{F}(t)\boldsymbol{x}(t) + \boldsymbol{G}(t)\boldsymbol{w}(t)
\end{aligned}$$

When $F(t)$ is a constant,

$$\begin{aligned}\Phi(t) &= \mathrm{e}^{Ft} \Rightarrow \\ \Phi(t, t_0) &= \mathrm{e}^{F\Delta t}, \text{ here} \Delta t = t - t_0\end{aligned} \tag{7.147}$$

In engineering practice, $F(t)$ is often a variable. In this case, an approximate method is to divide $[t_0,t]$ into several small time segments, for example, $[t_0,t_1,t_2,\ldots,t_N]$, where $t_N = t$. In each small time period $t \in [t_i, t_{i+1})$, it can be considered that $F(t) = F_i$ is a constant, then $\Phi(t_i, t_{i+1}) = \mathrm{e}^{F_i \delta t_i}$, here $\delta t_i = t_{i+1} - t_i$. Hence, throughout the time period $[t_0, t]$,

$$\Phi(t, t_0) \approx \prod_{i=0}^{N} \mathrm{e}^{F_i \delta t_i}, i = 0, 1, \cdots, N \tag{7.148}$$

The calculation step of Eq. (7.148) can be represented by Fig. 7.11. In the figure, $[t_0,t]$ is divided into N time periods. The approximation $F(t)$ can be considered as a constant in each time period, so that each time period Φ_i can be calculated according to Eq. (7.147), and the entire time span is the total $\Phi(t, t_0) = \Phi_0 \Phi_1 \cdots \Phi_{N-1}$.

Now, let's consider discrete time systems. All updates to the discrete time system occur at certain time intervals T_s. If T_s is small, $F(t)$ can be assumed to be constant. This assumption will greatly simplify later analysis. Therefore, the state vector t_{k+1} at the moment t_k has the following relationship:

$$\boldsymbol{x}(t_{k+1}) = \mathrm{e}^{\boldsymbol{F}T_s} \boldsymbol{x}(t_k) + \int_{t_k}^{t_{k+1}} \mathrm{e}^{\boldsymbol{F}T_s} \boldsymbol{G}(\lambda) \boldsymbol{w}(\lambda) \mathrm{d}\lambda \tag{7.149}$$

Equation (7.149) shows that after the state transition from time t_0 to t, the input noise variable also becomes:

$$\boldsymbol{w}_d = \int_{t_k}^{t_{k+1}} \mathrm{e}^{\boldsymbol{F}T_s} \boldsymbol{G}(\lambda) \boldsymbol{w}(\lambda) \mathrm{d}\lambda \tag{7.150}$$

Based on the assumption that $w(t)$ is a white noise process, the covariance matrix of w_d can be calculated as follows:

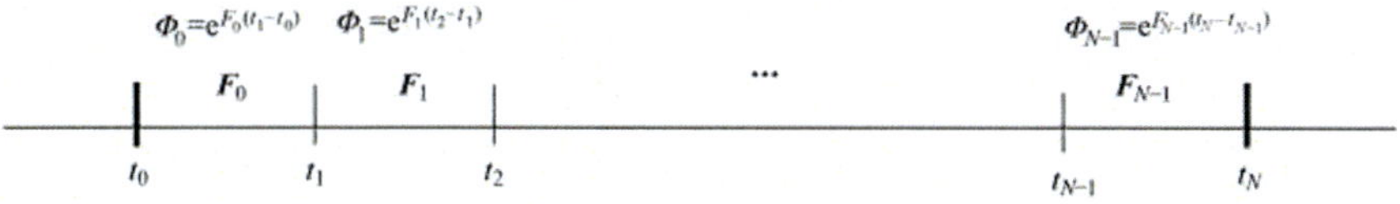

Fig. 7.11 The approximate method of calculating the $\Phi(t,t_0)$

$$
\begin{aligned}
QD_w &= E\{\boldsymbol{w}_d\boldsymbol{w}_d^{\mathrm{T}}\} \\
&= E\{\int_{t_k}^{t_{k+1}}\int_{t_k}^{t_{k+1}} \mathrm{e}^{\boldsymbol{F}(\lambda-t_k}\boldsymbol{G}(\lambda)\boldsymbol{w}(\lambda)\boldsymbol{w}^{\mathrm{T}}(\beta)\boldsymbol{G}^{\mathrm{T}}(\beta)(\mathrm{e}^{\boldsymbol{F}(\beta-t_k)})^{\mathrm{T}}\mathrm{d}\lambda\mathrm{d}\beta\} \\
&= \int_{t_k}^{t_{k+1}} \mathrm{e}^{\boldsymbol{F}(\lambda-t_k}\boldsymbol{G}(\lambda)\boldsymbol{Q}(\lambda)\boldsymbol{G}^{\mathrm{T}}(\lambda)(\mathrm{e}^{\boldsymbol{F}(\lambda-t_k)})^{\mathrm{T}}\mathrm{d}\lambda
\end{aligned} \tag{7.151}
$$

A rough approximation of Eq. (7.151) is:

$$
QD_w \approx \boldsymbol{GQG}^{\mathrm{T}}T_s \tag{7.152}
$$

The assumption T_s here is small enough that $\mathrm{e}^{\boldsymbol{F}(\lambda-t_k)} \approx \boldsymbol{I}$, $\forall\lambda \in [t_k, t_{k+1})$. A more precise approximation of Eq. (7.151) is first to use the Taylor series to expand $\mathrm{e}^{\boldsymbol{F}(\lambda-t_k)}$,

$$
\mathrm{e}^{\boldsymbol{F}T} = \boldsymbol{I} + \boldsymbol{F}T + \frac{1}{2!}\boldsymbol{F}^2T^2 + \frac{1}{3!}\boldsymbol{F}^3T^3 + \cdots
$$

and substitute into Eq. (7.151) to get

$$
\begin{aligned}
QD_w \approx \boldsymbol{Q}_G T + (\boldsymbol{F}\boldsymbol{Q}_G + \boldsymbol{Q}_G\boldsymbol{F}^{\mathrm{T}})\frac{T^2}{2!} \\
(\boldsymbol{F}^2\boldsymbol{Q}_G + 2\boldsymbol{F}\boldsymbol{Q}_G\boldsymbol{F}^{\mathrm{T}} + \boldsymbol{Q}_G(\boldsymbol{F}^{\mathrm{T}})^2)\frac{T^3}{3!} + \cdots
\end{aligned} \tag{7.153}
$$

Among them, $\boldsymbol{Q}_G = \boldsymbol{GQG}^{\mathrm{T}}$. Here, we only expand it to the third item in the Taylor series. Readers can expand the rest independently.

So far, we have understood the transition from continuous-time systems to discrete-time systems, including the transformation of system states and the conversion of covariance matrices of input noise variables. The following example in the BDS/GPS receiver will be used to help readers understand this section more deeply.

Consider a two-state linear system. The state vector is $\boldsymbol{x} = [x_1, x_2]^{\mathrm{T}}$. If we input the noise vector $\boldsymbol{w} = [w_1, w_2]^{\mathrm{T}}$, its continuous-time equation of state is:

$$
\begin{bmatrix} \dot{x}_1 \\ \dot{x}_2 \end{bmatrix} = \begin{bmatrix} 0 & 1 \\ 0 & 0 \end{bmatrix}\begin{bmatrix} x_1 \\ x_2 \end{bmatrix} + \begin{bmatrix} w_1 \\ w_2 \end{bmatrix} \tag{7.154}
$$

The state vector in Eq. (7.154) can be applied to the position and velocity in the receiver, that is, either $[P, v]^{\mathrm{T}}$ or the clock difference and the clock drift $[b, \dot{b}]^{\mathrm{T}}$. Note that P and v here are one-dimensional. Multidimensions can be treated similarly. As $[w_1, w_2]^{\mathrm{T}}$ is an input to the system, it is called the system handled noise of $[x_1, x_2]^{\mathrm{T}}$.

If we assume $\mathrm{var}\{w_1(t)\} = S_1$, $\mathrm{var}\{w_2(t)\} = S_2$, then:

$$
\boldsymbol{Q}(t) = \begin{bmatrix} S_1 & 0 \\ 0 & S_2 \end{bmatrix} \tag{7.155}
$$

At the same time, we know from Eq. (7.154) that:

$$\boldsymbol{F}(t)=\begin{bmatrix}0\ 1\\0\ 0\end{bmatrix},\boldsymbol{G}(t)=\begin{bmatrix}1\ 0\\0\ 1\end{bmatrix} \tag{7.156}$$

$F(t)$and $G(t)$ are constants. With calculation, it can be proven that $\boldsymbol{F}^n=\boldsymbol{0},\quad n\geq 2$. This property will simplify the subsequent derivation.

According to Eq. (7.147), we know that:

$$\begin{aligned}\Phi(t,t_0)&=\mathrm{e}^{\boldsymbol{F}(t-t_0)}\\&=\boldsymbol{I}+\boldsymbol{F}(t-t_0)\\&=\begin{bmatrix}1 & (t-t_0)\\0 & 1\end{bmatrix}\end{aligned} \tag{7.157}$$

The second line of the above equation is to expand the Taylor series $\mathrm{e}^{\boldsymbol{F}(t-t_0)}$ and use $\boldsymbol{F}^n=\boldsymbol{0},\quad n\geq 2$.

According to Eq. (7.153), QD_{w} is known:

$$\begin{aligned}\boldsymbol{QD}_w&\approx\boldsymbol{Q}_G\Delta t+(\boldsymbol{FQ}_G+\boldsymbol{Q}_G\boldsymbol{F}^{\mathrm{T}})\frac{\Delta t^2}{2!}+2\boldsymbol{FQ}_G\boldsymbol{F}^{\mathrm{T}}\frac{\Delta t^3}{3!}\\&=\begin{bmatrix}S_1\Delta t+S_2\frac{\Delta t^3}{3} & S_2\frac{\Delta t^2}{2}\\S_2\frac{\Delta t^2}{2} & S_2\Delta t\end{bmatrix}\end{aligned} \tag{7.158}$$

where $\Delta t=(t-t_0)$.

Then, the discrete time representation of the system is:

$$\begin{bmatrix}x_1(k)\\x_2(k)\end{bmatrix}=\begin{bmatrix}1 & T_s\\0 & 1\end{bmatrix}\begin{bmatrix}x_1(k-1)\\x_2(k-1)\end{bmatrix}+\begin{bmatrix}w_1(k-1)\\w_2(k-1)\end{bmatrix} \tag{7.159}$$

Here T_{s} is the time interval between two samples:

$$E\left\{\begin{bmatrix}w_1(k-1)\\w_2(k-1)\end{bmatrix}[w_1(k-1),w_2(k-1)]\right\}=\boldsymbol{QD}_w \tag{7.160}$$

QD_{w} is calculated by Eq. (7.158).

7.3.4 The Extended Kalman Filter

The Kalman filters described in the previous sections are all based on linear systems, but in practice, it cannot be guaranteed that the system state equations or observation equations will be linear. In such cases, the conventional linear Kalman filter is not applicable. Before applying Kalman filtering to a nonlinear system, it is often necessary to linearize it beforehand. This is known as the Extended Kalman Filter (EKF).

Suppose a nonlinear system is described by the following difference equations:

$$\dot{\boldsymbol{x}}(t) = \boldsymbol{f}(\boldsymbol{x}, \boldsymbol{u}, t) + \boldsymbol{g}(\boldsymbol{x}, t)\boldsymbol{w}(t) \tag{7.161}$$

$$\boldsymbol{y}(t) = \boldsymbol{h}(\boldsymbol{x}, t) + \boldsymbol{v}(t) \tag{7.162}$$

Equation (7.161) is the system state transition equation, and Eq. (7.162) the observation equation. $f(x,u,t)$ is a nonlinear equation about the state of the system x, the input quantity u, and t. $h(x,t)$ is a nonlinear equation about x and t. $w(t)$ and $v(t)$ are the processing noise and observed noise in the continuous time domain respectively.

If the state of the system $x(t)$ at the moment t is unknown, but we know its approximate range, we can choose a value $x^*(t)$. So long as we obtain the estimate of $\delta \boldsymbol{x}(t) = \boldsymbol{x}(t) - \boldsymbol{x}^*(t)$, we can correct $x^*(t)$ and get a more accurate estimate of $x(t)$.

When selecting $x^*(t)$, we need to ensure that the following equations are established:

$$\dot{\boldsymbol{x}}^*(t) = \boldsymbol{f}(\boldsymbol{x}^*, \boldsymbol{u}, t) \tag{7.163}$$

$$\boldsymbol{y}^*(t) = \boldsymbol{h}(\boldsymbol{x}^*, t) \tag{7.164}$$

Then, we use Eq. (7.161) and Eq. (7.163) to subtract. Equation (7.162) and Eq. (7.164) are subtracted. The Taylor series is used to expand, and the high-order term is omitted. We can obtain:

$$\delta\dot{\boldsymbol{x}}(t) = \boldsymbol{F}(t)\delta\boldsymbol{x}(t) + \boldsymbol{g}(\boldsymbol{x}, t)\boldsymbol{w}(t) \tag{7.165}$$

$$\delta\boldsymbol{y}(t) = \boldsymbol{H}(t)\delta\boldsymbol{x}(t) + \boldsymbol{v}(t) \tag{7.166}$$

where

$$\boldsymbol{F}(t) = \left.\frac{\partial \boldsymbol{f}(\boldsymbol{x},\boldsymbol{u},t)}{\partial \boldsymbol{x}}\right|_{\boldsymbol{x}=\boldsymbol{x}^*}, \boldsymbol{H}(t) = \left.\frac{\partial \boldsymbol{h}(\boldsymbol{x},t)}{\partial \boldsymbol{x}}\right|_{\boldsymbol{x}=\boldsymbol{x}^*}.$$

Equation (7.165) and Eq. (7.166) are consistent with the state equations and observation equations of the linear Kalman filter, so Kalman filtering can be implemented in a very similar way to the steps in Sect. 7.3.2. The only differences are the time update and prediction of the system state observation. At the same time, it should be noted that Eq. (7.165) and Eq. (7.166) are only applicable when $x^*(t)$ is close to the real state of the system. When the difference between $x^*(t)$ and the real state is too large, the linearization of the result of the first-order Taylor series expansion will contain a large error. Therefore, after the estimation of $\delta x(t)$ is completed by the Kalman filter, it is very important to update $x^*(t)$ in time.

Combining the above analysis with the principles of conventional linear Kalman filtering, the basic steps of extending the Kalman filter are as follows.

1. Time update of system status

Suppose that the estimation of the system state at the moment t_k is $\hat{\boldsymbol{x}}_{k|k}$, according to the equation $\dot{\boldsymbol{x}}^*(t) = \boldsymbol{f}(\boldsymbol{x}^*, \boldsymbol{u}, t)$, substitute $\hat{\boldsymbol{x}}_{k|k}$ to obtain $\dot{\boldsymbol{x}}^*(t)|_{\hat{\mathbf{x}}_{k|k}}$. Integrate it over time $[t_k, t_{k+1})$, and $\hat{\boldsymbol{x}}_{k+1|k}$ is obtained, that is:

$$\hat{\boldsymbol{x}}_{k+1|k} = \hat{\boldsymbol{x}}_{k|k} + \int_{t_k}^{t_{k+1}} \dot{\boldsymbol{x}}^*(t)|_{\hat{\mathbf{x}}_{k|k}} \mathrm{d}t \tag{7.167}$$

2. Calculating the observed residuals

Calculate the predicted observation according to the $\hat{\boldsymbol{x}}_{k+1|k}$ in the first step:

$$\widehat{y}_{k+1} = h(\widehat{x}_{k+1|k}, t_{k+1}) \tag{7.168}$$

Then calculate the observation residual between the actual observation and the predicted observation:

$$z_{k+1} = \tilde{\boldsymbol{y}}_{k+1} - \hat{\boldsymbol{y}}_{k+1} \tag{7.169}$$

This step shows that the predicted observation $\hat{\boldsymbol{y}}_{k+1}$ is obtained by state quantities $\hat{\boldsymbol{x}}_{k+1|k}$, which is based on nonlinear equations.

Time update of the error covariance matrix $\boldsymbol{P}_{k|k}$ and error state $\delta\hat{\boldsymbol{x}}_{k|k}$.

Substitute the obtained $\hat{\boldsymbol{x}}_{k+1|k}$ from step 1 into $\boldsymbol{F}(t) = \frac{\partial f}{\partial x}$ to obtain the matrix of $\boldsymbol{F}(t)|_{\hat{\boldsymbol{x}}_{k+1|k}}$ at the time t_{k+1}, and then use the method in Sect. 7.3.3 to calculate Φ_k and $\boldsymbol{QD}_w$. Perform time update to $\boldsymbol{P}_{k|k}$ and $\delta\hat{\boldsymbol{x}}_{k|k}$ and obtain:

$$\delta\hat{\boldsymbol{x}}_{k+1|k} = \Phi_k \delta\hat{\boldsymbol{x}}_{k|k} = \Phi_k \boldsymbol{0} = \boldsymbol{0} \tag{7.170}$$

$$\boldsymbol{P}_{k+1|k} = \Phi_k \boldsymbol{P}_{k|k} \Phi_k^T + \boldsymbol{QD}_w \tag{7.171}$$

The reason why $\delta\hat{\boldsymbol{x}}_{k|k} = \boldsymbol{0}$ will be illustrated in step 6.

Linearize the observation at time t_{k+1} and calculate the Kalman gain matrix $\boldsymbol{k}_{k+1}$.

The observation equation $\boldsymbol{h}(\boldsymbol{x}^*, t)$ is linearized when $\hat{\boldsymbol{x}}_{k+1|k}$. $\boldsymbol{H}_{k+1} = \left.\frac{\partial \boldsymbol{h}}{\partial \boldsymbol{x}}\right|_{\hat{\boldsymbol{x}}_{k+1|k}}$ is obtained, so the Kalman gain matrix is

$$\boldsymbol{k}_{k+1} = \boldsymbol{P}_{k+1|k} \boldsymbol{H}_{k+1}^{\mathrm{T}} [\boldsymbol{H}_{k+1} \boldsymbol{P}_{k+1|k} \boldsymbol{H}_{k+1}^{\mathrm{T}} + \boldsymbol{R}_{k+1}]^{-1} \tag{7.172}$$

Update the observation of the error state $\delta\hat{\boldsymbol{x}}_{k+1|k}$

$$\delta\hat{\boldsymbol{x}}_{k+1|k+1} = \delta\hat{\boldsymbol{x}}_{k+1|k} + \boldsymbol{k}_{k+1} z_{k+1}$$

$$= \boldsymbol{k}_{k+1} z_{k+1} \tag{7.173}$$

Here $\delta\hat{\boldsymbol{x}}_{k+1|k}$ is the time update of the error state $\delta\hat{\boldsymbol{x}}_{k|k}$ in the third step.

Update system status $\hat{\boldsymbol{x}}_{k+1|k+1}$ and $\boldsymbol{P}_{k+1|k+1}$.

Now that the error status $\delta\hat{\boldsymbol{x}}_{k+1|k+1}$ after the observation update is obtained, the time update of the system status $\hat{\boldsymbol{x}}_{k+1|k}$ has been obtained in the first step, so the observation update of $\hat{\boldsymbol{x}}_{k+1|k}$ can be performed:

$$\hat{\boldsymbol{x}}_{k+1|k+1} = \hat{\boldsymbol{x}}_{k+1|k} + \delta\hat{\boldsymbol{x}}_{k+1|k+1} \tag{7.174}$$

After this update, $\hat{\boldsymbol{x}}_{k+1|k+1} = E\{\boldsymbol{x}(t_{k+1)}\}$. Hence, the new information contained in $\delta\hat{\boldsymbol{x}}_{k+1|k+1}$ has already been used, so $\delta\hat{\boldsymbol{x}}_{k+1|k+1} = \boldsymbol{0}$ needs to be set, which explains why $\delta\hat{\boldsymbol{x}}_{k|k} = \boldsymbol{0}$ in step 3.

At the same time, the error co-variance matrix needs to be updated:

$$\boldsymbol{P}_{k+1|k+1} = (\boldsymbol{I} - \boldsymbol{k}_{k+1}\boldsymbol{H}_{k+1})\boldsymbol{P}_{k+1|k} \tag{7.175}$$

Steps 1 to 3 are time updates, and steps 4 to 6 are observation updates. The result of each processing $\boldsymbol{P}_{k+1|k+1}$ and $\hat{\boldsymbol{x}}_{k+1|k+1}$ are saved as the starting condition for the next iteration.

In Eq. (7.174), the estimation of the system state error is obtained by observing the residual and the Kalman gain matrix. It is a very critical step to use this estimate to correct the time update of the system's full state $\hat{\boldsymbol{x}}_{k+1|k}$. Only after this update can $\delta\hat{\boldsymbol{x}}_{k+1|k+1}$ be set to 0. This step also ensures that the system state $\hat{\boldsymbol{x}}_{k+1|k+1}$ is always in the vicinity of the near-value range of the real state of the system, thus ensuring the correctness of subsequent linearization.

In GPS receivers, pseudo-range observation is a nonlinear function of the system state. Therefore, the extended Kalman filter described in this section has been widely used in modern GPS receivers, in the BDS/GPS dual-mode receiver. The dual mode observation is slightly more complicated than the single GPS pseudo-range observation, but the extended Kalman filter principle described above is the same. Section 7.3.5 will analyze the models of several Kalman filters commonly used in receivers, and combine the steps in this section to analyze how to implement extended Kalman filters.

7.3.5 Several KF Models Commonly Used in Receivers

1. Static user: P model

When the receiver is in a static state, since the speed is constant to 0, it is only necessary to use the position coordinates and the clock as the system state, that is, $\boldsymbol{x} = [\boldsymbol{x}_{\mathrm{p}}^{\mathrm{T}}, \boldsymbol{x}_{\mathrm{c}}^{\mathrm{T}}]^{\mathrm{T}}$. The position state vector is $\boldsymbol{x}_{\mathrm{p}} = [x, y, z]^{\mathrm{T}}$; the clock state vector is

$\boldsymbol{x}_\mathbf{c} = [b, d]^\mathrm{T}$; the local clock difference b; and the local clock drift is d. In general, the clock vector must be included in the system state, so this KF model is called the P model for the sake of highlighting $\boldsymbol{x}_\mathbf{p}$.

In the P model, the position state is considered to be a random walk process. The system state equation of the P model is

$$\begin{bmatrix} \dot{x} \\ \dot{y} \\ \dot{z} \\ \dot{b} \\ \dot{d} \end{bmatrix} = \begin{bmatrix} 0\,0\,0\,0\,0 \\ 0\,0\,0\,0\,0 \\ 0\,0\,0\,0\,0 \\ 0\,0\,0\,0\,1 \\ 0\,0\,0\,0\,0 \end{bmatrix} \begin{bmatrix} x \\ y \\ z \\ b \\ d \end{bmatrix} + \begin{bmatrix} w_x \\ w_y \\ w_z \\ w_b \\ w_d \end{bmatrix} \tag{7.176}$$

$\boldsymbol{w}_\mathrm{p} = \begin{bmatrix} w_x \\ w_y \\ w_z \end{bmatrix}$ and $\boldsymbol{w}_c = \begin{bmatrix} w_b \\ w_d \end{bmatrix}$ are the processing noise vector of the position and the processing noise vector of the clock respectively.

As explained in Chap. 5, the observations include pseudo-range observations and Doppler observations. In the P model, since the system state does not include the velocity vector, it is generally sufficient to use pseudo-range observation. If you need to use Doppler observation, you must set its speed component to zero. The observation equation for single-mode pseudo-range observation is:

$$\rho_i = \sqrt{(x - x_{si})^2 + (y - y_{si})^2 + (z - z_{si})^2} + cb + n_i, i = 1, \cdots, m \tag{7.177}$$

In the above equation, c is the speed of light, $[x_\mathrm{si}, y_\mathrm{si}, z_\mathrm{si}]^\mathrm{T}$ is the position of the satellite, and n_i is the pseudo-range noise. If it is a BDS or GPS dual-mode receiver, and the method of considering T_GB as the state quantity of the system to be estimated, the observation equation of GPS and BDS pseudo-range observation is:

$$\tilde{\rho}_{\mathrm{G}i} = \sqrt{(x_\mathrm{u} - x_{\mathrm{Gs}_i})^2 + (y_\mathrm{u} - y_{\mathrm{Gs}_i})^2 + (z_\mathrm{u} - z_{\mathrm{Gs}_i})^2} + cb + n_{\rho \mathrm{G}i} \tag{7.178}$$

$$\tilde{\rho}_{\mathrm{B}i} = \sqrt{(x_\mathrm{u} - x_{\mathrm{Bs}_i})^2 + (y_\mathrm{u} - y_{\mathrm{Bs}_i})^2 + (z_\mathrm{u} - z_{\mathrm{Bs}_i})^2} + cb + cT_\mathrm{GB} + n_{\rho \mathrm{B}i} \tag{7.179}$$

At the same time, the equation of the system state should be adjusted to:

$$\begin{bmatrix} \dot{x} \\ \dot{y} \\ \dot{z} \\ \dot{T}_\mathrm{GB} \\ \dot{b} \\ \dot{d} \end{bmatrix} = \begin{bmatrix} 0\,0\,0\,0\,0\,0 \\ 0\,0\,0\,0\,0\,0 \\ 0\,0\,0\,0\,0\,0 \\ 0\,0\,0\,0\,0\,0 \\ 0\,0\,0\,0\,0\,1 \\ 0\,0\,0\,0\,0\,0 \end{bmatrix} \begin{bmatrix} x \\ y \\ z \\ T_\mathrm{GB} \\ b \\ d \end{bmatrix} + \begin{bmatrix} w_x \\ w_y \\ w_z \\ w_T \\ w_b \\ w_d \end{bmatrix} \tag{7.180}$$

The difference between Eq. (7.180) and Eq. (7.176) is that there is T_{GB} in the system state quantity of the former, with w_T during processing of noise vectors. That is, T_{GB} is also modeled as a random walk process. Since the T_{GB} changes little when time changes, the variance term w_T to which the Q matrix corresponds is set to a small value.

The equation for Doppler observation is:

$$f'_{d\_i} = cd + v_i, i = 1, \cdots m \tag{7.181}$$

Here, $f'_{d\_i}, i = 1, \cdots, m$ is the linearized Doppler observation in the equation, that is, the Doppler shift obtained from the NCO of the pseudo-code tracking loop minus the projection component of the satellite velocity on the direction cosine, and c is the speed of light. Since the receiver speed is 0, only the clock drift term is left in the $\boldsymbol{H}\boldsymbol{x}_v$.

Low dynamic users: PV model

When the user is in a low dynamic motion environment, the PV model should be used. Applications for such models include smooth driving of vehicles and boats, and walkers. In this model, the velocity component is considered to be a random walk, while the position component is the integral of the velocity component, and the clock model remains unchanged. The system state vector is $\boldsymbol{x} = [\boldsymbol{x}_{\mathbf{p}}^{\mathrm{T}}, \boldsymbol{x}_{\mathbf{v}}^{\mathrm{T}}, \boldsymbol{x}_{\mathbf{c}}^{\mathrm{T}}]^{\mathrm{T}}$, $\boldsymbol{x}_{\mathbf{v}} = [v_x, v_y, v_z]^{\mathrm{T}}$ is the velocity state vector, and the meaning of $\boldsymbol{x}_{\mathbf{p}}$ and $\boldsymbol{x}_{\mathbf{c}}$ remain the same.

The system state equation of the PV model is:

$$\begin{bmatrix} \dot{x} \\ \dot{y} \\ \dot{z} \\ \dot{v}_x \\ \dot{v}_y \\ \dot{v}_z \\ \dot{b} \\ \dot{d} \end{bmatrix} = \begin{bmatrix} 0&0&0&1&0&0&0&0 \\ 0&0&0&0&1&0&0&0 \\ 0&0&0&0&0&1&0&0 \\ 0&0&0&0&0&0&0&0 \\ 0&0&0&0&0&0&0&0 \\ 0&0&0&0&0&0&0&0 \\ 0&0&0&0&0&0&0&1 \\ 0&0&0&0&0&0&0&0 \end{bmatrix} \begin{bmatrix} x \\ y \\ z \\ v_x \\ v_y \\ v_z \\ b \\ d \end{bmatrix} + \begin{bmatrix} w_x \\ w_y \\ w_x \\ w_{v_x} \\ w_{v_y} \\ w_{v_z} \\ w_b \\ w_d \end{bmatrix} \tag{7.182}$$

Here, $\boldsymbol{w}_v = [w_{v_x}, w_{v_y}, w_{v_z}]^{\mathrm{T}}$ is the processing noise vector of the speed state. The processing noise vectors for the position and clock states $\boldsymbol{w}_p$ and $\boldsymbol{w}_c$ are the same as defined in the P model. In the PV model, the processing noise vector of the position state $\boldsymbol{w}_p = \mathbf{0}$ can be considered. That is, the position state is considered to have no processing noise, and the position state can be perfectly determined by the integral of the speed state.

The single-mode pseudo-range observation has no change compared with the P model. The BDS and GPS dual-mode pseudo-range observations are the same as Eq. (7.178) and Eq. (7.179), and the system state equation becomes Eq. (7.184). The dual-mode system has more instances of T_{GB}. The T_{GB} is also modeled as a random walk. Since the T_{GB} hardly changes with time changes, the corresponding variance w_{T} term in the Q matrix can be set to a small value. It should be noted that if the scheme of reading T_{GB} through system setting is adopted, the system state equation does not include T_{GB}, and the system state equation takes the form of Eq. (7.182). At this time, the mathematical model of dual-mode pseudo-range observation adopts the form of Eq. (7.177).

The Doppler observations under the PV model differ from Eq. (7.181). Since the velocity state vector is not zero at this time, the Doppler observation takes the form of (7.183):

$$f'_{d\_i} = h_{x\_i}v_x + h_{y\_i}v_y + h_{z\_i}v_z + cd + v_i, i = 1, \cdots m \tag{7.183}$$

$$\begin{bmatrix} \dot{x} \\ \dot{y} \\ \dot{z} \\ \dot{v}_x \\ \dot{v}_y \\ \dot{v}_z \\ \dot{T}_{\text{GB}} \\ \dot{b} \\ \dot{d} \end{bmatrix} = \begin{bmatrix} 0\,0\,0\,1\,0\,0\,0\,0\,0 \\ 0\,0\,0\,0\,1\,0\,0\,0\,0 \\ 0\,0\,0\,0\,0\,1\,0\,0\,0 \\ 0\,0\,0\,0\,0\,0\,0\,0\,0 \\ 0\,0\,0\,0\,0\,0\,0\,0\,0 \\ 0\,0\,0\,0\,0\,0\,0\,0\,0 \\ 0\,0\,0\,0\,0\,0\,0\,0\,0 \\ 0\,0\,0\,0\,0\,0\,0\,0\,1 \\ 0\,0\,0\,0\,0\,0\,0\,0\,0 \end{bmatrix} \begin{bmatrix} x \\ y \\ z \\ v_x \\ v_y \\ v_z \\ T_{\text{GB}} \\ b \\ d \end{bmatrix} + \begin{bmatrix} w_x \\ w_y \\ w_x \\ w_{v_x} \\ w_{v_y} \\ w_{v_z} \\ w_T \\ w_b \\ w_d \end{bmatrix} \tag{7.184}$$

The direction cosine vector $\boldsymbol{H} = [h_{x\_i}, h_{y\_i}, h_{z\_i}]^{\text{T}}$ in Eq. (7.183) is the same as defined in Sect. 7.1.3. Since the velocity vector is no longer zero at this time, the corresponding matrix elements in the Doppler observation equation are no longer zero.

High dynamic users: PVA model

In some applications, the user's motion acceleration varies widely, such as a high-speed aircraft. In this case, three acceleration components need to be added to the system state vector. This model is called a PVA model. Compared with the PV model, the PVA model has three more acceleration components and one more clock high-order term. Because the state component of the PVA model is relatively large, for the sake of simplicity, the expression of the block matrix will be used in the following equation.

In the PVA model, the system state vector is $\boldsymbol{x} = [\boldsymbol{x}_p^{\text{T}}, \boldsymbol{x}_v^{\text{T}}, \boldsymbol{x}_a^{\text{T}}, \boldsymbol{x}_c^{\text{T}}]^{\text{T}}$, where $\boldsymbol{x}_p$,$\boldsymbol{x}_v$, and $\boldsymbol{x}_c$ are the position state vector, the velocity state vector, and the clock state vector respectively. The definition of $\boldsymbol{x}_p$ and $\boldsymbol{x}_v$ are the same as the definition in the PV model, and $\boldsymbol{x}_c = [b, d, j]^{\text{T}}$ is the newly added acceleration state vector. At this time, $\boldsymbol{x}_c$ contains three quantities, namely clock bias, clock drift, and clock drift acceleration, that is, $x_{\text{c}} = [b,d,j]^{\text{T}}$ where j is the acceleration of the clock drift, and j can be modeled as a random walk process, then x_{c} can satisfy the following equation:

$$\dot{\boldsymbol{x}}_c = \boldsymbol{F}_c\boldsymbol{x}_c + \boldsymbol{w}_c \tag{7.185}$$

where $\boldsymbol{F}_c = \begin{bmatrix} 0\,1\,0 \\ 0\,0\,1 \\ 0\,0\,0 \end{bmatrix}, \boldsymbol{w}_c = \begin{bmatrix} w_b \\ w_d \\ w_j \end{bmatrix}$.

In the PVA model, the position state is the integral of the velocity state, the velocity state is the integral of the acceleration state, and the acceleration state can be approximated by the first-order Markov process, that is:

$$\begin{aligned} \dot{\boldsymbol{x}}_p &= \boldsymbol{x}_v \\ \dot{\boldsymbol{x}}_v &= \boldsymbol{x}_a \\ \dot{\boldsymbol{x}}_a &= \boldsymbol{D}\boldsymbol{x}_a \end{aligned} \tag{7.186}$$

where $\boldsymbol{D} = \mathrm{diag}\left(-\frac{1}{\tau_x}, -\frac{1}{\tau_y}, -\frac{1}{\tau_z}\right)$ is a diagonal matrix. τ_x, τ_y, and τ_z are the respective time-dependent constants of the three accelerations.

Based on the above analysis, the system state equation of the PVA model can be written as:

$$\begin{bmatrix} \dot{\boldsymbol{x}}_p \\ \dot{\boldsymbol{x}}_v \\ \dot{\boldsymbol{x}}_a \\ \dot{\boldsymbol{x}}_c \end{bmatrix} = \begin{bmatrix} \boldsymbol{0} & \boldsymbol{I} & \boldsymbol{0} & \boldsymbol{0} \\ \boldsymbol{0} & \boldsymbol{0} & \boldsymbol{I} & \boldsymbol{0} \\ \boldsymbol{0} & \boldsymbol{0} & \boldsymbol{D} & \boldsymbol{0} \\ \boldsymbol{0} & \boldsymbol{0} & \boldsymbol{0} & \boldsymbol{F}_c \end{bmatrix} \begin{bmatrix} \boldsymbol{x}_p \\ \boldsymbol{x}_v \\ \boldsymbol{x}_a \\ \boldsymbol{x}_c \end{bmatrix} + \begin{bmatrix} \boldsymbol{w}_p \\ \boldsymbol{w}_v \\ \boldsymbol{w}_a \\ \boldsymbol{w}_c \end{bmatrix} \tag{7.187}$$

where I is the 3×3 unit matrix.

For the case of BDS and GPS dual-mode observation, similar to the P model and the PV model, it is only necessary to add the T_{GB} term in the system state vector. The system state equation is as follows:

$$\begin{bmatrix} \dot{\boldsymbol{x}}_p \\ \dot{\boldsymbol{x}}_v \\ \dot{\boldsymbol{x}}_a \\ \dot{\boldsymbol{T}}_{\mathrm{GB}} \\ \dot{\boldsymbol{x}}_c \end{bmatrix} = \begin{bmatrix} \boldsymbol{0} & \boldsymbol{I} & 0 & 0 & \boldsymbol{0} \\ \boldsymbol{0} & \boldsymbol{0} & \boldsymbol{I} & 0 & \boldsymbol{0} \\ \boldsymbol{0} & \boldsymbol{0} & \boldsymbol{D} & 0 & \boldsymbol{0} \\ 0 & 0 & 0 & 0 & 0 \\ \boldsymbol{0} & \boldsymbol{0} & \boldsymbol{0} & 0 & \boldsymbol{F}_c \end{bmatrix} \begin{bmatrix} \boldsymbol{x}_p \\ \boldsymbol{x}_v \\ \boldsymbol{x}_a \\ \boldsymbol{T}_{\mathrm{GB}} \\ \boldsymbol{x}_c \end{bmatrix} + \begin{bmatrix} \boldsymbol{w}_p \\ \boldsymbol{w}_v \\ \boldsymbol{w}_a \\ \boldsymbol{w}_T \\ \boldsymbol{w}_c \end{bmatrix} \tag{7.188}$$

In the above equation, T_{GB} is still modeled as a random walk. The two-mode pseudo-range observation and Doppler observation did not change compared to the PV model. The Kalman filter which adopts Eq. (7.185) contains 12 state quantities, and the Kalman filter which adopts Eq. (7.188) 13 state quantities. When using single mode or dual mode pseudo-range observation and Doppler observation, the corresponding elements of the observation matrix are needed to be adjusted and the details will not be described here.

Latitude and longitude model

The three system models described above are all in the ECEF coordinate system. The latitude and longitude model represents the position coordinates in the geodetic coordinate system. At this time, the position state vector is $\boldsymbol{x_p} = [\phi, \lambda, h]^{\mathrm{T}}$, where ϕ, λ, and h are the longitude, latitude, and altitude of the receiver respectively. In the corresponding velocity state vector $\boldsymbol{x_v} = [v_n, v_e, v_d]^{\mathrm{T}}$, $\boldsymbol{x_v}$ contains the velocity component of the northeast direction. The clock state vector is $\boldsymbol{x_c} = [b, d]^{\mathrm{T}}$, b is the local clock difference, d is the local clock drift. So the state vector under this system model is $\boldsymbol{x} = [\boldsymbol{x}_p^{\mathrm{T}}, \boldsymbol{x}_v^{\mathrm{T}}, \boldsymbol{x}_c^{\mathrm{T}}]^{\mathrm{T}}$, which contains a total of 8 state variables.

The system state equation of the latitude and longitude model can be obtained according to Eqs. (1.39) and (1.40) in Chap. 1:

$$\begin{bmatrix} \dot{\phi} \\ \dot{\lambda} \\ \dot{h} \\ \dot{v}_n \\ \dot{v}_e \\ \dot{v}_d \\ \dot{b} \\ \dot{d} \end{bmatrix} = \begin{bmatrix} 0\,0\,0 & \frac{1}{R_M+h} & 0 & 0\;0\,0 \\ 0\,0\,0 & 0 & \frac{1}{(R_N+h)\cos\phi} & 0\;0\,0 \\ 0\,0\,0 & 0 & 0 & -1\,0\,0 \\ 0\,0\,0 & 0 & 0 & 0\;0\,0 \\ 0\,0\,0 & 0 & 0 & 0\;0\,0 \\ 0\,0\,0 & 0 & 0 & 0\;0\,0 \\ 0\,0\,0 & 0 & 0 & 0\;0\,1 \\ 0\,0\,0 & 0 & 0 & 0\;0\,0 \end{bmatrix} \begin{bmatrix} \phi \\ \lambda \\ h \\ v_n \\ v_e \\ v_d \\ b \\ d \end{bmatrix} + \begin{bmatrix} w_\phi \\ w_\lambda \\ w_h \\ w_{v_n} \\ w_{v_e} \\ w_{v_d} \\ w_b \\ w_d \end{bmatrix} \tag{7.189}$$

The definitions of RN and RM are as follows (refer to Chap. 1):

$$R_N = \frac{a}{\left[1 - e^2 \sin^2 \phi\right]^{1/2}}, R_M = \frac{a(1-e^2)}{\left[1 - e^2 \sin^2(\phi)\right]^{3/2}}$$

When using the latitude and longitude as the position state vector, the pseudo-range observation is difficult to express directly with the latitude and longitude, and the pseudo-range mathematical expression is still Eq. (7.177). Since both the satellite position and the receiver position in the equation are expressed in the ECEF coordinate system, the relationship between the pseudo-range observations and the $[\phi, \lambda, h]^{\mathrm{T}}$ cannot be obtained directly. In the process of using the EKF update step, there is pseudo-range observation residual in calculation Eq. (7.169):

$$\begin{aligned} \partial\rho_i &\approx H_{xi}\partial x + H_{yi}\partial y + H_{zi}\partial z + c\partial b \\ &= \left[h_{x\_i}\; h_{y\_i}\; h_{z\_i} \right] \begin{bmatrix} \partial x \\ \partial y \\ \partial z \end{bmatrix} + c\partial b \end{aligned} \tag{7.190}$$

where in $\boldsymbol{H}_{E\_i} = [h_{x\_i}, h_{y\_i}, h_{z\_i}]^{\mathrm{T}}$ is the corresponding satellite direction cosine vector measured for the i-th pseudo-range view expressed in the ECEF coordinate system. As seen in Sect. 1.2.4,

$$\begin{bmatrix} \partial x \\ \partial y \\ \partial z \end{bmatrix} = R_{t2e} \begin{bmatrix} \partial n \\ \partial e \\ \partial d \end{bmatrix} \tag{7.191}$$

If we substitute Eq. (7.191) into Eq. (7.190), we can obtain:

$$\begin{aligned} \partial \rho_i &\approx \begin{bmatrix} h_{n\_i} \; h_{e\_i} \; h_{d\_i} \end{bmatrix} \begin{bmatrix} \partial n \\ \partial e \\ \partial d \end{bmatrix} + c\partial b \\ &= \boldsymbol{H}_{T\_i} \begin{bmatrix} \partial n \\ \partial e \\ \partial d \end{bmatrix} + c\partial b \end{aligned} \tag{7.192}$$

where $\boldsymbol{H}_{T\_i} = [h_{n\_i}, h_{e\_i}, h_{d\_i}]^{\mathrm{T}}$ of Eq. (7.192) is still the direction cosine vector of the *i*-th satellite represented in the NED coordinate system. The relationship between $\boldsymbol{H}_{T\_i}$ and $\boldsymbol{H}_{E\_i}$ is:

$$\boldsymbol{H}_{T\_i} = \boldsymbol{H}_{E\_i} \boldsymbol{R}_{t2e} \tag{7.193}$$

where $\boldsymbol{R}_{t2e}$ is determined by Eq. (1.32).

$$\begin{bmatrix} \partial n \\ \partial e \\ \partial d \end{bmatrix} = \begin{bmatrix} (R_M + h) & 0 & 0 \\ 0 & (R_N + h)\cos(\phi) & 0 \\ 0 & 0 & -1 \end{bmatrix} \begin{bmatrix} \partial \phi \\ \partial \lambda \\ \partial h \end{bmatrix} \tag{7.194}$$

If we substitute Eq. (7.194) into Eq. (7.192), we can obtain:

$$\partial \rho_i \approx \begin{bmatrix} (R_M + h)\, h_{n\_i} \; (R_N + h)\cos(\phi) h_{e\_i} \; -h_{d\_i} \end{bmatrix} \begin{bmatrix} \partial \phi \\ \partial \lambda \\ \partial h \end{bmatrix} + c\partial b \tag{7.195}$$

The equation is the linear relationship between the pseudo-range residuals and $[\partial \phi, \partial \lambda, \partial h]^{\mathrm{T}}$, which is needed for the pseudo-range observation update.

The Doppler observation is simpler than the pseudo-range observation and can be easily derived.

$$\begin{aligned} f'_{d\_i} &= h_{n\_i} v_n + h_{e\_i} v_e + h_{d\_i} v_d + cd + v_i \\ &= \boldsymbol{H}_{T\_i} \begin{bmatrix} v_n \\ v_e \\ v_d \end{bmatrix} + cd + v_i, i = 1, \cdots m, \end{aligned} \tag{7.196}$$

where $\boldsymbol{H}_{T\_i}$ is calculated by Eq. (7.193).

The biggest advantage of the latitude and longitude height model is that the receiver's elevation is directly used as one of the system state quantities, meaning that it is easy to achieve high assistance using a barometer or altimeter. It is only necessary to add the height value of the sensor output as an observation to the Kalman filter measurement set.

The above analysis is based on the single-mode pseudo-range observation. In cases of BDS and GPS dual-mode observation, only the system state vector $\boldsymbol{x}$ needs to be expanded to be $[\boldsymbol{x}_p^{\mathrm{T}}, \boldsymbol{x}_v^{\mathrm{T}}, T_{\mathrm{GB}}, \boldsymbol{x}_c^{\mathrm{T}}]^{\mathrm{T}}$. This is similar to the processing in the P model, the PV model, and the PVA model. T_{GB} is modeled as a random walk, and readers can work it out according to the previous explanation.

Case Study: PV Model

Since the PV model is the most widely used in practical GPS receivers, it has practical significance, and we use it as an implementation example to summarize the principle of Kalman filtering described in this chapter. Hopefully readers can use this example to understand the principle of Kalman filtering and extend it to more practical applications.

The state vector of the PV model includes the position, velocity, and clock parameters of the receiver, so the applicable scenes include smooth moving cars, boats, pedestrians, and low-speed aircraft. Since it covers most of the moving scenes in daily life, the PV model is widely used in civilian receivers, which is why it is analyzed as an example.

Equation (7.182) shows that the system state equation of the PV model is linear and $F(t)$ is a constant matrix.

$$\boldsymbol{F}(t) = \begin{bmatrix} 0\,0\,0\,1\,0\,0\,0\,0 \\ 0\,0\,0\,0\,1\,0\,0\,0 \\ 0\,0\,0\,0\,0\,1\,0\,0 \\ 0\,0\,0\,0\,0\,0\,0\,0 \\ 0\,0\,0\,0\,0\,0\,0\,0 \\ 0\,0\,0\,0\,0\,0\,0\,0 \\ 0\,0\,0\,0\,0\,0\,0\,1 \\ 0\,0\,0\,0\,0\,0\,0\,0 \end{bmatrix} \tag{7.197}$$

However, since the pseudo-range observation is non-linear, the state vector estimation x requires the use of the extended Kalman filter described above.

First, the pseudo-range equation needs to be linearized, and the error state is taken according to the method in Sect. 7.3.4.

$$\delta\boldsymbol{x} = \boldsymbol{x} - \boldsymbol{x}^*$$

x in the above equation is an unknown system real state, and x^* is an estimate of x when it is close to the adjacent range. The task of the Kalman filtering process below is to find the best estimate of the difference between x^* and x, and then correct x^*.

Since the system state equation itself is linear, there is no need to linearize it. The pseudo-range equation is nonlinear, so linearization is required. Using the first-order Taylor series to develop the pseudo-range observation Equation

$$\delta \dot{\boldsymbol{x}}(t) = \boldsymbol{F}\delta \boldsymbol{x}(t) + \boldsymbol{w}(t), \tag{7.198}$$

$$\delta \rho = \boldsymbol{H}\delta \boldsymbol{x}(t) + v \tag{7.199}$$

In practice, the initialization of x^* is often the result of obtaining the PVT solution by the least-squares method, and then using the solution result to set the initial value of $\hat{\boldsymbol{x}}$, which is $\hat{\boldsymbol{x}}_0$. After Kalman enters the steady state, the time measurements come continuously. The system state is continuously updated, and must be near the neighboring value of x in the real state to ensure that linearization is correct.

Each row vector of H in Eq. (7.199) corresponds to the observation equation of a satellite:

$$\boldsymbol{H} = \begin{bmatrix} h_{x\_1} & h_{y\_1} & h_{z\_1} & 0 & 0 & 0 & 1 & 0 \\ h_{x\_2} & h_{y\_2} & h_{z\_2} & 0 & 0 & 0 & 1 & 0 \\ \vdots & \vdots & \vdots & \vdots & \vdots & \vdots & \vdots & \vdots \\ h_{x\_m} & h_{x\_m} & h_{x\_m} & 0 & 0 & 0 & 1 & 0 \end{bmatrix} \tag{7.200}$$

where $\left[h_{x\_i}, h_{y\_i}, h_{z\_i}\right] = \left[\left.\frac{\partial \rho_i}{\partial x}\right|_{\mathbf{x}^*}, \left.\frac{\partial \rho_i}{\partial y}\right|_{\mathbf{x}^*}, \left.\frac{\partial \rho_i}{\partial z}\right|_{\mathbf{x}^*}\right]$ is the direction cosine vector between the i-th satellite and x^*. It should be noted that as the position of the receiver changes, H also changes. Hence, each time the user position is updated, the matrix of H needs to be recalculated.

The co-variance matrix of the state processing noise $w(t)$ is represented as a Q matrix, as shown in Eq. (7.201), which contains only the diagonal elements. The non-diagonal elements are 0, indicating that the processing noise in each system state is irrelevant here. The actual Q matrix is not necessarily a diagonal matrix, but this does not affect the Kalman filter's update step.

$$\boldsymbol{Q} = \mathrm{var}\{\boldsymbol{w}(t)\} = \begin{bmatrix} \sigma_p & 0 & 0 & 0 & 0 & 0 & 0 & 0 \\ 0 & \sigma_p & 0 & 0 & 0 & 0 & 0 & 0 \\ 0 & 0 & \sigma_p & 0 & 0 & 0 & 0 & 0 \\ 0 & 0 & 0 & \sigma_v & 0 & 0 & 0 & 0 \\ 0 & 0 & 0 & 0 & \sigma_v & 0 & 0 & 0 \\ 0 & 0 & 0 & 0 & 0 & \sigma_v & 0 & 0 \\ 0 & 0 & 0 & 0 & 0 & 0 & \sigma_b & 0 \\ 0 & 0 & 0 & 0 & 0 & 0 & 0 & \sigma_d \end{bmatrix} \tag{7.201}$$

It is assumed here that the processing noise of each state quantity is independent of the others and is white noise, and the respective noise power spectral densities are respectively diag$\{\sigma_p, \sigma_p, \sigma_p, \sigma_v, \sigma_v, \sigma_v, \sigma_b, \sigma_d\}$. The Q matrix will be used later in the calculation of QD_w.

The F, Q, and H above are in the continuous time domain. Now, to implement Kalman filtering on a computer, the corresponding Φ and QD_w in the discrete time domain must be derived. According to the analysis in Sect. 7.3.3, it can be known that:

$$\Phi = e^{FT} = \begin{bmatrix} 1 & 0 & 0 & T & 0 & 0 & 0 & 0 \\ 0 & 1 & 0 & 0 & T & 0 & 0 & 0 \\ 0 & 0 & 1 & 0 & 0 & T & 0 & 0 \\ 0 & 0 & 0 & 1 & 0 & 0 & 0 & 0 \\ 0 & 0 & 0 & 0 & 1 & 0 & 0 & 0 \\ 0 & 0 & 0 & 0 & 0 & 1 & 0 & 0 \\ 0 & 0 & 0 & 0 & 0 & 0 & 1 & T \\ 0 & 0 & 0 & 0 & 0 & 0 & 0 & 1 \end{bmatrix} \tag{7.202}$$

$$\begin{aligned} QD_w &\approx QT + (FQ + QF^{\mathrm{T}})\frac{T^2}{2!} + 2FQF^{\mathrm{T}}\frac{T^3}{3!} \\ &= \begin{bmatrix} Q_p & 0 & 0 & \frac{\sigma_v T^2}{2} & 0 & 0 & 0 & 0 \\ 0 & Q_p & 0 & 0 & \frac{\sigma_v T^2}{2} & 0 & 0 & 0 \\ 0 & 0 & Q_p & 0 & 0 & \frac{\sigma_v T^2}{2} & 0 & 0 \\ \frac{\sigma_v T^2}{2} & 0 & 0 & \sigma_v T & 0 & 0 & 0 & 0 \\ 0 & \frac{\sigma_v T^2}{2} & 0 & 0 & \sigma_v T & 0 & 0 & 0 \\ 0 & 0 & \frac{\sigma_v T^2}{2} & 0 & 0 & \sigma_v T & 0 & 0 \\ 0 & 0 & 0 & 0 & 0 & 0 & Q_b & \frac{\sigma_d T^2}{2} \\ 0 & 0 & 0 & 0 & 0 & 0 & \frac{\sigma_d T^2}{2} & \sigma_d T \end{bmatrix} \end{aligned} \tag{7.203}$$

$$Q_p = \sigma_p T + \frac{\sigma_v T^3}{3}, \; Q_b = \sigma_b T + \frac{\sigma_d T^3}{3} \tag{7.204}$$

T is the interval between each update. Modern GPS receivers are generally $T = 1$ s. For certain special requirements of the receiver, such as receivers in large dynamic applications, the frequency of the positioning information required to output is higher. T will take a smaller value, but at this time the power of the processor has higher expectations.

The properties of F are applied in the derivation of Eq. (7.202) and Eq. (7.203): $\boldsymbol{F}^n = \mathbf{0}, \quad n \geq 2$. After completing the above preparations, a two-step update of the Kalman filter, namely time update and observation update, can be implemented.

If we assume that the system state at the time t_k is $\hat{\boldsymbol{x}}_{k|k}$, the co-variance matrix of the state is $\boldsymbol{P}_{k|k}$. First, a time update of the system state and state co-variance matrix is performed:

$$\hat{\boldsymbol{x}}_{k+1|k} = \Phi_k \hat{\boldsymbol{x}}_{k|k} \tag{7.205}$$

$$\boldsymbol{P}_{k+1|k} = \Phi_k \boldsymbol{P}_{k|k} \Phi_k^T + \boldsymbol{Q}\boldsymbol{D}_w \tag{7.206}$$

Here, since the state equation is linear, the time update of $\hat{\boldsymbol{x}}_{k|k}$ is also linear. Furthermore, for the PV model, the time update of $\hat{\boldsymbol{x}}_{k|k}$ is actually.

$$\begin{aligned} x_{k+1|k} &= x_{k|k} + v_{x,k|k}T \\ y_{k+1|k} &= y_{k|k} + v_{y,k|k}T \\ z_{k+1|k} &= z_{k|k} + v_{z,k|k}T \\ b_{k+1|k} &= b_{k|k} + d_{k|k}T \end{aligned}$$

Here, T is the time interval between this update and the last update, that is, $T = t_{k+1} - t_k$. The above equation can avoid a cumbersome matrix multiplication.

Then, based on the state of the system after the time update $\hat{\boldsymbol{x}}_{k+1|k}$, the observation is predicted:

$$\hat{\rho}_i = \sqrt{(x_{k+1|k} - x_{si})^2 + (y_{k+1|k} - y_{si})^2 + (z_{k+1|k} - z_{si})^2} + b_{k+1|k}, \ (i = 1, \cdots, m) \tag{7.207}$$

Since the observation equation is nonlinear, the prediction of the observations here is also a nonlinear function of the state parameters. With the advent of observation $\tilde{\boldsymbol{\rho}}_{k+1} = [\tilde{\rho}_1, \cdots, \tilde{\rho}_m]^{\mathrm{T}}$, the pseudo-range residual can be calculated:

$$\delta\boldsymbol{\rho}_{k+1} = \tilde{\boldsymbol{\rho}}_{k+1} - \hat{\boldsymbol{\rho}} \tag{7.208}$$

The calculation method of Kalman gain $\boldsymbol{k}_{k+1}$ is as follows:

$$\boldsymbol{k}_{k+1} = \boldsymbol{P}_{k+1|k}\boldsymbol{H}_{k+1}^{\mathrm{T}}[\boldsymbol{H}_{k+1}\boldsymbol{P}_{k+1|k}\boldsymbol{H}_{k+1}^{\mathrm{T}} + \boldsymbol{R}_{k+1}]^{-1} \tag{7.209}$$

Among them, $\boldsymbol{R}_{k+1}$ is the observation measurement variance matrix, and $\boldsymbol{H}_{k+1}$ is calculated by Eq. (7.210).

$$\hat{\boldsymbol{x}}_{k+1|k+1} = \hat{\boldsymbol{x}}_{k+1|k} + \boldsymbol{k}_{k+1}\delta\boldsymbol{\rho}_{k+1} \tag{7.210}$$

Then, we update the state co-variance matrix $\boldsymbol{P}_{k+1|k}$:

$$\boldsymbol{P}_{k+1|k+1} = (\mathbf{I} - \mathbf{k}_{k+1}\boldsymbol{H}_{k+1})\boldsymbol{P}_{k+1|k} \tag{7.211}$$

The updated $\hat{\boldsymbol{x}}_{k+1|k+1}$ and $\boldsymbol{P}_{k+1|k+1}$ are saved to prepare for the next update.

The above is the basic process of Kalman filtering of the user's position, velocity, and clock difference using the PV model inside the GPS receiver. BDS/GPS dual-mode observation can be extended from Eq. (7.184), for which the specific steps can be similar to the above process. Due to the limitations of space, we only describe the use of pseudo-range observations. Readers can deduce the processing steps of

Doppler observation according to the principles in this section and the previous sections. Within the actual receiver, completing a reliable and stable navigation algorithm is a very complicated and arduous task, and the work that needs to be done is far more than the above description. This section includes only a detailed explanation of the principle of the extended Kalman filter inside the receiver, which will help readers in practical work.

7.3.6 *Technical Processing in the Implementation of Kalman Filtering*

The basic equations of Kalman filtering are described in the above chapters, and are sufficient for understanding the principle. However, when Kalman filtering is used in engineering, problems occur that require special skills and techniques. These techniques do not change the basic properties of Kalman filtering in principle, but they are very important in practice. Some techniques can greatly reduce the amount of computation, some can stabilize the system, and some bring the noise model of the Kalman filter closer to the actual situation. The following section summarizes these issues.

When input is colored noise

From Eq. (7.115) and Eq. (7.116), the noise and observation noise that the Kalman filter needs to process are white noise. In cases where the input is not white noise, the subsequent Kalman filter process will encounter problems, especially with updates to the system state's co-variance matrix.

Consider the state equation and the observation equation of the system as shown in Eq. (7.212) and Eq. (7.213).

$$\boldsymbol{x}_m = \boldsymbol{\Phi}_{m-1}\boldsymbol{x}_{m-1} + \boldsymbol{G}_{m-1}\boldsymbol{u}_{m-1} + \boldsymbol{\Gamma}_{m-1}\boldsymbol{w}_{m-1} \tag{7.212}$$

$$\mathbf{y}_m = \boldsymbol{H}_m\boldsymbol{x}_m + \mathbf{v}_m \tag{7.213}$$

When the noise w_{m} is colored, it can be assumed that the w_{m} is fitted into Equation:

$$\boldsymbol{w}_m = \mathbf{B}_m\boldsymbol{w}_{m-1} + \varsigma_{m-1}. \tag{7.214}$$

$\varsigma_{\mathrm{m}-1}$ in Eq. (7.214) is zero-mean white noise.

At this point, we can augment w_{m} into the system state, that is, the new system state vector is $X_m = \begin{bmatrix} \boldsymbol{x}_m \\ \boldsymbol{w}_m \end{bmatrix}$. Then, the new system state equation and observation equations are:

$$\begin{bmatrix} \boldsymbol{x}_m \\ \boldsymbol{w}_m \end{bmatrix} = \begin{bmatrix} \boldsymbol{\Phi}_{m-1} & \boldsymbol{\Gamma}_{m-1} \\ \mathbf{0} & \mathbf{B}_{m-1} \end{bmatrix} \begin{bmatrix} \boldsymbol{x}_{m-1} \\ \boldsymbol{w}_{m-1} \end{bmatrix} + \begin{bmatrix} \boldsymbol{G}_{m-1} \\ \mathbf{0} \end{bmatrix} \boldsymbol{u}_{m-1} + \begin{bmatrix} \mathbf{0} \\ \boldsymbol{I} \end{bmatrix} \boldsymbol{\varsigma}_{m-1} \tag{7.215}$$

$$\boldsymbol{y}_m = \begin{bmatrix} \boldsymbol{H}_m & \mathbf{0} \end{bmatrix} \begin{bmatrix} \boldsymbol{x}_m \\ \boldsymbol{w}_m \end{bmatrix} + \boldsymbol{v}_m \tag{7.216}$$

The noise in Eqs. (7.215) and (7.216) is white noise, which is consistent with the noise requirements of the basic Kalman filtering equation.

In cases where the observed noise $\boldsymbol{v}_m$ is colored noise, the processing is slightly different. In such cases, the method of augmenting $\boldsymbol{v}_m$ to the system state cannot be adopted, because it will cause a lack of observation noise in the observation equation, and will lead to the inverse divergence of the matrix when calculating the Kalman gain.

Assuming that the observed noise $\boldsymbol{v}_m$ meets the following equation:

$$\boldsymbol{v}_m = \mathbf{T}_m \boldsymbol{v}_{m-1} + \boldsymbol{\zeta}_{m-1} \tag{7.217}$$

from Eq. (7.213), we can know that:

$$\begin{aligned} \boldsymbol{y}_{m+1} &= \boldsymbol{H}_{m+1}\boldsymbol{x}_{m+1} + \boldsymbol{v}_{m+1} \\ &= \boldsymbol{H}_{m+1}(\boldsymbol{\Phi}_m \boldsymbol{x}_m + \boldsymbol{G}_m \boldsymbol{u}_m + \boldsymbol{\Gamma}_m \boldsymbol{w}_m) + (\mathbf{T}_{m+1}\boldsymbol{v}_m + \boldsymbol{\zeta}_m) \\ &= \boldsymbol{H}_{m+1}(\boldsymbol{\Phi}_m \boldsymbol{x}_m + \boldsymbol{G}_m \boldsymbol{u}_m + \boldsymbol{\Gamma}_m \boldsymbol{w}_m) + [\mathbf{T}_{m+1}(\boldsymbol{y}_m - \boldsymbol{H}_m \boldsymbol{x}_m) + \boldsymbol{\zeta}_m] \\ &= (\boldsymbol{H}_{m+1}\boldsymbol{\Phi}_m - \mathbf{T}_{m+1}\boldsymbol{H}_m)\boldsymbol{x}_m + \boldsymbol{H}_{m+1}\boldsymbol{G}_m \boldsymbol{u}_m + \mathbf{T}_{m+1}\boldsymbol{y}_m \\ &\quad + (\boldsymbol{H}_{m+1}\boldsymbol{\Gamma}_m \boldsymbol{w}_m + \boldsymbol{\zeta}_m) \end{aligned} \tag{7.218}$$

If we slightly transform Eq. (7.218), we can get:

$$\begin{aligned} \boldsymbol{y}_{m+1} - \mathbf{T}_{m+1}\boldsymbol{y}_m = &(\boldsymbol{H}_{m+1}\boldsymbol{\Phi}_m - \mathbf{T}_{m+1}\boldsymbol{H}_m)\boldsymbol{x}_m + \boldsymbol{H}_{m+1}\boldsymbol{G}_m \boldsymbol{u}_m \\ &+ (\boldsymbol{H}_{m+1}\boldsymbol{\Gamma}_m \boldsymbol{w}_m + \boldsymbol{\zeta}_m) \end{aligned} \tag{7.219}$$

Let's define that:

$$\boldsymbol{Y}_k \triangleq \boldsymbol{y}_k - \mathbf{T}_k \boldsymbol{y}_{k-1},$$

$$\boldsymbol{H}_k^* \triangleq \boldsymbol{H}_k \boldsymbol{\Phi}_{k-1} - \mathbf{T}_k \boldsymbol{H}_{k-1},$$

$$\boldsymbol{V}_k \triangleq \boldsymbol{H}_k \boldsymbol{\Gamma}_{k-1} \boldsymbol{w}_{k-1} + \boldsymbol{\zeta}_{k-1}.$$

Then, we can rewrite Eq. (7.219) into:

$$\boldsymbol{Y}_{m+1} = \boldsymbol{H}_{m+1}^* \boldsymbol{x}_m + \boldsymbol{H}_{m+1}\boldsymbol{G}_m \boldsymbol{u}_m + \boldsymbol{V}_{m+1} \tag{7.220}$$

Equation (7.220) is a new measurement equation that can prove that the new observed noise V_k is zero-mean white noise, which is consistent with the requirements of the Kalman filter for observation noise. The only problem is that V_k and w_m are correlated, that is, measurement noise and processing noise are no longer irrelevant, and Eq. (7.117) is no longer valid. The Kalman filter for measuring noise and processing noise-related conditions is not described in detail in this book. Interested readers should refer to Reference [2] in Chap. 2.

Sequential (serial) update of observation

Assuming that at t_k time, m observations appear at the same time, these m observations are to be formed into $m \times 1$ vector according to the general processing in 7.3.2, and the Kalman gain and state correction will be obtained in a parallel processing method.

$$\begin{aligned} \boldsymbol{k}_k &= \boldsymbol{P}_k^- \boldsymbol{H}_k^{\mathrm{T}} (\boldsymbol{H}_k \boldsymbol{P}_k^- \boldsymbol{H}_k^{\mathrm{T}} + \boldsymbol{R}_k)^{-1} \\ \delta \hat{\boldsymbol{x}}_k &= \boldsymbol{k}_k (\tilde{\boldsymbol{y}}_k - \hat{\boldsymbol{y}}_k) \end{aligned}$$

It can be seen that in this kind of conventional parallel processing approach, the inverse of the matrix is essential. This is especially so when the number of observations is relatively large, as the dimension of $(\boldsymbol{H}_k \boldsymbol{P}_k^- \boldsymbol{H}_k^{\mathrm{T}} + \boldsymbol{R}_k)$ can be very large. Therefore, on many occasions, especially in the case of embedded applications, this method brings a very heavy computational burden.

In practice, when the observation covariance matrix R is a diagonal matrix, that is:

$$\boldsymbol{R} = \begin{bmatrix} R_1 & \cdots & 0 \\ \vdots & & \vdots \\ 0 & \cdots & R_m \end{bmatrix}.$$

the m observations can be regarded as in "sequential order", so that it can be processed in a serial manner, that is, for each observation $\tilde{y}_i$, assuming that the noise variance is R_i, $i = 1, \cdots, m$, we can calculate the Kalman gain k_i and the updated value of x_i caused by this observation.

First, two auxiliary vectors can be introduced:

$$\hat{\boldsymbol{x}}_0 = \boldsymbol{x}_k^-, \qquad \boldsymbol{P}_0 = \boldsymbol{P}_m^-$$

Here, $\boldsymbol{x}_k^-$ and $\boldsymbol{P}_m^-$ are the system state and state error variance matrix after the time update, and then the order i-th observation for the first observation, $i = 1, \cdots, m$. We can make the following calculation:

$$\boldsymbol{k}_i = \frac{\boldsymbol{P}_{i-1} \boldsymbol{h}_i^{\mathrm{T}}}{\boldsymbol{h}_i \boldsymbol{P}_{i-1} \boldsymbol{h}_i^{\mathrm{T}} + R_i} \tag{7.221}$$

$$\hat{\boldsymbol{x}}_i = \hat{\boldsymbol{x}}_{i-1} + \boldsymbol{k}_i (\tilde{y}_i - \boldsymbol{h}_i \hat{\boldsymbol{x}}_{i-1}) \tag{7.222}$$

$$\boldsymbol{P}_i = [\boldsymbol{I} - \boldsymbol{k}_i\boldsymbol{h}_i]\boldsymbol{P}_{i-1} \tag{7.223}$$

In the above equation, $\boldsymbol{h}_i$ is the row vector corresponding to the observation $\tilde{y}_i$ in the matrix $\boldsymbol{H}_k$. The processing of Eq. (7.221) to Eq. (7.223) is repeated until m observations have been processed. After processing m observations, the total system state and state variance matrix should be reset.

$$\boldsymbol{x}_k^+ = \hat{\boldsymbol{x}}_m, \qquad \boldsymbol{P}_m^+ = \boldsymbol{P}_m$$

When viewing observations are updated according to each observation, there is only one observation, so $(\boldsymbol{h}_i\boldsymbol{P}_{i-1}\boldsymbol{h}_i^{\mathrm{T}} + R_i)$ is a 1×1 "matrix", which is a number. Therefore, through this serial method, the inversion operation of a matrix with large dimensions is transformed into several numbers to find the inverse of the operation. k_i is also no longer a matrix, but a vector.

In this kind of processing, the system has been updated several times. This is because these observations come at the same time, that is, $\Delta t = 0$, so it is not necessary to perform a time update. The cumulative state updates of sequential processing $\delta\hat{\boldsymbol{x}}_k$ are the same as parallel processing. However, the final Kalman gain k_{m} is different from parallel processing because the k_{m} in parallel processing is based on all observations, whereas here, it is based on one observation at a time.

When the observation covariance matrix $\boldsymbol{R}$ is not a diagonal matrix, it is a positive definite matrix, so it can be decomposed into a form in which the lower triangular matrix is multiplied.

$$\boldsymbol{R} = \boldsymbol{E}\boldsymbol{E}^{\mathrm{T}} \tag{7.224}$$

Multiplying sides of the observation equations, we can obtain:

$$\boldsymbol{E}^{-1}\boldsymbol{y}_m = \boldsymbol{E}^{-1}\boldsymbol{H}_m\boldsymbol{x}_m + \boldsymbol{E}^{-1}\boldsymbol{v}_m \tag{7.225}$$

Defining:

$$\begin{aligned}
\overline{\boldsymbol{y}}_m &\triangleq \boldsymbol{E}^{-1}\boldsymbol{y}_m, \\
\overline{\boldsymbol{H}}_m &\triangleq \boldsymbol{E}^{-1}\boldsymbol{H}_m, \\
\overline{\boldsymbol{v}}_m &\triangleq \boldsymbol{E}^{-1}\boldsymbol{v}_m
\end{aligned}$$

The new observation equation is:

$$\overline{\boldsymbol{y}}_m = \overline{\boldsymbol{H}}_m\boldsymbol{x}_m + \overline{\boldsymbol{v}}_m \tag{7.226}$$

and the new observation co-variance matrix can be verified.

$$\overline{\boldsymbol{R}} = \boldsymbol{E}\{\overline{\boldsymbol{v}}_m\overline{\boldsymbol{v}}_m^{\mathrm{T}}\} = \boldsymbol{I}$$

Therefore, it is possible to continue to use the method of measuring the co-variance matrix as a diagonal matrix.

Observation validity test

In practical applications, for some reason, erroneous or invalid observations are often entered into the Kalman filter. Updating the system state based on these incorrect observations may result in errors. Fortunately, the Kalman filter provides a mechanism for screening these error observations.

Considering that the observation at a certain moment is $\tilde{y}$, and assuming that the observed variance is R and the system state covariance matrix is $\boldsymbol{P}$, then the predicted value $\hat{y}$ is observed according to the system state $\boldsymbol{x}$, and the observed residual is $\delta y = \tilde{y} - \hat{y}$, which is a random variable, and:

$$E\{\delta y\} = 0, \qquad \mathrm{var}\{\delta y\} = \boldsymbol{HPH}^{\mathrm{T}} + R \tag{7.227}$$

We can set a threshold α, making:

$$\text{概率}\{(\delta y)^2 > \alpha(\boldsymbol{HPH}^{\mathrm{T}} + R)\} \approx 0 \tag{7.228}$$

Therefore, an observation that satisfies the condition of the above equation can be regarded as an error or an invalid observation, and is thus eliminated without updating the Kalman filter. The value α can be determined according to the actual engineering situation. The basic principle is to eliminate the incorrect observation and not interfere with the correct observation update.

The method described here is often referred to as an Innovation Check. It should be noted that when the system state deviates far from the real state for some reason, even if the correct observation cannot be detected by the update, the practical strategy must consider how to deal with this situation. A simple and effective strategy is that if the signal quality is good enough but is removed by the update detection logic, it is necessary to carefully check the validity of the current system state. Clearly, the update detection can be applied naturally in the serial processing of the observation, and an update detection is performed each time an observation is updated. An interesting technique at this time is to observe the order of serial processing. Generally, it is necessary to sort the reliability of the observation, firstly to process the observation with high reliability, and to post the observation with low reliability, so that the normal observation will not be influenced by the incorrect update of the system state, which will pass update detection.

Lost or invalid observations

Sometimes, due to system failure or communication problems, observations are lost or cannot be obtained by the Kalman filter module. In this case, the time update can still be performed, while the measurement update cannot be performed.

From another point of view, when the observation is lost, the uncertainty of the observation can be considered to be infinite, that is, $\boldsymbol{R} = \infty$ according to the equation of the Kalman gain:

$$k = P^{-}H(HP^{-}H^{\mathrm{T}} + R)^{-1}.$$

Because of $R = \infty$, $k = 0$, the update amount of the system state is obviously 0, so there is no observation update to the system state at this time. However, the time update is continued, and at the same time, $P_{k+1|k} = \Phi_k P_{k|k} \Phi_k^{\mathrm{T}} + QD_w$. The system state covariance matrix is incremented to indicate the system state, and the uncertainty increases.

Maintaining the symmetry and positive definiteness of the matrix P

The co-variance matrix of the system state P is a symmetric matrix from its own definition, and it should also be a positive definite matrix. The equation that P^{-} obtains from the time update to P^{+} from the observation update is:

$$P^{+} = (I - kH)P^{-}$$

It seems that the symmetry of P^{+} is not visible from the form, but if the equation $k = P^{-}H^{\mathrm{T}}(HP^{-}H^{\mathrm{T}} + R)^{-1}$ of k is substituted, we can obtain:

$$P^{+} = P^{-} - P^{-}H^{\mathrm{T}}(HP^{-}H^{\mathrm{T}} + R)^{-1}HP^{-} \tag{7.229}$$

Equation (7.229) shows that in cases when P^{-} and R are symmetric arrays, theoretically P^{+} must also be a symmetric matrix, because at the next moment, when the P^{+} is time updated, the operation is also symmetrical. As a result, at any moment, P^{-} and P^{+} must be symmetric arrays.

In the specific implementation process, because the word length effect and rounding effect of the computer are inevitable, it will certainly lead **P** to gradually become asymmetrical, and at the same time lose its positive definiteness. When **P** becomes unsteady, the Kalman gain k cannot achieve a true weight value between the observation and the local state, which can cause serious errors in the updating of the system state. There are some ways to guarantee the symmetry of **P**, one of which is the average reset **P** with P^{T} and **P** after each observation update, that is:

$$P = \frac{1}{2}(P + P^{\mathrm{T}}) \tag{7.230}$$

This method is widely used in practice because it is simple and effective.

Another very simple method is to calculate only the elements of the diagonal and the elements of the upper part in the calculation process of P^{+}, that is, $\{P_{i,j}, i \leq j, i = 1, \cdots, n\}$, where n is the dimension of the matrix P. Then, according to the symmetry of the matrix, the upper part of its element can be used to place the elements of the lower part, that is:

$$P_{j,i} = P_{i,j} \qquad i < j$$

However, the above two methods can only guarantee the symmetry of the matrix $\boldsymbol{P}$. They cannot guarantee the positive definiteness of the matrix. Further methods can also use square root filtering and UDU decomposition filtering methods. The basic idea of square root filtering is to use a conclusion in matrix theory that a non-negative symmetric matrix $\boldsymbol{P}$ can always be decomposed into a form in which two triangular matrices are multiplied, that is:

$$\boldsymbol{P} = \Delta\Delta^{\mathrm{T}} \tag{7.231}$$

where Δ is the upper triangular or lower triangular matrix, which is called the square root matrix of $\boldsymbol{P}$. The Δ matrix can be obtained by performing Cholesky decomposition. The subsequent Kalman state co-variance matrix is updated based on the matrix Δ. Since Eq. (7.231) guarantees the positive definiteness and symmetry of $\boldsymbol{P}$, the square root filter has the advantage of being able to strictly guarantee the matrix $\boldsymbol{P}$'s positive definiteness while still using only half of the matrix $\boldsymbol{P}$. The word length guarantees the numerical accuracy of the filtering result. UDU decomposition filtering is the UD decomposition using the matrix $\boldsymbol{P}$, that is:

$$\boldsymbol{P} = \boldsymbol{U}\boldsymbol{D}\boldsymbol{U}^{\mathrm{T}} \tag{7.232}$$

where U is the upper triangular matrix; D is the diagonal matrix; and UDU decomposition filtering no longer transmits the matrix, making it easy to lose positive definiteness due to the word length effect, but transmitting the U and D matrix.

These methods change the update process of the matrix $\boldsymbol{P}$ to the square root matrix or the U and D matrix by square root decomposition Δ and UDU decomposition of the matrix. This can strongly guarantee the positive definiteness of the matrix $\boldsymbol{P}$, but at the cost of increasing the computation. Due to the limitations of space, the specific details of square root filtering and UDU filtering are not explained in detail here. Interested readers can consult related literature and books about the principle of Kalman filtering.

7.4 Summary of the Least-Squares and Kalman Filtering Solutions

The BDS and GPS receivers calculate the position, velocity, and time information of the user through pseudo-range observation and Doppler observation, which is referred to as the PVT solution. In the process of implementing this solution, the least-squares or Kalman filter algorithm can be used.

The positioning result obtained by the least-squares method is based on a set of observations of a certain time element, and has nothing to do with the observation of other time elements. When the receiver extracts the observation, the observations of different time elements are extracted independently of each other, so the observations from different time elements can be considered to be independent of each other. Thus, the error features of the least-squares positioning results based on different time elements can be considered to be similar to white noise, so the result of the least-squares positioning has a white noise-like jumping phenomenon in time.

Kalman filtering is a linear, recursive estimation method widely used in modern control, parameter estimation, and adaptive filtering systems. The Kalman filter theory is based on the premise that the system processing noise is Gaussian white noise. The Kalman filter in the receiver is based on observations of multiple time elements, and can be considered to share all of the information from observations of different time elements. Although the observation noise at different time scales can be considered as a white Gaussian distribution, the error features of the Kalman filtering positioning result are colored noise distribution due to the Kalman filtering information sharing for different time elements. The positioning error reflected in terms of time gradually and continuously varies. Figure 7.12 shows a two-dimensional error plot of the least-squares positioning result and the Kalman filter positioning result in the case of a static antenna. In the pursuit of fair comparison, the least-squares and Kalman filtering methods deal with the same intermediate frequency data. The left half is the positioning result of least-squares positioning, and the right half is the positioning result of Kalman filtering. The abscissa and the ordinate are the errors in the positive east and north directions of the user ENU coordinate system respectively, and the units are meters. The positioning error results of adjacent moments in the figure are connected by thin solid lines. Clearly, the error of the least-squares positioning is distributed in a disorderly manner, and the error of the Kalman filter exhibits a certain regular slow change. Overall, the positioning result of the Kalman filter is better than that of the least-squares positioning result. The error is smaller, which also reflects the low-pass filtering features of Kalman processing. The starting point in the Kalman filtering result is marked in the figure. Clearly, the update amount of the Kalman filter is large at the beginning of the iteration, which is reflected in the figure by the relatively large space between adjacent points. With the number

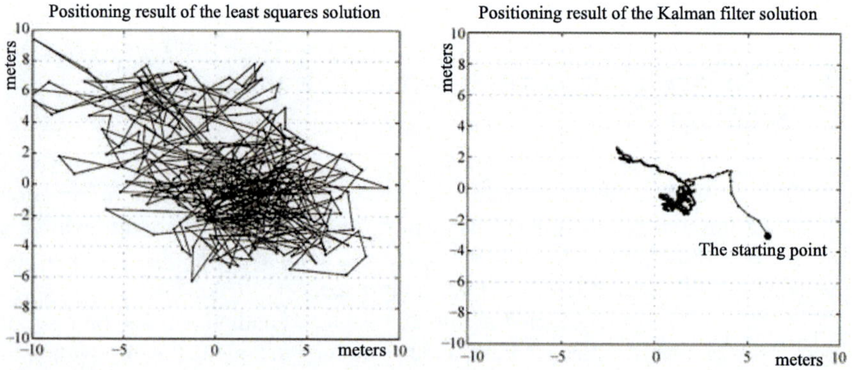

Fig. 7.12 Comparison of the positioning results of least-squares and Kalman filtering

of iterations increasing, the Kalman filter's P matrix converges. Then, the Kalman filter's gain matrix is more dependent on the local state, resulting in a smaller update of the Kalman filter, which is reflected in the figure where the space between adjacent points becomes smaller.

The following example illustrates the colored noise features of the Kalman filter positioning error more clearly. In this example, the receiver antenna is stationary, and the receiver's output contains the positioning results of the two methods, which lasted 500 s, as shown in Fig. 7.13. At the 200th second, the pseudo-range observation of a certain satellite is artificially introduced into a 50-km hop, and large positioning

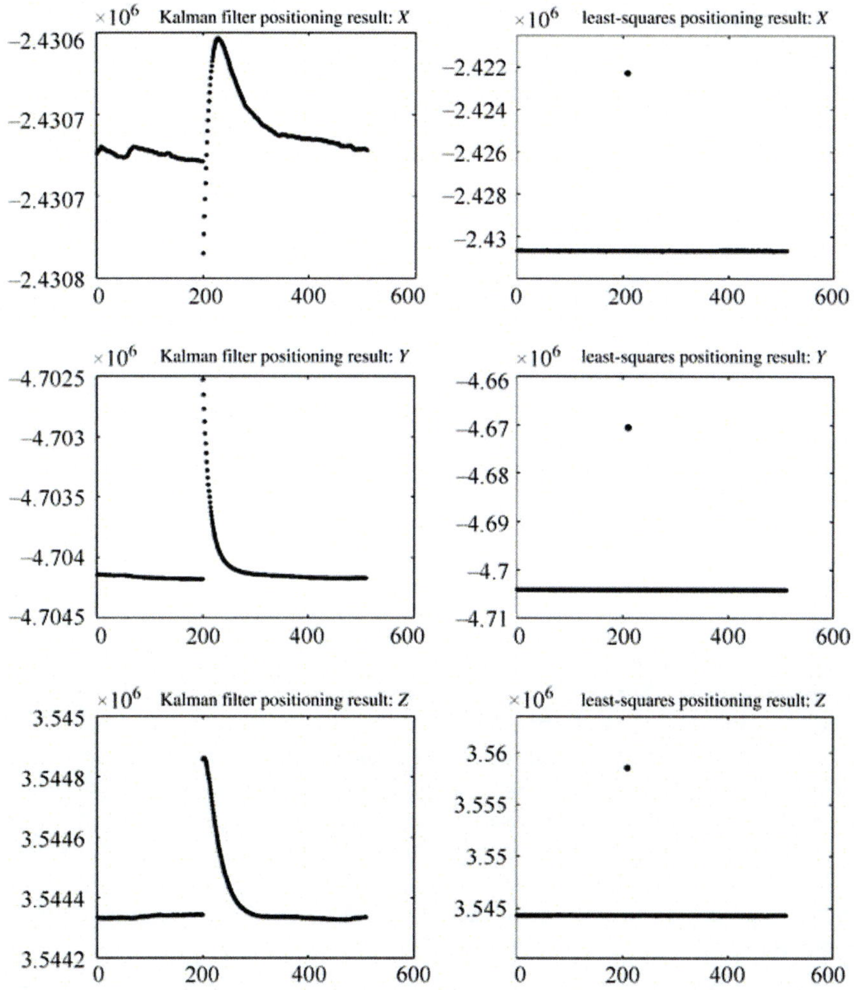

Fig. 7.13 Differences in the Kalman filter and the least-squares positioning result caused by a pseudo-range observation error introduced at the 200th second

errors will appear in both the Kalman filter and the least-squares algorithm. The three pictures on the left in the figure are the (x, y, z) coordinates of the ECEF coordinate system of the Kalman filter output, and the three pictures on the right are the (x, y, z) of the least-squares ECEF coordinate system. The least-squares method only deviates the positioning result at the 200th second, and then quickly returns to the correct positioning result, that is, the positioning result at the subsequent time is not affected by the error result of the previous moment. Meanwhile, in the Kalman filter, not only does the incorrect positioning result occur at the 200th second, but the error continues for a long period of time, before gradually returning to the correct positioning result. Therefore, the positioning results of Kalman filtering are correlated in time.

From the nature of the problem, the colored noise features of the Kalman filtering result are derived from the constraints of the system state equation. For example, in the PV model, there is a strict integral relationship between position and velocity, so between the current positioning result and that at the next moment, there is a physical constraint, which comes from the speed vector of the receiver itself. The time update in the Kalman filtering process is the embodiment of this physical relationship, and the subsequent observation update is only based on time. The residual between the update and the observation is corrected, which determines that the positioning results of the Kalman filter must be correlated in time, and the extent to which the time-correlation feature is applied marks the biggest difference between the Kalman filter and the least-squares method. This is why the positioning result of the Kalman filter is smoother than that of the least-squares method.

Judging from the perspective of optimal estimation, the least-squares method and the Kalman filtering algorithm both estimate the user position based on the least-squares principle, but in the least-squares algorithm, the least-squares estimation is performed for a set of observations of the current time element, while in the Kalman filtering algorithm, it is performed recursively based on observations from the power-on time of the receiver to the current time element. The statistical features of the system state and the observation are not required for the least-squares method, and the optimal index is merely the numerical accuracy of the measurement estimation, which is clearly shown in cost functions like Eqs. (7.2) and (7.15). The estimation accuracy of the least-squares method is not high, and when a large deviation appears in the observation, it will cause a similarly large deviation of the estimation result. However, it is still widely used in practice for the simple algorithm and wide application conditions. Although the co-variance matrix of the estimated error can be calculated according to Eqs. (7.10) and (7.20) in the case of a known covariance matrix of the observation vector, it is not necessary. This means that the result of least-squares estimation does not need to be performed based on the results of Eqs. (7.10) and (7.20). On the contrary, matrix $\boldsymbol{P}$, the covariance matrix of the estimation error, must be calculated in the process of obtaining the state estimation in the Kalman filtering algorithm.

Kalman filtering must first define a system state vector $\boldsymbol{x}$ and then solve a time-state equation based on $\boldsymbol{x}$, which must also be a linear or nonlinear equation of $\boldsymbol{x}$ for observation. If the time-state equation of the system state vector cannot be derived, the Kalman filtering method cannot be used. The time-state equation must also reflect the actual system operation, otherwise the Kalman filter will not achieve the desired results. For example, if a P model is used for a high-speed moving object, the velocity vector of the object cannot be accurately estimated.

At least four satellite observations are adopted in the least-squares algorithm to provide an equation for each satellite's observations. This solves four unknowns by nonlinear iterative methods, namely the three position coordinates of the user and a clock deviation. Unlike the least-squares algorithm, Kalman filtering obtains a time update of the current system state according to the system state transition equation, and has no requirements for the number of satellites. After the time update, with the arrival of the observation, the observation update is performed for the equation based on the time update. In the observation and updating process, there is still no requirement for the number of satellites, so even if there are observations from only one or two satellites, Kalman filtering can still provide a positioning result. Of course, this positioning result will gradually deviate from the correct position with the increase of time.

From the expression of Kalman gain k_{m},

$$\boldsymbol{k}_m = \boldsymbol{P}_m^- \boldsymbol{H}_m^{\mathrm{T}} (\boldsymbol{H}_m \boldsymbol{P}_m^- \boldsymbol{H}_m^{\mathrm{T}} + \boldsymbol{R}_m)^{-1}$$

When an element on the diagonal of $\boldsymbol{R}$ is large, the corresponding element in $\boldsymbol{k}_m$ will be small, indicating that $\delta \boldsymbol{x}$ derived from the corresponding observation will also be small in value. Meanwhile, if an element of $\boldsymbol{P}$ is small, the corresponding element of $\boldsymbol{k}_m$ will be small, indicating that the value of $\delta \boldsymbol{x}$ at this time is small too. From the above analysis, we can see that the update of the system state through Kalman filtering is based on a compromise between the uncertainty of the current system state and the uncertainty of the observation. If the uncertainty of the current system state estimation is zero (that is, if the current system state estimation is very reliable), then the update caused by the observation will be very small. Conversely, if the uncertainty of the observation is very low (that is, if the observation is credible), the update of the system state will be mainly determined by the observation. So Kalman filtering is an adaptive process, and the matrix $\boldsymbol{P}$ provides a variance estimate of the system state error, that is,

$$\boldsymbol{P} = \mathrm{cov}\{\delta \boldsymbol{x} \delta^{\mathrm{T}} \boldsymbol{x}\}$$

In practice, the matrix $\boldsymbol{P}$ is often used to evaluate the state estimation error of $\boldsymbol{x}$. If the elements in the matrix $\boldsymbol{P}$ are relatively large, the corresponding system state quantity is less reliable. Otherwise, the corresponding system state quantity is more reliable.

At the same time, the matrix $\boldsymbol{P}$ can also be used to evaluate the credibility of the observations. We saw how to use the observation validity test in Sect. 7.3.6. In the least-squares method, there is no intuitive way to evaluate the effectiveness of observations.

The Kalman filtering method functions as a way of data fusion, so for system designers, it is the only choice in some applications. In the system design of integrated navigation, in addition to the satellite pseudo-range and Doppler observation of the BDS and GPS receivers themselves, other sensors can provide more observations. For example, accelerometers and gyroscopes provide the acceleration and angular velocity in three directions; the altimeter provides the current level; and the digital compass provides the current course angle. To fully and effectively utilize the observations from these sensors, we must resort to the Kalman filtering method to integrate all of this information and make an optimal estimate of the state of the system based on the system state equation and the observational equation. More details can be found in References [1, 3, 4].

For BDS and GPS receiver designers, the least-squares method must be understood as a conventional positioning solution. With the advancement of technology, the price of sensors is declining, and inexpensive receivers are beginning to use sensors to assist navigation. Therefore, a thorough comprehension of the principles and methods of Kalman filtering is becoming increasingly important for the analysis of the actual system and the design of an acceptable filtering model. It is also a very important theoretical basis for learning about other non-linear filtering methods such as the new filtering principles of the Unscented Kalman Filter (UKF) [5, 6, 7, 8] and particle filtering. Hopefully, this chapter has offered a preliminarily introduction to this field as a foundation for further study.

References

1. Farrell JA (2008) Aided navigation, GPS with high rate sensors. McGraw Hill
2. Qin Y (2012) Kalman filter and integrated navigation principle, 2nd edn. Northwest Polytechnic University Press, Xi'an
3. Jay AF, Barth M (1999) The global positioning system and inertial navigation. McGraw-Hill
4. Grewal M, Weill L, Andrews A (2001) Global positioning systems, inertial navigation, and integration. Wiley Inc
5. Julier Simon J, Uhlmann Jeffrey K (1997) "A new extension of the Kalman filter to nonlinear systems." Proceedings of AeroSense, the 11th international symposium on aerospace/defense sensing, simulation and controls.
6. Wan Eric A, Rudolph VD Merwe. The unscented Kalman filter for nonlinear estimation
7. https://www.cs.unc.edu/~welch/kalman/
8. Arthur G (May 15, 1974) Applied optimal estimation. The MIT Press
9. Grover Brown R (1996) Receiver autonomous integrity monitoring. In: Parkinson B, Spiker J, Axelrad P, Enge P (eds), Global positioning system: theory and applications, vol 2
10. Kaplan ED (2006) Understanding GPS principles and applications, 2nd edn. Artech House Publishers
11. China Satellite Navigation System Management Office (Dec 2013) Beidou satellite navigation system space signal interface control document (Open Service Signal) Version 2.0
12. Misra P, Enge P (2008) Global positioning system—signal, measurement and performance, 2nd edn. Electronic Industry Press, Beijing
13. Changxu G (2010) GPS: Theory, Algorithms and Applications, 2nd edn. Springer
14. Shouxin Z (1996) The theory and application of GPS satellite surveying and positioning. National University of Defense Science and Technology Press, Beijing

15. Sturza MA, Brown AK (Sep 19–21, 1990) Comparison of fixed and variable threshold RAIM algorithms." Proceedings of the third international technical meeting of the satellite division of the institute of navigation (ION GPS-90), Colorado Springs, CO. pp 437–43
16. Lee YC (June 24–26, 1986) "Analysis of range and position comparison methods as a means to provide GPS integrity in the user receiver." Proceedings of the annual meeting of the institute of navigation. Seattle, WA, pp l–4
17. Parkinson BW, Axelrad P (1987) "A basis for the development of operational algorithms for simplified GPS integrity checking." Proceedings of the satellite division first technical meeting, the institute of navigation, Colorado Springs, CO. pp 269–76
18. Sturza MA (1988–89) "Navigation system integrity monitoring using redundant measurements." Navig, J Inst Navig 35(4, Winter):483–501
19. Brown RG, Chin GY (1997) "GPS RAIM: calculation of threshold and protection radius using chi-square methods a geometric approach," Global positioning system, navigation. J Inst Navig, 5:155–178
20. Brown RG (1992) "A baseline GPS RAIM scheme and a note on the equivalence of three RAIM methods." Navig, J Inst Navig 39(3). Fall
21. Pakinson BW, Axelrad PA (1988) "Autonomous GPS integrity monitoring using the pseudorange residual." Navig, J Inst Navig 35(2, Summer)
22. Chin GY, Kraemer JH (1992–1993) "GPS RAIM screen out bad geometries under worst-case bias conditions." Navig, J Inst Navig 39(4, Winter)
23. Brown RG, Hwang PYC (1992) Introduction to random signals and applied Kalman filtering, 2nd edn. Wiley, New York
24. Pervan BS et al (Apr 1996) "Parity space methods for autonomous fault detection and exclusion algorithms using carrier phase." Proceedings of PLANS 96 symposium. Altanta, GA
25. Kelly RJ (1998) "The linear model, RNP, and the near-optimum fault detection and exclusion algorithm." Invited paper of GPS red book, vol 5
26. Brenner M (Sep 1990) "Implementation of a RAIM monitor in a GPS receiver and an integrated GPS/IRS." Proceedings of ION GPS-90. Colorado Springs, CO
27. Pullen S, Enge P (Sep 2004) "Satellite integrity monitoring concepts for GPS/Galileo augmentation systems." Proceedings of ION GNSS 17th ITM of the satellite division. Long Beach, CA
28. Brown RG, McBurney PW (1987) "Self-contained GPS integrity check using maximum solution separation as the test statistic." Proceedings of the satellite division first technical meeting, ION, Colorado Springs, CO
29. Grewal M, Andrews A (2001) Kalman filtering: theory and practice using matlab, 2nd edn. Wiley Inc
30. Haykin S (1996) Adaptive filter theory, 3rd edn. Prentice-Hall, Inc
31. Kalman RE (1960) "A new approach to linear filtering and prediction problems." Trans ASME, J Basic Eng 82
32. Thornton CL (1976) Triangular covariance factorizations for Kalman filtering. NASA Ph.D. Thesis
33. Bar-Shalom Y, Li XR, Thiagalingarm K (2001) Estimation with applications to tracking and navigation. Wiley Inc
34. Bierman Gerald J (1977) "Factorization methods for discrete sequential estimation." Mathematics in science and engineering, vol 128. Academic Press, NY
35. Eric AW, Rudolph VD (2001) Merwe. "The unscented Kalman filter." Kalman filtering and neural networks. Wiley Inc
36. Tine L, Herman B (2002) Comment on "New method for the nonlinear transformation of means and covariances in filters and estimators." IEEE transactions of automatic control
37. Papoulis A (1991) Probability, random variables and stochastic processes. McGraw Hill
38. Oppenheim AV, Willsky AS, Nawab SH (1997) Signals and systems, 2nd edn. Prentice-Hall, Inc

Chapter 8
RF Front-End

The setup of the RF front-end plays a pivotal role in the satellite receiver, especially for software receivers, in which the RF front-end may be the only hardware in the system. In conventional hardware receivers where ASIC chips form the main structure, the RF front-end is the source of subsequent baseband signal processing, which is vital for the performance of the receiver. The importance of the RF front-end can be interpreted from the following aspects:

① The signal bandwidth, noise figure, insertion loss, and RF impedance of the RF front-end directly affect the strength and noise properties of the received signal, which will thereby affect the performance of the subsequent baseband processing, including the performance of the tracking loop and the precision of the navigation positioning result.

② The frequency scheme setting of the RF front-end determines the theoretical intermediate frequency value f_c of the received satellite signal. In physics,it is the theoretical value of the carrier frequency of the received satellite signal without Doppler shift and local clock drift that directly determines the frequency search range of the signal acquisition algorithm, and it is an indispensable parameter for subsequent signal processing.

③ With the implementation of multi-mode satellite navigation systems, signals from multiple satellite navigation systems need to be received and processed simultaneously in multi-mode multi-frequency receivers. For example, the new civilian signals L2C and L5 in GPS; B1, B2, and B3 signals in BDS; and G1 and G2 signals in GLONASS. Proper RF settings will enable the receiver to simultaneously receive GNSS signals in multiple frequency bands and multiple systems with fewer RF hardware overheads. This chapter will elaborate on these three points.

In modern satellite navigation receivers, the digital signal processing scheme is extensively used to process the satellite signal. The satellite signal enters the RF front-end initially in the form of an analog signal, and is finally processed in the baseband

Y. Lu, *BDS/GPS Dual-Mode Software Receiver*, Navigation: Science and Technology 10, https://doi.org/10.1007/978-981-16-1075-2_8

in the form of a digital signal, so the analog-digital conversion (hereafter referred to as ADC) will be an indispensable part of the RF front-end. Later in this chapter, we will find that the finite quantization bit width of the ADC will bring loss to the signal energy, and the establishment of the quantization level is also related to the setting of the AGC. These factors must be considered during the design of the RF front-end. Since satellite navigation signals tend to be narrowband signals, bandpass sampling is widely adopted in the RF front-end of GNSS receivers, and deliberate spectral aliasing is often used to convert narrowband signals in higher frequency bands into digital baseband signals. For better comprehension of the sampling process, this section also explains the basic principle of bandpass sampling.

Two common RF front-end schemes are explained in this chapter—the traditional IF sampling scheme and the RF sampling scheme commonly used in software radio—of which the pros and cons will also undergo comparison. The strength of the signal received by the antenna relative to the noise is often represented by the carrier-to-noise ratio CN_0, which is closely related to the subsequent signal processing performance. This chapter explains the relationship between signal-to-noise ratio S/N and carrier-to-noise ratio from the perspective of the RF signal. At the end of this section, a GPS RF chip commonly seen in the market is taken as an example to offer a detailed analysis of the specific frequency scheme based on the principle of the RF front-end explained in this section. This will help GPS receiver engineers to understand the most important parameters of the RF front-end, which are essential for the design of subsequent signal processing algorithms.

8.1 Transmission and Reception of Satellite Signals

The signal power transmitted by BDS and GPS satellites is limited by space-borne devices. For example, the transmission power of the C/A code signal of the GPS satellite on the L1 carrier is only about 27 W if it is in decibels, that is, $10\lg 27 = 14.3$ dBW. When the satellite transmits a signal uniformly in all directions in the space, the speed at which the signal attenuates during the transmission in free space is proportional to $\frac{1}{R^2}$, where R is the linear distance from the satellite to the receiver. More specifically, the signal power space density at a place whose linear distance to the satellite is R is

$$P_D = \frac{P_T}{4\,\pi\,R^2} \tag{8.1}$$

where P_T is the transmit power of the signal, and P_D is the power intensity per unit area, whose unit is $\mathrm{W/m^2}$.

In engineering, P_D is more commonly written in the unit of dB, so

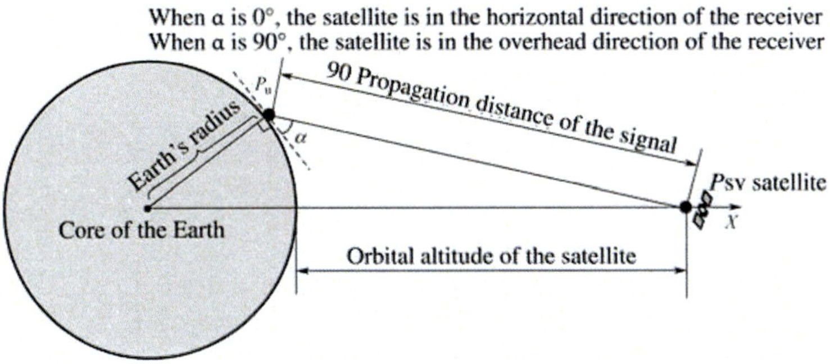

Fig. 8.1 Satellite signal propagation distance in space

$$\begin{aligned} P_{D,\mathrm{dB}} &= P_{\mathrm{T,dB}} - 10\lg(4\,\pi) - 20\lg R \\ &= P_{\mathrm{T,dB}} - 11 - 20\lg R \quad \mathrm{dB}/m^2 \end{aligned} \tag{8.2}$$

where $P_{\mathrm{T,dB}}$ is the satellite transmit power, expressed in $\mathrm{dB/m^2}$. In Eq. (8.2),

$$L_{R,dB} = 11 + 20\lg R \tag{8.3}$$

is the signal loss caused by the path.

The propagation distance of satellite signals in space R can be calculated according to Fig. 8.1. In the figure, α is the elevation angle of the receiver on the surface of the earth. When the elevation angle is 90°, the satellite is in the overhead direction of the receiver, and when the elevation angle is 0°, the satellite is in the horizontal direction of the receiver. As seen in Fig. 8.1, if the satellite's elevation angle is known, according to the Earth's radius and the orbital height of the satellite, the signal propagation distance can be calculated. The specific steps can be deduced from the correlation triangles, which are skipped here.

In terms of GPS signal reception, when the user is on the surface of the Earth, R is a function related to the satellite elevation angle α. When the satellite is right above the user, i.e. $\alpha \approx 90^{\circ}$, $R \approx 20\ 190$ km, according to Eq. (8.3), the path loss at this time is about −157.1 dB. Meanwhile, when the satellite is in the horizontal direction, such as when the elevation angle is 5 °, $R \approx 25\ 240$ km, the path loss at this time is approximately −159 dB.

For the BDS MEO satellite, when the satellite is over the user, $R \approx 21\ 500$ km. The path loss at this time is about −157.6 dB, and when the satellite is on the horizon, the satellite's elevation angle is still 5 °, then $R \approx 25\ 688$ km. The path loss is about −159.5 dB, which shows that in each case, the path loss of the MEO satellite is 0.5 dB more than the GPS satellite.

For the IGSO/GEO satellite of BDS, the propagation distances of the signal in the above two cases ($\alpha \approx 90^{\circ}$ and $\alpha \approx 5^{\circ}$) are $R \approx 35\ 780$ km and $R \approx 41\ 120$ km respectively. According to Eq. (8.3), the path losses are −162 dB and −163.3 dB

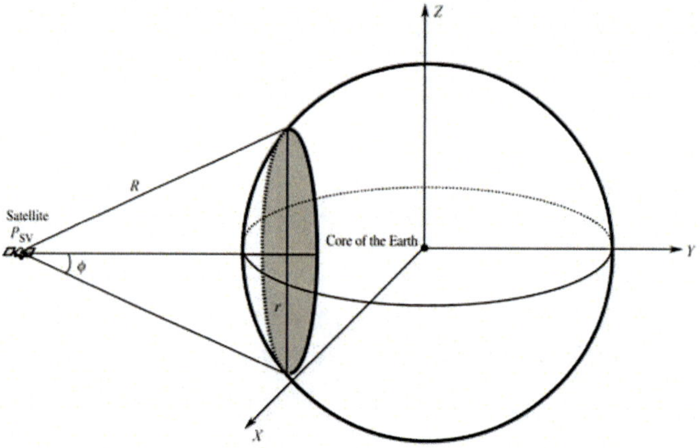

Fig. 8.2 Satellite transmission of antenna gain

respectively, which is nearly 4 dB more than the path loss in the BDS MEO satellite. Thus, the higher satellite orbit of the IGSO/GEO satellite brings about 4 dB of additional signal loss to the receiver on the Earth's surface as compared to the MEO satellite.

The above analysis is based on the omnidirectional transmit antenna that can distribute signals evenly in all directions of space. Meanwhile, in practice, the BDS and GPS satellites will transmit signals towards the Earth. Therefore, the signal received on the surface of the earth is enhanced as compared with those in the free space analyzed above. The enhancement is determined by the transmit antenna, so it is called the transmit antenna gain. Figure 8.2 shows this principle.

According to Fig. 8.2, when the antenna transmits signals within a range whose angle is 2ϕ, as indicated by the shaded portion of the figure, the signals will appear within a spherical cap with a radius of r. So, compared with the spherical surface formed by the omnidirectional antenna, the antenna gain is approximately

$$G_T(\phi) \approx \frac{4\,\pi\,R^2}{\pi\,r^2} = \frac{4}{\sin^2\phi} \tag{8.4}$$

The antenna gain is a function of ϕ. The sign "≈" is used in Eq. (8.4) because the area of the actual spherical cap is not $\pi\,r^2$, but slightly larger than $\pi\,r^2$, and the smaller ϕ is, the lager the approximation will be. In specific GPS implementations, the satellite antenna transmits the signal at an angle of $\pm\,21.3\,°$, and the antenna gain calculated through Eq. (8.4) is approximately $10\lg G_T(21.3^{\circ}) \approx 14.8\,\text{dB}$.

In this way, if the satellite signal is uniformly emitted inside the entire spherical cap, the signal received by the user at the vertex of the spherical cap region is slightly stronger, and the satellite signal received bY the user at the edge of the spherical cap is slightly weaker. The reason is obvious: there is a small distance loss at the vertex of the spherical cap. In order to ensure that the signal received by users at different locations on the surface of the Earth is consistent in strength, the transmitting power of the GPS satellite antenna is slightly lower when the emission angle is 0 ° and slightly higher at the edge of the spherical cap. As mentioned in Chap. 2, the "hot spot", i.e. the local signal enhancement function, will be added to the Block-IIIC satellite of the modernized GPS, mainly through the spot beam. This will raise the signal intensity of local hotspots on the Earth so that the anti-interference capability of GPS receivers in the region can be improved. This special function is achieved through adjusting the coverage area of the beam.

Satellite signals travel long distances before reaching the Earth, and the path loss is already very weak. The power of the signal obtained by the receiver is equal to the product of the signal power density P_D and the effective area of the antenna. Theoretically, the relationship between the effective area of the antenna and the signal wavelength λ and the gain of the receiving antenna G_{R} is

$$S_{\mathrm{Ant}} = G_{\mathrm{R}}\lambda^2/4\,\pi \tag{8.5}$$

where λ is the wavelength of the signal carrier for GPS L1 signals. The carrier wavelength $\lambda \approx 0.19$ m and the carrier wavelength of BDS signals can be calculated by readers independently. G_{R} is the gain of the receiving antenna, indicating the antenna's ability to capture signals intensively from a certain direction. Considering the design of the BDS and GPS satellite constellations, a satellite's signal may come from any direction. Therefore, the receiver's antenna is usually designed to be omnidirectional, that is, the antenna gain is identical for satellites in any direction. Otherwise, the receiver would not generate good GDOPs if the antenna does not have fine reception of signals from certain directions.

The BDS and GPS satellite signals will be very weak after a long transmission distance, so the specific signal power received by the antenna is determined by several factors, including the initial transmitting power of the satellite, the transmitting antenna gain, the path loss, and the effective area of the receiving antenna. Let's take the GPS signal as an example for a rough estimate.

We assume that the transmitting power of the GPS satellite signal at L1 frequency is 27 W, which is 14.3 dBW. According to the above calculation, the transmit antenna gain is 14.8 dB and the path loss is −159 dB. The power density of the signal that reaches the Earth's surface is −130 dBW/m^2. If we assume that the effective area of the antenna is $S_{\mathrm{Ant}} = \lambda^2/4\pi$, and for the L1 signal, $\lambda = 0.19$ m, then $S_{\mathrm{Ant}} \approx 2.872 \times 10^{-}3$ m^2, which is −25.4 dB, so the received signal power is about −156 dBW. It is worth noting that this value is only a rough estimate; the actual value will vary depending on the satellite's elevation angle and the different factors of the receiving antenna. However, one thing we can be sure of is that the BDS and GPS signals received by users on the surface of the Earth are already extremely weak.

8.2 Noise Figure of the Cascaded System

Starting from the antenna, the RF front-end of the GNSS receiver has multiple RF modules, including low noise amplifiers, mixers, and filters. So, the entire RF front-end can be regarded as a cascade of multiple modules, each having an important indicator called the noise figure, which is written as F here. The noise figure is defined as

$$F = \frac{\text{Input SNR}}{\text{Output SNR}} \tag{8.6}$$

The noise figure is always greater than 1, because a module will always generate noise itself. At any temperature above absolute zero (0 K), a semiconductor or conductor will inevitably generate thermal noise inside, so the signal-to-noise ratio after the signal passes through the module is always less than the input signal-to-noise ratio. Furthermore, if the noise power in the input signal is N_0, the gain of the module is G, and the noise power generated by the module itself is N_{Local}, then

$$F = 1 + \frac{N_{\text{Local}}}{GN_0} \tag{8.7}$$

Equation (8.7) can be understood from Fig. 8.3. The input signal of the module in the figure is $S + N_{\text{in}}$, where S is the signal, N_{in} is the input noise, G_1 is the gain of the module, and the noise generated by the module itself is N_{Local}. Then, the signal outputted by the module plus noise is $\text{G}_1 S + \text{G}_1 N_{\text{in}} + N_{\text{Local}}$, so

$$\text{Input SNR} = \frac{S}{N_{\text{in}}}, \quad \text{Output SNR} = \frac{G_1 S}{G_1 N_{\text{in}} + N_{\text{local}}} \tag{8.8}$$

The result of Eq. (8.7) can be obtained via substituting Eq. (8.8) into Eq. (8.6).

According to Eq. (8.7), if the noise figure is known, the local system noise can be written as

$$N_{\text{Local}} = (F - 1)GN_{\text{In}} \tag{8.9}$$

According to the definition of the noise figure, it measures the extent to which a module degrades the input signal-to-noise ratio. The larger the noise figure is, the worse the output signal-to-noise ratio will be. Conversely, the smaller the noise figure is, the closer the output signal-to-noise ratio will be to the input signal-to-noise ratio.

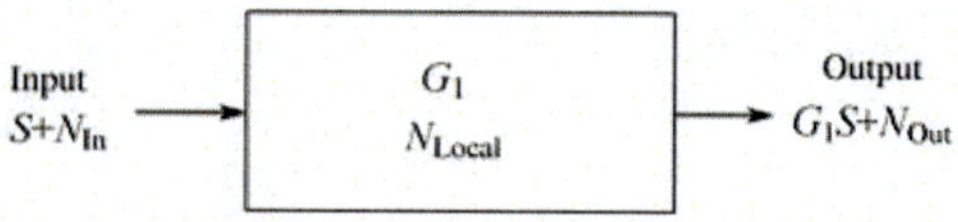

Fig. 8.3 Calculation of the noise figure

Fig. 8.4 Calculation of the noise figure of N cascaded systems

When the module itself does not generate any noise, the output signal-to-noise ratio is equal to the input signal-to-noise ratio, and the noise figure is 1. Of course, this is an ideal situation that cannot be achieved in practice. Some interesting conclusions can also be drawn from Eq. (8.7). If the two modules have the same noise power, the module with larger gain has the smaller noise figure. For passive devices such as cables, connectors, and passive filters, since the gain $G < 1$, it attenuates the signal. It can be proven that the noise figure of the passive device is equal to its insertion loss, for which the larger the attenuation of the passive device is, the larger the noise figure will be.

The above is the analysis for a single RF module. When multiple modules, such as N linear systems, are cascaded, the noise figure and gain of each system and the local noise generated by each module are assumed to be $F_i, \quad G_i, \quad i = 1, 2, \cdots, N$ and $N_{\mathrm{L},i}$, as shown in Fig. 8.4.

Based on the analysis of single modules, the following conclusions can be derived in a similar way:

- The signal outputted by the first-stage module is $G_1 S$, and the noise is $G_1 N_{\mathrm{In}} + N_{\mathrm{L},1}$;
- The signal outputted by the second-stage module is $G_1 G_2 S$, and the noise is $G_1 G_2 N_{\mathrm{In}} + G2 N_{\mathrm{L},1} + N_{\mathrm{L},2}$; ……

By analogy, the signal outputted by the N-stage module is $\prod_{i=1}^{n} G_i S$, and the noise is

$$\prod_{i=1}^{n} G_i N_{\mathrm{In}} + \prod_{i=2}^{n} G_i N_{\mathrm{L},1} + \cdots + G_N N_{\mathrm{L},N-1} + N_{\mathrm{L},N}$$

If the N-stage cascaded system is regarded as a module, the total noise figure is

$$\begin{aligned}
F_{\mathrm{All}} &= \frac{\prod_{i=1}^{n} G_i N_{\mathrm{In}} + \prod_{i=2}^{n} G_i N_{\mathrm{L},1} + \cdots + G_N N_{\mathrm{L},N-1} + N_{\mathrm{L}_N}}{\prod_{i=1}^{n} G_i S} \frac{S}{N_{\mathrm{In}}} \\
&= \frac{\prod_{i=1}^{n} G_i N_{\mathrm{In}} + \prod_{i=2}^{n} G_i N_{\mathrm{L},1} + \cdots + G_N N_{\mathrm{L},N-1} + N_{\mathrm{L}_N}}{\prod_{i=1}^{n} G_i N_{In}} \\
&= 1 + \frac{N_{\mathrm{L},1}}{G_1 N_{\mathrm{In}}} + \frac{N_{\mathrm{L},2}}{G_1 G_2 N_{\mathrm{In}}} + \cdots + \frac{N_{L,N}}{G_1 G_2 \cdots G_N N_{\mathrm{In}}} \\
&= F_1 + \frac{F_2 - 1}{G_1} + \cdots + \frac{F_N - 1}{G_1 G_2 \cdots G_{N-1}}
\end{aligned} \tag{8.10}$$

Equation (8.10) is the Friis Formula that is often adopted in RF front-ends. This formula reveals a useful property of the RF front-end, namely that when G_1 is large, the total noise figure of the entire cascaded system is basically determined by the noise system of the first-stage module. Therefore, in actual engineering design, the first module behind the satellite antenna is often a first-stage low-noise amplifier. The low-noise amplifier has a relatively lower noise figure and higher gain, so it determines the noise figure of the entire RF front-end. Therefore, in such a cascaded system, the noise figure of subsequent passive devices does not require much consideration. Some satellite antennae have built-in low-noise amplifiers. This type of antenna is called an active antenna. The antenna must be fed by the outside environment to ensure that it outputs a normal RF signal. In this case, regardless of how the subsequent RF front-end in the receiver is designed, the total noise figure is basically determined by the low noise amplifier in the antenna.

8.3 Bandpass Sampling

Bandpass sampling is widely used in the RF front-end of GNSS receivers. Especially in the application of Intentional Frequency Aliasing, the bandpass sampling process can be regarded as a combination of the digital quantization of the analog signal and the frequency conversion of the narrowband signal. Since bandpass sampling is so widely applied in the RF front-end of GNSS receivers, this section will explain the principles in detail. We will now review the principles of baseband sampling.

If the bandwidth of a signal $s(t)$ is B, the discrete sequence sampled at a frequency of $f_s > 2B$ will contain all the information of $s(t)$, which means that $s(t)$ can be reconstructed without distortion by the discrete sequence. Assuring that $f_s > 2B$ is the key for the sampled discrete sequence to reconstruct the original signal, and $2B$ is called the Nyquist frequency of $s(t)$. The reason why a sampling rate above two times the signal bandwidth shall be selected is to avoid spectral aliasing. In order to understand this, a theoretical analysis of the entire sampling principle is required.

For the convenience of theoretical analysis, it is assumed that the sampling waveform is a periodic impulse function $p(t)$ as shown in Fig. 8.5. Assume that the sampling period is T_s, obviously, $T_s = \frac{1}{f_s}$. The mathematical expression of the sampling pulse is

$$p(t) = \sum_{n=-\infty}^{\infty} \delta(t - nT_s) \tag{8.11}$$

whose Fourier transform is

$$P(f) = \frac{1}{T_s} \sum_{n=-\infty}^{\infty} \tag{8.12}$$

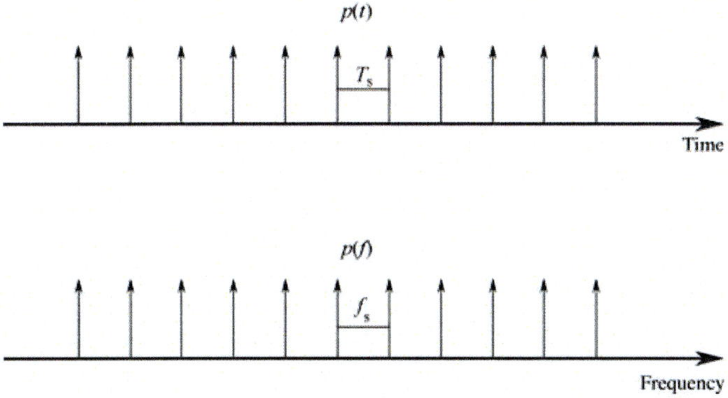

Fig. 8.5 Sampling the waveform and its spectrum [the arrows stand for the impulse function $\delta(t)$]

The signal is sampled by the sampling pulse described above, which can be regarded as the multiplication of the sampling waveform $p(t)$ and $s(t)$. If the signal after sampling is recorded as $\hat{s}(t)$, then

$$\hat{s}(t) = s(t) \sum_{n=-\infty}^{\infty} \delta(t - nT_s) \tag{8.13}$$

We perform a Fourier transform on $\hat{s}(t)$ to obtain its spectrum $\hat{S}(f)$

$$\begin{aligned}\hat{S}(f) &= S(f) \otimes P(f) \\ &= \frac{1}{T_s} \sum_{n=-\infty}^{\infty} S(f - \frac{n}{T_s})\end{aligned} \tag{8.14}$$

where $\otimes$ stands for the convolution.

After the sampling of the periodic impulse function, the spectrum of the obtained discrete signal is equivalent to the spectrum of the original signal after periodically shifting and superimposing, and the period is the sampling frequency f_s. This process is shown in Fig. 8.6. The upper part of the figure is the spectrum $S(f)$ of the original signal. The abscissa axis indicates the frequency, and the shaded part is the spectrum range of the signal. Since the sampling frequency $f_s > 2B$, the signal spectrum is between 0 and $0.5 f_s$. The lower half of Fig. 8.6 is the spectrum $\hat{S}(f)$ of the sample sequence, which is the expression in Eq. (8.14).

As seen in Fig. 8.6, if the signal bandwidth exceeds $f_s/2$, there must be an aliasing between the high frequency and the low frequency parts of the signal after sampling, for which the sampling frequency must be higher than two times the bandwidth. Figure 8.6 shows that the fundamental way to avoid signal aliasing is to ensure that the spectrum of the entire signal is distributed within $[kf_s, (k + 1/2) f_s]$, where $k = 0, 1, \cdots$, or any integer. Also, there may be a suspicion that the signal spectrum

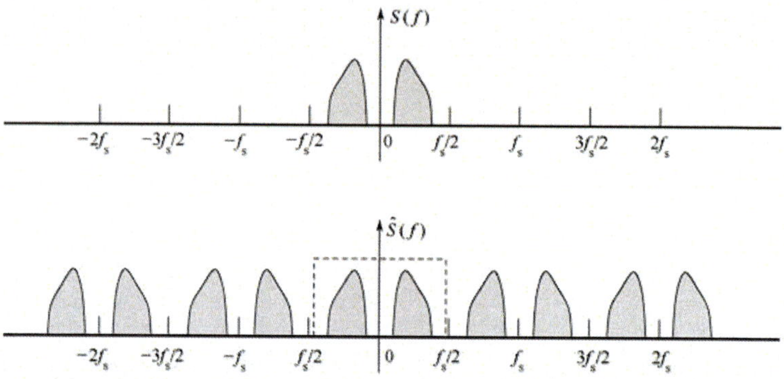

Fig. 8.6 Original signal spectrum $S(f)$ and signal spectrum $\hat{S}(f)$ after sampling

is in the negative frequency domain. In fact, if $s(t)$ is a real signal, its negative spectrum must be symmetrical with the positive spectrum with respect to the zero frequency, so it must also fall within $[-(k+1/2)f_s, -kf_s]$. Therefore, when $k = 0$, the sampling is baseband sampling. Meanwhile, when $k > 0$, since the highest frequency component in the signal is already higher than the sampling rate, the sampling is bandpass sampling.

It should be noted that when the spectrum of the signal is distributed between $[(k+1/2)f_s, (k+1)f_s]$, spectrum aliasing can also be avoided. However, at this time, "spectral inversion" occurs, which means that the high-frequency component in the original signal corresponds to the low frequency component in the sampled signal, while the low frequency component of the original signal corresponds to the high frequency component of the sampled signal. It is a situation that worthy of attention during the design of the system. Figure 8.7 shows a case of spectrum inversion.

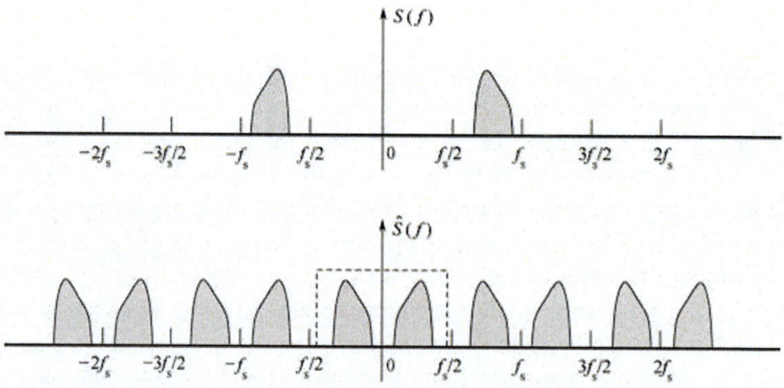

Fig. 8.7 The original signal spectrum $S(f)$ and the signal spectrum inversion after it is sampled

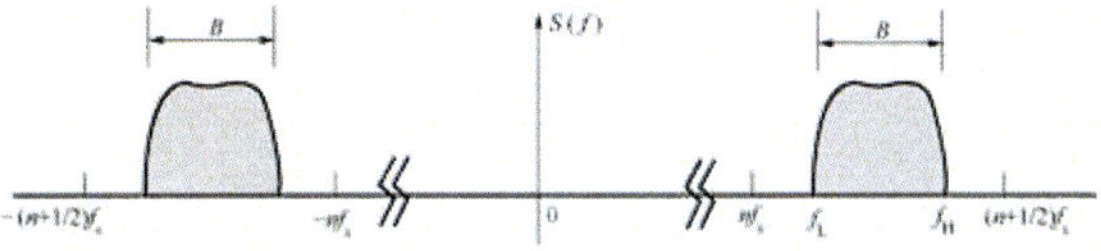

Fig. 8.8 Conditions for the selection of f_s

Bandpass sampling is often performed on narrowband signals, which are signals whose bandwidth is relatively smaller than their carrier frequency. The signal bandwidth of the C/A code in GPS is 2 MHzm. That of the P/Y code is only 20 MHz, which can be considered quite small compared to its carrier frequency of 1.57 GHz. Thus, the GPS signal is a standard narrowband signal. The BDS signal can also be considered as a narrowband signal judging from its relatively small bandwidth compared to its carrier frequency. According to the above analysis of the sampling principle, to avoid spectral aliasing, f_s must be selected based on the following three conditions during the sampling of the narrowband signal:

$$\begin{aligned} &f_s > 2B; \\ &f_L > nf_s/2; \\ &f_H < (n+1)f_s/2 \end{aligned} \tag{8.15}$$

Figure 8.8 illustrates Eq. (8.15).

In Fig. 8.8, B is the signal bandwidth, and f_L and f_H are the lowest frequency and the highest frequency component in the signal spectrum respectively. Obviously, $B = f_H - f_L$. In fact, only two of the three conditions in Eq. (8.15) are necessary, and the third one can be derived from any two of them. However, for the sake of clarity, all three of them are listed here. The value of n is not unique. It can be anything that meets the above conditions, and its value is only unique when there are other additional requirements. For example, in practice, the sampling rate often needs to be as low as possible on the premise that spectral aliasing does not occur. At this time, the value of n can be determined. It should be noted that when the value of n is even, the signal after sampling has no spectral inversion, while when n is odd, the spectrum after sampling is reversed.

The bandpass sampling principle can be represented by Fig. 8.9. The signal spectrum in the figure is between $[nf_s, (n+1/2)f_s]$, so the signal after sampling is shown in the lower part of the figure. Obviously, after sampling, the spectrum of the signal that was originally at a high frequency is linearly "moved" to the digital baseband, i.e. $[0, 1/2f_s]$. From this perspective, bandpass sampling can be considered as the combination of digital quantization and down-conversion.

An important premise of this processing is that the signal energy is concentrated within $[nf_s, (n+1/2)f_s]$ and the signal components are zero in other bands, because the out-of-band noise is also "moved" to the digital baseband during the down-conversion of the signal to the digital baseband, where the "out-of-band" noise includes noise in all frequency bands from DC to infinity. There are high requirements for bandpass sampling on the filter. It must be guaranteed that the flow of the

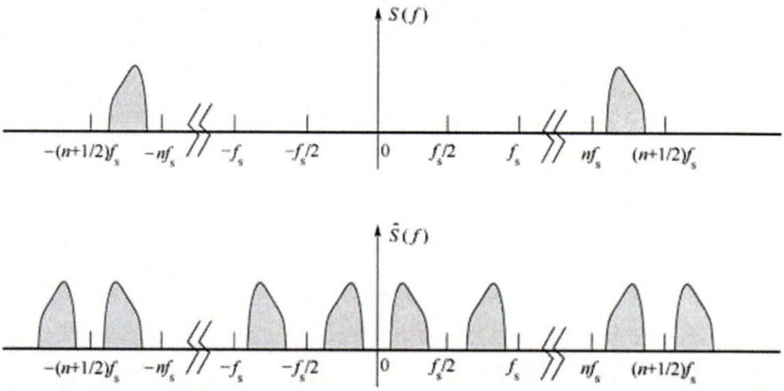

Fig. 8.9 Signal spectrum of the narrowband signal after bandpass sampling

amplitude-frequency response of the filter is unimpeded in the signal band and noise, and that unnecessary signals outside the signal band are intensively filtered out.

Since the BDS and GPS signal spectrum must be symmetrical with respect to its carrier frequency f_c, in practical engineering, when the value of n is determined, f_s can be selected so that

$$f_c = nf_s + 1/4f_s \tag{8.16}$$

Then, the center frequency of the signal after sampling must be at $0.25f_s$ on the digital baseband. This processing is the safest for the sake of avoiding spectral aliasing. Of course, it is not the only option.

Theoretically, there is no special requirement for the spectral distribution of the input signal in bandpass sampling as long as the three conditions in Eq. (8.15) are met. Therefore, in the design of the BDS and GPS receivers, bandpass sampling is applicable to the intermediate-frequency signal after down-conversion in the RF front-end and the RF signal after direct amplification, so there are two sampling schemes: IF sampling and RF sampling.

8.4 IF Sampling and RF Sampling Schemes

8.4.1 IF Sampling Scheme

The IF sampling scheme is currently a widely used RF front-end scheme in GNSS receivers. The basic configuration of this scheme is to down-convert the RF signal to the intermediate frequency through the local oscillator and the mixer, and sample the IF signal through the ADC to obtain the signal in the discrete time domain, and then send it to the subsequent signal processing software for baseband signal processing.

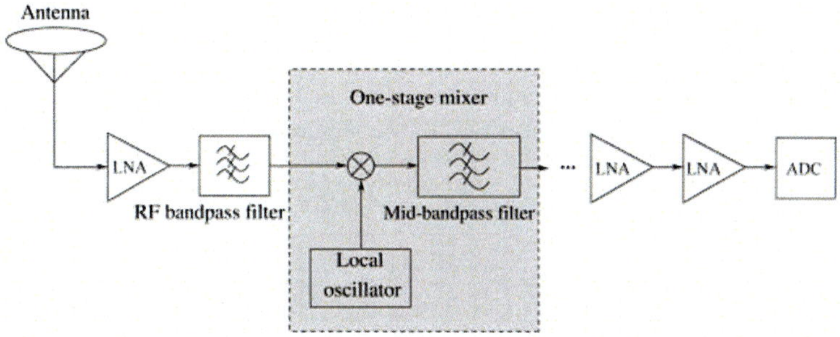

Fig. 8.10 IF sampling scheme diagram

In Sect. 5.3, a conventional RF solution is briefly explained in order to illustrate the extraction of Doppler observations, as shown in Fig. 5.5, where the RF scheme described is an IF sampling scheme. In general, the principle of the IF sampling scheme is shown in Fig. 8.10.

The RF signal in the figure is received by the antenna at first, and then amplified by a first-stage low noise amplifier (LNA), which is the first RF module behind the antenna here. This design is for the convenience of analyzing the noise figure of the cascaded system described in Sect. 8.2. After the LNA, unserviceable out-of-band signals and noise are filtered through the RF bandpass filter, followed by the first stage mixer. The mixer multiplies the RF signal and the local carrier of the local oscillator to obtain an intermediate frequency signal, which is represented by IF_1, where IF is the abbreviation of Intermediate Frequency. Some RF schemes have multi-stage mixers, for which there will be multiple intermediate frequencies, like IF_2, and IF_3. The values of the intermediate frequencies of each stage decrease successively. Besides, the out-of-band noise and interference are also gradually eliminated or reduced. Multi-stage and first-stage mixing schemes are all applicable in practice. The mixed signal is further filtered out of the out-of-band signal and noise by the mid-band pass filter to obtain the final down-converted signal. If the signal is not sufficiently amplified, there will be one or more stages of LNA in the mid-band to generate the necessary gain. Finally, the IF signal that is sufficiently amplified is sampled by the ADC to obtain a digital IF signal, which is sent to the subsequent module for processing.

Figure 8.10 contains only the basic functional units of the RF front-end. In fact, more are required for a practical GNSS RF front-end, such as automatic gain control (AGC). AGC is to ensure that the analog signal level input to the ADC is within the proper range and does not change with external signals and noise. If AGC is included, the LNA in Fig. 8.10 needs to be replaced by a controllable gain amplifier (PGA). In addition, modern RF front-ends typically include flexible local frequency synthesizers to accommodate different frequency inputs. In-phase and quadrature-channel signal sampling are also frequently included in ADC. Besides, clock and reset logic are also essential. At the same time, since the RF front-end often needs flexible configuration, peripheral configuration interfaces such as UART, IIC, or SPI bus interfaces are also necessary.

The filter in the RF front-end is an easily overlooked but essential part. According to the description of the bandpass sampling in Sect. 8.3, if there is no filter, noise, or other signals outside, the signal band will be "moved" into the digital baseband, which will affect the signal-to-noise ratio, signal capture, and tracking. If the filter works directly in the RF band, the passband width ($\approx 2\,\text{MHz}$) of the filter is very narrow compared to its operating frequency ($= 1.57$ GHz). Such a filter is regarded as one with a high value of Q, so it is hard to manufacture. The one commonly used today is the surface acoustic wave (SAW) filter, which works on the IF frequency. It is easier to manufacture and the cost is lower. Therefore, in terms of filter selection, the IF sampling scheme is more advantageous, and sometimes the common LC or RC filter can meet the requirements.

The above process is the entire signal flow of the IF RF scheme. Actually, there are many options for the implementation of the scheme, and Fig. 8.10 is just one of them. For example, the LNA in the mid-band can also be installed in the RF section, i.e. before the mixer. The performance and properties of different devices, such as the working frequency band of the amplifier, the impedance matching of each device during the cascade, and the noise figure of different devices should be taken into consideration in the specific implementations.

Since the RF signal received by the antenna is a real signal, its spectrum is symmetrical with respect to the RF carrier. If the local mixing signal is single in the carrier, then the mixing will only change the carrier frequency and not the signal spectrum. So, in the final stage of the mixer, the signal spectrum must be symmetrical with respect to the IF value, while the IF signal is used as the input signal to the ADC to complete the sampling process. The sampling scheme can be baseband sampling or bandpass sampling as mentioned in Sect. 8.3, depending on the final IF value f_{IF} and the sampling frequency f_s. If the IF value falls within the digital fundamental frequency, baseband sampling should be adopted, and if the IF value falls outside the digital fundamental frequency, the scheme should be bandpass sampling. The overall process can be expressed as follows:

$$\text{Sampling scheme} = \begin{cases} \text{Baseband sampling}, & When \quad f_{\text{IF}} \in [0, 0.5f_s] \\ \text{Bandpass sampling}, & When \quad f_{IF} \in [0.5f_s, \infty] \end{cases}$$

For subsequent work in the receiver, the two most important technical indicators that need to be known from the RF front-end are the sampling frequency and the theoretical intermediate frequency. The sampling frequency is important for two reasons. First, the operating frequency of the carrier NCO and the pseudo-code NCO in the subsequent tracking loop comes from the sampling frequency, so it must be accurate to generate the correct local carrier and pseudo-code. Second, the operating frequency of the software receiver is the sampling frequency. The arrival of each sample point indicates the beginning of a new clock cycle, and the local time comes from the integration of the sampling clock. The principle of the theoretical IF value has been explained in Sect. 5.3, and will not be repeated here. What needs to be mentioned is that the theoretical IF value is determined by the setting of the specific

RF front-end, and there is no general method of calculation. Basically, in the entire IF sampling scheme, there are two modules that affect the theoretical IF value. One is obviously the mixer, and the other is the ADC, which is easy to understand if we look back at the explanation of the sampling principle in Sect. 8.3. In the bandpass sampling process, what is actually performed is the down-conversion of the IF signal. Therefore, in this process, the theoretical intermediate frequency value will inevitably be affected. Hence, the detailed RF frequency scheme and the sampling frequency of the final ADC is indispensable for the confirmation of the theoretical IF value.

It can be seen from the above analysis that the working clock is needed for the local oscillator, ADC, and other modules of the RF front-end. The signal of the clock is critical, because the clock's performance directly affects the theoretical IF value f_{IF} of the signal after sampling and the stability of the sampling frequency f_{s}. It will also directly impact the subsequent signal processing, for example, the performance of the carrier tracking loop and the pseudo-code tracking loop. Therefore, the reference clock used in the RF front-end of the GNSS receiver generally has a temperature compensated crystal oscillator (TCXO). This quartz crystal oscillator can achieve a frequency stability of 10^{-6}~10^{-7} through the negative feedback control of the temperature compensation loop.

The following is an example of the effect of clock bias on f_{IF} and f_{s}.

Let's assume that there is an RF solution with IF sampling as shown in Fig. 8.10. The intermediate frequency is obtained from the first-stage mixer and sent to the ADC through two stages of LNA. Let's assume that the theoretical frequency of the local oscillator in the mixer is 1 571.0 MHz, then the theoretical IF value $f_{\mathrm{IF}} = 1\ 575.42 - 1\ 571.0 = 4.42$ MHz. However, due to the deviation of the local clock, the actual frequency of the local oscillator will not be 1,571.0 MHz, but 1,571.010 MHz. This is 10 kHz higher than the theoretical IF value, so the actual IF value obtained will also have a corresponding deviation of 10 kHz. Due to the relative motion between the satellite and the user, the received signal must have a Doppler shift, so the value of the final intermediate frequency is determined by both the Doppler shift and the frequency deviation of the local oscillator. If the deviation between the true IF value and the theoretical IF value is too large, the complexity of the signal acquisition algorithm will increase. For example, it will take a long time to cover all possible frequency ranges. More seriously, the excessively large frequency deviation will make it impossible to capture the signal with the acquisition algorithm. For GPS signals at the L1 frequency and BDS signals at the B1 frequency, the theoretical intermediate frequency will deviate by approximately 1.5 kHz for every 1 ppm increase in the frequency deviation of the TCXO. Therefore, the worse the frequency stability of the TCXO is, the more the Doppler frequency search range is to be increased when the signal is captured.

As mentioned earlier, the deviation of f_{s} during bandpass sampling will affect the actual IF value of the Doppler observation. However, the deviation of the clock will only cause common deviation to all Doppler observations, but will not affect the accuracy of the estimation of the receiver's speed. To understand this, we only need to review Sect. 7.4. In addition, the clock input of the subsequent signal processing module comes from the sampling frequency f_{s}, and the local time of the receiver is maintained by integrating the f_s clock. The extraction of the pseudo-range observation requires local time, so the deviation of the sampling frequency will directly affect the deviation of the pseudo-range observation. Of course, similar to the Doppler observation, the deviation caused by the clock is the common deviation of the pseudo-range observations, which can be estimated with the positioning algorithm, as can the clock drift. Hence, it can be deduced that there is an integral relationship between the clock deviation and the clock drift calculated through the PVT, which has been explained in detail in the description of the PVT algorithm in Chap. 7.

8.4.2 RF Sampling Scheme

The biggest difference between the RF sampling scheme and the IF sampling mentioned above is that there is no mixer and IF involved, and all processing is done at the RF end. The RF sampling scheme is the best choice for software radio systems, as its simple hardware setup conforms to the principles of software radio. It is necessary for the entire RF front-end to have a certain gain, which is an amplification of the signal in the RF band. The RF signal, amplified to a certain extent, will be sent directly to the ADC for sampling and quantization. It can be seen that the ADC also works in the RF band instead of the IF band. The workflow of the RF sampling scheme is shown in Fig. 8.11. Compared with the IF sampling scheme in Fig. 8.10, the RF sampling scheme is much simpler in structure. However, this does not mean that the RF sampling scheme is easier to implement. The main RF front-end chips on the market are made using the IF sampling scheme.

In Fig. 8.11, the first stage after the GPS antenna is the LNA, followed by the RF bandpass filter, considering the noise figure. Since one stage of the LNA may not be enough to achieve the required gain, one or more LNAs are needed. There will be one or more stages of filters to remove out-of-band noise or interference, and the sampling is finally completed by the ADC. Because the RF frequency of the GPS L1 signal is L1 = 1575.42 MHz, it is obvious that if the baseband sampling is adopted, the sampling rate should be higher than $2 \times \text{L1} = 3\,150.84$ MHz, which is very difficult to achieve using current technology. Therefore, the bandpass sampling

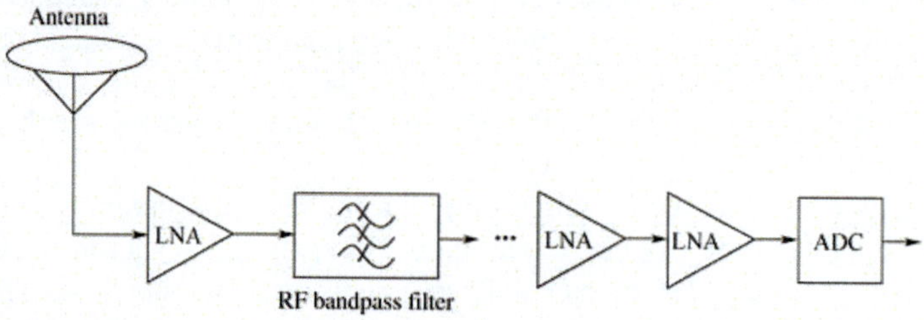

Fig. 8.11 The principle of the RF sampling scheme

principle is generally applied for the ADC in the RF sampling scheme. During the bandpass sampling, the signal band is transformed from the radio frequency to the digital baseband, where deliberate spectral aliasing is performed to complete the down-conversion. Detailed theoretical analysis can be seen in Sect. 8.3. In order to give readers a clearer understanding of this process, here is an example.

Let's assume that the sampling rate $f_s = 100$ MHz, as in the analysis in Sect. 7.2, for the GPS signal

$$f_c = 1\,575.4\,\text{MHz},\ B \approx 2\,\text{MHz} \Rightarrow f_L = 1\,574.4\,\text{MHz},\ f_H = 1\,576.4\,\text{MHz}$$

According to Eq. (8.15), for f_L and f_H of the GPS signal

$$f_L > 31 f_s/2, \quad and \quad f_H < (31 + 1) f_s/2$$

So, when $f_s = 100$ MHz, $n = 31$, the BDS or GPS signal is moved from the RF band, i.e. [1,550 MHz, 1,600 MHz], to the digital baseband [0,50 MHz] through bandpass sampling. Since n is an odd number at this time, the sampled signal undergoes spectrum inversion.

The above process is cumbersome and unintelligible in words and formulas, so some scholars (including Akos and Poppe) skillfully use the mirror steps shown in Fig. 8.12 to describe it.

The lowermost range in Fig. 8.12 is the digital baseband, i.e. $\left[0, \frac{f_s}{2}\right]$. The frequencies above the digital baseband are all reflected on the upper ladder, and each of the

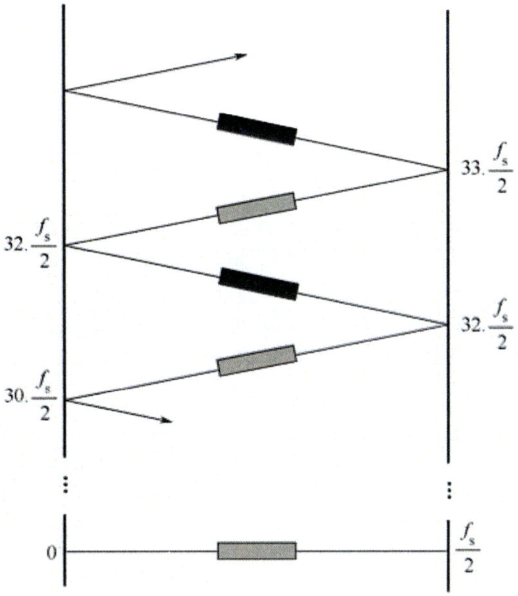

Fig. 8.12 The steps of the RF sampling image frequency, where $f_s = 100$ MHz; at this point the GPS signal falls between $\left[31 \times \frac{f_s}{2}, 32 \times \frac{f_s}{2}\right]$

fold lines corresponds to a band of frequency $\left[i\frac{f_s}{2}, (i+1)\frac{f_s}{2}\right]$, $i = 1, 2, \cdots$. The GPS signal is within $\left[31 \times \frac{f_s}{2}, 32 \times \frac{f_s}{2}\right]$. The shaded area in the middle of each fold line in the figure corresponds to the signal spectrum. According to Eq. (8.16), the center frequency of the signal is preferably located at the center point of each fold line. Careful readers will notice that some of the signal spectrum of the fold line is lighter in color, and some is darker. The shallower signal bands indicate that no spectrum inversion occurs, and the deeper ones indicate the opposite. It can be seen that spectrum inversion occurs in the line where the GPS signal is located.

The effect of the filter on the signal-to-noise ratio of the sampled signal is also easy to see in Fig. 8.12. Because the noise in each step will be moved to the digital baseband, a filter is necessary to filter out noises outside the signal band. Otherwise, there will be a lot of out-of-band noise in the signal after sampling.

Figure 8.12 also illustrates the effect of the deviation of the sampling frequency on the theoretical intermediate frequency of the signal after sampling. Let's assume that the actual sampling rate is 100.001 MHz, which is 1 kHz higher than the theoretical sampling rate. Because the step where the GPS signal is located corresponds to $n = 31$, 16 complete $[0, f_s]$ frequency bands need to be "folded" for it to reach the digital baseband. So, the GPS signal after sampling will have a deviation of 16 kHz, from which we can tell that the RF sampling process requires high frequency accuracy and stability of the sampling clock.

In the RF sampling scheme with bandpass sampling, although the operating frequency of the ADC is not high, such as 100 MHz in the above example, the designer must always be aware that the signal frequency is in the RF band, that is, L1 = 1575.42 MHz. Therefore, the ADC must not cause unnecessary attenuation to such high-frequency signals, and the distribution parameters of the device should also be considered during the design of the circuit. For example, the pins of the device and the cables among them are equivalent to a low-pass network, as shown in Fig. 8.13. These factors must be carefully considered in both device selection and system hardware design.

Another advantage of the RF sampling scheme is the ability to simultaneously sample multiple GNSS signals. Taking GPS signals as an example, the current GPS signals are mainly L1 and L2 signals. With the modernization of GPS systems, new civilian signals L2C and L5 will enter the civilian receiver market, hopefully allowing multiple GPS signals to be sampled with just one RF front-end. Theory and practice have proved that if the sampling rate is carefully selected so that GPS signals of multiple frequencies fall within a certain frequency band of the digital

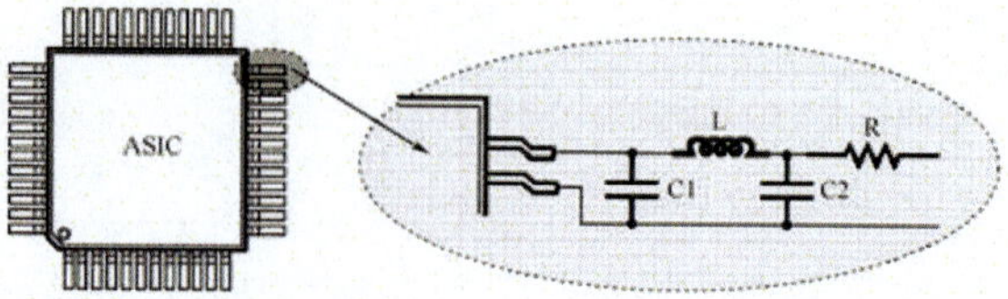

Fig. 8.13 Equivalent circuit of the ADC input pin

Table 8.1 Signal properties of future GPS L1, L2C, and L5 signals

Signal type	L1	L2C	L5
Signal bandwidth	≈2 MHz	≈2 MHz	≈20 MHz
Chip rate of the signal	1.023 Mchip/s	1.023 Mchip/s	10.23 Mchip/s
Carrier frequency	1 575.42 MHz	1 227.6 MHz	1 176.45 MHz
$[f_L, f_H]$/MHz	[1 574.42,1 576.42]	[1 226.6,1 228.6]	[1 171.45, 1 181.45]

baseband without overlapping, these signals can be down-converted through one bandpass sampling. The following is an example to illustrate the implementation of this conception.

Future GPS civilian signals will include L1 C/A, L2C, and L5 signals, whose signal bandwidth and carrier frequency are shown in Table 8.1. The last column of the table shows the spectrum range of the three signals.

The sum of the three signal bandwidths is approximately 24 MHz, or $\approx (2+2+20)$ MHz, so according to the Nyquist sampling principle, the required sampling frequency is at least 50 MHz. After the sampling frequency is selected, according to the above description, each of the signals can be bandpassed and moved from the RF band to the digital baseband. Instead of single frequency, there are now multiple signals, and it must be ensured that the spectrum of each signal does not overlap in the digital baseband. Several sampling frequencies as shown in Table 8.2 are obtained through correlated calculation, allowing the spectrum of the three signals after bandpass sampling to satisfy the above conditions on the digital baseband. The sampling frequency and the corresponding center frequency f_c of the three GPS signals at the frequency of the digital baseband, as well as the range of the spectrum $[f_c - 0.5B, fc + 0.5B]$, are also given in Fig. 8.2.

Table 8.2 Selection of the sampling frequency scheme for simultaneous bandpass sampling of GPS L1, L2C, and L5 signals

Sample frequency f_s	L1		L2C		L5	
	f_c	$f_c \pm 0.5B$	f_c	$f_c \pm 0.5B$	f_c	$f_c \pm 0.5B$
64.5	27.42	(26.39, 28.44)	2.1	(1.077, 3.123)	15.45	(5.22, 25.68)
166	1.42	(80.397,82.443)	65.6	(64.577,66.623)	14.45	(4.22, 24.68)
166.5	76.92	(75.897,77.943)	62.1	(61.077,63.123)	10.95	(0.72,21.18)
190	55.42	(54.397,56.443)	87.6	(86.577,88.623)	36.45	(26.22,46.48)
191.5	43.42	(42.397,44.443)	78.6	(77.577,79.623)	27.45	(17.22,37.68)
192	39.42	(38.397,40.443)	75.6	(74.577,76.623)	24.45	(14.22,34.68)
Note: All the units above are MHz						

The following is an example of how the results in Table 8.2 can be obtained when the sampling frequency $f_s = 64.5$ MHz.

For the GPS L1 signal, according to Eq. (8.15)

$$f_{L1} > nf_s/2, \quad and f_{H1} < (n+1)f_s/2, \quad \Rightarrow n = 48$$

where $f_{\mathrm{L1}} = 1\ 574.42$, $f_{\mathrm{H1}} = 1\ 576.42$, the theoretical intermediate frequency on the digital baseband is

$$f_{\mathrm{C1}} = 1\ 575.42 - nf_s/2 = 27.42 \text{ MHz}$$

If we analyze the L2C signal in a similar way, then

$$f_{L2} > nf_s/2, \quad and f_{\mathrm{H2}} < (n+1)f_s/2, \quad \Rightarrow \quad n = 38$$

where $f_{\mathrm{L2}} = 1\ 226.6$, $f_{\mathrm{H2}} = 1\ 228.6$, and the theoretical intermediate frequency on the digital baseband is

$$f_{\mathrm{C2}} = 1\ 227.6 - nf_s/2 = 2.1 \text{ MHz}$$

Likewise, for L5 signals,

$$f_{L3} > nf_s/2, \quad and f_{\mathrm{H3}} < (n+1)f_s/2, \quad \Rightarrow \quad n = 36$$

where $f_{\mathrm{L3}} = 1\,171.45$, $f_{\mathrm{H3}} = 1\,181.45$, 1,171.45, and 1,181.45, so the theoretical intermediate frequency on the digital baseband is

$$f_{\mathrm{C3}} = 1\ 176.45 - nf_s/2 = 15.45 \text{ MHz}$$

For other sampling frequency schemes, readers can perform independent analysis.

8.5 Automatic Gain Control (AGC) and Quantization Bit Width

The environment in which the RF front-end of the GNSS receiver operates varies. In addition to GNSS signals, various noises and interference signals will enter the antenna, including blocking signals, some of which are accidentally captured, and some of which are intentionally sent, such as malicious blocking signals sent to enemy receivers in hostile states. The main noises and interference signals that civil receivers may encounter include cosmic background noise, antenna thermal noise, continuous wave interference, and electromagnetic compatibility interference. For instance, for GNSS receivers in mobile phones, the interference of the image

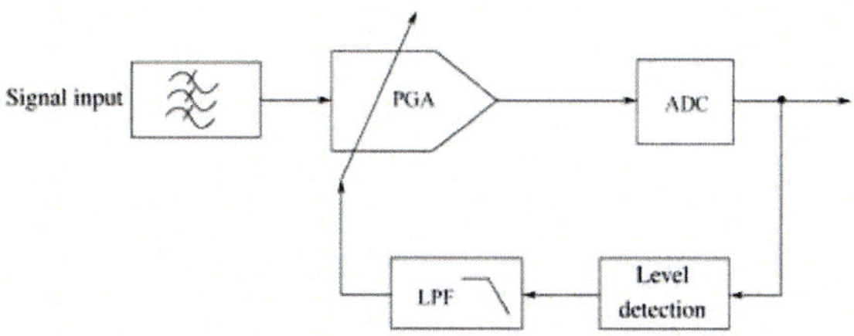

Fig. 8.14 The basic principle of automatic gain control

frequency or harmonic frequency of the communication frequency band (the working frequency band of communication systems such as GSM, TDS-CDMA, WCDMA, and cdma2000) must be taken into consideration. The presence of these interference and noise signals causes the signal received by the antenna to fluctuate drastically in intensity when the signal arrives at the ADC after being amplified and filtered. The main purpose of the automatic gain control module is to ensure that the input level of the ADC remains stable over a certain range. Without automatic gain control, the output of the ADC will be saturated in the presence of a strong interference signal, and the signal will not be output normally.

Figure 8.14 shows the principle of an AGC that is widely used in the GNSS RF front-end, in which the automatic gain control is composed of a level detection unit, a low-pass filter unit, and a programmable gain amplifier (PGA). The level detecting unit performs level detection on the digital sample outputted by the ADC, and generates a control signal according to the magnitude of the level value, which is filtered by the low-pass filter to remove the high frequency and then used to control the amplification factor of the PGA. Thereby, the gain of the PGA is small when the sampled data of the ADC output is too high, and the gain of the PGA becomes large when the sampled level of the ADC output is too low, so the sampled data outputted by the ADC remains in a stable range of levels. The input signal of the level detection unit in Fig. 8.14 is from the output of the ADC, i.e. the digital sampling signal. At this time, the level detection is performed by the digital circuit. Of course, the input signal of the level detection can also be extracted from the input of the ADC, i.e. the analog signal. At this time the level detection is performed by the analog circuit.

The factors affecting the function of AGC mainly include the gain control range of the PGA, the valve time of the control signal, and the algorithm of level detection. The gain control range of the PGA is written in dB, indicating the range of amplifier gain that the control signal can control. Currently, the general GNSS RF chip can achieve a gain control range of 40 to 60 dB [4–6]. The larger the gain control range is, the larger the dynamic range of interference and noise that the AGC can handle will be. However, excessive gain can also cause problems for the stability of the amplifier. The valve time of the control signal is a time constant for the generation of the control signal. In the level detection algorithm, the level in a certain period of time needs to be analyzed to obtain the control signal, and the valve time is the length of time period here, so the shorter the valve time is, the more sensitive the control signal will be. However, if the valve time is too short, too much noise will be contained in the signal. The longer the valve time is, the more the control signal will lag behind the input interference and the change of noise, so the general valve time should be kept accurate to the order of milliseconds. The level detection is another important factor affecting the performance of the AGC. In general, it includes peak-to-peak detection, average leveling, power calculation, and sampled data distribution analysis. For the scheme where the sampling level comes from the ADC output sampling, the sampled data distribution analysis is extensively used in the current GNSS RF front-end.

The sampled data distribution analysis relies on a basic assumption that the noise and the interference in the input signal of the antenna are distributed as Gaussian white noise, and the sampling level distribution of the ADC output should follow the same distribution pattern. There is actually an implicit premise for this method that the main component of the antenna input signal is the noise component, and the intensity of the GNSS signal is so weak that it does not affect the strength of the noise component. This is often in line with the actual situation, because when satellite signals reach the ground, they are generally obliterated by noise. To prove this, consider that even in open sky, the power of the GPS signal is about 19 dB lower than the noise power. Therefore, the adjustment of AGC in the GNSS RF front-end is different from other near-field communication terminals. The AGC of the GNSS RF front-end generates the control signal according to the power of the noise level, while the general near-field communication terminals generate it according to the power of the signal intensity.

Figure 8.15 shows the distribution of the sampling point outputted by the ADC after the AGC controls the PGA in the two-bit quantization case. Because of the two-bit quantization, the digital sample points outputted by the ADC have four levels: $\pm 1, \pm 3$. The levels are $+1$ and $+3$ when the input signal is positive, and are -1 and -3 when the input signal is negative. The value of the quantization level V_T determines whether the absolute value of the ADC output is 1 or 3. The AGC will generate a control signal to adjust the gain of the PGA, so it actually controls the analog level of the ADC input, not the quantization level of the ADC. However, the distribution of the input level of the ADC can be considered as fixed, and the quantization level of the ADC can be adjusted accordingly, so the AGC control can be demonstrated by the relationship between the distribution curve of the input level and the value of the quantization level V_T in Fig. 8.15.

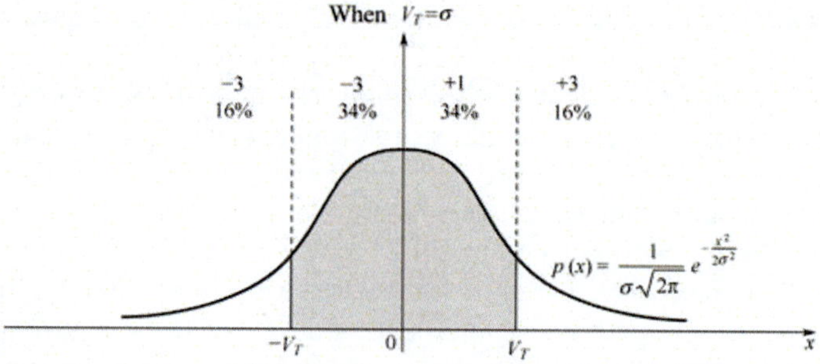

Fig. 8.15 The influence of quantization level on sampling point distribution after AGC control

If we assume that the distribution probability density function of the input signal is

$$p(x) = \frac{1}{\sigma\sqrt{2\pi}} e^{-\frac{x^2}{2\sigma^2}} \tag{8.17}$$

then the distribution probabilities of the four sampling points are

$$P(\pm 3) = \int_{V_T}^{\infty} \frac{1}{\sigma\sqrt{2\pi}} e^{-\frac{x^2}{2\sigma^2}} \mathrm{d}x \tag{8.18}$$

$$P(\pm 1) = \int_{0}^{V_T} \frac{1}{\sigma\sqrt{2\pi}} e^{-\frac{x^2}{2\sigma^2}} \mathrm{d}x \tag{8.19}$$

Substitute $V_T = \sigma$ into Eq. (8.18) and (8.19), then $\mathrm{P}\{-3, -1, +1, +3\} = [16\%, 34\%, 34\%, 16\%]$, which is also currently the most widely adopted sampled data distribution under the two-bit quantization case. The level detecting unit of the AGC checks the distribution of sampled data 1 and 3 in the valve time. If the distribution of the sample value 1 is greater than 34%, it indicates that the gain of the PGA is too low and needs to be made higher. Otherwise, the PGA gain is too high and needs to be lowered.

The above analysis is based on the two-bit quantization case, and similar analysis can be conducted in cases of multi-bit quantization. Figure 8.16 shows the sampling level distribution in the three types of PGA gains when the quantization width is 6 bits. Since the quantization bit is 6, the range of samples at this time is [−63, +63]. Figure 8.16a shows the distribution of sampled points when the gain is too small. It can be seen that the sampling points are mostly distributed in the range with low-value sampling points. Figure 8.16b is the case where the gain is appropriate. At this time, the distribution of sampling points exhibits a Gaussian pattern over the entire range. In Fig. 8.16c, where the gain is too large, the sample points are abnormally distributed in a range with high-value sampling points.

In cases where the quantization bit width is limited, quantization loss in the ADC will be inevitable during quantization, so the wider the quantization bit is, the smaller the quantization loss will be. Since the spread gain of the GPS or BDS signal is large (for example, the spread gain of the GPS CA code is 30 dB, while that of the BDS D1 code is 33 dB), the coherent integration time of baseband processing is longer, and the quantization loss of high-bit quantization tends to decrease gradually. According to References [9, 10 and 12], the minimum quantization losses with the quantization bit width of 1 bit, 2 bits, 3 bits, 4 bits are shown in Table 8.3 on the premise that the noise is in Gaussian distribution and the RF bandwidth is infinite. If the bandwidth is limited, the actual quantization losses will be slightly larger than those in Table 8.3, by 0.4 to 0.5 dB.

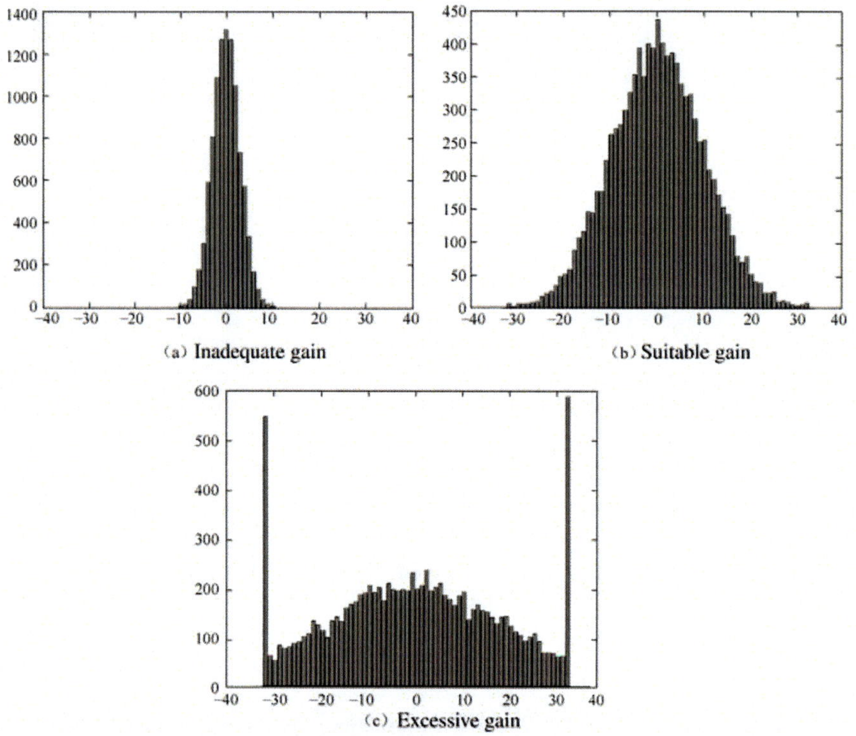

Fig. 8.16 Three types of AGC control in bit quantization

Table 8.3 Relationship between minimum quantization loss and bit width

Quantized bit	Minimum quantization loss/dB
1	−1.96
2	−0.55
3	−0.17
4	−0.05

According to Table 8.3, when the number of quantization bits is larger than four, the quantization loss can almost be negligible. By this time, the main purpose of increasing the quantization bits is not to reduce the quantization loss, but to improve the dynamic range of the RF front-end. Because the dynamic range can be increased by 6 dB with each additional bit, there will be an additional 60 dB in the range with a sampling of 12 bits as compared to that of 2 bits. With the adjustable gain of 40–60 dB brought by the PGA, the total dynamic range can reach 120 dB, which is highly advantageous when dealing with signal blocking and interference. Therefore, high-bit sampling schemes are widely adopted in the RF front-ends of anti-blocking receivers.

8.6 The Relationship Between RF Carrier-to-Noise Ratio and Baseband Signal-to-Noise Ratio

In order to indicate the quality of GNSS satellite signals, in the RF segment, the carrier-to-noise ratio CN_0 is often used to indicate the relationship between signal strength and noise, and the baseband is often expressed by the signal-to-noise ratio S/N. It is important to understand the relationship between the two, because one of the tasks of the receiver is to estimate CN_0 based on the S/N of the baseband. The meaning of CN_0 has been roughly introduced in Sect. 4.2.5 of this book. This section will detail the meaning of CN_0 and the relationship between CN_0 and S/N. In order to understand the relationship between the two, the meaning of CN_0 should be understood first.

Noise is included in the signal received by the antenna. A reasonable assumption is that the noise here is white noise, that is, the power spectral density of the noise is flat and does not vary with frequency. Based on this assumption, in a system with a signal bandwidth B, if the single sideband power spectral density of the input noise is N_0, the total noise power is

$$P_n = BN_0 \tag{8.20}$$

This equation shows that the unit of noise power spectral density N_0 should be watts/hertz (W/Hz).

The power spectral density of thermal noise is proportional to its equivalent noise temperature,

$$N_0 = kT_e \tag{8.21}$$

$k = 1.38 \times 10^{-23}$ is the Boltzmann constant, and T_e is the equivalent noise temperature whose unit is kelvins. The thermal noise comes from the conductors in the device and the electron motion inside the semiconductor, so thermal noise is inevitable as long as the device's temperature is above absolute zero. It can be calculated from Eq. (8.21), when the equivalent noise temperature is 290 K, $N_0 =$ –204 dBW/Hz. For the BDS and GPS antennae, the equivalent noise temperature is 70–100 K due to the main reception of signals from space.

As analyzed in Sect. 8.1, the receiver on the Earth's surface receives a signal power which is approximately −156 dBW. The noise power spectral density at the RF front-end of the receiver is primarily determined by the antenna and the first-stage LNA, mostly around −201 dBW/Hz. Therefore, the carrier-to-noise ratio of the RF front-end of the receiver is roughly

$$CN_0 \approx (-156) - (-201) = 45 \text{ dB/Hz} \tag{8.22}$$

Similar to the calculation of signal power, the calculation of the carrier-to-noise ratio is also a rough estimate. The actual value is related to many factors, such as the elevation angle of the satellite, the circumstance that the user use in, and the performance parameters of the receiver antenna. The carrier-to-noise ratio CN_0 is an important parameter in the signal processing of GNSS receivers. Many signal processing performance and results are directly related to the carrier-to-noise ratio. For example, in the initial signal search process, the carrier-to-noise ratio directly determines whether the capture is successful or not; in the tracking process of the signal, the phase noise of the tracking loop is also directly related to the carrier-to-noise ratio. Therefore, it is important to know the carrier-to-noise ratio information for the GNSS receiver. The specific calculation method is described in detail in Sect. 4.2.5.

It is difficult to estimate the carrier-to-noise ratio directly at the RF front-end because the signal is annihilated in the noise and cannot be separated from the signal without special processing. Beginners may have questions about this. Given that the input signal-to-noise ratio of the RF signal is about 45 dB/Hz, isn't this signal stronger than noise? This doubt is due to a shallow understanding of the meaning of the carrier-to-noise ratio. This ratio measures the relationship between signal power and noise power per unit bandwidth, so the system noise bandwidth must be considered to truly measure the relationship between signal power and noise power. Specific to the GPS C/A code signal, since the signal bandwidth is 2 MHz, the bandwidth of the RF front-end must be greater than 2 MHz to allow the signal power to pass without loss. It is assumed here that the bandwidth of the RF front-end is 2 MHz, and the signal power and noise power spectral density are calculated in the value of Eq. (8.22), that is, about 201 dBW/Hz. Then, the noise power is $P_n(\mathrm{dB}) = -201 + 10\,\mathrm{lg}(2 \times 10^6) = -138$ dBW. Hence, the signal-to-noise ratio is $[-156 - (-138)] = -18$ dB. This shows that the signal is indeed annihilated in the noise.

Figure 8.17 shows the relationship between the carrier-to-noise ratio and the signal-to-noise ratio. The thick line above in the figure indicates the noise power

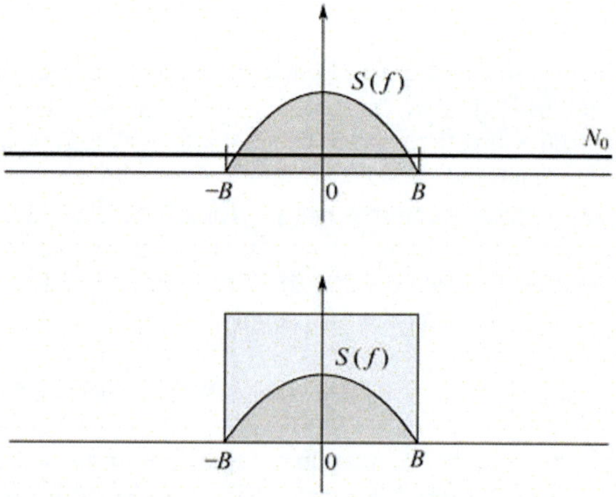

Fig. 8.17 The relationship between the carrier-to-noise ratio and signal-to-noise ratio

spectral density. The rectangular shaded area below can be considered as the noise power in cases where the RF front-end noise bandwidth is B.

According to the physics of the carrier-to-noise ratio, if the bandwidth of the RF front-end is B, the signal-to-noise ratio of the baseband processing is

$$S/N_{\text{in}} = \frac{CN_0}{B} \tag{8.23}$$

It is assumed that the coherent integration of the integration time T is performed during the de-spreading process, so the resulting spread gain is

$$G_{\text{d}} = \frac{T}{T_{\text{c}}} \tag{8.24}$$

Here, T_{c} is the chip width of the spreading code; for the GPS signal C/A code, $T_{\text{c}} = \frac{1}{1\,023}$ ms, and for the BDS D1 code, $T_{\text{c}} = \frac{1}{2\,046}$ ms. Based on the above analysis, and without considering other losses, the relationship between S/N and CN_0 can be obtained:

$$S/N = S/N_{\text{in}}G_{\text{d}} = CN_0\frac{T}{BT_{\text{c}}} \tag{8.25}$$

The above equation can be written in logarithmic form:

$$CN_0(\text{dB}) = S/N(\text{dB}) + 10\,\text{lg}(\text{B}) - 10\,\text{lg}(T/T_{\text{c}}) \tag{8.26}$$

Considering that the noise figure and other losses of the RF front-end are not zero, the above equation can be adjusted to

$$CN_0(\text{dB}) = S/N(\text{dB}) + 10\,\text{lg}(B) - 10\,\text{lg}(T/T_{\text{c}}) + \beta \tag{8.27}$$

where β is the sum of the noise figure of the receiver RF front-end and the baseband processing loss. Equation (8.27) shows the relationship between the carrier-to-noise ratio and the baseband signal-to-noise ratio.

8.6.1 Case Analysis of the RF Front-End Frequency Scheme

GP2015 is a GPS RF front-end chip from Zarlink that provides a low-power, low-cost, and highly integrated GPS L1 band RF front-end solution for device-constrained applications. GP2015 integrates a frequency synthesizer, a three-stage mixer, an AGC control unit, and a two-bit ADC. It requires only a few peripheral components to form a complete RF front-end. GP2015 was designed more than ten years ago, and has been overtaken by contemporary mainstream products in terms of the performance

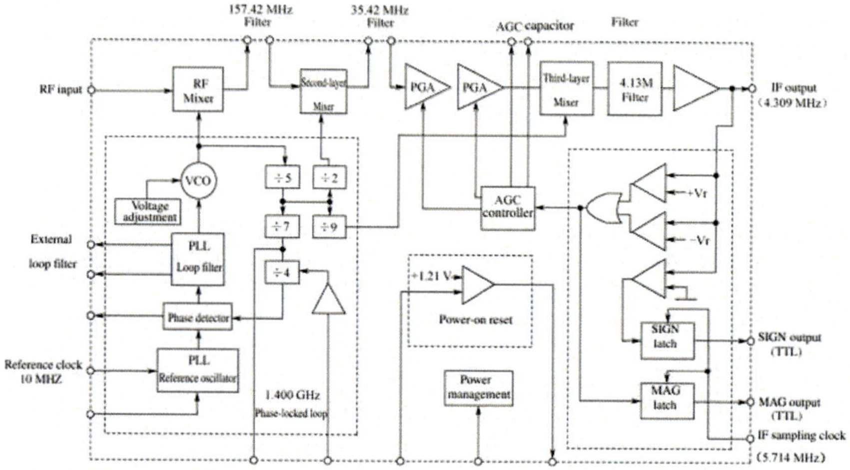

Fig. 8.18 Block diagram of GP2015

of the indicator and the RF front-end. However, its clear system structure, comprehensive functional unit, and rich product documentation make it a good example to demonstrate the principle and structure of the GFSS RF front-end. So, in this section, with GP2015 as an example, the principles explained earlier in this chapter will be integrated with actual engineering products to deepen readers' understanding of the RF front-end of GNSS receivers.

Figure 8.18 is a diagram of GP2015, showing just a few key modules that determine the RF frequency scheme, including local frequency synthesizers, mixers, automatic gain control modules, and ADC sampling modules. The rest of the devices that are significant but not directly related to the frequency scheme are omitted or simply listed.

Briefly speaking, the function of GP2015 is to amplify the L1 signal of the GPS satellite received by the antenna. After mixing with the local oscillator, it is converted to the intermediate frequency band. In GP2015, there are three intermediate frequencies, which are 175.42 MHz. 35.42 MHz, and 4.3 MHz respectively. Finally, a digital intermediate frequency signal with a theoretical IF of 1.405 MHz is sampled by an ADC with a sampling rate of 5.714 MHz.

The signal carrier frequency received by the antenna is L1 = 1,575.42 MHz, and enters the GP2015 mixer. The mixer multiplies the input signal by the carrier generated by the local oscillator, and filters out the high frequency component to obtain the difference frequency component. GP2015 contains a total of three mixers, so there are three intermediate frequencies, which are represented by IF1, IF2, and IF3 respectively. The local oscillator signals of the three-stage mixer are respectively derived from the three-carrier output of the frequency synthesizer.

The first-stage local oscillator carrier frequency is 1,400 MHz, which is obtained from the frequency synthesizer through multiplying the input fundamental frequency

(10 MHz) by 140 with a phase-locked loop. The second-stage local oscillator carrier frequency is 140 MHz, which is obtained through dividing the first-stage local oscillator carrier signal frequency by 10, while the third-stage local oscillator carrier frequency is obtained through dividing the first-stage local oscillator carrier signal by 45, that is, 31.111 MHz. Then, after the first stage mixer, the first intermediate frequency IF1 is (1 575.42 – 14 000) = 175.42 MHz. The second intermediate frequency IF2 after the second-stage mixing is (175.42 – 140) = 35.42 MHz. After the third-stage mixing, IF2 becomes the third intermediate frequency IF3, which is (35.42 – 31.11) = 4.31 MHz.

After the third intermediate frequency signal passes through a bandpass filter with a center frequency of 4.3 MHz, the sampling rate $f_s = 5.714$ MHz is converted by the ADC into a two-bit digital signal. Since the value of IF_3 is within $[f_s/2, f_s]$, the bandpass sampling described in Sect. 8.3 and the spectral inversion occur during AD sampling. The final center carrier frequency f_c is 5.714–4.31 $\approx$ 1.405 MHz. The sampling clock of the ADC is obtained through dividing the 40 MHz clock by 7, while the exact value of IF_3 is 4.308 888 MHz, so the exact value of its f_c should be (40/7-4.308 888) = 1.405 396 MHz. It should be noted that these values are only theoretical. This means that when the local reference clock is exactly 10 MHz, and the input GPS signal carrier is exactly 1575.42 MHz, any deviation will cause the actual received signal frequency to deviate from the theoretical intermediate frequency.

The significance of analyzing the theoretical intermediate frequency value is to determine the search space of the Doppler frequency in the subsequent signal acquisition process, and also to make readers aware of the influence of the frequency scheme of the RF front-end on the value of the theoretical intermediate frequency. The theoretical IF value of the IF signal is closely related to the specific setting of the RF scheme. There is no general equation to obtain it, and readers should analyze a variety of specific circumstances.

According to the GP2015 chip manual, the three-stage mixer also generates certain gains upon completion of the mixing. The three-stage gains are written as $G1$, $G2$ and $G3$, respectively. The total gain of the three-stage mixer is $(G1 + G2 + G3)$. To determine the value of the total gain is also a problem that chip designers must solve.

The input level of the GP2015's on-chip ADC is required to be at approximately 100 mV, and with 50 Ω of the RF impedance, the power is 0.12/50 = 2×10-3 W, expressed in decibels as −37 dBW or −7 dBm. Considering factors such as the gain and noise figure of the low noise amplifier, the attenuation of the peripheral SAW filter, and the signal bandwidth of 2 MHz, the following inequality can be obtained

$$\begin{aligned} &-7\,\text{dBm} < -174\,\text{dBm/Hz} + 19\,\text{dB} \\ &+ (G1 + G2 + G3) - 21\,\text{dB} + 63\,\text{dB}+ \\ &\Rightarrow \quad (G1 + G2 + G3) > 106\,\text{dB} \end{aligned} \tag{8.28}$$

The result of this inequality shows that the total gain of the three-stage mixer must be greater than 106 dB. −174 dBm/Hz is the background noise power spectral density of the RF input, in which 63 dB is from the 2 MHz GPS signal bandwidth,

Table 8.4 Probability of ADC sampling distribution in GP2015

Sampled value	Percentage/(%)
3	15
−1	35
1	35
3	15

and −21 dB is the attenuation of the peripheral SAW filter (mainly the 175 MHz and 35 MHz bandpass filters), and 19 dB is the combined result of the gain and noise figure provided by the low noise amplifier before the ADC. The above relationships and parameters are cited from the GP2015 data sheet. [4]

The input of the automatic gain control unit in GP2015 is the sampling result of the ADC output. The signal strength of the sampling result is determined according to a certain level of detection logic to control the controllable gain amplifier in the third-stage mixer. The AGC's gain control range is 60 dB, which covers the variation range of the signal in most applications. In the description of the AGC principle in Sect. 8.5, the method of determining the AGC control signal according to the sample value distribution has been emphasized. The control signal of the AGC in GP2015 is adjusted according to this method.

The AGC adjusts the gain by judging the numerical distribution of the output data stream. The final effect is that 70% of the sampled data falls within [−1,1] and 30% of the sampled data within [−3,3]. Specifically, Table 8.4 can be used to indicate its distribution. Obviously, the sampled data after the AGC control must have no DC component, and the sampling level is near the noise averaging square root.

The valve time of the GP2015 AGC is determined by the external capacitor. According to the GP2015 data sheet, the recommended capacitor value is 100 nF, and the valve time is 2 ms.

Other similar GNSS RF front-end chips include SE4110/4120 from the SiGe Semiconductor and MAX2769 from Maxim-IC. Domestic manufacturers include Zhongke Microelectronics, Southwest Microelectronics, Runxin Information Technology Co., Ltd., and Jiameixinxin Communication Technology. Ltd. The working principles of their products are similar, and readers can work out their frequency schemes and key parameters according to the chip data manual based on their understanding of this section.

References

1. Lindsey WC, Simon MK (1973) Telecommunication systems engineering. Prentice-Hall, Englewood Cliffs, NJ
2. John G Proakis. Digital communications. McGraw-Hill Inc
3. Changxin Fan et al (1995) Communication principles, 4th edn. National Defense Industry Press, Beijing

4. Datasheet: GP2015, GPS Receiver RF Front End. ISSUE 3.1, Zarlink Semiconductor, Feb 2002
5. Datasheet: SE4110L, GPS Receiver IC, Rev 6.4, SiGe Semiconductor, May 2009
6. Datesheet: Max2769, Universal GPS Receiver, Rev 2, Maxim-IC, Jul 2010
7. Kaplan ED (2006) Understanding GPS principles and applications, 2nd edn. Artech House Publishers
8. Misra Pratap, Enge Per (2008) Global positioning system—signal, measurement and performance, 2nd edn. Electronic Industry Press, Beijing
9. van Diggelen F (2009) A-GPS: Assisted GPS, GNSS, and SBAS. Artech House
10. van Dierendonck AJ (1996) "GPS receivers."In: Parkinson B, Spiker J, Axelrad P, Enge P (eds) Global positioning system: theory and applications, vol. I
11. Tsui JBY (2008) Fundamentals of global positioning receivers: a software approach, 2nd edn. Wiley
12. Sturza MA (1996) Digital direct-sequence spread-spectrum receiver design considerations. Proc. of the Fourth Annual WIRELESS Symposium, Santa Clara, California
13. Vaughan RG, Scott NL, White DR (1991) The theory of bandpass sampling. IEEE Trans Sig Proc 39(9):1973–1984
14. Akos DM (1997) A software radio approach to global navigation satellite receiver design. Ph.D. Dissertation, Department of Electrical Engineering and Computer Science, Ohio University
15. Peterson RL, Ziemer RE, Borth DE (1995) Introduction to spread-spectrum communications. Prentice-Hall
16. Navstar GPS Space Segment/Navigation User Interfaces, IS-GPS-200G, September 5, 2012
17. Navstar GPS Space Segment/User Segment L5 Interfaces, IS-GPS-705, Rev. A, June 8, 2010
18. Navstar GPS Space Segment/User Segment L1C Interfaces, IS-GPS-800, Rev. A, June 8, 2010
19. China Satellite Navigation System Management Office (Dec 2013) Beidou satellite navigation system space signal interface control document (Open Service Signal) Version 2.0

Chapter 9
Implementation of BDS/GPS Dual-Mode Software Receivers

This chapter will discuss the implementation of the BDS and GPS dual-mode receivers in software according to the principles of the GNSS receiver that were explained in previous chapters. This implementation is the software radio receiver solution mentioned in Chap. 2. Compared with the hardware solution, the software solution is considerably more flexible, and is easier to reconfigure and upgrade, so it is ideal for beginners to learn principles of the GNSS receiver. For engineers, it can also be used as a development platform for the GNSS receiver, for related signal processing and algorithms.

The basic principles of the implemented software receivers in this chapter have been covered in the previous chapters. The essential task of software receivers is to implement the most important receiver principles on a computer through a specific programming language. In the implementation process of specific software, the choice of programming language should be carefully considered. In principle, any programming language can be used to implement the principles of GNSS software receivers. However, considering that the readership of this book mainly consists of students and researchers, MathWorks' Matlab scripting language (USA) has obvious advantages, as it is widely used in algorithm research, data visualization, data analysis, and numerical calculation.

This chapter will be divided into three parts. The first part explains the signal source (input signal) of the software receiver, which often takes the form of data files. The data stored in the signal source comes from the AD sampled quantized data of the RF front-end, and is stored as a data file in a certain format, so the principles of the RF front-end introduced in Chap. 8 will be referred to in this part.

The second part of this chapter deals with the code implementation of the software receiver. It explains the Matlab code module with a focus on the main functions of each source file and the signal processing principles involved, which are related to the principles of BDS and GPS signal, signal acquisition and tracking, telegraphy demodulation, satellite position and velocity calculation, and PVT solution. This will

Y. Lu, *BDS/GPS Dual-Mode Software Receiver*, Navigation: Science and Technology 10, https://doi.org/10.1007/978-981-16-1075-2_9

give readers a clearer and more practical understanding of the theoretical knowledge mentioned in the previous chapter.

The third part of this chapter presents the data processing result of the software receiver. The software receiver explained in the second part is applied to process typical BDS and GPS dual-mode IF data, and analyzes the signal processing results. The results of signal acquisition and tracking, and the results of the PVT positioning solution will be given in the form of graphs, offering readers a more visualized understanding of the internal principles of the GNSS receiver, signal processing flow, and various functional modules. At the same time, it can also help readers to understand the principles of GNSS receivers discussed in the first half of the book.

The source code of the software receiver in this chapter is given on the website that accompanies this book (https://www.gnssbook.cn). At the same time, there is a BDS and GPS dual-mode IF data file lasting about 90 s. This data file contains the signals from 15 BDS and GPS satellites. The time period of 90 s ensures that the GPS and BDS satellites can at least contain a complete set of ephemeris data. Readers can develop their own BDS and GPS software receivers and corresponding signal processing algorithms based on their understanding of the receiver's source code.

9.1 Signal Source for Dual-Mode Software Receivers

The source of the dual-mode software receiver comes from the IF data file. Since the IF data file comes from the sampled quantized data of the BD/GPS dual-mode RF front-end, the signal source of the dual-mode software receiver is also from the BD/GPS dual mode RF front-end. In fact, the logical relationship between the three is shown in Fig. 9.1.

The BD/GPS dual-mode RF front-end in Fig. 9.1 contains two RF signal channels, which process the BDS and GPS signals respectively, and complete a series of signal processing tasks such as mixing, filtering, down-conversion, and AD sampling

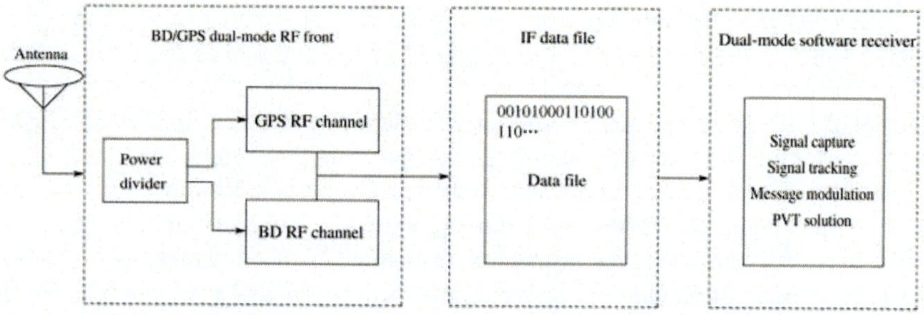

Fig. 9.1 The relationship between the signal input, IF data files, and dual-mode RF front-ends for dual-mode software receivers

quantization. The local clock ensures that the clocks of the BDS and GPS signals are strictly aligned. The two AD-quantized digital signals are stored as intermediate-frequency data files through the high-speed data interface, and then the IF data files are processed by the dual-mode software receiver. Figure 9.1 shows that the dual-mode software receiver still processes the signal of the hardware RF front-end, which is only processing the stored data file. It is not a real-time processing method. Signal processing algorithms have obvious advantages, and are widely used in the early stages of product development for receivers, as well as in scientific research.

In the implementation of the software receiver code, the format of the IF data file needs to be analyzed first to obtain the data sample value. The specific format of the IF data is determined by the AD quantization format of the RF front-end, and the AD output quantization formats of different RF front-ends are different from each other. Hence, only the BDS and GPS dual-mode IF data file formats attached to this book are described here. Readers can analyze data files stored by digital sampling output from other RF front-ends according to their hardware manuals.

The data file attached to this book is in binary format. Each byte contains two sets of sample data. Each set of sample data contains one sample of BDS and GPS. The quantization bit width of each sampled data is 2 bits. The detailed data format is shown in Fig. 9.2.

The dual-mode data format in each byte is shown in Fig. 9.2, where the data is stored in chronological order from high bit (MSB) to low bit (LSB), which readers should note. The reason is that when reading data at the software receiver, the data needs to match the order in which it was stored. Each data sample (GPS and BDS) is two-bit quantization. The high bit is the Sign bit (Sign) and the low bit is the Amplitude bit (Mag). For the specific quantization principle, please refer to Sect. 8.5. If the Sign is 0, the sample value is positive. Otherwise, the sample value is negative. If Mag is 0, the amplitude should be 1, otherwise the amplitude is 3. The specific meaning is shown in Table 9.1.

The IF data file in the software receiver has two main functions. The first is to provide IF sampling data, and the second is to provide the working clock of the core processing module. The first function can be easily understood, but the second is

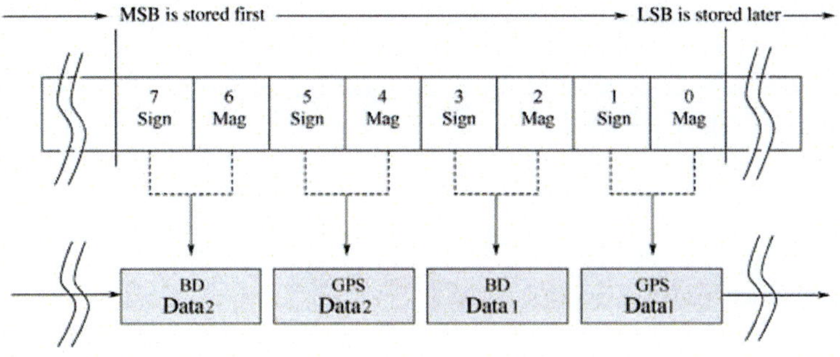

Fig. 9.2 IF data file format

Table 9.1 Mapping table of two-bit quantization and sample values

Sampling value	Sign bit	Mag bit
+1	0	0
+3	0	1
−1	1	0
−3	1	1

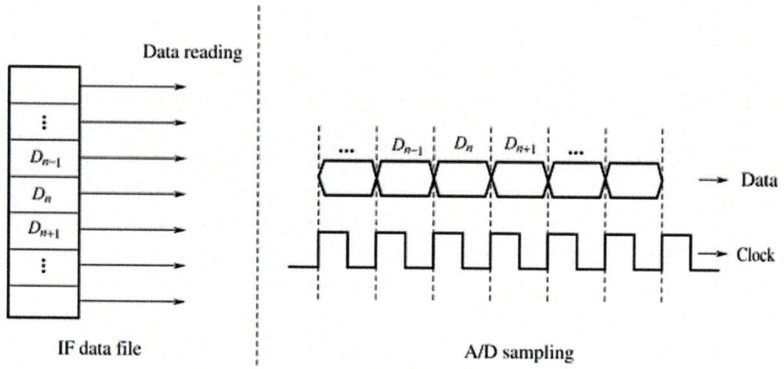

Fig. 9.3 Correspondence between the reading of data files and AD sampling

less simple. To understand it, we only need to grasp the specific process of ADC sampling. Each sampling clock leads to the generation of a sample. In the practical use of BDS and GPS receivers, this sampled data is sent to the subsequent hardware processing stage. However, in the software receiver, the sampled data is stored in the data file. When undertaking the signal processing of the software receiver, each time sample data is read from the data file, it also means the arrival of a sampling clock. This clock will cause each state update of the correlator. The integration of the clock is the local time change, which is used to update the local time.

The above analysis is shown in Fig. 9.3. The left side of the dashed line in the figure is the representation of the IF data in the computer file. Each reading of the data corresponds to the sampling process of the AD to the intermediate frequency signal in the hardware, which is shown to the right of the dotted line in the figure. For the dual-mode software receiver in this book, each piece of data in the figure contains BDS data and GPS data.

9.2 Software Modules and Program Interfaces for Dual-Mode Receivers

As mentioned in the previous section, the code for the software receiver in this book is the Matlab scripting language, so readers should ensure that they understand it independently. There are many reference books and technical materials explaining it, as well as a variety of free tutorials on the Internet for reference.

In general, the dual-mode software receiver code can be divided into five parts, namely the graphical interface, the signal acquisition, the signal tracking, the telegram demodulation, and the PVT solution. The functions belong to five subdirectories in accordance with their respective functions, namely GUI, Acquisition, Tracking, Navmsg, and PVT. The functions performed by the code modules in each subdirectory are shown in Table 9.2.

The following is a description of the source code files in each subdirectory. It should be noted that this is only a brief introduction to the basic functions of each source code file. For the details and principles of how it works, please read the source code.

1. **GUI Subdirectory**

main_fig.m: Completes initialization of the main program interface.

initial_mainfig.m: Updates the main program interface according to the information from the initial configuration. It only operates once during the running of the program.

GuiUpdate.m: Updates different information areas during the running of the program, including the data file information area, processing progress information area, processing message area, baseband signal processing information area, least-squares result area, and Kalman filtering result area.

pushInfoMsg.m: Shows the key information and intermediate variables during the running of the program in the processing message area.

Table 9.2 Module divisions of software receiver functions

Subdirectory	Function
GUI	Generates the graphical interface. Updates according to the results of data processing during the running of the program
Tracking	Allocates the tracking channel according to the results of signal acquisition, to realize signal tracking bit synchronization and observation measurement of GPS and BD
Acquisition	Receiver initialization, GPS and BD pseudo-random code generation, generation of various LUTs, and capture of GPS and BD signals
Navmsg	Sub-frame synchronization of GPS and BD signals, GPS data check and BCH check of BD data, and demodulation of ephemeris data
PVT	PVT solution, least-squares solution, Kalman filtering solution, ECEF to LLH coordinate transformation, GPS and BD satellite positioning, and velocity calculation

2. **Acquisition Subdirectory**

ReadGnssConfig.m: Reads the name, the IF sampling frequency, the theoretical IF frequency, and the signal format of the input configuration file that users need to modify to suit the specific file information.

gnssInit.m: Completes global variable initialization, including the generation of GPS and BD pseudo-random codes, sine/cosine and signal mapping look-up tables, initialization of tracking channels, least-squares, and initialization of Kalman filtering results.

GpsCodeGen.m: Generates the pseudo-random code of 32 GPS satellites.

BdCodeGen.m: Generates the pseudo-random codes of 32 BDS satellites.

DownSampling.m: Down-samples the original input signal to satisfy the software FFT requirements.

createValueMapping.m: Establishes a signal map look-up table.

BDsearchNH.m: Searches the starting position of the NH code in the D1 code of the BD, which is required for the capture of the BDS IGSO/MEO satellite signal.

AcquisitionByFFT.m: Captures the signal by the time domain parallel FFT algorithm, which is the core file of the signal capture in the software receiver.

AcquisitionEngine.m: Down-samples the original input signal, and then signal acquisition is performed by the FFT algorithm. It is the upper function interface for the main program to call signal capture.

3. **Tracking Subdirectory**

AllocateTrackingChannel.m: Allocates and initializes the tracking channel according to the result of the capture module.

ResetChannel.m: Resets the tracking channel.

BDGeoTrkLoop.m: Updates loops for BDS GEO satellites, including the code loop update and carrier ring update.

BDMeoTrkLoop.m: Updates loops for BDS MEO/IGSO satellites, including the code loop update and carrier ring update.

GPSTrkLoop.m: Updates loops for GPS satellites, including the code loop update and carrier ring update.

SignalTracking.m: The main function entry of the signal tracking, which performs the I and Q integration of the E, P, and L branches for all the assigned tracking channels, and uses their respective loop update functions to conduct loop updates.

SignalTrackingByC.c: Correlates the I and Q integrals of the E, P, and L branches. It is the only C language file in all codes. Its main purpose is to improve the running speed of the program.

TicMeasurement.m: Extracts pseudo-range and Doppler observations after the signal tracking channel completes the sub-frame synchronization.

UpdateTrackingLoop.m: Calls loops to update their respective functions based on the difference in signal systems.

4. **Navmsg Subdirectory**

BdBchDecode.m: Performs BCH decoding of BDS telegram data bits.

BdGeoDecodeEph.m: Performs the ephemeris data decoding of the BDS GEO satellite.

BdMeoDecodeEph.m: Performs the ephemeris data decoding of the BDS MEO/IGSO satellite.

BdNavProcess.m: The main function entry of the BDS baseband data processing.

BdSearchPreamble.m: Performs the synchronization word search and sub-frame synchronization of the BDS message, upon the completion of which the SOW and the current sub-frame number can be known.

GpsDecodeEph.m: Performs the decoding of the ephemeris data of the GPS satellite.

GpsParityCheck.m: Performs the Hamming code verification of GPS data bits.

GpsSearchPreamble.m: Performs the synchronization word search and sub-frame synchronization of the GPS data. The Z-Count and the sub-frame number can be known after the step is completed.

GpsNavProcess.m: The main function entry for GPS baseband data processing.

5. **PVT Subdirectory**

Ecef2Llh.m: Converts ECEF coordinates to LLH coordinates (latitude and longitude coordinate system).

Llh2Ecef.m: Converts LLH coordinates to ECEF coordinates.

gnssPvt.m: The main function entry of the PVT solution.

initKF.m: Initializes the Kalman filter state with the result of the least-squares solution.

kalmanFix.m: Implements Kalman filtering, including time update and observation update, and the generation of rotation matrix of ECEF to LLH coordinate system and calculation of DOP value. What the Kalman filter model selects is the PV model explained in Sect. 7.2.5.

lsfix_double.m: Performs a least-squares solution (5 state) for dual mode observation.

lsfix_single.m: Performs a least-squares solution (4 states) for single mode observation.

sv_pos_eph.m: Calculates the position, velocity, and clock correction of GPS satellites and BDS MEO/IGSO satellites.

sv_pos_eph_geo.m: Calculates the position, speed, and clock correction of the BDS GEO satellite.

In the source code root directory, the main program script file is startRun.m. In the Matlab environment, the current working directory is set to the source code root directory. Then, the main function (startRun) is run in the command line window, and the main program interface appears. This is shown in Fig. 9.4.

Click the start button in the upper left corner of the interface to start the program. You can stop it by clicking the stop button while the program is running. In the process of running the program, in order to visually display the results of data processing and the status of the program, the program running interface is divided into six parts: the data file information area, processing progress information area, processing message area, baseband signal processing information area, least-squares result zone, and Kalman filter result zone.

The data file information area mainly shows the current data file name, IF sampling frequency value, theoretical IF frequency value, and GPS/BD data file mode (single-mode data or dual-mode data), as shown in Fig. 9.5.

The processing progress information area shows the total duration of the data file currently being processed and the sampling time being processed, as well as the percentage progress of the current data processing, which is shown in Fig. 9.6. The total duration and the time being processed are in milliseconds.

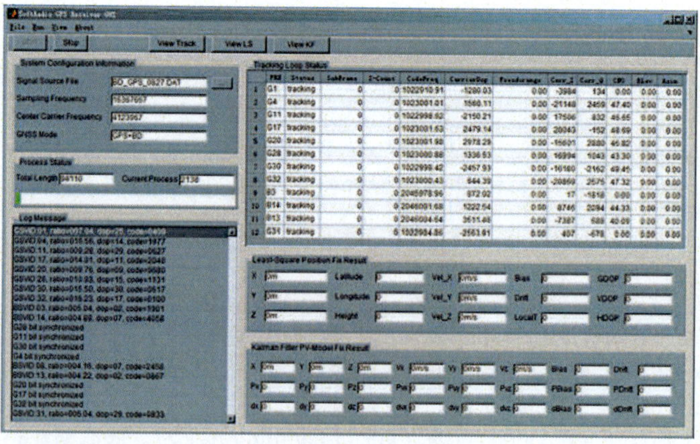

Fig. 9.4 The main program interface of the software receiver

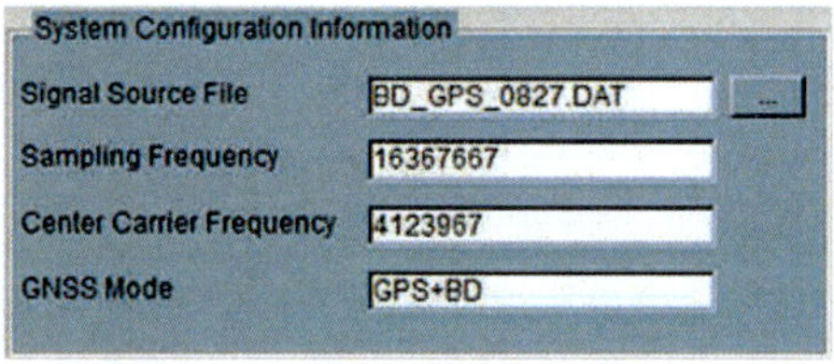

Fig. 9.5 Data file information area

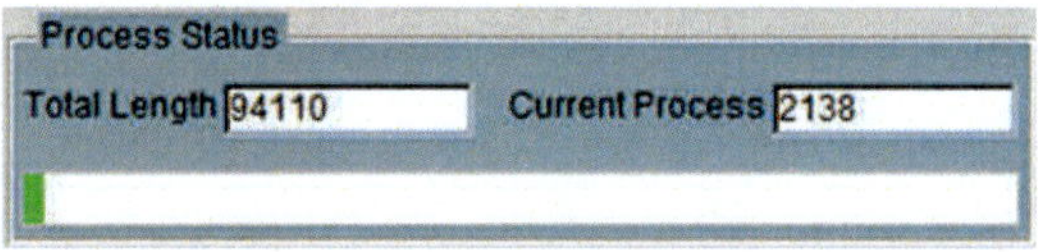

Fig. 9.6 Data processing progress information area

During the running of the program, key information in BDS and GPS signal processing is generated, such as signal acquisition, signal tracking loop status, bit synchronization, sub-frame synchronization results, local time initialization, ephemeris data demodulation results, and PVT solution results. This information is crucial for judging whether the software receiver is working normally, so the signal processing message area is specifically allocated in the main interface to display the information, as shown in Fig. 9.7.

The state quantity during signal tracking is displayed in the baseband signal processing information area, including satellite number, satellite type (G indicates GPS satellite, B indicates BDS satellite), sub-frame synchronization result, Z-Count (or TOW) value, code tracking frequency, carrier tracking frequency, pseudo-range observation, Doppler observation, IQ integral value, CN0, and satellite elevation and amplitude, as shown in Fig. 9.8.

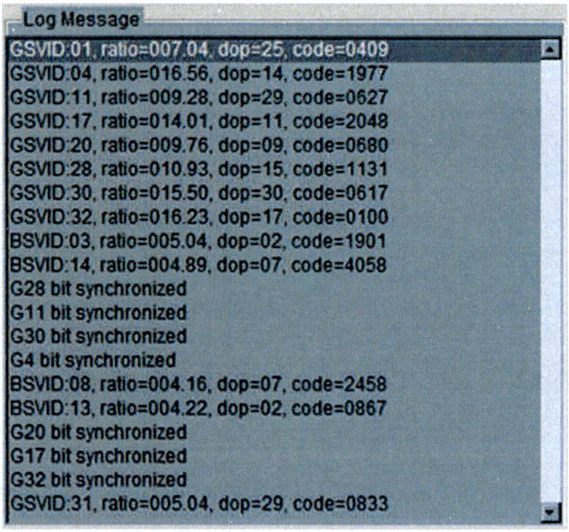

Fig. 9.7 Signal processing message areas

Tracking Loop Status

	PRN	Status	SubFrame	Z-Count	CodeFreq	CarrierDop	Pseudorange	Corr_I	Corr_Q	CNO	Elev	Azim
1	G1	tracking	0	0	1022910.91	-1280.03	0.00	-3984	134	0.00	0.00	0.00
2	G4	tracking	0	0	1023001.01	1560.11	0.00	-21148	2459	47.40	0.00	0.00
3	G11	tracking	0	0	1022998.62	-2150.21	0.00	17506	832	46.65	0.00	0.00
4	G17	tracking	0	0	1023001.63	2479.14	0.00	20043	-152	48.69	0.00	0.00
5	G20	tracking	0	0	1023001.98	2978.29	0.00	-15601	2880	45.82	0.00	0.00
6	G28	tracking	0	0	1023000.88	1336.53	0.00	16994	1043	43.30	0.00	0.00
7	G30	tracking	0	0	1022998.42	-2457.93	0.00	-16180	-2162	49.45	0.00	0.00
8	G32	tracking	0	0	1023000.43	644.39	0.00	-20869	2575	47.32	0.00	0.00
9	B3	tracking	0	0	2045978.96	872.02	0.00	17	-1818	0.00	0.00	0.00
10	B14	tracking	0	0	2046001.68	1222.54	0.00	8746	2094	44.33	0.00	0.00
11	B13	tracking	0	0	2046004.64	3511.48	0.00	-7387	688	40.09	0.00	0.00
12	G31	tracking	0	0	1022984.86	-2553.91	0.00	407	-678	0.00	0.00	0.00

Fig. 9.8 The baseband signal processing information area

With the completion of the ephemeris demodulation, the determination of the local time, and the extraction of the pseudo-range and carrier frequency observations, the receiver software starts the PVT solution. The initial solution algorithm is completed by the least-squares method, and the program will adopt either single-mode or dual-mode algorithms based on the combination of observation measurements. The results of least-squares calculation include information in the ECEF coordinate system and the latitude and longitude coordinate system, the speed of the ECEF coordinate system, clock difference and clock drift, and the geometric accuracy factor. A further PVT solution is completed by Kalman filtering. The Kalman filter in the software receiver is in system state, which is [*x, y, z, b, vx, vy, vz, dr, Tgb*], where [*x, y, z*] are the positions in the ECEF coordinate system, [*vx, vy, vz*] are the amounts of velocity in the ECEF coordinate system, and [*b, dr, Tgb*] are the clock difference, the clock drift, and the system time deviation of GPST-BDT respectively. The results of least-squares and Kalman filtering are shown in the least-squares result area and the Kalman filter result area, as in Fig. 9.9.

During the running of the program, you can click the three View buttons at the top of the interface to implement the function similar to "Oscilloscope". The View Track button can show the I/Q integral and E, P, and L branch integration results of each tracking channel. Thus, we can make a real-time observation of the entire process of transient response and loop update during signal tracking, which is shown in Fig. 9.10.

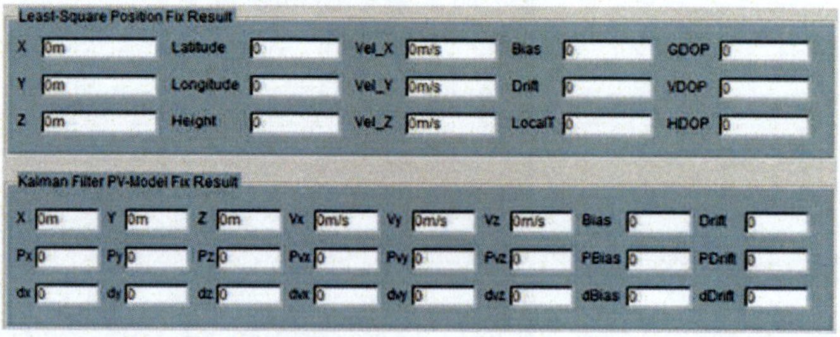

Fig. 9.9 The least-squares result area and Kalman filter results area

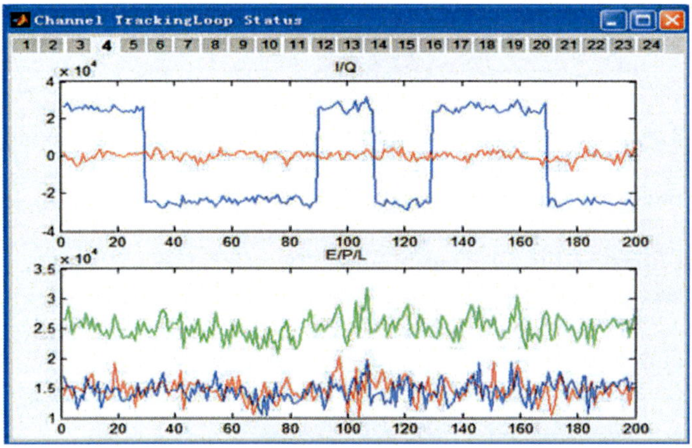

Fig. 9.10 I\Q integral and E, P, and L integration results of the tracking channel

The View LS button shows the curve of the least-squares positioning result on the time axis. The View KF button shows the value of the system state of the Kalman filter, the correction amount, and the curve of the covariance matrix element on the time axis. They are shown in the left and right halves of Fig. 9.11 respectively. In the least-squares result, there is information such as receiver position, speed, clock drift, GPST-BDT system time difference, and DOP value. They are the values of nine system state quantities in the Kalman filter result display, and display their respective correction amount and variance value.

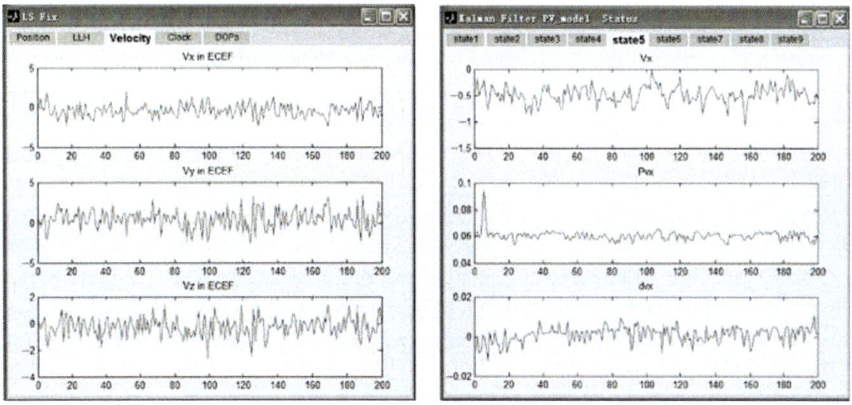

Fig. 9.11 Least-squares results (left) and Kalman filter results (right)

9.3 Data Processing in Dual-Mode Receivers

As well as the Matlab code for BDS and GPS dual-mode software receivers, a section on the BDS and GPS dual-mode IF data file is also provided in this book. This file comes from the UTREK210 satellite belonging to Beijing Jetstar Quanta Technology Co., Ltd. The data acquisition system obtains the actual satellite signal acquisition, with a sampling frequency of 16.367 6 MHz, a theoretical intermediate frequency of 4.130 4 MHz, and two-bit quantization. The data storage format is consistent with the description in Fig. 9.2. Combined with the source code of the dual-mode software receiver and the dual-mode IF data file, readers can directly debug and modify the implementation of the software receiver. At the same time, the software algorithm can be programmed to process the actual satellite signal, which can deepen the theoretical understanding of the GNSS receiver.

In this section, the software receiver is used to process the above IF data files, including key processing such as signal acquisition, tracking, message demodulation, and navigation solution. The results are presented in the form of graphical curves. Readers can run the software receiver code independently, and analyze and compare the intermediate results and variables that are of interest.

Figure 9.12 shows the result of signal acquisition for dual-mode IF data files. Due to the characteristics of the software receiver, there is no a priori information at the beginning of the software receiver operation. Therefore, the signal acquisition method adopts the "cold start blind search" strategy. Figures 9.12 and 9.13 show the results of capturing all GPS and BDS satellites. The specific signal acquisition algorithm uses the time domain parallel FFT acquisition algorithm from Sect. 4.1.4. The figure only shows the capture results of the satellites with stronger signals, from which the obvious signal peaks can be seen.

According to the results of Figs. 9.12 and 9.13, the pseudo-code phase and Doppler frequency corresponding to the satellite signal peak after successful acquisition can be calculated. Then, the information is submitted to the signal tracking channel through a certain logic conversion, and the carrier channel of the tracking channel is initialized. The pseudo-code NCO is set to enter the tracking processing of the signal.

Figure 9.14 shows the coherent integration result of the in-phase path (I way) and the orthogonal path (Q way) of the signal tracking loop output. The integral length is 1 ms, so the horizontal axis in the figure is the time axis in ms. The axis is the integral amplitude value. The results in the figure are from GPS satellite signals and have similar patterns to the BDS satellite signal, except that the data hopping period of the BDS GEO satellite is 2 ms. The upper part of the figure is the I way integral result, and the lower part is the Q way integral result. From the curves of the I and Q way integral results with time, the following two points can be seen:

① At the beginning of the loop (within 500 ms), the signal energy gradually increases, which means that $\sqrt{I^2 + Q^2}$ gradually increases. This is because the loop adjustment makes the carrier frequency difference smaller, so the energy loss caused by the frequency difference becomes smaller.

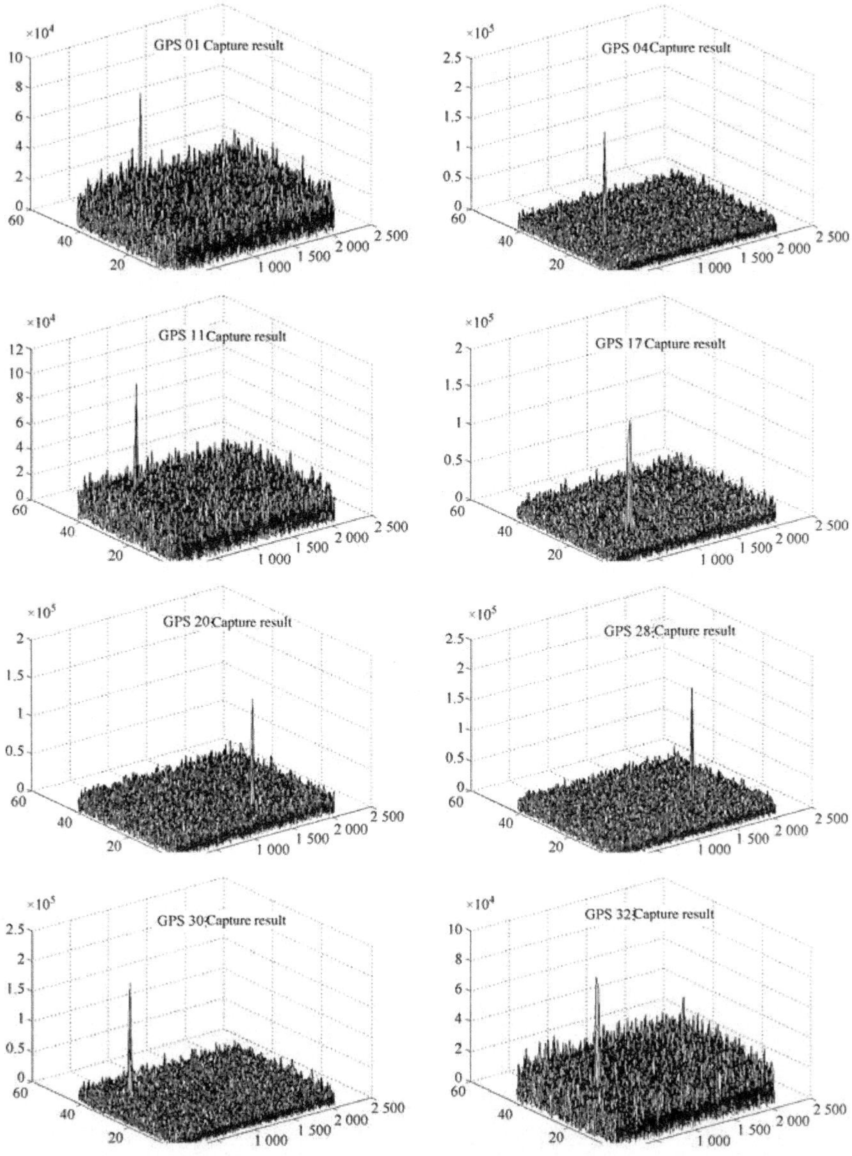

Fig. 9.12 Results of signal acquisition for all GPS and BDS satellites

② After the loop enters stable tracking (after 500 ms), the integral amplitude of the I path is stable, and the integral amplitude of the Q channel becomes smaller. This is because the phase lock is started when the frequency difference is small to a certain degree. After the phase lock is realized, the energy is concentrated

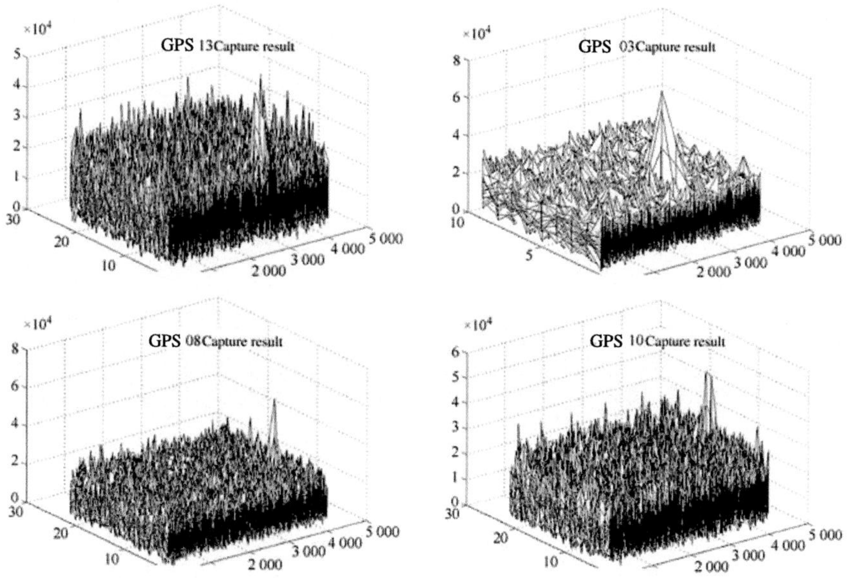

Fig. 9.13 Results of signal acquisition for all BDS satellites

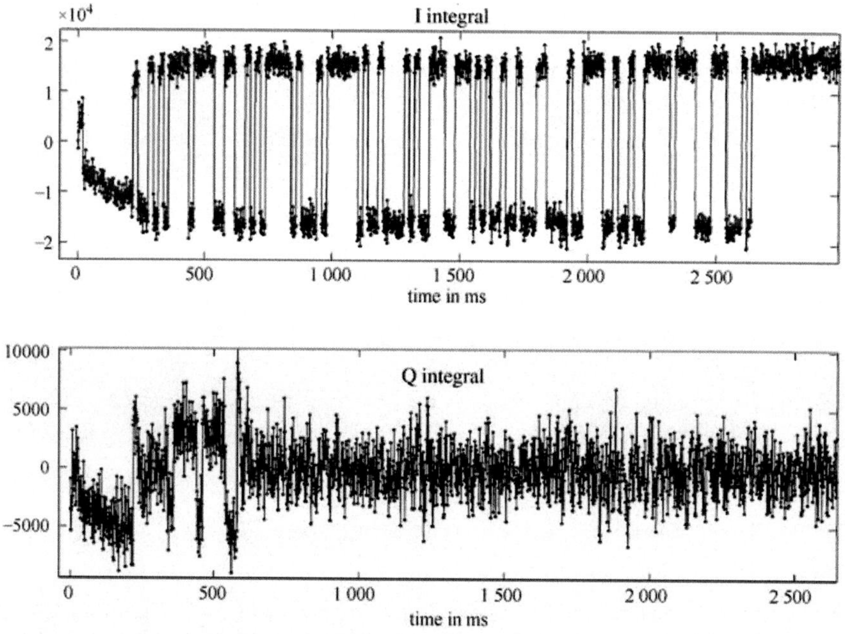

Fig. 9.14 The I and Q integral results of the signal tracking loop

in the I path, and the Q path only contains the noise component. At this time, the navigation signal can be demodulated by the integral value of the I path.

Figure 9.15 shows the NCO frequency value of the carrier tracking loop and the pseudo-code tracking loop in the signal tracking process. The upper part is the carrier tracking frequency value, and the lower part is the pseudo-code tracking frequency value. From Figs. 9.15 and 9.14, corresponding conclusions can be made. At the beginning of the loop, the variance of the carrier frequency and the pseudo-code frequency is large, indicating that a relatively drastic adjustment is being made at this time. The variation of the carrier frequency and the pseudo-code frequency after entering the stable tracking is much smaller than in the initial stage. In particular, the pseudo-code frequency is greater due to the role of carrier assistance, resulting in a very smooth pseudo-code frequency.

For loop update strategies at different stages of loop tracking, please refer to the source code to understand the details. Adjust the parameters and strategies to improve the loop performance.

After the tracking loop enters a steady state, the sub-frame synchronization and the demodulation of the navigation message can be started. Synchronization of the BDS and GPS sub-frames can be realized by referring to the methods described in

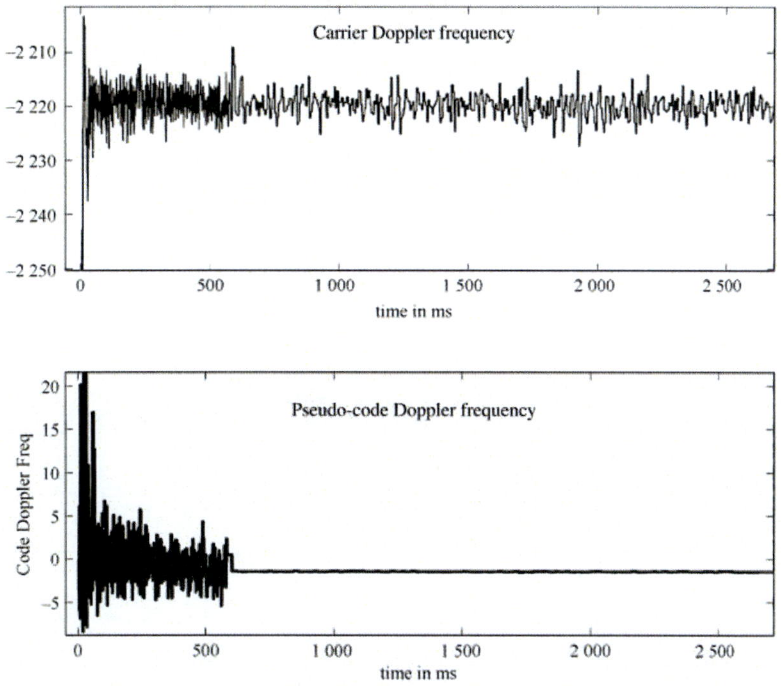

Fig. 9.15 Carrier frequency of the carrier tracking loop, and pseudo-code frequency of the pseudo-code tracking loop

Sects. 3.3.2 and 4.2.7 of this book. Comprehensive navigation in telemetry demodulation includes TOW demodulation, ephemeris data demodulation, almanac data demodulation, UTC parameter demodulation, and ionospheric parameter demodulation. Due to limitations of the data file length, the software receiver code only implements TOW demodulation with ephemeris data, but this is sufficient for subsequent PVT solvers.

After the TOW time of the satellite signal is acquired, the observation measurement can be performed. Here, the pseudo-range observation and the Doppler frequency observation are extracted. The PVT solution can be achieved after obtaining more than four satellite observations and pieces of ephemeris data.

The PVT solution strategy is to perform the least-squares positioning solution first, then initialize the Kalman filter state quantity with the result of the least-squares method, and then start the Kalman filter positioning solution. The least-squares method in the specific implementation will either adopt the 4-state model or the 5-state model according to the situation of single-mode or dual-mode observation, wherein the 4-state model is for the single-mode (single GPS or single BDS) and 5-state model is for the dual-mode observation. The 5-state model has one GPST-BDT system time deviation more than the 4-state model. For the specific principle, refer to Chap. 7 of this book.

Figure 9.16 shows the receiver position coordinates calculated by the least-squares method, expressed in the ECEF coordinate system. In order to display the range of variation of the coordinates, the difference between the calculated coordinate result and its mean value is shown in the figure. Since the software receiver code extracts the observation every 100 ms, the horizontal axis of Fig. 9.16 is the time axis in units

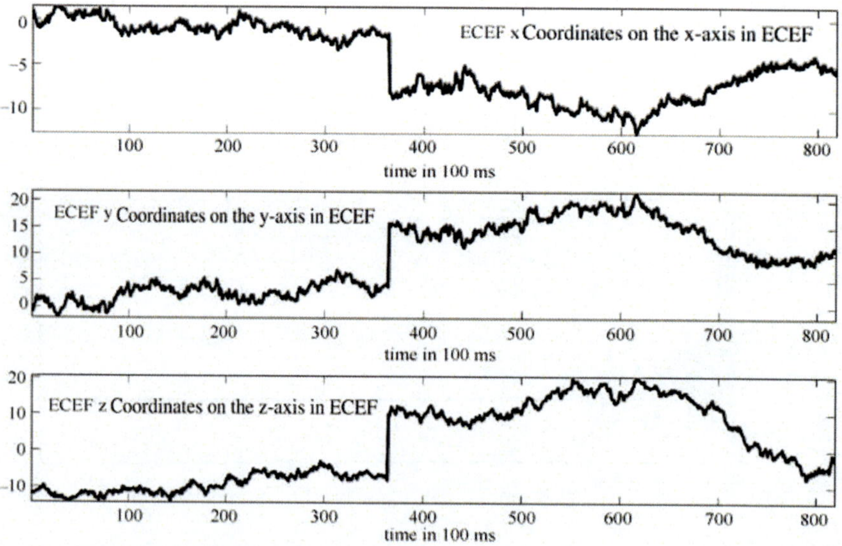

Fig. 9.16 Location calculated by the least-squares method

of 100 ms, and the vertical axis is in meters. It can be seen that the positioning result of the least-squares method has large jumping phenomenon.

In contrast, Fig. 9.17 shows the position coordinates calculated by the Kalman filter, in which the mean value is also deducted for the same reason. It can be seen from the comparison of the position results of the least-squares method that the position result of the Kalman filter is smoother, and the position jump range is much smaller in the entire positioning time. The reader can adjust the R and Q matrix parameters to obtain different positioning results based on the understanding of the Kalman filter source code. The reader can also consider the effect that the result of the Kalman filter obtains from the R matrix and the Q matrix.

The velocity component results of the Kalman filter are given in Fig. 9.18. The upper, middle, and lower curves are the velocity components in the X-axis, Y-axis, and Z-axis directions respectively. Since it is a stationary antenna, the velocity component has a mean value of 0, and the velocity deviation in the Z-axis direction is slightly larger. The three curves in Fig. 9.19 are the local clock error, local clock drift, and GPST-BDT system time deviation, where the local clock and system time deviation are in meters, and the local clock drift is in m/s. The local clock difference and the system time deviation can be converted to the speed of light in seconds, and the speed of light at the local clock can be converted to ppm.

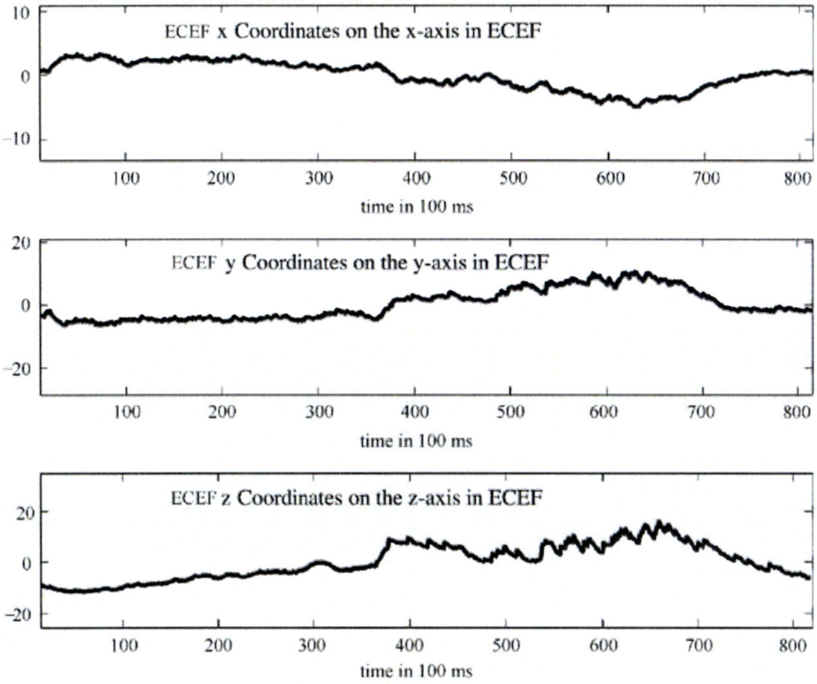

Fig. 9.17 Location calculated by Kalman filtering

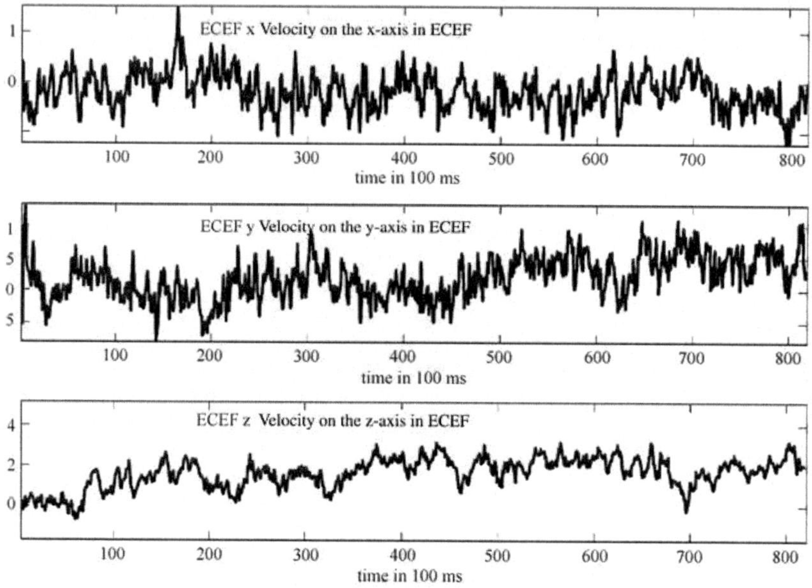

Fig. 9.18 Speed of Kalman filtering

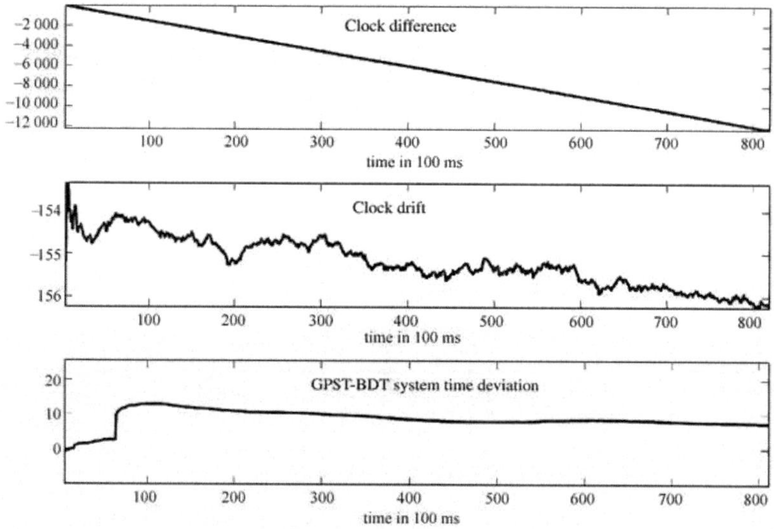

Fig. 9.19 Clock time difference and clock drift of Kalman filtering, and GPST-BDT system time deviation

The local clock difference is the integral of the local clock drift. This conclusion is determined by the physical nature of the clock difference and the clock drift. This can also be easily verified from the values in the curve in Fig. 9.19. It can be seen from Fig. 9.19 that the value of the GPST-BDT system time deviation is around 0 (if expressed in seconds), which means that BDT and GPST are basically synchronized. If the observation error is subtracted, the value of the system time deviation will be smaller. The figure shows that the variation of the time deviation of the GPST-BDT system is small and remains basically unchanged during the observation.

Figures 9.20, 9.21 and 9.22 show the respective curves of the variance of the various system states of the Kalman filter over time. The Kalman filter in the software receiver uses the PV model explained in Sect. 7.2.5. The system state vector is $[x, y, z, b, vx, vy, vz, d, T_{GB}]$, which consists of three position quantities, three speed quantities, and three clock quantities. The P matrix is the covariance matrix of the nine system state variables, while Figs. 9.20, 9.21, and 9.22 give the time curve of the nine diagonal elements of the *P* matrix. The unit of the vertical axis in Fig. 9.20 is m, the unit of the vertical axis in Fig. 9.21 is m/s, and the unit of the vertical axis in Fig. 9.22 is m (top), m/s (middle), and m (bottom). The horizontal axis in these three figures is the time in 100 ms.

Figures 9.20, 9.21, and 9.22 show that the time curves of the diagonal elements of the *P* matrix of the Kalman filter are consistent. The initial values are large. They gradually become smaller as time passes, and finally converge near a stable value. This is because when the Kalman filter is initialized by using the result of the least-squares method, the *P* matrix is generally initialized to a larger value, which

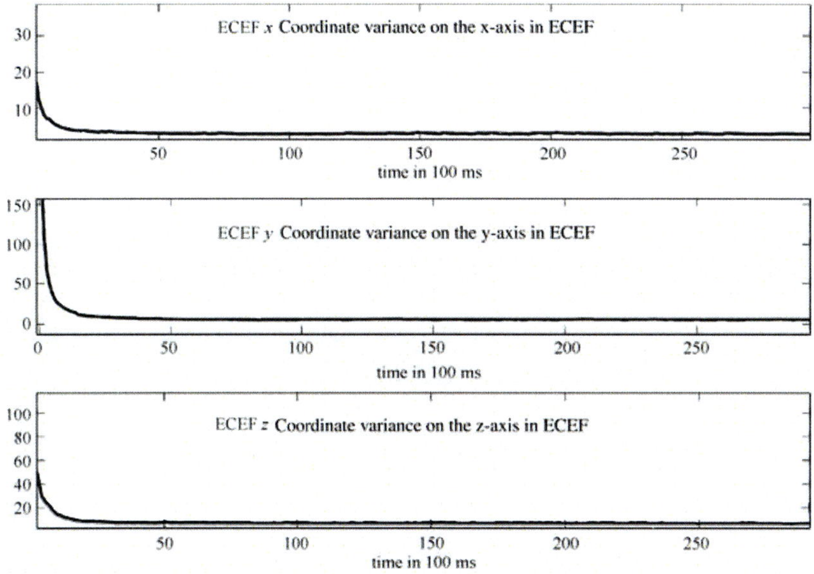

Fig. 9.20 The variance of the positional state of the Kalman filter output

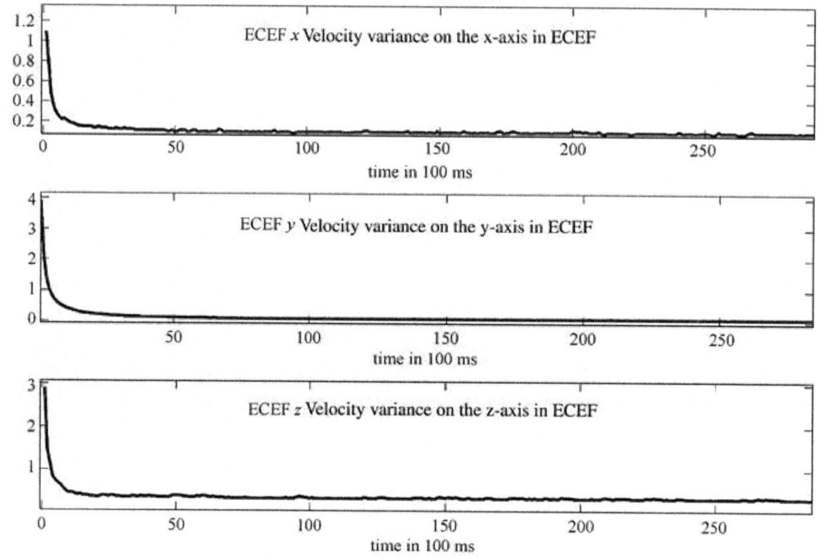

Fig. 9.21 The variance of the velocity state of the Kalman filter output

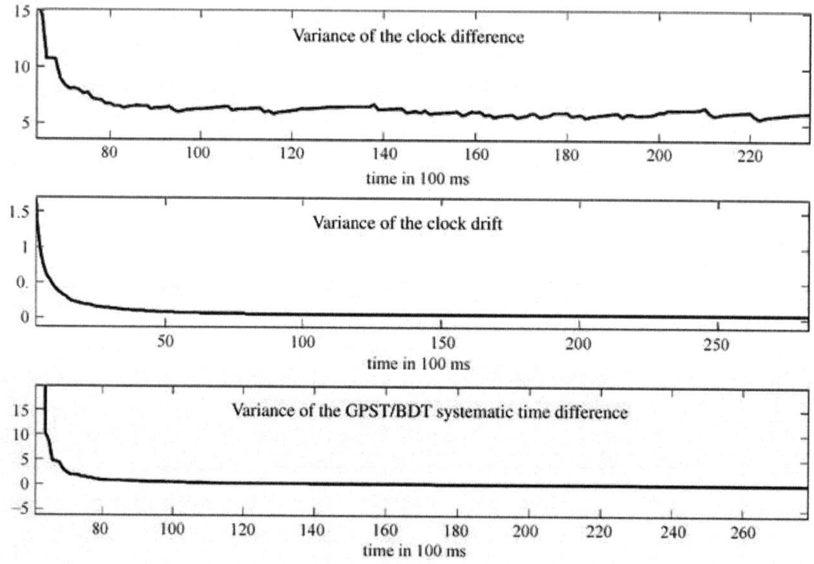

Fig. 9.22 The variance of the time state of the Kalman filter output

allows the observation update to quickly correct the local state. As time passes and the observation quantity update continues, the covariance matrix of the local system state will gradually converge. At this time, the observed noise variance and the P matrix jointly determine the Kalman gain matrix K. The correction for the local system state will be the weighted result of the local state variance and the observed noise variance. The final P matrix will also stabilize near a convergence value. In the subsequent update of the Kalman filter, the observation noise situation and the system processing noise matrix Q will affect the convergence value of the P matrix, such as when there are no observations for a period of time (a typical scenario in which this happens is when the receiver enters an underground garage or tunnel), or if there is a decrease in the quality of the measurement results in an increase in the P matrix.

A simple explanation has been offered on the processing of the actual IF data file by the dual-mode software receiver in this book. This analysis covers only a small part of the intermediate result in the software receiver, and some implementation details and debugging processes have been skipped. We hope that readers will understand the software receiver code implemented in this book, and grasp its theoretical principles. Modifying and optimizing the existing code, and observing the changes in output will improve future work and research.

Appendix A
Basic Matrices and Vector Operations

Calculations of matrices and vectors appear extensively in this book, so some basic knowledge is essential for readers. Due to the limitations of space, the specific derivation process will not be given. Interested readers can refer to books on linear algebra and matrix theory for a more in-depth understanding.

The notation format of this book is as follows:

- Scalars are represented by variables in regular font, such as x;
- Vectors and matrices are represented by variables in bold, such as $\boldsymbol{x}$, $\boldsymbol{X}$;
- The real number field set is written as $\mathbb{R}$;
- The set of real n-dimensional vectors is written as $\mathbb{R}^n$;
- The set of real $m \times n$ matrices is written as $\mathbb{R}^{m\times n}$;
- The set of real $n \times n$ squares is written as $\mathbb{R}^{n\times n}$;
- $\mathrm{diag}\{a_1, a_2, \ldots, a_n\}$ represents an n-th order matrix with diagonal elements as $\{a_1, a_2, \ldots, a_n\}$ and other elements as 0;
- $\det(\boldsymbol{A})$ or $|\boldsymbol{A}|$ represents the determinant of the matrix;
- $r(\boldsymbol{A})$ represents the rank of the matrix.

A.1 The Inverse Matrix and Its Properties

The necessary and sufficient condition for the reversibility of matrix $\boldsymbol{A} \in \mathbb{R}^{n\times n}$ is $r(\boldsymbol{A}) = n$, which means that $\boldsymbol{A}$ is in full rank. So,

$$\det(\mathrm{A}) \neq 0 \leftrightharpoons \text{A Full rank} \leftrightharpoons \text{AReversible}$$

When calculating the inverse matrix of the n-th order matrix $\boldsymbol{A}$, we can convert it into a matrix with a lower order and perform the inversion through the partitioning of the matrix. Specifically, if $\boldsymbol{A}$ can be written as

Y. Lu, *BDS/GPS Dual-Mode Software Receiver*, Navigation: Science and Technology 10, https://doi.org/10.1007/978-981-16-1075-2

$$A = \begin{bmatrix} A_{11} & A_{12} \\ A_{21} & A_{22} \end{bmatrix}$$

and A_{11} is reversible, then

$$A^{-1} = \begin{bmatrix} A_{11}^{-1}(I + A_{12}W^{-1}A_{21}A_{11}^{-1}) & -A_{11}^{-1}A_{12}W^{-1} \\ -W^{-1}A_{21}A_{11}^{-1} & W^{-1} \end{bmatrix} \tag{A.1}$$

where $W = A_{22} - A_{21}A_{11}^{-1}A_{12}$.

When A_{22} is a one-dimensional matrix, i.e. a scalar, the inversion of the n-th order matrix is converted to that of the n-1-th order matrix. At this point, the calculations of $-A_{11}^{-1}A_{12}W^{-1}$ and $-W^{-1}A_{21}A_{11}^{-1}$ are relatively simple vector multiplications.

The following is a useful matrix inversion lemma:

If matrix $E = A + BCD$ and the inverse matrices of A, C, and E exist, then

$$E^{-1} = A^{-1} - A^{-1}B\left(DA^{-1}B + C^{-1}\right)^{-1}DA^{-1} \tag{A.2}$$

This lemma is very useful in the calculation of the gain k_{m} matrix of a Kalman filter. The derivation of Eq. (7.132) using a matrix inversion lemma is given below.

According to Eq. (7.129)

$$(P_m^+)^{-1} = (P_m^-)^{-1} + H_m^{\mathrm{T}}R_m^{-1}H_m \tag{A.3}$$

Compared with Eq. (A.2), $A = (P_m^-)^{-1}$, $B = H_m^{\mathrm{T}}$, $C = R_m^{-1}$, $D = H_m$, and based on the matrix inversion lemma,

$$P_m^+ = P_m^- - P_m^-H_m^{\mathrm{T}}\left(H_mP_m^-H_m^{\mathrm{T}} + R_m\right)^{-1}H_mP_m^- \tag{A.4}$$

By substituting the above equation into Eq. (7.130), we obtain

$$\begin{aligned} k_m &= P_m^+H_m^{\mathrm{T}}R_m^{-1} \\ &= P_m^-H_m^{\mathrm{T}}R_m^{-1} - P_m^-H_m^{\mathrm{T}}\left(H_mP_m^-H_m^{\mathrm{T}} + R_m\right)^{-1}H_mP_m^-H_m^{\mathrm{T}}R_m^{-1} \\ &= P_m^-H_m^{\mathrm{T}}\left[R_m^{-1} - \left(H_mP_m^-H_m^{\mathrm{T}} + R_m\right)^{-1}H_mP_m^-H_m^{\mathrm{T}}R_m^{-1}\right] \\ &= P_m^-H_m^{\mathrm{T}}\left[I - \left(H_mP_m^-H_m^{\mathrm{T}} + R_m\right)^{-1}H_mP_m^-H_m^{\mathrm{T}}\right]R_m^{-1} \\ &= P_m^-H_m^{\mathrm{T}}\left(H_mP_m^-H_m^{\mathrm{T}} + R_m\right)^{-1}R_mR_m^{-1} \\ &= P_m^-H_m^{\mathrm{T}}\left(H_mP_m^-H_m^{\mathrm{T}} + R_m\right)^{-1} \end{aligned} \tag{A.5}$$

If we substitute Eqs. (A.5) to (A.4), then

$$P_m^+ = P_m^- - k_mH_mP_m^-$$

$$= (\boldsymbol{I} - \boldsymbol{k}_m \boldsymbol{H}_m)\boldsymbol{P}_m^- \tag{A.6}$$

A.2 Eigenvalues and Eigenvectors of the Matrix

For matrix $\boldsymbol{A} \in \mathbb{R}^{n \times n}$, if there is a non-zero vector $\boldsymbol{x} \in \mathbb{R}^n$ that makes

$$\boldsymbol{A}\boldsymbol{x} = \lambda \boldsymbol{x} \tag{A.7}$$

then λ is called the eigenvalue of $\boldsymbol{A}$, and the corresponding vector $\boldsymbol{x}$ is called the eigenvector of $\boldsymbol{A}$ with respect to the feature value λ.

From the definition of the eigenvalues of the matrix, we can see that λ satisfies the homogeneous linear equation.

$$(\lambda \boldsymbol{I} - \boldsymbol{A})\boldsymbol{x} = \boldsymbol{0} \tag{A.8}$$

To give the equation a non-zero solution,

$$\det(\lambda \boldsymbol{I} - \boldsymbol{A}) = 0 \tag{A.9}$$

$\det(\lambda \boldsymbol{I} - \boldsymbol{A})$ is the n-th order equation of λ, so the root of the equation is the eigenvalue of $\boldsymbol{A}$.

If $\boldsymbol{A} = [a_{ij}]$, then

$$f(\lambda) = \det(\lambda \boldsymbol{I} - \boldsymbol{A}) \tag{A.10}$$

This is called the characteristic polynomial of $\boldsymbol{A}$, and

$$\begin{bmatrix} \lambda - a_{11} & -a_{12} & \cdots & -a_{1n} \\ -a_{21} & \lambda - a_{22} & \cdots & -a_{2n} \\ \vdots & \vdots & \ddots & \vdots \\ -a_{n1} & -a_{n2} & \cdots & \lambda - a_{nn} \end{bmatrix} \tag{A.11}$$

is called the characteristic matrix of $\boldsymbol{A}$.

If we assume that λ is an eigenvalue of $\boldsymbol{A}$, the following conclusions can be established:

- $k\lambda$ is the eigenvalue of $k\boldsymbol{A}$, and k is an arbitrary constant;
- λ^m is the eigenvalue of $\boldsymbol{A}^m$, and m is any positive integer;
- If $\boldsymbol{A}$ is reversible, then $\frac{1}{\lambda}$ is the eigenvalue of $\boldsymbol{A}^{-1}$;
- If $\boldsymbol{A}$ is a real symmetric matrix or a Hermite matrix, then λ must be a real number;

- If $\lambda = 0$, then A must be a singular matrix, i.e. $\det(\mathbf{A}) = 0$;
- λ is also the eigenvalue of A^{T}, which means that A and A^{T} have the same eigenvalue.

Assume that the n eigenvalues of $A = [a_{ij}] \in \mathbb{R}^{n\times n}$ are $\lambda_1, \lambda_2, \cdots, \lambda_n$, then

$$(i)\ \sum_{i=1}^{n} \lambda_i = \sum_{i=1}^{n} a_{ii} \tag{A.12}$$

$$(ii)\ \prod_{i=1}^{n} \lambda_i = det(\mathbf{A}) \tag{A.13}$$

where, $\sum_{i=1}^{n} a_{ii}$ is generally written as $Tr(\mathbf{A})$, which is called the trace of A.

There is a close relationship between the eigenvectors and the diagonalization of the matrix. More specifically, the necessary condition for the diagonalization of the matrix is that there is a linearly independent eigenvector. Since the eigenvectors corresponding to different eigenvalues are linearly independent, if there are n different eigenvalues of A, A must be diagonalizable. The inverse proposition of this conclusion is not true, and the diagonalizable matrix A does not necessarily have n different eigenvalues.

A special case is when A is a real symmetric matrix. In this case, the eigenvectors corresponding to different eigenvalues are not only linearly independent, but also orthogonal. The real symmetric matrix must be diagonalizable, that is, there is an orthogonal matrix $P \in \mathbb{R}^{n\times n}$ that makes

$$P^{-1}AP = \mathrm{diag}\{\lambda_1, \lambda_2, \ldots, \lambda_n\} \tag{A.14}$$

where, $P^{\mathrm{T}}P = I$.

A.3 Quadratic Form and Fixed Matrices

The quadratic homogeneous polynomial of the n-ary variable $x_1, x_2, \cdots, x_n$

$$\begin{aligned} &a_{11}x_1^2 + a_{12}x_1x_2 + \cdots + a_{1n}x_1x_n + \\ &a_{21}x_2x_1 + a_{22}x_2^2 + \cdots + a_{2n}x_2x_n + \\ &\vdots \\ &a_{n1}x_nx_1 + a_{n2}x_nx_2 + \cdots + a_{nn}x_n^2 \end{aligned} \tag{A.15}$$

is called an n-ary quadratic form, referred to as a quadratic form.

If

$$\boldsymbol{A} = \begin{bmatrix} a_{11} & a_{12} & \cdots & a_{1n} \\ a_{21} & a_{22} & \cdots & a_{2n} \\ \vdots & \vdots & \ddots & \vdots \\ a_{m1} & a_{m2} & \cdots & a_{mn} \end{bmatrix}, \mathbf{x} = [x_1, x_2, \ldots, x_n]^{\mathrm{T}}$$

then the quadratic form can be written as

$$\sum_{i=1}^{n}\sum_{j=1}^{n} a_{ij}x_i x_j = \boldsymbol{x}^{\mathrm{T}}\boldsymbol{A}\boldsymbol{x} \tag{A.16}$$

Matrix $\boldsymbol{A}$ can be written as the sum of the symmetrical part $\boldsymbol{A}_{\mathrm{s}}$ and the asymmetric part $\boldsymbol{A}_{\mathrm{u}}$, so

$$\boldsymbol{x}^{\mathrm{T}}\boldsymbol{A}\boldsymbol{x} = \boldsymbol{x}^{\mathrm{T}}\boldsymbol{A}_{\mathrm{s}}\boldsymbol{x} + \boldsymbol{x}^{\mathrm{T}}\boldsymbol{A}_{\mathrm{u}}\boldsymbol{x} \tag{A.17}$$

It can be proven that

$$\begin{aligned} &\boldsymbol{x}^{\mathrm{T}}\boldsymbol{A}_{\mathrm{u}}\boldsymbol{x} = 0 \\ &\Rightarrow \qquad \boldsymbol{x}^{\mathrm{T}}\boldsymbol{A}\boldsymbol{x} = \boldsymbol{x}^{\mathrm{T}}\boldsymbol{A}_{\mathrm{s}}\boldsymbol{x} \end{aligned}$$

Therefore, in the analysis of the quadratic form in the real number domain, only the real symmetric matrix needs to be considered.

Through a certain non-degenerate linear transformation $\boldsymbol{x} = \boldsymbol{C}\boldsymbol{y}$, where $\boldsymbol{C} \in \mathbb{R}^{n\times n}$ is an invertible matrix, $\boldsymbol{y} = [y_1, y_2, \cdots, y_n]^{\mathrm{T}}$, the quadratic form of $\boldsymbol{x}$ in Eq. (A.15) can be transformed into a quadratic form for $\boldsymbol{y}$ with only mixed but not quadratic terms, i.e.

$$b_1 y_1^2 + b_2 y_2^2 + \cdots + b_n y_n^2 \tag{A.18}$$

It is called a standard quadratic form. Obviously, the conversion from general quadratic form to standard quadratic form is to find the reversible matrix $\boldsymbol{C}$ to make $\boldsymbol{C}^{\mathrm{T}}\boldsymbol{A}\boldsymbol{C}$ a diagonal matrix.

For the real symmetric matrix $\boldsymbol{A}$, no matter how the non-degenerate linear transformation matrix $\boldsymbol{C}$ is selected, the number of positive square terms in the standard quadratic form obtained is constant, and so is the number of negative square terms. They are called the positive and the negative inertia indexes of $\boldsymbol{A}$ respectively.

When the positive inertia index of the real symmetric matrix $\boldsymbol{A}$ is n, for any non-zero vector x, there is always

$$\sum_{i=1}^{n}\sum_{j=1}^{n} a_{ij}x_i x_j = \boldsymbol{x}^{\mathrm{T}}\boldsymbol{A}\boldsymbol{x} < 0 \tag{A.19}$$

At this time, $\boldsymbol{x}^{\mathrm{T}}\boldsymbol{A}\boldsymbol{x}$ is called a positive definite quadratic form, and $\boldsymbol{A}$ is a positive definite matrix. If $\boldsymbol{x}^{\mathrm{T}}\boldsymbol{A}\boldsymbol{x} \geq 0$ in the above equation, then $\boldsymbol{x}^{\mathrm{T}}\boldsymbol{A}\boldsymbol{x} \geq 0$ is called a semi-positive definite matrix.

Similarly, when the negative inertia index of the real symmetric matrix $\boldsymbol{A}$ is n, for any non-zero vector $\boldsymbol{x}$, there is always

$$\sum_{i=1}^{n}\sum_{j=1}^{n} a_{ij}x_i x_j = \boldsymbol{x}^{\mathrm{T}}\boldsymbol{A}\boldsymbol{x} < 0 \tag{A.20}$$

At this time, $\boldsymbol{x}^{\mathrm{T}}\boldsymbol{A}\boldsymbol{x}$ is called a negative quadratic form, and $\boldsymbol{A}$ is a negative fixed matrix.

For the positive definite symmetric matrix $\boldsymbol{A} = [a_{ij}]$, it has the following properties:

- The diagonal elements of $\boldsymbol{A}a_{ii} > 0, i = 1, 2, \cdots, n$;
- The characteristic value of $\boldsymbol{A}\lambda_i > 0, i = 1, 2, \cdots, n$;
- $\det(\boldsymbol{A}) > 0$;
- There is a reversible matrix $\boldsymbol{P}$, making $\boldsymbol{A} = \boldsymbol{P}^{\mathrm{T}}\boldsymbol{P}$;
- $\boldsymbol{A}$ has an inverse matrix, and $\boldsymbol{A}^{-1}$ is also a positive definite matrix.

The eigenvalue of the semi-definite matrix is greater than or equal to 0 and is not less than 0, so its determinant is greater than or equal to 0. The positive definite matrix must be of full rank. Meanwhile, the semi-positive definite matrix is not necessarily of full rank, so the semi-positive definite matrix does not necessarily have an inverse matrix. The case of a semi-negative fixed matrix can be analyzed similarly.

It is mentioned in Sect. 7.2.6 that matrix $\boldsymbol{P}$ in the Kalman filter is positive definite. In actual engineering, the positive definiteness can be determined according to the above conditions.

A.4 Several Important Matrix Decompositions

1. Matrix eigenvalue decomposition

We assume that matrix $\boldsymbol{A} \in \mathbb{R}^{n\times n}$ has n linearly independent eigenvectors $\boldsymbol{x}_i, i = 1, 2, \cdots, n$ and the corresponding eigenvalue is λ_i, i.e.

$$\boldsymbol{A}\boldsymbol{x}_i = \lambda_i \boldsymbol{x}_i \tag{A.21}$$

We arrange the n feature vectors into a matrix $\boldsymbol{P}$ of $n \times n$,

$$P = [x_1, x_2, \cdots, x_n]$$

Then, we multiply A left-ward by P,

$$AP = P\begin{bmatrix} \lambda_1 & 0 & \cdots & 0 \\ 0 & \lambda_2 & \cdots & 0 \\ \vdots & \vdots & \ddots & 0 \\ 0 & 0 & 0 & \lambda_n \end{bmatrix} \tag{A.22}$$

Because x_i is linearly independent, P is of full rank and P^{-1} exists, so there is

$$A = P\begin{bmatrix} \lambda_1 & 0 & \cdots & 0 \\ 0 & \lambda_2 & \cdots & 0 \\ \vdots & \vdots & \ddots & 0 \\ 0 & 0 & 0 & \lambda_n \end{bmatrix}P^{-1} \tag{A.23}$$

Then, A is decomposed by the eigenvalue into the diagonal matrix $\text{diag}\{\lambda_1, \lambda_2, \cdots, \lambda_n\}$.

2. **SVD decomposition**

SVD decomposition means the singular value decomposition of the matrix. It can be performed on non-diagonal matrices.

If $A \in \mathbb{R}^{m\times n}$, then the matrix $A^{\mathrm{T}}A$ is an n-order square matrix whose positive square root $\sigma = \sqrt{\lambda}$ of n eigenvalues λ is called the singular value of A. Since $A^{\mathrm{T}}A$ is a real symmetric matrix and must be semi-positive, if its eigenvalues are listed by size, there must be

$$\lambda_1 > \lambda_2 > \cdots > \lambda_r > \lambda_{r+1} = \cdots = \lambda_n = 0 \tag{A.24}$$

where r is the rank of $A^{\mathrm{T}}A$, $r \leq min(m, n)$.

An interesting conclusion is that in the calculation of the singular value of A, if a similar process is performed for AA^{T}, the same result will be obtained. AA^{T} is an m-th order square matrix, which is also semi-positive. If its eigenvalues are calculated and sorted, the resulting sequence is

$$\mu_1 > \mu_2 > \cdots > \mu_r > \mu_{r+1} = \cdots = \mu_m = 0 \tag{A.25}$$

then $\mu_i = \lambda_i, i = 1, 2, \ldots, r$.

For A, there must be orthogonal matrices $U \in \mathbb{R}^{m\times m}$ and $V \in \mathbb{R}^{n\times n}$ that make

$$U^{\mathrm{T}}AV = \begin{bmatrix} \mathbf{S} & \mathbf{0} \\ \mathbf{0} & \mathbf{0} \end{bmatrix}_{(m\times n)} \tag{A.26}$$

where $S = \text{diag}\{\sigma_1, \sigma_2, \cdots, \sigma_r\}$.

It can be derived from Eq. (A.26) and the orthogonal characteristics of U and V that

$$\begin{aligned} AA^{\mathrm{T}}U &= UU^{\mathrm{T}}AVV^{\mathrm{T}}A^{\mathrm{T}}U \\ &= U(U^{\mathrm{T}}AV)(U^{\mathrm{T}}AV)^{\mathrm{T}} \\ &= U\begin{bmatrix} S^2 & 0 \\ 0 & 0 \end{bmatrix}_{(m\times m)} \end{aligned} \tag{A.27}$$

$$\begin{aligned} A^{\mathrm{T}}AV &= VV^{\mathrm{T}}A^{\mathrm{T}}UU^{\mathrm{T}}AV \\ &= V(U^{\mathrm{T}}AV)^{\mathrm{T}}(U^{\mathrm{T}}AV) \\ &= V\begin{bmatrix} S^2 & 0 \\ 0 & 0 \end{bmatrix}_{(n\times n)} \end{aligned} \tag{A.28}$$

So, it can be seen that U and V are respectively composed of linearly independent eigenvectors of AA^{T} and $A^{\mathrm{T}}A$.

When A is a real symmetric matrix, the singular value of its **SVD** decomposition is the absolute value of its eigenvalue, i.e.

$$\sigma_i = |\lambda_i|, i = 1, 2, \cdots, r \tag{A.29}$$

Furthermore, when A is a real symmetric positive definite matrix, its singular value is equal to its eigenvalue, that is, its SVD decomposition is its matrix eigenvalue decomposition.

3. **LU decomposition**

For matrix $A \in \mathbb{R}^{n\times n}$, its LU decomposition is the multiplication of a matrix of the lower triangular L and an upper triangular matrix U, i.e. $A = LU$. The LU decomposition of the matrix is proposed along with the solution of equation $Ax = b$. If LU decomposition is performed for A, then the equation can be solved into two steps.

$$\begin{aligned} &LUx = b \Rightarrow \\ &Ly = b \quad Ux = y \end{aligned} \tag{A.30}$$

Thereby, the solution can be greatly simplified, and the generally adopted method is Gaussian elimination.

First, we generate matrix L_k, $k = 1, 2, \ldots, n-1$ according to the following rules:

$$
\boldsymbol{L}_k = \begin{bmatrix} 1 & \cdots & 0 & 0 & \cdots & 0 \\ \vdots & \ddots & \vdots & \vdots & \ddots & \vdots \\ 0 & & 1 & 0 & & 0 \\ 0 & & -\tau_{k+1} & 1 & & 0 \\ \vdots & \vdots & \vdots & \vdots & \ddots & \vdots \\ 0 & \cdots & -\tau_n & 0 & \cdots & 1 \end{bmatrix} \tag{A.31}
$$

where $\tau_{k+1} = \frac{a_{k+1,k}}{a_{k,k}}, \ldots, \tau_n = \frac{a_{n,k}}{a_{k,k}}$.

It is easy to prove that

$$
\boldsymbol{L}_k^{-1} = \begin{bmatrix} 1 & \cdots & 0 & 0 & \cdots & 0 \\ \vdots & \ddots & \vdots & \vdots & \ddots & \vdots \\ 0 & & 1 & 0 & & 0 \\ 0 & & \tau_{k+1} & 1 & & 0 \\ \vdots & \vdots & \vdots & \vdots & \ddots & \vdots \\ 0 & \cdots & \tau_n & 0 & \cdots & 1 \end{bmatrix} \tag{A.32}
$$

That is, the inversion of $\boldsymbol{L}_k$ is to reverse the sign of τ_k.

If we multiply $\boldsymbol{L}_k$ left-ward by matrix $\boldsymbol{A}$ and turn all elements below $a_{k,k}$, and the element at the k-th column and the k-th row of $\boldsymbol{A}$ into 0, then the upper triangular matrix $\boldsymbol{U}$ can be obtained through multiplying elements from $k = 1$ to $n-1$ left-ward by $\boldsymbol{L}_k$, i.e.

$$
\boldsymbol{L}_{n-1} \cdots \boldsymbol{L}_2 \boldsymbol{L}_1 \boldsymbol{A} = \boldsymbol{U} \tag{A.33}
$$

Note: the order of left multiplication by $\boldsymbol{L}_k$ should not be confused. Multiply both sides of the above equation by $\boldsymbol{L}_k^{-1}$ successively to get

$$
\boldsymbol{A} = \boldsymbol{L}_1^{-1} \boldsymbol{L}_2^{-1} \cdots \boldsymbol{L}_{n-1}^{-1} \boldsymbol{U} \tag{A34}
$$

We define $\boldsymbol{L} = \boldsymbol{L}_1^{-1} \boldsymbol{L}_2^{-1} \cdots \boldsymbol{L}_{n-1}^{-1}$, and obviously $\boldsymbol{L}$ is still a lower triangular matrix, so

$$
\boldsymbol{A} = \boldsymbol{L}\boldsymbol{U} \tag{A.35}
$$

where $\boldsymbol{L}_k^{-1}$ is determined by Eq. (A.32).

4. **QR decomposition**

For full rank matrix $\boldsymbol{A} \in \mathbb{R}^{n \times n}$, if it is written in the form of column vector, i.e.

$$A = [a_1, a_2, \cdots, a_n]$$

then, $a_i, i = 1, 2, \cdots, n$, is n linearly independent vectors, and we can perform Gram–Schmidt on these vectors. The process needs to be completed in n steps as follows.

Step 1: Make $u_1 = a_1, \quad y_1 = \|u_1\|^{-1} u_1$;

Step 2: Make $u_2 = a_2 - (a_2 \cdot y_1) y_1, \quad y_2 = \|u_2\|^{-1} u_2$;

...

Step n: Make $u_n = a_n + \sum_{i=1}^{n-1} [-(a_n \cdot y_i) y_i], \quad y_n = \|u_n\|^{-1} u_n$.

where $(x \cdot y)$ is the point multiplication of vector x and vector y.

The resulting vectors y_i must be orthogonal to each other, i.e.

$$y_i \cdot y_j = \begin{cases} 0 \text{ when } i \neq j \\ 1 \text{ when } i = j \end{cases}$$

If we reverse the above process,

$$\begin{aligned} a_1 &= \|u_1\| y_1 \\ a_2 &= (a_2 \cdot y_1) y_1 + |u_2| y_2 \\ &\vdots \\ a_n &= \sum_{i=1}^{n-1} [(a_n \cdot y_i) y_i] + \|u_n\| y_n \end{aligned}$$

and write them in the form of matrix, then

$$A = QR \tag{A.36}$$

where

$$R = \begin{bmatrix} \|u_1\| & (a_2 \cdot y_1) & \cdots & (a_n \cdot y_1) \\ 0 & \|u_2\| & \cdots & (a_n \cdot y_2) \\ \vdots & \vdots & \ddots & \vdots \\ 0 & 0 & \cdots & \|u_n\| \end{bmatrix}, \quad Q = [y_1, y_2, \cdots, y_n] \tag{A.37}$$

A similar decomposition can easily be applied to long square matrices. For the full rank matrix $A \in \mathbb{R}^{m \times n}, \quad (m > n)$, there is an orthogonal matrix $U \in \mathbb{R}^{m \times m}$ and an upper triangular matrix $D \in \mathbb{R}^{n \times n}$ that make

$$\boldsymbol{A} = \boldsymbol{U}\begin{bmatrix} \boldsymbol{D} \\ \boldsymbol{O} \end{bmatrix} \tag{A.38}$$

where $\boldsymbol{o} \in \mathbb{R}^{m-n \times n}$.

The QR decomposition of the matrix is useful in the solution of an overdetermined equation

$$\boldsymbol{A}\boldsymbol{x} = \boldsymbol{b} \tag{A.39}$$

where, $\boldsymbol{A} \in \mathbb{R}^{m \times n}$, \$m > n\$, rank$(\boldsymbol{A}) = n$, $b \in \mathbb{R}^m$, $\mathbf{x} \in \mathbb{R}^n$.

For a pending solution $\boldsymbol{x}$, we define the cost function as

$$J(\boldsymbol{x}) = \|\boldsymbol{A}\boldsymbol{x} - \boldsymbol{b}\|^2 \tag{A.40}$$

Obviously,

$$\begin{aligned} J(\boldsymbol{x}) &= \left\|\boldsymbol{U}^{\mathrm{T}}\boldsymbol{A}\boldsymbol{x} - \boldsymbol{U}^{\mathrm{T}}\boldsymbol{b}\right\|^2 \\ &= \left\|\begin{bmatrix} \boldsymbol{D}\boldsymbol{x} \\ \boldsymbol{O} \end{bmatrix} - \begin{bmatrix} \boldsymbol{b1} \\ \boldsymbol{b2} \end{bmatrix}\right\|^2 \\ &= \|\boldsymbol{D}\boldsymbol{x} - \boldsymbol{b1}\|^2 + \|\boldsymbol{b2}\|^2 \end{aligned} \tag{A.41}$$

where $\boldsymbol{U}^{\mathrm{T}}\boldsymbol{b} = \begin{bmatrix} \boldsymbol{b1} \\ \boldsymbol{b2} \end{bmatrix}$.

The minimum solution $\boldsymbol{x}_{\mathrm{o}}$ of the cost function can be determined by $\boldsymbol{D}\boldsymbol{x}_{\mathrm{o}} = \boldsymbol{b1}$. Since $\boldsymbol{D}$ is reversible, there is

$$\boldsymbol{x}_{\mathrm{o}} = \boldsymbol{D}^{-1}\boldsymbol{b1} \tag{A.42}$$

At this time, the cost function

$$J(\boldsymbol{x}) = \|\boldsymbol{b2}\|^2 \tag{A.43}$$

It can be proved that the solution obtained at this time is the least-squares solution, i.e.

$$\boldsymbol{x}_{\mathrm{LS}} = (\boldsymbol{A}^{\mathrm{T}}\boldsymbol{A})^{-1}\boldsymbol{A}^{\mathrm{T}}\boldsymbol{b} \tag{A.44}$$

and

$$\boldsymbol{x}_{\mathrm{o}} = \boldsymbol{x}_{\mathrm{LS}} \tag{A.45}$$

A.5 Preliminary Matrix Analysis

The matrix function is a type of function in which the independent variable as well as the range are both matrices. It is similar to the polynomial of real number field

$$f(\lambda) = a_0 + a_1\lambda + a_2\lambda^2 + \cdots + a_n\lambda^n (a_m \neq 0)$$

The matrix polynomial of matrix $\boldsymbol{A} \in \mathbb{R}^{n\times n}$ can be defined as follows:

$$f(\boldsymbol{A}) = a_0 + a_1\boldsymbol{A} + a_2\boldsymbol{A}^2 + \cdots + a_n\boldsymbol{A}^n (a_m \neq 0) \tag{A.46}$$

The relationship between the two matrix polynomials $f_1(\boldsymbol{A})$ and $f_2(\boldsymbol{A})$ is

$$f_1(\boldsymbol{A}) + f_2(\boldsymbol{A}) = f_2(\boldsymbol{A}) + f_1(\boldsymbol{A}) \tag{A.47}$$

$$f_1(\boldsymbol{A})f_2(\boldsymbol{A}) = f_2(\boldsymbol{A})f_1(\boldsymbol{A}) \tag{A.48}$$

$$f(\boldsymbol{P}^{-1}\boldsymbol{A}\boldsymbol{P}) = \boldsymbol{P}^{-1}f(\boldsymbol{A})\boldsymbol{P} \text{ (When P is a nonsingular matrix)} \tag{A.49}$$

If the characteristic polynomial of $\boldsymbol{A}$ is

$$\Gamma(\lambda) = |\lambda\boldsymbol{I} - A| = \lambda^n + \alpha_1\lambda^{n-1} + \cdots + \alpha_{n-1}\lambda + \alpha_n \tag{A.50}$$

then

$$\Gamma(\boldsymbol{A}) = \boldsymbol{0} \tag{A.51}$$

That is, $\Gamma(\boldsymbol{A})$ is a zero polynomial of $\boldsymbol{A}$.

Based on this property, higher-order matrix polynomials can be transformed into low-cost ($\leq n$) matrix polynomials. For example, for $f(\boldsymbol{A})$ below,

$$f(\boldsymbol{A}) = \Gamma(\boldsymbol{A})\phi(\boldsymbol{A}) + s(\boldsymbol{A}) \tag{A.52}$$

Only $s(\boldsymbol{A})$ needs to be solved.

With the matrix polynomial concept, some special functions can be solved. For example, the exponential function of a matrix that appears many times in this book can be solved with the matrix power series.

$$e^{\boldsymbol{F}t} = \boldsymbol{I} + \boldsymbol{F}t + \frac{(\boldsymbol{F}t)^2}{2!} + \frac{(\boldsymbol{F}t)^3}{3!} + \cdots \tag{A.53}$$

where $\boldsymbol{F} \in \mathbb{R}^{n\times n}$.

The definition of a function matrix is a matrix with functions of the independent variable x as the matrix elements, i.e.

$$\boldsymbol{A}(x) = \begin{bmatrix} a_{11}(x) & a_{12}(x) & \cdots & a_{1n}(x) \\ a_{21}(x) & a_{22}(x) & \cdots & a_{2n}(x) \\ \vdots & \vdots & \ddots & \vdots \\ a_{m1}(x) & a_{m2}(x) & \cdots & a_{mn}(x) \end{bmatrix} \tag{A.54}$$

where, $a_{ij}(x)$ are all functions of x. Please note that the definition of function matrices should not be confused with that of matrix functions.

If all functions $a_{ij}(x)$ in the function matrix $\boldsymbol{A}(x)$ are divisible in (s, t), then the derivative of $\boldsymbol{A}(x)$ in (s, t) can be defined as

$$\boldsymbol{A}'(x) = \lim_{x \to x_0} \frac{\boldsymbol{A}(x_0 + \Delta x) - \boldsymbol{A}(x_0)}{\Delta x} \tag{A.55}$$

If the function $f(\boldsymbol{A})$ takes the matrix $\boldsymbol{A} \in \mathbb{R}^{m \times n}$ as its independent variable and the result is a scalar, then define its derivative with respect to the matrix as $\boldsymbol{A}$.

$$\frac{\mathrm{d}f(\boldsymbol{A})}{\mathrm{d}\boldsymbol{A}} = \begin{bmatrix} \frac{\partial f}{\partial a_{11}} & \frac{\partial f}{\partial a_{12}} & \cdots & \frac{\partial f}{\partial a_{1n}} \\ \frac{\partial f}{\partial a_{21}} & \frac{\partial f}{\partial a_{22}} & \cdots & \frac{\partial f}{\partial a_{2n}} \\ \vdots & \vdots & \ddots & \vdots \\ \frac{\partial f}{\partial a_{m1}} & \frac{\partial f}{\partial a_{m2}} & \cdots & \frac{\partial f}{\partial a_{mn}} \end{bmatrix} \tag{A.56}$$

Obviously, $\frac{\mathrm{d}f(\boldsymbol{A})}{\mathrm{d}\boldsymbol{A}}$ is a matrix.

A simple example can be given here for a better understanding of this deduction. If $f(\boldsymbol{A}) = \mathrm{tr}(\boldsymbol{A})$, then $\frac{\mathrm{d}f(\boldsymbol{A})}{\mathrm{d}\boldsymbol{A}} = \mathbf{I}$,

When $f(\boldsymbol{v})$ is a scalar function with vector $\boldsymbol{v} \in \mathbb{R}^m$ as an argument, similarly,

$$\frac{\mathrm{d}f(\boldsymbol{v})}{\mathrm{d}\boldsymbol{v}} = \begin{bmatrix} \frac{\partial f}{\partial v_1} \\ \frac{\partial f}{\partial v_2} \\ \vdots \\ \frac{\partial f}{\partial v_m} \end{bmatrix} \tag{A.57}$$

Therefore, it is safe to conclude that the derivative of the scalar function to the vector is still a vector, which can be regarded as a special form of the above derivative of the matrix.

Here are some useful relationships, in which $\boldsymbol{A} \in \mathbb{R}^{n \times n}$, $\boldsymbol{X} \in \mathbb{R}^{n \times n}$, $\boldsymbol{v} \in \mathbb{R}^n$, $\boldsymbol{u} \in \mathbb{R}^n$.

$$\frac{\mathrm{d}}{\mathrm{d}\boldsymbol{v}}(\boldsymbol{u} \cdot \boldsymbol{v}) = \frac{\mathrm{d}}{\mathrm{d}\boldsymbol{v}}(\boldsymbol{v} \cdot \boldsymbol{u}) = \boldsymbol{u} \tag{A.58}$$

$$\frac{\mathrm{d}}{\mathrm{d}\boldsymbol{v}}(\boldsymbol{A}\boldsymbol{v}) = \boldsymbol{A} \tag{A.59}$$

$$\frac{\mathrm{d}}{\mathrm{d}\boldsymbol{v}}(\boldsymbol{v}^{\mathrm{T}}\boldsymbol{A}) = \boldsymbol{A}^{\mathrm{T}} \tag{A.60}$$

$$\frac{\mathrm{d}}{\mathrm{d}\boldsymbol{v}}(\boldsymbol{v}^{\mathrm{T}}\boldsymbol{A}\boldsymbol{v}) = \boldsymbol{v}^{\mathrm{T}}(\boldsymbol{A} + \boldsymbol{A}^{\mathrm{T}}) \tag{A.61}$$

$$\frac{\mathrm{d}}{\mathrm{d}\boldsymbol{X}}\mathrm{tr}(\boldsymbol{A}\boldsymbol{X}) = \boldsymbol{A}^{\mathrm{T}} \tag{A.62}$$

$$\frac{\mathrm{d}}{\mathrm{d}\boldsymbol{X}}\mathrm{tr}(\boldsymbol{X}^{\mathrm{T}}\boldsymbol{A}\boldsymbol{X}) = \boldsymbol{A}\boldsymbol{X} + \boldsymbol{A}^{\mathrm{T}}\boldsymbol{X} \tag{A.63}$$

Appendix B
Conversion and Rotation of the Cartesian Coordinate System

Suppose that there are two Cartesian coordinate systems *XYZ* and $X'''Y'''Z'''$ as shown in Fig. B.1. The origins of the two coordinate systems coincide, but the three axes in each of them point to different directions.

Suppose that the coordinates *XYZ* of point $\boldsymbol{P}$ in the coordinate system are (x_1, y_1, z_1), the vector pointing from the origin of the coordinate system to point $\boldsymbol{P}$ can be expressed as

$$\boldsymbol{OP} = x_1\boldsymbol{I}_1 + y_1\boldsymbol{J}_1 + z_1\boldsymbol{K}_1 = \begin{bmatrix} \boldsymbol{I}_1 & \boldsymbol{J}_1 & \boldsymbol{K}_1 \end{bmatrix} \begin{bmatrix} x_1 \\ y_1 \\ z_1 \end{bmatrix} \tag{B.1}$$

Here $\boldsymbol{I}_1$, $\boldsymbol{J}_1$, and $\boldsymbol{K}_1$ are the respective unit vectors that the *X*-axis, *Y*-axis, and *Z*-axis point to.

Similarly, the coordinates of point $\boldsymbol{P}$ in the coordinate system $X'''Y'''Z'''$ is (x_2, y_2, z_2), and the representation of vector $\boldsymbol{OP}$ in the coordinate system $X'''Y'''Z'''$ is

$$\boldsymbol{OP} = x_2\boldsymbol{I}_2 + y_2\boldsymbol{J}_2 + z_2\boldsymbol{K}_2 = \begin{bmatrix} \boldsymbol{I}_2 & \boldsymbol{J}_2 & \boldsymbol{K}_2 \end{bmatrix} \begin{bmatrix} x_2 \\ y_2 \\ z_2 \end{bmatrix} \tag{B.2}$$

$\boldsymbol{I}_2$, $\boldsymbol{J}_2$, and $\boldsymbol{K}_2$ are the respective unit vectors that the X'''-axis, Y'''-axis, and Z'''-axis point to.

(x_1, y_1, z_1) and (x_2, y_2, z_2) in Eqs. (B.1) and (B.2) are the coordinates of point $\boldsymbol{P}$ in two different coordinate systems. There must be an inevitable connection between these two coordinates that reveals the relationship between the two coordinate systems. In order to figure out this conversion, let's analyze the representation of vectors $\boldsymbol{I}_2$, $\boldsymbol{J}_2$, and $\boldsymbol{K}_2$ of $X'''Y'''Z'''$ in XYZ.

First is the representation of the X'''-axis in XYZ as shown in Fig. B.2.

Y. Lu, *BDS/GPS Dual-Mode Software Receiver*, Navigation: Science and Technology 10, https://doi.org/10.1007/978-981-16-1075-2

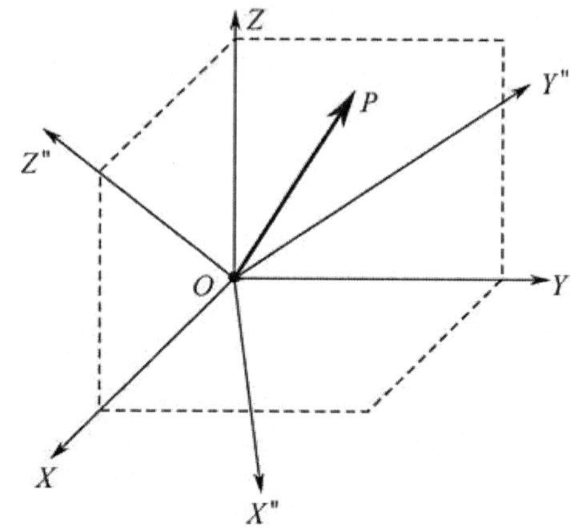

Fig. B.1 Representation of point P in two Cartesian coordinates

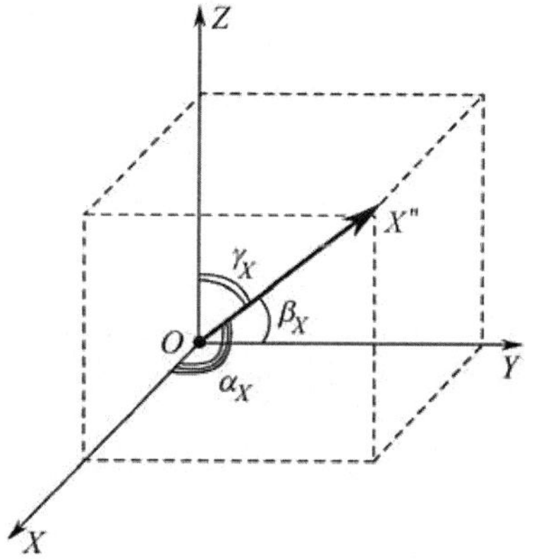

Fig. B.2 Representation of the X'''-axis in XYZ

In the figure, the angles between the X'''-axis and the X-axis, and the Y-axis and the Z-axis are α_X, β_X, and γ_X respectively. The unit vector $\boldsymbol{I}_2$ that the X'''-axis points to is represented in XYZ as

$$\boldsymbol{I}_2 = \cos\alpha_X \boldsymbol{I}_1 + \cos\beta_X \boldsymbol{J}_1 + \cos\gamma_X \boldsymbol{K}_1 = \begin{bmatrix} \boldsymbol{I}_1 & \boldsymbol{J}_1 & \boldsymbol{K}_1 \end{bmatrix} \begin{bmatrix} \cos\alpha_X \\ \cos\beta_X \\ \cos\gamma_X \end{bmatrix} \tag{B.3}$$

Hence, the representation of $\boldsymbol{J}_2$ and $\boldsymbol{K}_2$ in XYZ can be obtained through similar analysis.

$$\boldsymbol{J}_2 = \cos\alpha_Y \boldsymbol{I}_1 + \cos\beta_Y \boldsymbol{J}_1 + \cos\gamma_Y \boldsymbol{K}_1 = \begin{bmatrix} \boldsymbol{I}_1 & \boldsymbol{J}_1 & \boldsymbol{K}_1 \end{bmatrix} \begin{bmatrix} \cos\alpha_Y \\ \cos\beta_Y \\ \cos\gamma_Y \end{bmatrix} \tag{B.4}$$

$$\boldsymbol{K}_2 = \cos\alpha_Z \boldsymbol{I}_1 + \cos\beta_Z \boldsymbol{J}_1 + \cos\gamma_Z \boldsymbol{K}_1 = \begin{bmatrix} \boldsymbol{I}_1 & \boldsymbol{J}_1 & \boldsymbol{K}_1 \end{bmatrix} \begin{bmatrix} \cos\alpha_Z \\ \cos\beta_Z \\ \cos\gamma_Z \end{bmatrix} \tag{B.5}$$

α_Y, β_Y, and γ_Y are the angles between the Y'''-axis and the X-axis, the Y-axis, and the Z-axis, while α_Z, β_Z, and γ_Z are the angles between the Z'''-axis and the X-axis, the Y-axis and the Z-axis.

According to (B.3)–(B.5),

$$\begin{bmatrix} \boldsymbol{I}_2 & \boldsymbol{J}_2 & \boldsymbol{K}_2 \end{bmatrix} = \begin{bmatrix} \boldsymbol{I}_1 & \boldsymbol{J}_1 & \boldsymbol{K}_1 \end{bmatrix} \begin{bmatrix} \cos\alpha_X & \cos\alpha_Y & \cos\alpha_Z \\ \cos\beta_X & \cos\beta_Y & \cos\beta_Z \\ \cos\gamma_X & \cos\gamma_Y & \cos\gamma_Z \end{bmatrix} \tag{B.6}$$

If we substitute (B.6) into (B.2), then

$$\boldsymbol{OP} = [\boldsymbol{I}_1 \ \boldsymbol{J}_1 \ \boldsymbol{K}_1] \begin{bmatrix} \cos\alpha_X & \cos\alpha_Y & \cos\alpha_Z \\ \cos\beta_X & \cos\beta_Y & \cos\beta_Z \\ \cos\gamma_X & \cos\gamma_Y & \cos\gamma_Z \end{bmatrix} \begin{bmatrix} x_2 \\ y_2 \\ z_2 \end{bmatrix} \tag{B.7}$$

If we compare (B.7) and (B.1),

$$\begin{bmatrix} x_1 \\ y_1 \\ z_1 \end{bmatrix} = \begin{bmatrix} \cos\alpha_X & \cos\alpha_Y & \cos\alpha_Z \\ \cos\beta_X & \cos\beta_Y & \cos\beta_Z \\ \cos\gamma_X & \cos\gamma_Y & \cos\gamma_Z \end{bmatrix} \begin{bmatrix} x_2 \\ y_2 \\ z_2 \end{bmatrix} \tag{B.8}$$

Define

$$\boldsymbol{R}_{2\to 1} = \begin{bmatrix} \cos\alpha_X & \cos\alpha_Y & \cos\alpha_Z \\ \cos\beta_X & \cos\beta_Y & \cos\beta_Z \\ \cos\gamma_X & \cos\gamma_Y & \cos\gamma_Z \end{bmatrix} \tag{B.9}$$

This matrix is called the rotation matrix from $X'''Y'''Z'''$ to the coordinate system XYZ. $\boldsymbol{R}_{2-1}$ contains nine unknown quantities $\{\alpha_X, \beta_X, \gamma_X, \alpha_Y, \beta_Y, \gamma_Y, \alpha_Z, \beta_Z, \gamma_Z\}$. However, in fact, there is a constraint as follows:

First of all, for $\alpha_i, \beta_i, \gamma_i, \ i = \{ x, \ y, z\}$

$$\cos^2\alpha_X + \cos^2\beta_X + \cos^2\gamma_X = 1 \tag{B.10}$$

$$\cos^2\alpha_Y + \cos^2\beta_Y + \cos^2\gamma_Y = 1 \tag{B.11}$$

$$\cos^2\alpha_Z + \cos^2\beta_Z + \cos^2\gamma_Z = 1 \tag{B.12}$$

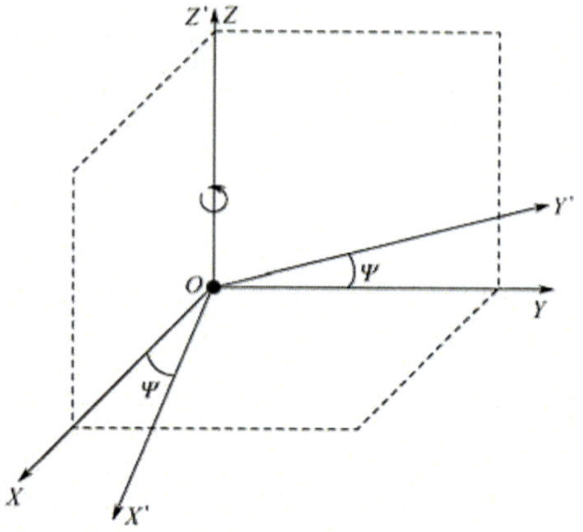

Fig. B.3 Step 1: Rotate the coordinate system around the Z-axis by angle ψ to get the $X'Y'Z'$ coordinate system

Second, because of the orthogonality between the coordinate axes, there are

$$\cos\alpha_X\cos\alpha_Y+\cos\beta_X\cos\beta_Y+\cos\gamma_X\cos\gamma_Y=0 \tag{B.13}$$

$$\cos\alpha_Y\cos\alpha_Z+\cos\beta_Y\cos\beta_Z+\cos\gamma_Y\cos\gamma_Z=0 \tag{B.14}$$

$$\cos\alpha_X\cos\alpha_Z+\cos\beta_X\cos\beta_Z+\cos\gamma_X\cos\gamma_Z=0 \tag{B.15}$$

There are six constraints in Eqs. (B.10)–(B.15), so there are only three degrees of freedom for the nine unknowns in the rotation $\boldsymbol{R}_{2\text{-}1}$ matrix. That is to say, there are only three angles that need to be determined.

Now, another method can be used to derive the rotation matrix. From the initial XYZ coordinate system, three coordinate axes are rotated by a certain angle to obtain the final $X'''Y'''Z'''$ coordinate system. This method fully embodies the "rotation" characteristics of the rotation matrix.

The first step is shown in Fig. B.3. We rotate the coordinate system around the Z-axis by angle ψ to get the $X'Y'Z'$ coordinate system. It is easy to see that after this step, the Z'-axis of the new coordinate system still coincides with that of the original system, but the X-axis and the Y-axis no longer coincide. The new X'-axis and Y'-axis are still within the XY plane, but differ from the X-axis and the Y-axis by an angle of ψ.

The $Z^{'}$-axis coordinates in the new coordinate system are the same as the Z-axis coordinates in the original coordinate system. The initial coordinates of the X-axis and the Y-axis undergo a two-dimensional rotation to become the coordinates of the X'-axis and the Y'-axis, so the new coordinate system and the rotation matrix of the coordinate system are

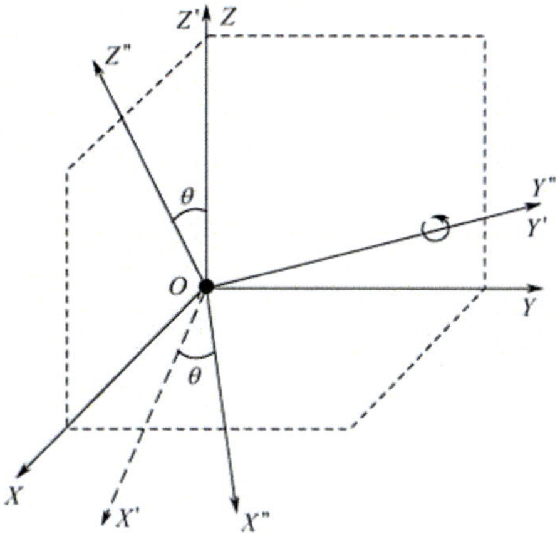

Fig. B.4 Step 2: Rotate the coordinate system around the Y'-axis by an angle of $-\theta$ to get the $X''Y''Z''$ coordinate system

$$\boldsymbol{R}_1 = \begin{bmatrix} \cos\psi & \sin\psi & 0 \\ -\sin\psi & \cos\psi & 0 \\ 0 & 0 & 1 \end{bmatrix} \tag{B.16}$$

Figure B.4 shows the second step. We rotate the coordinate system around the Y'-axis by an angle of $-\theta$ to get the $X''Y''Z''$ coordinate system. The rotation in this step is based on the $X'Y'Z'$ coordinate system in the last system. Because it is rotated around the Y'-axis, the coordinates of the Y'-axis will not change, and the X''-axis and the Z''-axis differ from the X'-axis and the Z'-axis by an angle of $-\theta$.

Thus, the rotation matrix from $X'Y'Z'$ to $X''Y''Z''$ is

$$\boldsymbol{R}_2 = \begin{bmatrix} \cos\theta & 0 & -\sin\theta \\ 0 & 1 & 0 \\ \sin\theta & 0 & \cos\theta \end{bmatrix} \tag{B.17}$$

Figure B.5 shows the third step. We rotate the coordinate system around the X''-axis by an angle of ϕ to get the $X'''Y'''Z'''$ coordinate system. The rotation in this step is based on the $X'''Y'''Z'''$ coordinate system in the last step. Because it is rotated around the X''-axis, the coordinates of this axis will remain unchanged. A similar analysis shows the rotation matrix from $X''Y''Z''$ to $X'''Y'''Z'''$

$$\boldsymbol{R}_3 = \begin{bmatrix} 1 & 0 & 0 \\ 0 & \cos\phi & \sin\phi \\ 0 & -\sin\phi & \cos\phi \end{bmatrix} \tag{B.18}$$

$X'''Y'''Z'''$ is the final coordinate system. Integrating the above steps, we can get the rotation matrix from the initial coordinate system to $X'''Y'''Z'''$.

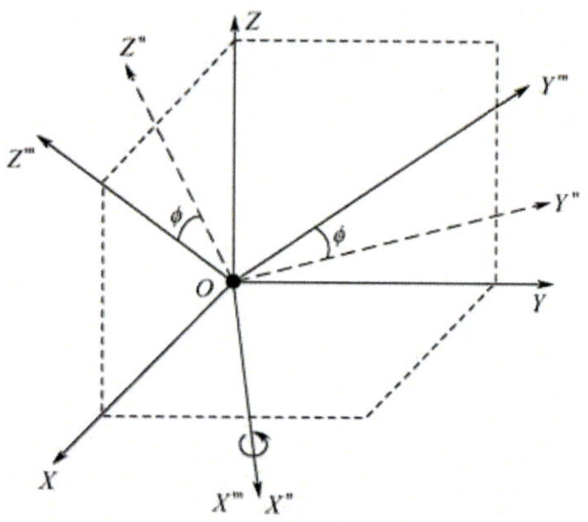

Fig. B.5 Rotate the coordinate system around the X'''-axis by an angle of ϕ to get the $X'''Y'''Z'''$ coordinate system

$$R_{1\to 2} = R_3 R_2 R_1$$
$$= \begin{bmatrix} c(\theta)c(\psi) & c(\theta)s(\psi) & -s(\theta) \\ -c(\phi)s(\psi)+s(\phi)s(\theta)c(\psi) & c(\phi)c(\psi)+s(\phi)s(\theta)s(\psi) & c(\theta)s(\phi) \\ s(\phi)s(\psi)+c(\phi)s(\theta)c(\psi) & -s(\phi)c(\psi)+c(\phi)s(\theta)s(\psi) & c(\theta)c(\phi) \end{bmatrix} \tag{B.19}$$

where $c(a) = \cos(a)$, $s(a) = \sin(a)$.

Equation (B.19) gives the rotation matrix from XYZ to $X'''Y'''Z'''$, and $R_{2\to 1}$ in Eq. (B.9) is the rotation matrix from $X'''Y'''Z'''$ to XYZ. The relationship between the two is

$$R_{2\to 1} = R_{1\to 2}^{\mathrm{T}} \tag{B.20}$$

and

$$R_{2\to 1} R_{1\to 2} = I \tag{B.21}$$

It can be verified that the rotation matrix obtained from Eq. (B.19) still satisfies Eqs. (B.10)–(B.15).

It can be seen from the above steps that only three angles are required to determine the rotation matrix and the relationship between the two coordinate systems. This conclusion is also consistent with that of the previous three degrees of freedom.

So far, our analysis is still based on cases where the origins of the two coordinate systems coincide, which means that the two coordinate systems only rotate without translation. Now, let's consider a more general case, as shown in Fig. B.6.

In Fig. B.6, the origins of $X'''Y'''Z'''$ and XYZ no longer coincide, and a translation occurs. In this case, we assume that the coordinates of point P in the $X'''Y'''Z'''$ coordinate system are (x_2, y_2, z_2), i.e.

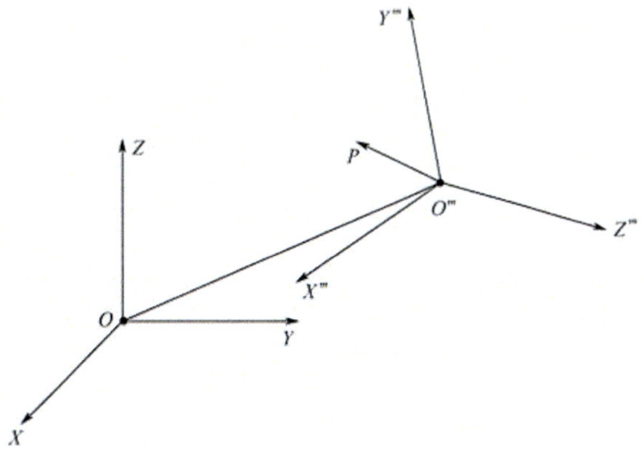

Fig. B.6 A case where both rotation and translation occur to the two coordinate systems

$$\boldsymbol{O'''P} = \begin{bmatrix} \boldsymbol{I}_2 & \boldsymbol{J}_2 & \boldsymbol{K}_2 \end{bmatrix} \begin{bmatrix} x_2 \\ y_2 \\ z_2 \end{bmatrix} \tag{B.22}$$

In this case, point P is expressed in XYZ as

$$\begin{aligned} \boldsymbol{OP} &= \boldsymbol{OO'''} + [\boldsymbol{O'''P}]_1 \\ &= \begin{bmatrix} x_o \\ y_o \\ z_o \end{bmatrix} + \boldsymbol{R_{2\to 1}} \begin{bmatrix} x_2 \\ y_2 \\ z_2 \end{bmatrix} \end{aligned} \tag{B.23}$$

where $[x_o, y_o, z_o]^T$ is the vector translation of the origin of $X'''Y'''Z'''$ relative to the origin of XYZ, and the subscript 1 in $[\boldsymbol{O'''P}]_1$ is the representation of vector $\boldsymbol{O'''P}$ in XYZ.

Knowing the conversion of the coordinates of a point in different coordinate systems, we can derive the conversion of a vector in different coordinate systems. As shown in Fig. B.7, if $\boldsymbol{QP} = [x_2, y_2, z_2]^{\mathrm{T}}$ in system $X'''Y'''Z'''$, how can we convert its coordinates to system XYZ?

According to the knowledge of vectors, in the $X'''Y'''Z'''$ coordinate system,

$$\boldsymbol{QP} = \boldsymbol{O'''P} - \boldsymbol{O'''Q} \tag{B.24}$$

Writing the above formula in XYZ, with the result of Eq. (B.23), we can deduce that

$$[\boldsymbol{QP}]_1 = [\boldsymbol{O'''P}]_1 - [\boldsymbol{O'''Q}]_1 = \left(\begin{bmatrix} x_o \\ y_o \\ z_o \end{bmatrix} + \boldsymbol{R_{2\to 1}} \begin{bmatrix} x_P \\ y_P \\ z_P \end{bmatrix} \right)$$

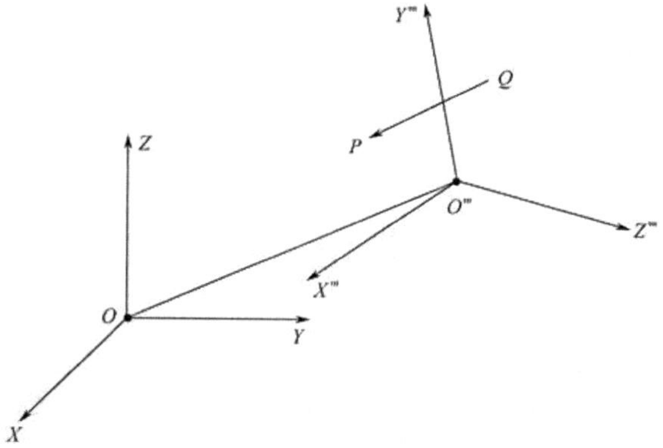

Fig. B.7 Representation of vector $\overrightarrow{QP}$ in two coordinate systems

$$
\begin{aligned}
&- \left(\begin{bmatrix} x_o \\ y_o \\ z_o \end{bmatrix} + \boldsymbol{R}_{2\to 1} \begin{bmatrix} x_Q \\ y_Q \\ z_Q \end{bmatrix} \right) \\
&= \boldsymbol{R}_{2\to 1} \left(\begin{bmatrix} x_P \\ y_P \\ z_P \end{bmatrix} - \begin{bmatrix} x_Q \\ y_Q \\ z_Q \end{bmatrix} \right) \\
&= \boldsymbol{R}_{2\to 1} \begin{bmatrix} x_2 \\ y_2 \\ z_2 \end{bmatrix}
\end{aligned} \tag{B.25}
$$

Therefore, for vectors, what needs to be considered is only the rotation of the coordinate systems, not their translation.

Appendix C
Mean and Variance of NBP and WBP

This section aims to prove Eqs. (4.202)–(4.206), which is mainly about probability theory.

According to Eqs. (4.200) and (4.201) in Chap. 4,

$$\text{WBP} = \sum_{k=1}^{M} (S_{I,k}^2 + S_{Q,k}^2) \tag{C.1}$$

$$\text{NBP} = \left(\sum_{k=1}^{M} S_{I,k}\right)^2 + \left(\sum_{k=1}^{M} S_{Q,k}\right)^2 \tag{C.2}$$

$$\begin{aligned} S_{I,k} &= A\cos(\Phi) + n_{I,k} \\ S_{Q,k} &= A\sin(\Phi) + n_{Q,k} \end{aligned} \tag{C.3}$$

where A is the signal amplitude and Φ is the initial phase difference, which can be considered as constants or minor changes in the coherent integration time period. $n_{I,k}$ and $n_{Q,k}$ are noise terms that obey the distribution of $N(0, \sigma^2)$. n_I and n_Q at different integration moments can be considered to be independent of each other.

Substituting Eq. (C.3) into Eqs. (C.1) and (C.2), we can get

$$\begin{aligned} \text{WBP} = A^2M + 2A\cos\Phi\left(\sum_{k=1}^{M} n_{I,k}\right) + 2A\sin\Phi\left(\sum_{k=1}^{M} n_{Q,k}\right) \\ + \left(\sum_{k=1}^{M} n_{I,k}^2\right) + \left(\sum_{k=1}^{M} n_{Q,k}^2\right) \end{aligned} \tag{C.4}$$

$$\text{NBP} = A^2M^2 + 2AM\cos\Phi\left(\sum_{k=1}^{M} n_{I,k}\right) + 2AM\sin\Phi\left(\sum_{k=1}^{M} n_{Q,k}\right)$$

Y. Lu, *BDS/GPS Dual-Mode Software Receiver*, Navigation: Science and Technology 10, https://doi.org/10.1007/978-981-16-1075-2

$$+\left(\sum_{k=1}^{M} n_{I,k}\right)^2 + \left(\sum_{k=1}^{M} n_{Q,k}\right)^2 \tag{C.5}$$

Let $N_{I,M} = \sum_{k=1}^{M} n_{I,k}, \mathrm{X}_{I,M} = \sum_{k=1}^{M} n_{I,k}^2$. Obviously $N_{I,M}$ and $N_{Q,M}$ fit into the distribution of $N(0, M\sigma^2)$. Let $\mathrm{X}_{I,M} = \sum_{k=1}^{M} n_{I,k}^2$, $\mathrm{X}_{Q,M} = \sum_{k=1}^{M} n_{Q,k}^2$, then $\mathrm{X}_{I,M}$ and $\mathrm{X}_{Q,M}$ need to fit into the distribution of M degrees of freedom $\chi^2(M)$. The mean of $\mathrm{X}_{I,M}$ and $\mathrm{X}_{Q,M}$ is $M\sigma^2$ and the variance of them is $2M\sigma^4$.

From this, the mean values of NBP and WBP can be derived. The results are shown in Eqs. (C.6) and (C.7). The results are consistent with those in Eqs. (4.200) and (4.203) in Chap. 4.

$$\begin{aligned} E[\mathrm{WBP}] &= A^2M + 2A\cos\Phi \cdot E\left(\sum_{k=1}^{M} n_{I,k}\right) + 2A\sin\Phi \cdot E\left(\sum_{k=1}^{M} n_{Q,k}\right) \\ &\quad + E\left(\sum_{k=1}^{M} n_{I,k}^2\right) + E\left(\sum_{k=1}^{M} n_{Q,k}^2\right) \\ &= A^2M + 2\sigma^2 M \end{aligned} \tag{C.6}$$

$$\begin{aligned} E[\mathrm{NBP}] &= A^2M^2 + 2AM\cos\Phi \cdot E\left(\sum_{k=1}^{M} n_{I,k}\right) + 2AM\sin\Phi \cdot E\left(\sum_{k=1}^{M} n_{Q,k}\right) \\ &\quad + E\left[\left(\sum_{k=1}^{M} n_{I,k}\right)^2\right] + E\left[\left(\sum_{k=1}^{M} n_{Q,k}\right)^2\right] \\ &= A^2M^2 + 2\sigma^2 M \end{aligned} \tag{C.7}$$

To derive the following Eqs. (4.204) and (4.205), $E\left[(\mathrm{WBP})^2\right]$ and $E\left[(\mathrm{NBP})^2\right]$ need to be derived first.

$$\begin{aligned} E\left[(\mathrm{WBP})^2\right] &= E\left[\left(A^2M + 2A\cos\Phi N_{I,M} + 2A\sin\Phi N_{Q,M} + \mathrm{X}_{I,M} + \mathrm{X}_{Q,M}\right)^2\right] \\ &= A^4M^2 + 4A^2\cos^2\Phi \cdot E\left[N_{I,M}^2\right] + 4A^2\sin^2\Phi \cdot E\left[N_{Q,M}^2\right] \\ &\quad + E\left[(\mathrm{X}_{I,M} + \mathrm{X}_{Q,M})^2\right] + 2A^2M \cdot E\left[\mathrm{X}_{I,M} + \mathrm{X}_{Q,M}\right] \\ &= A^4M^2 + 4A^2M\sigma^2 + 4A^2M^2\sigma^2 + 4M\sigma^4 + 4M^2\sigma^4 \end{aligned} \tag{C.8}$$

The following conclusions are used in the derivation of Eq. (C.8):

$$E\left[\mathrm{X}_{I,M}\right] = E\left[\mathrm{X}_{Q,M}\right] = M\sigma^2 \tag{C.9}$$

$$E\left[(\mathrm{X}_{I,M})\right]^2 = E\left[\left(\mathrm{X}_{Q,M}\right)^2\right] = 2M\sigma^4 + M^2\sigma^4 \tag{C.10}$$

$$
\begin{aligned}
E\Big[(\text{NBP})^2\Big] &= E\Big[\Big(A^2M^2 + 2AM\cos\Phi N_{I,M} + 2AM\sin\Phi N_{Q,M} + N_{I,M}^2 + N_{Q,M}^2\Big)^2\Big] \\
&= A^4M^4 + 4A^2M^2\cos^2\Phi \cdot E\Big[N_{I,M}^2\Big] + 4A^2M^2\sin^2\Phi \cdot E\Big[N_{Q,M}^2\Big] \\
&\quad + E\Big[(N_{I,M}^2 + N_{Q,M}^2)^2\Big] + 2A^2M^2 \cdot E\Big[N_{I,M}^2 + N_{Q,M}^2\Big] \\
&= A^4M^4 + 8A^2M^3\sigma^2 + 8M^2\sigma^4
\end{aligned} \tag{C.11}
$$

The following conclusions are used in the derivation of Eq. (C.11):

$$
E\big[N_{I,M}^2\big] = E\big[N_{Q,M}^2\big] = M\sigma^2 \tag{C.12}
$$

$$
E\big[N_{I,M}^2 + N_{Q,M}^2\big] = 2M\sigma^2 \tag{C.13}
$$

$$
E\big[(N_{I,M}^2 + N_{Q,M}^2)^2\big] = 8M^2\sigma^4 \tag{C.14}
$$

$(N_{I,M}^2 + N_{Q,M}^2)$ is subject to the square root distribution of the two degrees of freedom $\chi^2(2)$, with the mean $2M\sigma^2$ and the variance $4M^2\sigma^4$. According to the nature of the square root distribution, Eq. (C.14) can be derived easily.

Then, according to Eqs. (C.8) and (C.14), we can get

$$
\begin{aligned}
\text{var}[\text{WBP}] &= E\big[(\text{WBP})^2\big] - (E[\text{WBP}])^2 \\
&= 4M\sigma^4 + 4A^2M\sigma^2 \\
&= 4M\sigma^4\left(\frac{A^2}{\sigma^2} + 1\right)
\end{aligned} \tag{C.15}
$$

At the same time, according to Eqs. (C.11) and (C.7), we can get

$$
\begin{aligned}
\text{var}[\text{NBP}] &= E\big[(\text{NBP})^2\big] - (E[\text{NBP}])^2 \\
&= 4M^2\sigma^4 + 4A^2M^3\sigma^2 \\
&= 4M^2\sigma^4\left(\frac{MA^2}{\sigma^2} + 1\right)
\end{aligned} \tag{C.16}
$$

Equations (4.204) and (4.205) are proven.

$$
\begin{aligned}
\text{cov}[\text{WBP}, \text{NBP}] &= E[(\text{WBP} - E[\text{WBP}]) \cdot (\text{NBP} - E[\text{NBP}])] \\
&= E[(\text{WBP}) \cdot (\text{NBP})] - E[\text{WBP}] \cdot E[\text{NBP}]
\end{aligned} \tag{C.17}
$$

From Eq. (C.17) we can see that if we want to know cov[WBP, NBP], the key is to find $E[(\text{WBP}) \cdot (\text{NBP})]$.

$$
\begin{aligned}
(\mathrm{WBP})\cdot(\mathrm{NBP}) &= \left(\sum_{k=1}^{M}(S_{I,k}^2+S_{Q,k}^2)\right)\left[\left(\sum_{k=1}^{M}S_{I,k}\right)^2+\left(\sum_{k=1}^{M}S_{Q,k}\right)^2\right]\\
&= \left(\sum_{k=1}^{M}(S_{I,k}^2+S_{Q,k}^2)\right)\\
&\quad\times\left[\sum_{k=1}^{M}(S_{I,k}^2+S_{Q,k}^2)+\sum_{\substack{i,j\\ i\neq j}}^{1..M}2S_{I,i}S_{I,j}+\sum_{\substack{i,j\\ i\neq j}}^{1..M}2S_{Q,i}S_{Q,j}\right]\\
&= (\mathrm{WBP})^2+(\mathrm{WBP})\cdot\left(\sum_{\substack{i,j\\ i\neq j}}^{1..M}2S_{I,i}S_{I,j}+\sum_{\substack{i,j\\ i\neq j}}^{1..M}2S_{Q,i}S_{Q,j}\right)
\end{aligned}
\tag{C.18}
$$

According to Eq. (C.18), we can get

$$
\begin{aligned}
E[(\mathrm{WBP})\cdot(\mathrm{NBP})] &= E\left[(\mathrm{WBP})^2\right]+E[\mathrm{WBP}]\\
&\quad\cdot E\left(\sum_{\substack{i,j\\ i\neq j}}^{1..M}2S_{I,i}S_{I,j}+\sum_{\substack{i,j\\ i\neq j}}^{1..M}2S_{Q,i}S_{Q,j}\right)\\
&= A^4M^3+4A^2M\sigma^2+2A^2M^2\sigma^2\\
&\quad+2A^2M^3\sigma^2+4M\sigma^4+4M^2\sigma^4
\end{aligned}
\tag{C.19}
$$

Substituting Eqs (C.19), (C.6), and (C.7) into the Eq. (C.17), we get

$$
\begin{aligned}
\mathrm{cov}[\mathrm{WBP},\mathrm{NBP}] &= 4A^2M\sigma^2+4M\sigma^4\\
&= 4M\sigma^4\left(\frac{A^2}{\sigma^2}+1\right)
\end{aligned}
\tag{C.20}
$$

Equation (4.205) is proven.

Appendix D
Derivations Related to the Elliptical Orbit of the Satellite

In this section, Eq. (1.15) in Sect. 1.2.3 and Eqs. (6.14), (6.21)–(6.23) in Sect. 6.1 will be derived.

First, based on Fig. D.1, assuming the point coordinates of point N is (x_0, y_0), and according to the elliptic curve equation

$$y_0 = b\sqrt{1 - \frac{x^2}{a^2}} \tag{D.1}$$

To pass through point N and construct the tangent line of the ellipse, the slope is

$$k_{\|} = -\frac{bx_0}{a\sqrt{a^2 - x_0^2}} \tag{D.2}$$

Then, the slope of line NT is

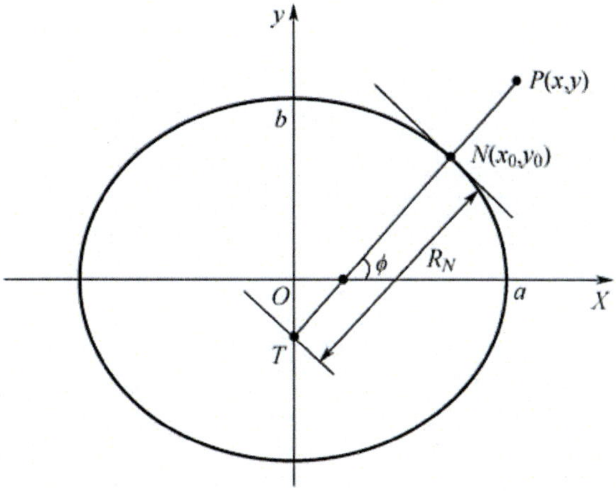

Fig. D.1 The meaning of RN

Y. Lu, *BDS/GPS Dual-Mode Software Receiver*, Navigation: Science and Technology 10, https://doi.org/10.1007/978-981-16-1075-2

$$k_{\perp} = -1/k_{\parallel} = \frac{a\sqrt{a^2 - x_0^2}}{bx_0} \tag{D.3}$$

Since line NT passes through point N, with the knowable coordinates of the point N, the linear equation of NT is

$$y = k_{\perp}x + \left(y_0 - \frac{a\sqrt{a^2 - x_0^2}}{b}\right) \tag{D.4}$$

To set x in the above equation to 0, the coordinate of T is,

$$\left(0, y_0 - \frac{a\sqrt{a^2 - x_0^2}}{b}\right)$$

$\boldsymbol{R}_{\mathrm{N}}$ is

$$\begin{aligned} R_N = \|\overline{NT}\| &= \sqrt{x_0^2 + \frac{a^2(a^2 - x_0^2)}{b^2}} \\ &= \sqrt{\frac{a^4}{b^2} + (1 - \frac{a^2}{b^2})x_0^2} \end{aligned} \tag{D.5}$$

Based on the relational equation,

$$\tan\phi = \frac{a\sqrt{a^2 - x_0^2}}{bx_0} \tag{D.6}$$

we can derive

$$x_0^2 = \frac{a^4\cos^2\phi}{b^2\sin^2\phi + a^2\cos^2\phi} \tag{D.7}$$

Substituting Eq. (D.7) into Eq. (D.5), we can get

$$R_N = \frac{a^2}{\sqrt{b^2\sin^2\phi + a^2\cos^2\phi}} \tag{D.8}$$

The result of Eq. (1.15) is proven.

To prove Eqs. (6.21)–(6.23), let's first look at Fig. D.2.

The curve of point S in Fig. 9 is an elliptic curve. The long semi-axis of the ellipse is a and the short semi-axis is b. The curve where point S' is located is concentric

Fig. D.2 The relationship between the eccentric anomaly and the true anomaly

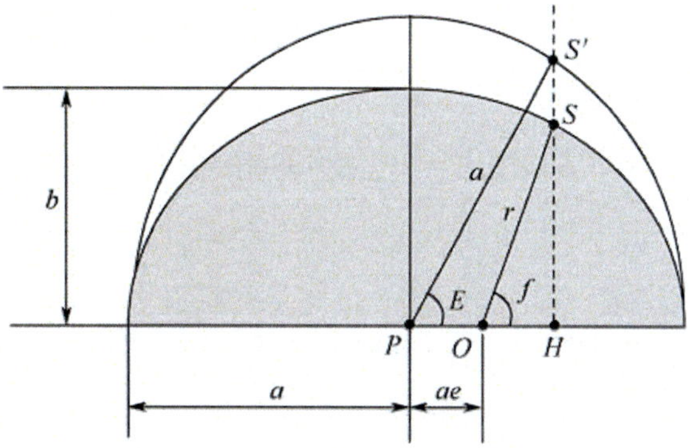

with the elliptic curve, and the radius of the circular curve is the semi-axis of the ellipse.

A significant Equation in Fig. 9 is

$$\frac{\overline{\|S'H\|}}{\overline{\|SH\|}} = \frac{a}{b} = \frac{1}{\sqrt{1-e^2}} \tag{D.9}$$

Equation (D.9) is easily derived by knowledge of analytical geometry. In fact, it is clear when the equations for the circumference and the elliptic curve are listed. Assuming that the abscissa of point $S^{'}$ and point S is x,

$$\overline{\|S'H\|} = a\sqrt{1-\frac{x^2}{a^2}} \tag{D.10}$$

$$\overline{\|S'H\|} = b\sqrt{1-\frac{x^2}{a^2}} \tag{D.11}$$

Because

$$\overline{\|S'H\|} = r\sin f \tag{D.12}$$

To combine the above Equation with Eq. (D.9),

$$\overline{\|S'H\|} = \frac{r\sin f}{\sqrt{1-e^2}} \tag{D.13}$$

because

$$\overline{\|PH\|} = a\cos E \tag{D.14}$$

$$\overline{\|PO\|} = ae \tag{D.15}$$

Therefore,

$$\overline{\|OH\|} = r\cos f = a\cos E - ae \tag{D.16}$$

$\overline{\|S'H\|}$ can be expressed as

$$\overline{\|S'H\|} = a\sin E \tag{D.17}$$

According to Eq. (D.9), we can get

$$\overline{\|SH\|} == r\sin f = a\sqrt{1-e^2}\sin E \tag{D.18}$$

Equations (6.21)–(6.23) are proven.

The proof of Eq. (6.14) is more complicated, and requires the following known conditions.

From Eqs. (D.16) and (D.18), the coordinates of the satellite in the orbit can be expressed as

$$\vec{r} = \begin{bmatrix} a\cos E - ae \\ a\sqrt{1-e^2}\sin E \end{bmatrix} \tag{D.19}$$

Deriving Eq. (D.19), the velocity vector of the satellite in the orbit can be expressed as

$$\dot{\vec{r}} = \begin{bmatrix} -a\sin E \\ a\sqrt{1-e^2}\cos E \end{bmatrix} \frac{\mathrm{d}E}{\mathrm{d}t} \tag{D.20}$$

According to Eqs. (6.12) and (6.13), we can get

$$\left|\vec{r} \times \dot{\vec{r}}\right| = \sqrt{\mu a(1-e^2)} \tag{D.21}$$

Substituting Eqs. (D.19) and (D.20) into Eq. (D.21), we can get

$$\frac{\mathrm{d}E}{\mathrm{d}t} = \frac{-\sqrt{\mu}}{a^{\frac{3}{2}}(1-e\cos E)} \tag{D.22}$$

Therefore, the satellite speed is

$$v^2 = \left\{(-a\sin E)^2 + \left[a\sqrt{1-e^2}\cos E\right]^2\right\}\left(\frac{\mathrm{d}E}{\mathrm{d}t}\right)^2$$

$$= a^2\left(1 - e^2 \cos^2 E\right)\left(\frac{\mathrm{d}E}{\mathrm{d}t}\right)^2 \tag{D.23}$$

Substituting Eq. (D.22) into Eq. (D.23), we can get

$$v^2 = \frac{\mu(1 + e \cos E)}{a(1 - e \cos E)} \tag{D.24}$$

Substituting the relational expression $r = a(1 - e \cos E)$ into the above Equation, we can get

$$v(r) = \sqrt{\mu\left(\frac{2}{r} - \frac{1}{a}\right)} \tag{D.25}$$

Equation (D.25) is Eq. (6.14).

Appendix E
The Klobuchar Model with Ionospheric Delay

At present, the ionospheric delay models of BDS and GPS navigation message propagation are both Klobuchar models, and can provide about 50% ionospheric delay correction globally. The Klobuchar model uses eight parameter constraints, and its theoretical mathematical model can be expressed as the following equation:

$$D_{\text{iono}} = F\left[b + A\cos\left(\frac{2\pi(t-\varsigma)}{\text{Per}}\right)\right] \tag{E.1}$$

where F is the tilt factor; b is the vertical delay constant; A is the amplitude of the cosine curve; and ς is the initial phase moment, which is generally the moment of the cosine curve, that is, when local time is 14:00. It is 50,400 s when it is expressed in seconds. *Per* is the cosine curve period.

Equation (E.1) shows that the Klobuchar model regards the ionospheric delay as a cosine function of a certain period. The ionospheric parameters of satellite propagation include eight parameters (α_1, α_2, α_3, α_4, β_1, β_2, β_3, β_4), where α_i is used to calculate A, and β_i is used to calculate *Per*. This model is only used in single-frequency independent receivers, and should not be used in differential mode and dual-frequency receivers.

The calculation of D_{iono} requires the following known quantities.

- E: Elevation angle of the receiver relative to the satellite;
- Z: Azimuth of the receiver relative to the satellite;
- ϕ_u: Receiver's WGS-84 latitude;
- λ_u: Receiver's WGS-84 longitude;
- Tgps: The time of the GPS system

The calculation process is as follows.

① Calculate the receiver position and the geocentric angle ψ where the ionospheric pierce point is between the projections of the Earth

Y. Lu, *BDS/GPS Dual-Mode Software Receiver*, Navigation: Science and Technology 10, https://doi.org/10.1007/978-981-16-1075-2

$$\psi = \frac{0.0137}{E + 0.11} - 0.022 \tag{E.2}$$

② Calculate the latitude ϕ_i and longitude λ_i of the ionospheric pierce point at the projected position of the Earth

$$\phi_i = \begin{cases} \phi_u + \psi \cos(Z) & |\phi_i| \le 0.416 \\ 0.416 & \phi_i > 0.416 \\ -0.416 & \phi_i < -0.416 \end{cases} \tag{E.3}$$

$$\lambda_i = \lambda_u + \frac{\psi \sin(Z)}{\cos(\phi_i)} \tag{E.4}$$

③ Calculate ϕ_m, the magnetic latitude of the ionospheric pierce point

$$\phi_m = \phi_i + 0.064 \cos(\lambda_i - 1.617) \tag{E.5}$$

④ Calculate *Per* using the ionospheric parameter β_i

$$\text{Per} = \begin{cases} \sum_{i=0}^{3} \beta_i (\phi_m)^i & \text{Per} > 72,000 \\ 72,000 & \text{Per} \le 72,000 \end{cases} \tag{E.6}$$

⑤ Calculate A using the ionospheric parameter α_i

$$A = \begin{cases} \sum_{i=0}^{3} \alpha_i (\phi_m)^i & A \ge 0 \\ 72,000 & A < 0 \end{cases} \tag{E.7}$$

⑥ Calculate the local time t and cosine term parameter x of the ionospheric pierce point

$$t = 4.32 \times 10^4 \lambda_i + T_{\text{gps}} \tag{E.8}$$

$$x = \frac{2\pi(t - 50\,400)}{\text{Per}} \tag{E.9}$$

⑦ Calculate the slope factor F

$$F = 1.0 + 16(0.53 - E)^3 \tag{E.10}$$

⑧ Calculate $\boldsymbol{D}_{\text{iono}}$, the ionospheric delay correction of L1

$$D_{\text{iono}} = \begin{cases} F\left[5 \times 10^{-9} + A\left(1 - \frac{x^2}{2} + \frac{x^4}{24}\right)\right] & |x| < \frac{\pi}{2} \\ F \times 5 \times 10^{-9} & |x| \geq \frac{\pi}{2} \end{cases} \tag{E.11}$$

In Eq. (E.11), the Taylor series approximation is performed on the cosine function $\cos(x)$, and the vertical delay constant $\text{b} = 5 \times 10^{-9}$.

If the ionospheric correction at the L2 frequency needs to be calculated, then

$$D_{\text{iono2}} = \frac{77^2}{60^2} D_{\text{iono}} \tag{E.12}$$

It can be found that the coefficient in Eq. (E.12) is the ratio of the square of the L1 and L2's carrier frequencies.

The Klobuchar parameter for calculating the ionospheric delay of the BDS satellite is given in the BDS interface control document. The principle is similar to the one described in this section. Readers can refer to the BDS ICD for specific processing details and steps.